Springer Texts in Statistics

Springer
New York
Berlin
Heidelberg
Hong Kong
London
Milan
Paris
Tokyo

Springer Texts in Statistics

Alfred: Elements of Statistics for the Life and Social Sciences
Berger: An Introduction to Probability and Stochastic Processes
Bilodeau and Brenner: Theory of Multivariate Statistics
Blom: Probability and Statistics: Theory and Applications
Brockwell and Davis: Introduction to Times Series and Forecasting, Second Edition
Chow and Teicher: Probability Theory: Independence, Interchangeability, Martingales, Third Edition
Christensen: Advanced Linear Modeling: Multivariate, Time Series, and Spatial Data—Nonparametric Regression and Response Surface Maximization, Second Edition
Christensen: Log-Linear Models and Logistic Regression, Second Edition
Christensen: Plane Answers to Complex Questions: The Theory of Linear Models, Third Edition
Creighton: A First Course in Probability Models and Statistical Inference
Davis: Statistical Methods for the Analysis of Repeated Measurements
Dean and Voss: Design and Analysis of Experiments
du Toit, Steyn, and Stumpf: Graphical Exploratory Data Analysis
Durrett: Essentials of Stochastic Processes
Edwards: Introduction to Graphical Modelling, Second Edition
Finkelstein and Levin: Statistics for Lawyers
Flury: A First Course in Multivariate Statistics
Jobson: Applied Multivariate Data Analysis, Volume I: Regression and Experimental Design
Jobson: Applied Multivariate Data Analysis, Volume II: Categorical and Multivariate Methods
Kalbfleisch: Probability and Statistical Inference, Volume I: Probability, Second Edition
Kalbfleisch: Probability and Statistical Inference, Volume II: Statistical Inference, Second Edition
Karr: Probability
Keyfitz: Applied Mathematical Demography, Second Edition
Kiefer: Introduction to Statistical Inference
Kokoska and Nevison: Statistical Tables and Formulae
Kulkarni: Modeling, Analysis, Design, and Control of Stochastic Systems
Lehmann: Elements of Large-Sample Theory
Lehmann: Testing Statistical Hypotheses, Second Edition
Lehmann and Casella: Theory of Point Estimation, Second Edition
Lindman: Analysis of Variance in Experimental Design
Lindsey: Applying Generalized Linear Models
Madansky: Prescriptions for Working Statisticians

(continued after index)

Helge Toutenburg

Statistical Analysis of Designed Experiments

Second Edition

With Contributions by Thomas Nittner

Springer

Helge Toutenburg
Institut für Statistik
Universität München
Akademiestrasse 1
80799 München
Germany
toutenb@stat.uni-muenchen.de

Library of Congress Cataloging-in-Publication Data
Toutenburg, Helge.
Statistical analysis of designed experiments / Helge Toutenburg.—2nd ed.
p. cm — (Springer texts in statistics)
Includes bibliographical references and index.
ISBN 0-387-98789-4 (alk. paper)
1. Experimental design. I. Title. II. Series.
QA279 .T88 2002
519.5—dc21 2001058976

Printed on acid-free paper.

Production managed by Timothy Taylor; manufacturing supervised by Jacqui Ashri.
Photocomposed copy prepared from the author's LaTeX files.
Printed and bound by Sheridan Books, Inc., Ann Arbor, MI.
Printed in the United States of America.

9 8 7 6 5 4 3 2 1

ISBN 0-387-98789-4 SPIN 10715322

Springer-Verlag New York Berlin Heidelberg
A member of BertelsmannSpringer Science+Business Media GmbH

Preface

This book is the second English edition of my German textbook that was originally written parallel to my lecture "Design of Experiments" which was held at the University of Munich. It is thought to be a type of resource/reference book which contains statistical methods used by researchers in applied areas. Because of the diverse examples it could also be used in more advanced undergraduate courses, as a textbook.

It is often called to our attention, by statisticians in the pharmaceutical industry, that there is a need for a summarizing and standardized representation of the design and analysis of experiments that includes the different aspects of classical theory for continuous response, and of modern procedures for a categorical and, especially, correlated response, as well as more complex designs as, for example, cross–over and repeated measures. Therefore the book is useful for non statisticians who may appreciate the versatility of methods and examples, and for statisticians who will also find theoretical basics and extensions. Therefore the book tries to bridge the gap between the application and theory within methods dealing with designed experiments.

In order to illustrate the examples we decided to use the software packages SAS, SPLUS, and SPSS. Each of these has advantages over the others and we hope to have used them in an acceptable way. Concerning the data sets we give references where possible.

Staff and graduate students played an essential part in the preparation of the manuscript. They wrote the text in well–tried precision, worked–out examples (Thomas Nittner), and prepared several sections in the book (Ulrike Feldmeier, Andreas Fieger, Christian Heumann, Sabina Illi, Christian Kastner, Oliver Loch, Thomas Nittner, Elke Ortmann, Andrea Schöpp, and Irmgard Strehler).

Especially I would like to thank Thomas Nittner who has done a great deal of work on this second edition. We are very appreciative of the efforts of those who assisted in the preparation of the English version. In particular, we would like to thank Sabina Illi and Oliver Loch, as well as V.K. Srivastava (1943–2001), for their careful reading of the English version.

This book is constituted as follows. After a short Introduction, with some examples, we want to give a compact survey of the comparison of two samples (Chapter 2). The well–known linear regression model is discussed in Chapter 3 with many details, of a theoretical nature, and with emphasis on sensitivity analysis at the end. Chapter 4 contains single–factor experiments with different kinds of factors, an overview of multiple regressions, and some special cases, such as regression analysis of variance or models with random effects. More restrictive designs, like the randomized block design or Latin squares, are introduced in Chapter 5. Experiments with more than one factor are described in Chapter 6, with some basics such as, e.g., effect coding. As categorical response variables are present in Chapters 8 and 9 we have put the models for categorical response, though they are more theoretical, in Chapter 7. Chapter 8 contains repeated measure models, with their whole versatility and complexity of designs and testing procedures. A more difficult design, the cross–over, can be found in Chapter 9. Chapter 10 treats the problem of incomplete data. Apart from the basics of matrix algebra (Appendix A), the reader will find some proofs for Chapters 3 and 4 in Appendix B. Last but not least, Appendix C contains the distributions and tables necessary for a better understanding of the examples.

Of course, not all aspects can be taken into account, specially as development in the field of generalized linear models is so dynamic, it is hard to include all current tendencies. In order to keep up with this development, the book contains more recent methods for the analysis of clusters.

To some extent, concerning linear models and designed experiments, we want to recommend the books by McCulloch and Searle (2000), Wu and Hamada (2000), and Dean and Voss (1998) for supplying revised material.

Finally, we would like to thank John Kimmel, Timothy Taylor, and Brian Howe of Springer–Verlag New York for their cooperation and confidence in this book.

Universität München
March 25, 2002

Helge Toutenburg
Thomas Nittner

Contents

1
Introduction

This chapter will give an overview and motivation of the models discussed within this book. Basic terms and problems concerning practical work are explained and conclusions dealing with them are given.

1.1 Data, Variables, and Random Processes

Many processes that occur in nature, the engineering sciences, and biomedical or pharmaceutical experiments cannot be characterized by theoretical or even mathematical models.

The analysis of such processes, especially the study of the cause effect relationships, may be carried out by drawing inferences from a finite number of samples. One important goal now consists of designing sampling experiments that are productive, cost effective, and provide a sufficient data base in a qualitative sense. Statistical methods of experimental design aim at improving and optimizing the effectiveness and productivity of empirically conducted experiments.

An almost unlimited capacity of hardware and software facilities suggests an almost unlimited quantity of information. It is often overlooked, however, that large numbers of data do not necessarily coincide with a large amount of information. Basically, it is desirable to collect data that contain a high level of information, i.e., *information–rich data*. Statistical methods of experimental design offer a possibility to increase the proportion of such information–rich data.

As data serve to understand, as well as to control processes, we may formulate several basic ideas of experimental design:

- Selection of the appropriate variables.
- Determination of the optimal range of input values.
- Determination of the optimal process regime, under restrictions or marginal conditions specific for the process under study (e.g., pressure, temperature, toxicity).

Examples:

(a) Let the response variable Y denote the flexibility of a plastic that is used in dental medicine to prepare a set of dentures. Let the binary input variable X denote if silan is used or not. A suitably designed experiment should:

(i) confirm that the flexibility increases by using silan (cf. Table 1.1); and

(ii) in a next step, find out the optimal dose of silan that leads to an appropriate increase of flexibility.

PMMA 2.2 Vol% quartz without silan	PMMA 2.2 Vol% quartz with silan
98.47	106.75
106.20	111.75
100.47	96.67
98.72	98.70
91.42	118.61
108.17	111.03
98.36	90.92
92.36	104.62
80.00	94.63
114.43	110.91
104.99	104.62
101.11	108.77
102.94	98.97
103.95	98.78
99.00	102.65
106.05	
$\bar{x} = 100.42$	$\bar{y} = 103.91$
$s_x^2 = 7.9^2$	$s_y^2 = 7.6^2$
$n = 16$	$m = 15$

TABLE 1.1. Flexibility of PMMA with and without silan.

(b) In metallurgy, the effect of two competing methods (oil, A; or salt water, B), to harden a given alloy, had to be investigated. Some metallic pieces were hardened by Method A and some by Method B. In both samples the average hardness, $\bar{x}_A$ and $\bar{x}_B$, was calculated and interpreted as a measure to assess the effect of the respective method (cf. Montgomery, 1976, p. 1).

In both examples, the following questions may be of interest:

- Are all the explaining factors incorporated that affect flexibility or hardness?
- How many workpieces have to be subjected to treatment such that possible differences are statistically significant?
- What is the smallest difference between average treatment effects that can be described as being substantial?
- Which methods of data analysis should be used?
- How should treatments be randomized to units?

1.2 Basic Principles of Experimental Design

This section answers parts of the above questions by formulating kinds of basic principles for designed experiments.

We shall demonstrate the basic principles of experimental design by the following example in dental medicine. Let us assume that a study is to be planned in the framework of a prophylactic program for children of preschool age. Answers to the following questions are to be expected:

- Are different intensity levels of instruction in dental care for pre–school children different in their effect?
- Are they substantially different from situations in which no instruction is given at all?

Before we try to answer these questions we have to discuss some topics:

(a) Exact definition of *intensity levels of instruction* in medical care.

Level I:	Instruction by dentists and parents and instruction to the kindergarten teacher by dentists.
Level II:	as Level I, but without instruction of parents.
Level III:	Instruction by dentists only.

Additionally, we define:

Level IV:	No instruction at all (control group).

(b) How can we measure the effect of the instruction?
As an appropriate parameter, we chose the increase in caries during the period of observation, expressed by the difference in carious teeth.

Obviously, the most simple plan is to give instructions to one child whereas another is left without advice. The criterion to quantify the effect is given by the increase in carious teeth developed during a fixed period:

Treatment	Unit	Increase in carious teeth
A (without instruction)	1 child	Increase (a)
B (with instruction)	1 child	Increase (b)

It would be unreasonable to conclude that instruction will definitely reduce the increase in carious teeth if (b) is smaller than (a), as only one child was observed for each treatment. If more children are investigated and the difference of the average effects (a) – (b) still continues to be large, one may conclude that instruction definitely leads to improvement.

One important fact has to be mentioned at this stage. If more than one unit per group is observed, there will be some variability in the outcomes of the experiment in spite of the homogeneous experimental conditions. This phenomenon is called *sampling error* or *natural variation.*

In what follows, we will establish some basic principles to study the sampling error. If these principles hold, the chance of getting a data set or a design which could be analyzed, with less doubt about structural nuisances, is higher as if the data was collected arbitrarily.

Principle 1 Fisher's Principle of Replication. The experiment has to be carried out on several units (children) in order to determine the sampling error.

Principle 2 Randomization. The units have to be assigned *randomly* to treatments. In our example, every level of instruction must have the same chance of being assigned. These two principles are essential to determine the sampling error correctly. Additionally, the conditions under which the treatments were given should be comparable, if not identical. Also the units should be similar in structure. This means, for example, that children are of almost the same age, or live in the same area, or show a similar sociological environment. An appropriate set–up of a correctly designed trial would consist of blocks (defined in Principle 3), each with, for example (the minimum of), four children that have similar characteristics. The four levels of instruction are then randomly distributed to the children such that, in the end, all levels are present in every group. This is the reasoning behind the following:

Principle 3 Control of Variance. To increase the sensitivity of an experiment, one usually stratifies the units into groups with similar (homogeneous) characteristics. These are called blocks. The criterion for stratifying is often given by age, sex, risk exposure, or sociological factors.

For Convenience. The experiment should be balanced. The number of units assigned to a specific treatment should nearly be the same, i.e., every instruction level occurs equally often among the children. The last principle ensures that every treatment is given as often as the others.

Even when the analyst follows these principles to the best of his ability there might still occur further problems as, for example, the scaling of variables which influences the amount of possible methods. The next two sections deal with this problem.

1.3 Scaling of Variables

In general, the applicability of the statistical methods depends on the scale in which the variables have been measured. Some methods, for example, assume that data may take any value within a given interval, whereas others require only an ordinal or ranked scale. The measurement scale is of particular importance as the quality and goodness of statistical methods depend to some extent on it.

Nominal Scale (Qualitative Data)

This is the most simple scale. Each data point belongs uniquely to a specific category. These categories are often coded by numbers that have no real numeric meaning.

Examples:

- Classification of patients by sex: two categories, *male* and *female*, are possible;
- classification of patients by blood group;
- increase in carious teeth in a given period. Possible categories: 0 (no increase), 1 (1 additional carious tooth), etc;
- profession;
- race; and
- marital status.

These types of data are called nominal data. The following scale contains substantially more information.

Ordinal or Ranked Scale (Quantitative Data)

If we intend to characterize objects according to an ordering, e.g., grades or ratings, we may use an ordinal or ranked scale. Different categories now symbolize different qualities. Note that this does not mean that differences between numerical values may be interpreted.

Example: The *oral hygiene index* (OHI) may take the values 0, 1, 2, and 3. The OHI is 0 if teeth are entirely free of dental plaque and the OHI is 3 if more than two–thirds of teeth are attacked. The following classification serves as an example for an ordered scale:

Group 1	0–1	Excellent hygiene
Group 2	2	Satisfactory hygiene
Group 3	3	Poor hygiene

Further examples of ordinal scaled data are:

- age groups (< 40, < 50, < 60, ≥ 60 years);
- intensity of a medical treatment (low, average, high dose); and
- preference rating of an object (low, average, high).

Metric or Interval Scale

One disadvantage of a ranked scale consists of the fact that numerical differences in the data are not liable to interpretation. In order to measure differences, we shall use a metric or interval scale with a defined origin and equal scaling units (e.g., temperature). An interval scale with a natural origin is called a ratio scale. Length, time, or weight measurements are examples of such ratio scales. It is convenient to consider interval and ratio scales as one scale.

Examples:

- Resistance to pressure of material.
- p_H–Value in dental plaque.
- Time to produce a workpiece.
- Rates of return in per cent.
- Price of an item in dollars.

Interval data may be represented by an ordinal scale and ordinal data by a nominal scale. In both situations, there is a loss of information. Obviously, there is no way to transform data from a lower scale into a higher scale.

Advanced statistical techniques are available for all scales of data. A survey is given in Table 1.2.

	Appropriate measures	Appropriate test procedures	Appropriate measures of correlation
Nominal scale	Absolute and relative frequency mode	χ^2–Test	Contingency coefficient
Ranked scale	Frequencies, mode, ranks, median, quantiles, rank variance	χ^2–Test, nonparametric methods based on ranks	Rank correlation coefficient
Interval scale	Frequencies, mode, ranks, quantiles, median, skewness, $\bar{x}, s, s^2$	χ^2–Test, nonparametric methods, parametric methods (e.g., under normality) χ^2–, t–, F–Tests, variance, and regression analysis	Correlation coefficient

TABLE 1.2. Measurement scales and related statistics.

It should be noted that all types of measurement scales may occur simultaneously if more than one variable is observed from a person or an object.

Examples: Typical data on registration at a hospital:

- Sex (nominal).
- Deformities: congenital/transmitted/received (nominal).
- Age (interval).
- Order of therapeutic steps (ordinal).
- OHI (ordinal).
- Time of treatment (interval).

1.4 Measuring and Scaling in Statistical Medicine

We shall discuss briefly some general measurement problems that are typical for medical data. Some variables are *directly measurable*, e.g., height, weight, age, or blood pressure of a patient, whereas others may be observed only via *proxy* variables. The latter case is called *indirect measurement*. Results for the variable of interest may only be derived from the results of a proxy.

Examples:

- Assessing the health of a patient by measuring the effect of a drug.

- Determining the extent of a cardiac infarction by measuring the concentration of transaminase.

An indirect measurement may be regarded as the sum of the actual effect and an additional random effect. To quantify the actual effect may be problematic. Such an indirect measurement leads to a metric scale if:

- the indirect observation is metric;
- the actual effect is measurable by a metric variable; and
- there is a unique relation between both measurement scales.

Unfortunately, the latter case arises rarely in medicine.

Another problem arises by introducing *derived scales* which are defined as a function of metric scales. Their statistical treatment is rather difficult and more care has to be taken in order to analyze such data.

Example: Heart defects are usually measured by the ratio

$$\frac{\text{strain duration}}{\text{time of expulsion}}.$$

For most biological variables $Z = X \mid Y$ is unlikely to have a normal distribution.

Another important point is the scaling of an interval scale itself. If measurement units are chosen unnecessarily wide, this may lead to identical values (ties) and therefore to a loss of information.

In our opinion, it should be stressed that real interval scales are hard to justify, especially in biomedical experiments.

Furthermore, metric data are often derived by transformations such that parametric assumptions, e.g., normality, have to be checked carefully.

In conclusion, statistical methods based on rank or nominal data assume new importance in the analysis of bio medical data.

1.5 Experimental Design in Biotechnology

Data represent a combination of *signals and noise.* A signal may be defined as the effect a variable has on a process. Noise, or experimental errors, cover the natural variability in the data or variables.

If a biological, clinical, or even chemical trial is repeated several times, we cannot expect that the results will be identical. Response variables always show some variation that has to be analyzed by statistical methods.

There are two main sources of uncontrolled variability. These are given by a pure experimental error and a measurement error in which possible interactions (joint variation of two factors) are also included. An *experimental error* is the variability of a response variable under exactly the same experimental conditions. *Measurement errors* describe the variability of a response if repeated measurements are taken. Repeated measurements mean observing values more than once for a given individual.

In practice, the experimental error is usually assumed to be much higher than the measurement error. Additionally, it is often impossible to separate both errors, such that noise may be understood as the sum of both errors. As the measurement error is negligible, in relation to the experimental error, we have

$$\text{noise} \approx \text{experimental error}.$$

One task of experimental design is to separate signals from noise under marginal conditions given by restrictions in material, time, or money.

Example: If a response is influenced by two variables, A and B, then one tries to quantify the effect of each variable. If the response is measured only at low or high levels of A and B, then there is no way to isolate their effects. If measurements are taken according to the following combinations of levels, then individual effects may be separated:

- A low, B low.
- A low, B high.
- A high, B low.
- A high, B high.

1.6 Relative Importance of Effects—The Pareto Principle

The analysis of models of the form

$$\text{response} = f(X_1, \ldots, X_k),$$

where the X_i symbolize exogeneous influence variables, is subject to several requirements:

- Choice of the functional dependency $f(\cdot)$ of the response on $X_1, \ldots, X_k$.
- Choice of the factors X_i.
- Consideration of interactions and hierarchical structures.
- Estimation of effects and interpretation of results.

A Pareto chart is a special form of bar graph which helps to determine the importance of problems. Figure 1.1 shows a Pareto chart in which influence variables and interactions are ordered according to their relative importance. The theory of loglinear regression (Agresti, 1990; Fahrmeir and Tutz, 2001; Toutenburg, 1992a) suggests that a special coding of variables as dummies yields estimates of the effects that are independent of measurement units. Ishihawa (1976) has also illustrated this principle by a Pareto chart.

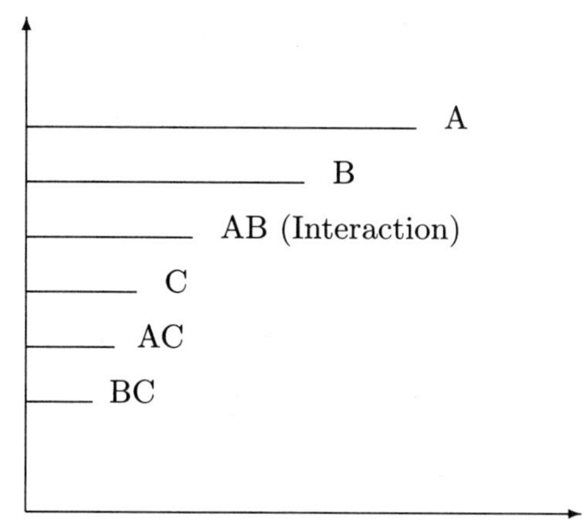

FIGURE 1.1. Typical Pareto chart of a model: response $= f(A, B, C)$.

1.7 An Alternative Chart

The results of statistical analyses become strictly more apparent if they are accompanied by the appropriate graphs and charts. Based on the Pareto principle, one such chart has been presented in the previous section. It helps to find and identify the main effects and interactions. In this section, we will illustrate a method developed by Heumann, Jacobsen and Toutenburg (1993), where bivariate cause effect relationships for ordinal data are investigated by loglinear models. Let the response variable Y take two values

$$Y = \begin{cases} 1 & \text{if response is a success,} \\ 0 & \text{otherwise.} \end{cases}$$

Let the influence variables A and B have three ordinal factor levels (low, average, high).
The loglinear model is given by

$$\ln(n_{1jk}) = \mu + \lambda_1^{\text{success}} + \lambda_j^A + \lambda_k^B + \lambda_{1j}^{\text{success}/A} + \lambda_{1k}^{\text{success}/B} . \tag{1.1}$$

Data is taken from Table 1.3.

Y	Factor A	Factor B		
		low	average	high
0	low	40	10	20
	average	60	70	30
	high	80	90	70
1	low	20	30	5
	average	60	150	20
	high	100	210	50

TABLE 1.3. Three–dimensional contingency table.

The loglinear model with interactions (1.1)

$$Y \ / \ \text{Factor A}, \qquad Y \ / \ \text{Factor B},$$

yields the following parameter estimates for the main effects (Table 1.4).

Parameter	Standardized estimate
$Y = 0$	0.257
$Y = 1$	–0.257
Factor A low	–13.982
Factor A average	4.908
Factor A high	14.894
Factor B low	2.069
Factor B average	10.515
Factor B high	–10.057

TABLE 1.4. Main effects in model (1.1).

The estimated interactions are given in Table 1.5.

The interactions are displayed in Figures 1.2 and 1.3. The effects are shown proportional to the highest effect. Note that a comparison of the main effects (shown at the border) and interactions is not possible due to different scaling. Solid circles correspond to a positive interaction, non-solid circles to a negative interaction. The standardization was calculated according to

$$\text{area effect}_i = \pi r_i^2 \tag{1.2}$$

with

$$r_i = \sqrt{\frac{\text{estimation of effect}_i}{\max_i\{\text{estimation of effect}_i\}}} \cdot r,$$

where r denotes the radius of the maximum effect.

Parameter	Standardized estimate
$Y = 0$/Factor A low	3.258
$Y = 0$/Factor A average	-1.963
$Y = 0$/Factor A high	-2.589
$Y = 1$/Factor A low	-3.258
$Y = 1$/Factor A average	1.963
$Y = 1$/Factor A high	2.589
$Y = 0$/Factor B low	1.319
$Y = 0$/Factor B average	-8.258
$Y = 0$/Factor B high	5.432
$Y = 1$/Factor B low	-1.319
$Y = 1$/Factor B average	8.258
$Y = 1$/Factor B high	-5.432

TABLE 1.5. Estimated interactions.

Interpretation. Figure 1.2 shows that (A low)/failure and (A high)/success are positively correlated, such that a recommendation to control is given by "A high". Analogously, we extract from Figure 1.3 the recommendation "B average".

Note. Interactions are to be assessed only within one figure and not between different figures, as standardization is different. A Pareto chart for the effects of positive response yields Figure 1.4, where the negative effects are shown as thin lines and the positive effects are shown as thick lines.

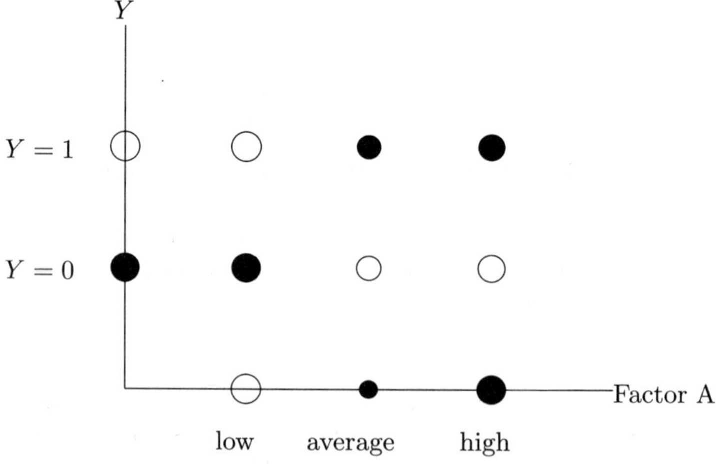

FIGURE 1.2. Main effects and interactions of Factor A.

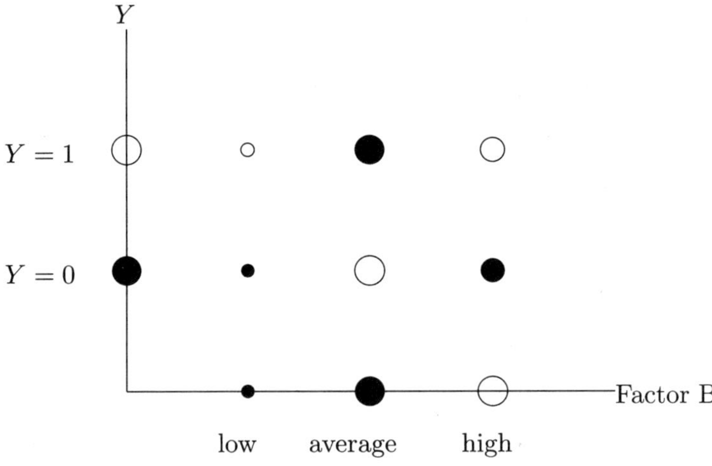

FIGURE 1.3. Main effects and interactions of Factor B.

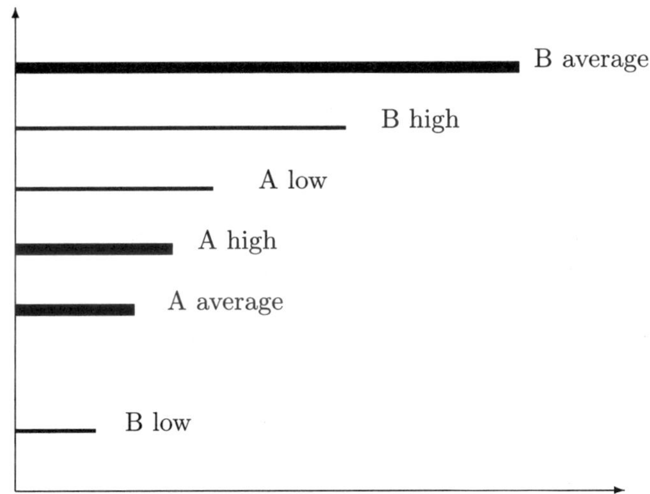

FIGURE 1.4. Simple Pareto chart of a loglinear model.

Example 1.1. To illustrate the principle further, we focus our attention on the cause effect relationship between smoking and tartar. The loglinear model related to Table 1.6 is given by

$$\ln(n_{ij}) = \mu + \lambda_i^{\text{Smoking}} + \lambda_j^{\text{Tartar}} + \lambda_{ij}^{\text{Smoking/Tartar}}, \tag{1.3}$$

with $\lambda_i^{\text{Smoking}}$ as main effect of the three levels nonsmoker, light smoker, and heavy smoker, $\lambda_j^{\text{Tartar}}$ as main effect of the three levels (low/average/high) of tartar, and $\lambda_{ij}^{\text{Smoking/Tartar}}$ as interaction smoking/tartar.

Parameter estimates are given in Table 1.7.

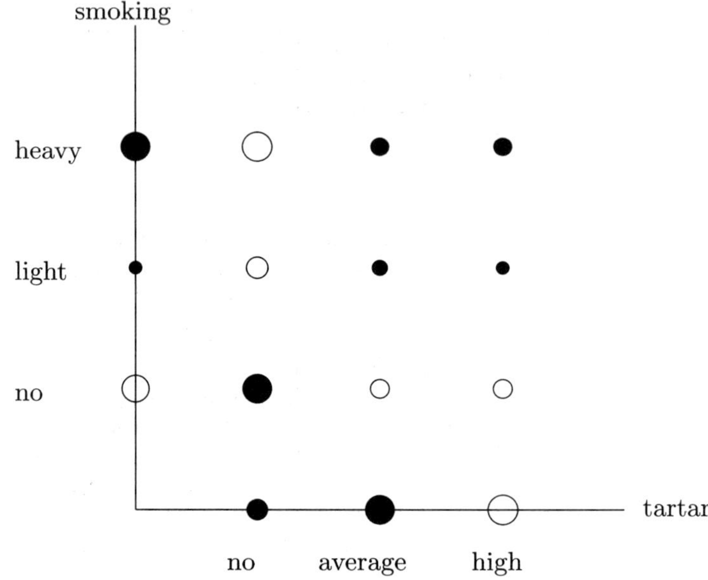

FIGURE 1.5. Effects in a loglinear model (1.3) displayed proportional to size.

		No tartar	Medium tartar	High–level tartar	
	i	j 1	2	3	$n_{i\cdot}$
Nonsmoker	1	284	236	48	568
Smoker, less than 6.5 g per day	2	606	983	209	1798
Smoker, more than 6.5 g per day	3	1028	1871	425	3324
$n_{\cdot j}$		1918	3090	682	5690

TABLE 1.6. Contingency table: consumption of tobacco / tartar.

Basically, Figure 1.5 shows a diagonal structure of interactions, where positive values are located on the main diagonal. This indicates a positive relationship between tartar and smoking.

Standardized parameter estimates	Effect
-25.93277	smoking(non)
7.10944	smoking(light)
32.69931	smoking(heavy)
11.70939	tartar(no)
23.06797	tartar(average)
-23.72608	tartar(high)
7.29951	smoking(non)/tartar(no)
-3.04948	smoking(non)/tartar(average)
-2.79705	smoking(non)/tartar(high)
-3.51245	smoking(light)/tartar(no)
1.93151	smoking(light)/tartar(average)
1.17280	smoking(light)/tartar(high)
-7.04098	smoking(heavy)/tartar(no)
2.66206	smoking(heavy)/tartar(average)
3.16503	smoking(heavy)/tartar(high)

TABLE 1.7. Estimations in model (1.3).

1.8 A One–Way Factorial Experiment by Example

To illustrate the theory of the preceding section, we shall consider a typical application of experimental design in agriculture. Let us assume that $n_1 = 10$ and $n_2 = 10$ plants are randomly collected out of n (homogeneous) plants. The first group is subjected to a fertilizer A and the second to a fertilizer B. After a period of growth, the weight (response) y of all plants is measured.

Suppose, for simplicity, that the response variable in the population is distributed according to $Y \sim N(\mu, \sigma^2)$. Then we have, for both subpopulations (fertilizers A and B),

$$Y_A \sim N(\mu_A, \sigma^2)$$

and

$$Y_B \sim N(\mu_B, \sigma^2),$$

where the variances are assumed to be equal.

These assumptions include the following one–way factorial model, where the factor fertilizer is imposed on two levels, A and B. For the actual response values we have

$$y_{ij} = \mu_i + \epsilon_{ij} \qquad (i = 1, 2, \quad j = 1, \ldots, n_i) \tag{1.4}$$

with

$$\epsilon_{ij} \sim N(0, \sigma^2)$$

and ϵ_{ij} independent, for all $i \neq j$. The null hypothesis is given by

$$\mathrm{H}_0 : \mu_1 = \mu_2 \qquad (\text{i.e., } \mathrm{H}_0 : \mu_A = \mu_B).$$

The alternative hypothesis is

$$\mathrm{H}_1 : \mu_1 \neq \mu_2 .$$

The one–way analysis of variance is equivalent to testing the equality of the expected values of two samples by the t–test under normality. The test statistic, in the case of independent samples of size n_1 and n_2, is given by

$$t = \frac{\bar{x} - \bar{y}}{s} \sqrt{\frac{n_1 \cdot n_2}{n_1 + n_2}} \sim t_{n_1+n_2-2}\,, \tag{1.5}$$

where

$$s^2 = \frac{\sum_{i=1}^{n_1}(x_i - \bar{x})^2 + \sum_{j=1}^{n_2}(y_j - \bar{y})^2}{n_1 + n_2 - 2} \tag{1.6}$$

is the pooled estimate of the variance (experimental error). H_0 will be rejected, if

$$|t| > t_{n_1+n_2-2;1-\alpha/2}, \tag{1.7}$$

where $t_{n_1+n_2-2;1-\alpha/2}$ stands for the $(1-\alpha/2)$–quantile of the $t_{n_1+n_2-2}$–distribution. Assume that the data from Table 1.8 was observed.

	Fertilizer A		Fertilizer B	
i	x_i	$(x_i - \bar{x})^2$	y_i	$(y_i - \bar{y})^2$
1	4	1	5	1
2	3	4	4	4
3	5	0	6	0
4	6	1	7	1
5	7	4	8	4
6	6	1	7	1
7	4	1	5	1
8	7	4	8	4
9	6	1	5	1
10	2	9	5	1
$\sum$	50	26	60	18

TABLE 1.8. One–way factorial experiment with two independent distributions.

We calculate $\bar{x} = 5$, $\bar{y} = 6$, and

$$\begin{aligned} s^2 = \frac{26+18}{10+10-2} = \frac{44}{18} &= 1.56^2\,, \\ t_{18} = \frac{5-6}{1.56}\sqrt{\frac{100}{20}} &= -1.43\,, \\ t_{18;0.975} &= 2.10\,, \end{aligned}$$

such that $H_0 : \mu_A = \mu_B$ cannot be rejected.

The underlying assumption of the above test is that both subpopulations can be characterized by identical distributions which may differ only in location. This assumption should be checked carefully, as (insignificant) differences may come from inhomogeneous populations. This inhomogeneity leads to an increase in experimental error and makes it difficult to detect different factor effects.

Pairwise Comparisons (Paired t–Test)

Another experimental set–up that arises frequently in the analysis of biomedical data is given if two factor levels are subjected, consecutively, to the same object or person. After the first treatment a *wash–out period* is established, in which the response variable is traced back to its original level.

Consider, for example, two alternative pesticides, A and B, which should reduce lice attack on plants. Each plant is treated initially by Method A before the concentration of lice is measured. Then, after some time, each plant is treated by Method B and again the concentration is measured. The underlying statistical model is given by

$$y_{ij} = \mu_i + \beta_j + \epsilon_{ij}, \qquad \begin{cases} i = 1, 2, \\ j = 1, \ldots, J, \end{cases} \tag{1.8}$$

where:

y_{ij} is the concentration in plant j after treatment i;
μ_i is the effect of treatment i;
β_j is the effect of the jth replication; and
ϵ_{ij} is the experimental error.

A comparison of the treatments is possible by inspecting the individual differences

$$d_j = y_{1j} - y_{2j}, \qquad j = 1, \ldots, J, \tag{1.9}$$

of concentrations on one specific plant. We derive

$$\begin{aligned} \mu_d := \mathrm{E}(d_j) &= \mathrm{E}(y_{1j} - y_{2j}) \\ &= \mu_1 + \beta_j - \mu_2 - \beta_j \\ &= \mu_1 - \mu_2. \end{aligned}$$

Testing $H_0 : \mu_1 = \mu_2$ is therefore equivalent to testing for the significance of $H_0 : \mu_d = 0$. In this situation, the paired t–test for one sample may be applied, assuming $d_i \sim N(0, \sigma_d^2)$,

$$t_{n-1} = \frac{\bar{d}}{s_d}\sqrt{n} \tag{1.10}$$

with

$$s_d^2 = \frac{\sum(d_i - \bar{d})^2}{n-1} .$$

H_0 is rejected if

$$|t_{n-1}| > t_{n-1;1-\alpha/2}.$$

Let us assume that the data shown in Table 1.9 was observed (i.e., the same data as in Table 1.8). We get

j	y_{1j}	y_{2j}	d_j	$(d_j - \bar{d})^2$
1	4	5	-1	0
2	3	4	-1	0
3	5	6	-1	0
4	6	7	-1	0
5	7	8	-1	0
6	6	7	-1	0
7	4	5	-1	0
8	7	8	-1	0
9	6	5	1	4
10	2	5	-3	4
$\sum$			-10	8

TABLE 1.9. Pairwise experimental design.

$$\begin{aligned} \bar{d} &= -1 , \\ s_d^2 = \frac{8}{9} &= 0.94^2 , \\ t_9 = \frac{-1}{0.94}\sqrt{10} &= -3.36 , \\ t_{9;0.975} &= 2.26 , \end{aligned}$$

such that $H_0 : \mu_1 = \mu_2$ (i.e., $\mu_A = \mu_B$) is rejected, which confirms that Method A is superior to Method B.

If we compare the two experimental designs a loss in degrees of freedom becomes apparent in the latter design. The respective confidence intervals are given by

$$\begin{aligned} (\bar{x} - \bar{y}) &\pm t_{18;0.975}\, s\sqrt{\frac{n_1 + n_2}{n_1 n_2}} , \\ -1 &\pm 2.10 \cdot 1.56\sqrt{\frac{20}{100}} , \\ -1 &\pm 1.46 , \end{aligned}$$

$$[-2.46; +0.46] ,$$

and

$$\begin{aligned} \bar{d} &\pm t_{9;0.975}\frac{s_d}{\sqrt{n}}, \\ -1 &\pm 2.26\frac{0.94}{\sqrt{10}}, \\ -1 &\pm 0.67, \end{aligned}$$

$$[-1.67; -0.33].$$

We observe a smaller interval in the second experiment. A comparison of the respective variances, $s^2 = 1.56^2$ and $s_d^2 = 0.94^2$, indicates that a reduction of the experimental error to $(0.94/1.56) \cdot 100 = 60\%$ was achieved by blocking with the paired design.

Note that these positive effects of blocking depend on the homogeneity of variances within each block. In Chapter 4 we will discuss this topic in detail.

1.9 Exercises and Questions

1.9.1 Describe the basic principles of experimental design.

1.9.2 Why are control groups useful?

1.9.3 To what type of scaling do the following data belong?

- Male/female.
- Catholic, Protestant.
- Pressure.
- Temperature.
- Tax category.
- Small car, car in the middle range, luxury limousine.
- Age.
- Length of stay of a patient in a clinical trial.
- University degrees.

1.9.4 What is the difference between direct and indirect measurements?

1.9.5 What are ties and their consequences in a set of data?

1.9.6 What is a Pareto chart?

1.9.7 Describe problems occurring in experimental set–ups with paired observations.

2
Comparison of Two Samples

2.1 Introduction

Problems of comparing two samples arise frequently in medicine, sociology, agriculture, engineering, and marketing. The data may have been generated by observation or may be the outcome of a controlled experiment. In the latter case, randomization plays a crucial role in gaining information about possible differences in the samples which may be due to a specific factor. Full nonrestricted randomization means, for example, that in a controlled clinical trial there is a constant chance of every patient getting a specific treatment. The idea of a blind, double blind, or even triple blind set–up of the experiment is that neither patient, nor clinician, nor statistician, know what treatment has been given. This should exclude possible biases in the response variable, which would be induced by such knowledge. It becomes clear that careful planning is indispensible to achieve valid results.

Another problem in the framework of a clinical trial may consist of the fact of a systematic effect on a subgroup of patients, e.g., males and females. If such a situation is to be expected, one should stratify the sample into homogeneous subgroups. Such a strategy proves to be useful in planned experiments as well as in observational studies.

Another experimental set–up is given by a *matched–pair design*. Subgroups then contain only one individual and pairs of subgroups are compared with respect to different treatments. This procedure requires pairs to be homogeneous with respect to all the possible factors that may exhibit an influence on the response variable and is thus limited to very special situations.

2.2 Paired t–Test and Matched–Pair Design

In order to illustrate the basic reasoning of a matched–pair design, consider an experiment, the structure of which is given in Table 2.1.

	Treatment		
Pair	1	2	Difference
1	y_{11}	y_{21}	$y_{11} - y_{21} = d_1$
2	y_{12}	y_{22}	$y_{12} - y_{22} = d_2$
$\vdots$	$\vdots$	$\vdots$	$\vdots$
n	y_{1n}	y_{2n}	$y_{1n} - y_{2n} = d_n$
			$\bar{d} = \sum d_i/n$

TABLE 2.1. Response in a matched–pair design.

We consider the linear model already given in (1.8). Assuming that

$$d_i \overset{i.i.d.}{\sim} N\left(\mu_d, \sigma_d^2\right), \tag{2.1}$$

the best linear unbiased estimator of μ_d, $\bar{d}$, is distributed as

$$\bar{d} \sim N(\mu_d, \frac{\sigma_d^2}{n}). \tag{2.2}$$

An unbiased estimator of σ_d^2 is given by

$$s_d^2 = \frac{\sum_{i=1}^{n}(d_i - \bar{d})^2}{n-1} \sim \frac{\sigma_d^2}{n-1}\,\chi_{n-1}^2 \tag{2.3}$$

such that under $H_0 : \mu_d = 0$ the ratio

$$t = \frac{\bar{d}}{s_d}\sqrt{n} \tag{2.4}$$

is distributed according to a (central) t–distribution.
A two–sided test for $H_0 : \mu_d = 0$ versus $H_1 : \mu_d \neq 0$ rejects H_0, if

$$|t| > t_{n-1;1-\alpha(\text{two–sided})} = t_{n-1;1-\alpha/2}\,. \tag{2.5}$$

A one–sided test $H_0 : \mu_d = 0$ versus $H_1 : \mu_d > 0$ ($\mu_d < 0$) rejects H_0 in favor of $H_1 : \mu_d > 0$, if

$$t > t_{n-1;1-\alpha}\,. \tag{2.6}$$

H_0 is rejected in favor of $H_1 : \mu_d < 0$, if

$$t < -t_{n-1;1-\alpha}\,. \tag{2.7}$$

Necessary Sample Size and Power of the Test

We consider a test of H_0 versus H_1 for a distribution with an unknown parameter θ. Obviously, there are four possible situations, two of which

Decision	Real situation	
	H_0 true	H_0 false
H_0 accepted	Correct decision	False decision
H_0 rejected	False decision	Correct decision

TABLE 2.2. Test decisions.

lead to a correct decision. The probability

$$P_\theta(\text{reject } H_0 \mid H_0 \text{ true}) = P_\theta(H_1 \mid H_0) \leq \alpha \quad \text{for all} \quad \theta \in H_0 \tag{2.8}$$

is called the probability of a *type I error.* α is to be fixed before the experiment. Usually, $\alpha = 0.05$ is a reasonable choice. The probability

$$P_\theta(\text{accept } H_0 \mid H_0 \text{ false}) = P_\theta(H_0 \mid H_1) \geq \beta \quad \text{for all} \quad \theta \in H_1 \tag{2.9}$$

is called the probability of a *type II error.* Obviously, this probability depends on the true value of θ such that the function

$$G(\theta) = P_\theta(\text{reject } H_0) \tag{2.10}$$

is called the *power* of the test. Generally, a test on a given α aims to fix the type II error at a defined level or beyond. Equivalently, we could say that the power should reach, or even exceed, a given value. Moreover, the following rules apply:

(i) the power rises as the sample size n increases, keeping α and the parameters under H_1 fixed;

(ii) the power rises and therefore β decreases as α increases, keeping n and the parameters under H_1 fixed; and

(iii) the power rises as the difference δ between the parameters under H_0 and under H_1 increases.

We bear in mind that the power of a test depends on the difference δ, on the type I error, on the sample size n, and on the hypothesis being one–sided or two–sided. Changing from a one–sided to a two–sided problem reduces the power.

The comparison of means in a matched–pair design yields the following relationship. Consider a one–sided test ($H_0 : \mu_d = \mu_0$ versus $H_1 : \mu_d = \mu_0 + \delta$, $\delta > 0$) and a given α. To start with, we assume σ_d^2 to be known. We now try to derive the sample size n that is required to achieve a fixed power of $1 - \beta$ for a given α and known σ_d^2. This means that we have to settle n

in a way that $H_0 : \mu_d = \mu_0$, with fixed α, is accepted with probability β, although the true parameter is $\mu_d = \mu_0 + \delta$. We define

$$u := \frac{\bar{d} - \mu_0}{\sigma_d/\sqrt{n}}.$$

Then, under $H_1 : \mu_d = \mu_0 + \delta$, we have

$$\tilde{u} = \frac{\bar{d} - (\mu_0 + \delta)}{\sigma_d/\sqrt{n}} \sim N(0, 1). \tag{2.11}$$

$\tilde{u}$ and u are related as follows:

$$u = \tilde{u} + \frac{\delta}{\sigma_d}\sqrt{n} \sim N\left(\frac{\delta}{\sigma_d}\sqrt{n}, 1\right). \tag{2.12}$$

The null hypothesis $H_0 : \mu_d = \mu_0$ is accepted erroneously if the test statistic u has a value of $u \leq u_{1-\alpha}$. The probability for this case should be $\beta = P(H_0 \mid H_1)$. So we get

$$\begin{aligned} \beta &= P(u \leq u_{1-\alpha}) \\ &= P\left(\tilde{u} \leq u_{1-\alpha} - \frac{\delta}{\sigma_d}\sqrt{n}\right) \end{aligned}$$

and, therefore,

$$u_\beta = u_{1-\alpha} - \frac{\delta}{\sigma_d}\sqrt{n},$$

which yields

$$\begin{aligned} n &\geq \frac{(u_{1-\alpha} - u_\beta)^2 \sigma_d^2}{\delta^2} && (2.13) \\ &= \frac{(u_{1-\alpha} + u_{1-\beta})^2 \sigma_d^2}{\delta^2}. && (2.14) \end{aligned}$$

For application in practice, we have to estimate σ_d^2 in (2.13). If we estimate σ_d^2 using the sample variance, we also have to replace $u_{1-\alpha}$ and $u_{1-\beta}$ by $t_{n-1;1-\alpha}$ and $t_{n-1;1-\beta}$, respectively. The value of δ is the difference of expectations of the two parameter ranges, which is either known or estimated using the sample.

2.3 Comparison of Means in Independent Groups

2.3.1 Two–Sample t–Test

We have already discussed the two–sample problem in Section 1.8. Now we consider the two independent samples

$$\begin{array}{lll} A : & x_1, \ldots, x_{n_1}, & x_i \sim N(\mu_A, \sigma_A^2), \\ B : & y_1, \ldots, y_{n_2}, & y_i \sim N(\mu_B, \sigma_B^2). \end{array}$$

Assuming $\sigma_A^2 = \sigma_B^2 = \sigma^2$, we may apply the linear model. To compare the two groups A and B we test the hypothesis $\mathrm{H}_0 : \mu_A = \mu_B$ using the statistic, i.e.,

$$t_{n_1+n_2-2} = (\bar{x} - \bar{y})/s\sqrt{(n_1 n_2)/(n_1 + n_2)}\,.$$

In practical applications, we have to check the assumption that $\sigma_A^2 = \sigma_B^2$.

2.3.2 Testing $H_0 : \sigma_A^2 = \sigma_B^2 = \sigma^2$

Under H_0, the two independent sample variances

$$s_x^2 = \frac{1}{n_1 - 1} \sum_{i=1}^{n_1} (x_i - \bar{x})^2$$

and

$$s_y^2 = \frac{1}{n_2 - 1} \sum_{i=1}^{n_2} (y_i - \bar{y})^2$$

follow a χ^2–distribution with $n_1 - 1$ and $n_2 - 1$ degrees of freedom, respectively, and their ratio follows an F–distribution

$$F = \frac{s_x^2}{s_y^2} \sim F_{n_1-1, n_2-1}\,. \tag{2.15}$$

Decision

Two–sided:

$$\mathrm{H}_0 : \sigma_A^2 = \sigma_B^2 \text{ versus } \mathrm{H}_1 : \sigma_A^2 \neq \sigma_B^2\,.$$

H_0 is rejected if

$$F > F_{n_1-1, n_2-1; 1-\alpha/2}$$

or

$$F < F_{n_1-1, n_2-1; \alpha/2} \tag{2.16}$$

with

$$F_{n_1-1, n_2-1; \alpha/2} = \frac{1}{F_{n_1-1, n_2-1; 1-\alpha/2}}\,. \tag{2.17}$$

One–sided:

$$\mathrm{H}_0 : \sigma_A^2 = \sigma_B^2 \text{ versus } \mathrm{H}_1 : \sigma_A^2 > \sigma_B^2 \,. \tag{2.18}$$

If

$$F > F_{n_1-1,n_2-1;1-\alpha}\,, \tag{2.19}$$

then H_0 is rejected.

Example 2.1. Using the data set of Table 1.8, we want to test $\mathrm{H}_0 : \sigma_A^2 = \sigma_B^2$. In Table 1.8 we find the values $n_1 = n_2 = 10$, $s_A^2 = \frac{26}{9}$, and $s_B^2 = \frac{18}{9}$. This yields

$$F = \frac{26}{18} = 1.44 < 3.18 = F_{9,9;0.95}$$

so that we cannot reject the null hypothesis $\mathrm{H}_0 : \sigma_A^2 = \sigma_B^2$ versus $\mathrm{H}_1 : \sigma_A^2 > \sigma_B^2$ according to (2.19). Therefore, our analysis in Section 1.8 was correct.

2.3.3 *Comparison of Means in the Case of Unequal Variances*

If $\mathrm{H}_0 : \sigma_A^2 = \sigma_B^2$ is not valid, we are up against the so–called Behrens Fisher problem, which has no exact solution. For practical use, the following correction of the test statistic according to *Welch* gives sufficiently good results

$$t = \frac{|\bar{x} - \bar{y}|}{\sqrt{(s_x^2/n_1) + (s_y^2/n_2)}} \sim t_v \tag{2.20}$$

with degrees of freedom approximated by

$$v = \frac{\left(s_x^2/n_1 + s_y^2/n_2\right)^2}{(s_x^2/n_1)^2/(n_1+1) + (s_y^2/n_2)^2/(n_2+1)} - 2 \tag{2.21}$$

(v is rounded). We have $\min(n_1 - 1, n_2 - 1) < v < n_1 + n_2 - 2$.

Example 2.2. In material testing, two normal variables, A and B, were examined. The sample parameters are summarized as follows:

$$\begin{array}{rrrrrrl} \bar{x} = & 27.99, & s_x^2 = & 5.98^2, & n_1 = & 9 & , \\ \bar{y} = & 1.92, & s_y^2 = & 1.07^2, & n_2 = & 10 & . \end{array}$$

The sample variances are not equal

$$F = \frac{5.98^2}{1.07^2} = 31.23 > 3.23 = F_{8,9;0.95}\,.$$

Therefore, we have to use *Welch's test* to compare the means

$$t_v = \frac{|27.99 - 1.92|}{\sqrt{5.98^2/9 + 1.07^2/10}} = 12.91$$

with $v \approx 9$ degrees of freedom. The critical value of $t_{9;0.975} = 2.26$ is exceeded and we reject $H_0 : \mu_A = \mu_B$.

2.3.4 Transformations of Data to Assure Homogeneity of Variances

We know from experience that the two–sample t–test is more sensitive to discrepancies in the homogeneity of variances than to deviations from the assumption of normal distribution. The two–sample t–test usually reaches the level of significance if the assumption of normal distributions is not fully justified, but sample sizes are large enough ($n_1, n_2 > 20$) and the homogeneity of variances is valid. This result is based on the central limit theorem. Analogously, deviations from variance homogeneity can have severe effects on the level of significance.

The following transformations may be used to avoid the inhomogeneity of variances:

- logarithmic transformation $\ln(x_i)$, $\ln(y_i)$; and
- logarithmic transformation $\ln(x_i + 1)$, $\ln(y_i + 1)$, especially if x_i and y_i have zero values or if $0 \le x_i, y_i \le 10$ (Woolson, 1987, p. 171).

2.3.5 Necessary Sample Size and Power of the Test

The necessary sample size, to achieve the desired power of the two–sample t–test, is derived as in the paired t–test problem. Let $\delta = \mu_A - \mu_B > 0$ be the one–sided alternative to be tested against $H_0 : \mu_A = \mu_B$ with $\sigma_A^2 = \sigma_B^2 = \sigma^2$. Then, with $n_2 = a \cdot n_1$ (if $a = 1$, then $n_1 = n_2$), the minimum sample size to preserve a power of $1 - \beta$ (cf. (2.14)) is given by

$$n_1 = \sigma^2(1 + 1/a)(u_{1-\alpha} + u_{1-\beta})^2/\delta^2 \tag{2.22}$$

and

$$n_2 = a \cdot n_1 \quad \text{with } n_1 \text{ from (2.22).}$$

2.3.6 Comparison of Means without Prior Testing $H_0 : \sigma_A^2 = \sigma_B^2$; Cochran–Cox Test for Independent Groups

There are several alternative methods to be used instead of the two–sample t–test in the case of unequal variances. The test of Cochran and Cox (1957) uses a statistic which approximately follows a t–distribution. The Cochran–Cox test is conservative compared to the usually used t–test. Substantially, this fact is due to the special number of degrees of freedom that have to be used. The degrees of freedom of this test are a weighted average of $n_1 - 1$ and $n_2 - 1$. In the balanced case ($n_1 = n_2 = n$) the Cochran–Cox test has $n - 1$ degrees of freedom compared to $2(n - 1)$ degrees of freedom used in the two–sample t–test. The test statistic

$$t_{c-c} = \frac{\bar{x} - \bar{y}}{s_{(\bar{x}-\bar{y})}} \tag{2.23}$$

with

$$s^2_{(\bar{x}-\bar{y})} = \frac{s_x^2}{n_1} + \frac{s_y^2}{n_2}$$

has critical values at:

two–sided: (2.24)

$$t_{c-c(1-\alpha/2)} = \frac{s_x^2/n_1 \, t_{n_1-1;1-\alpha/2} + s_y^2/n_2 \, t_{n_2-1;1-\alpha/2}}{s^2_{(\bar{x}-\bar{y})}}, \tag{2.25}$$

one–sided: (2.26)

$$t_{c-c(1-\alpha)} = \frac{s_x^2/n_1 \, t_{n_1-1;1-\alpha} + s_y^2/n_2 \, t_{n_2-1;1-\alpha}}{s^2_{(\bar{x}-\bar{y})}}. \tag{2.27}$$

The null hypothesis is rejected if $|t_{c-c}| > t_{c-c}(1 - \alpha/2)$ (two–sided) (resp., $t_{c-c} > t_{c-c}(1 - \alpha)$ (one–sided, H$_1$: $\mu_A > \mu_B$)).

Example 2.3. (Example 2.2 continued).
We test H$_0$: $\mu_A = \mu_B$ using the two–sided Cochran–Cox test. With

$$\begin{aligned} s^2_{(\bar{x}-\bar{y})} &= \frac{5.98^2}{9} + \frac{1.07^2}{10} \\ &= 3.97 + 0.11 = 4.08 = 2.02^2 \end{aligned}$$

and

$$\begin{aligned} t_{c-c(1-\alpha/2)} &= \frac{3.97 \cdot 2.31 + 0.11 \cdot 2.26}{4.08} \\ &= 1.86\,, \end{aligned}$$

we get $t_{c-c} = |27.99 - 1.92|/2.02 = 12.91 > 2.31$, so that H$_0$ has to be rejected.

2.4 Wilcoxon's Sign–Rank Test in the Matched–Pair Design

Wilcoxon's test for the differences of pairs is the nonparametric analog to the paired t–test. This test can be applied to a continuous (not necessarily normal distributed) response. The test allows us to check whether the differences $y_{1i} - y_{2i}$ of paired observations (y_{1i}, y_{2i}) are symmetrically distributed with median $M = 0$.

In the two–sided test problem, the hypothesis is given by

$$\mathrm{H}_0 : M = 0 \quad \text{or, equivalently,} \quad \mathrm{H}_0 : P(Y_1 < Y_2) = 0.5\,, \tag{2.28}$$

versus

$$\mathrm{H}_1 : M \neq 0 \tag{2.29}$$

and in the one–sided test problem

$$\mathrm{H}_0 : \quad M \leq 0 \quad \text{versus} \quad \mathrm{H}_1 : \quad M > 0\,. \tag{2.30}$$

Assuming $Y_1 - Y_2$ being distributed symmetrically, the relation $f(-d) = f(d)$ holds for each value of the difference $D = Y_1 - Y_2$, with $f(\cdot)$ denoting the density function of the difference variable. Therefore, we can expect, under H_0, that the ranks of absolute differences $|d|$ are equally distributed amongst negative and positive differences. We put the absolute differences in ascending order and note the sign of each difference $d_i = y_{1i} - y_{2i}$. Then we sum over the ranks of absolute differences with positive sign (or, analogously, with negative sign) and get the following statistic (cf. Büning and Trenkler, 1978, p. 187):

$$W^+ = \sum_{i=1}^{n} Z_i R(|d_i|) \tag{2.31}$$

with

$$\begin{aligned} d_i &= y_{1i} - y_{2i}\,, \\ R(|d_i|) &: \text{rank of } |d_i|, \\ Z_i &= \begin{cases} 1, & d_i > 0\,, \\ 0, & d_i < 0\,. \end{cases} \end{aligned} \tag{2.32}$$

We also could sum over the ranks of negative differences (W^-) and get the relationship $W^+ + W^- = n(n+1)/2$.

Exact Distribution of W^+ under H_0

The term W^+ can also be expressed as

$$W^+ = \sum_{i=1}^{n} i Z_{(i)} \quad \text{with} \quad Z_{(i)} = \begin{cases} 1, & D_j > 0\,, \\ 0, & D_j < 0\,. \end{cases} \tag{2.33}$$

In this case D_j denotes the difference for which $r(|D_j|) = i$ for given i. Under $H_0 : M = 0$ the variable W^+ is symmetrically distributed with center

$$E(W^+) = E\left(\sum_{i=1}^{n} i\, Z_{(i)}\right) = \frac{n(n+1)}{4}.$$

The sample space may be regarded as a set L of all n–tuples built of 1 or 0. L itself consists of 2^n elements and each of these has probability $1/2^n$ under H_0. Hence, we get

$$P(W^+ = w) = \frac{a(w)}{2^n} \tag{2.34}$$

with $a(w)$: number of possibilities to assign + signs to the numbers from 1 to n in a manner that leads to the sum w.

Example: Let $n = 4$. The exact distribution of W^+ under H_0 can be found in the last column of the following table:

w	Tuple of ranks	$a(w)$	$P(W^+ = w)$
10	(1 2 3 4)	1	1/16
9	(2 3 4)	1	1/16
8	(1 3 4)	1	1/16
7	(1 2 4), (3 4)	2	2/16
6	(1 2 3), (2 4)	2	2/16
5	(1 4), (2 3)	2	2/16
4	(1 3), (4)	2	2/16
3	(1 2), (3)	2	2/16
2	(2)	1	1/16
1	(1)	1	1/16
0		1	1/16
			$\sum$: 16/16 = 1

For example, $P(W^+ \geq 8) = 3/16$.

Testing

Test A:

$H_0 : M = 0$ is rejected versus $H_1 : M \neq 0$, if $W^+ \leq w_{\alpha/2}$ or $W^+ \geq w_{1-\alpha/2}$.

Test B:

$H_0 : M \leq 0$ is rejected versus $H_1 : M > 0$, if $W^+ \geq w_{1-\alpha}$.

The exact critical values can be found in tables (e.g., Table H, p. 373 in Büning and Trenkler, 1978). For large sample sizes ($n > 20$) we can use the following approximation

$$Z = \frac{W^+ - \mathrm{E}(W^+)}{\sqrt{\mathrm{Var}(W^+)}} \overset{\mathrm{H}_0}{\sim} N(0,1)\,,$$

i.e.,

$$Z = \frac{W^+ - n(n+1)/4}{\sqrt{n(n+1)(2n+1)/24}}\,. \tag{2.35}$$

For both tests, H_0 is rejected if $|Z| > u_{1-\alpha/2}$ (resp., $Z > u_{1-\alpha}$).

Ties

Ties may occur as *zero–differences* ($d_i = y_{1i} - y_{2i} = 0$) and/or as *compound–differences* ($d_i = d_j$ for $i \neq j$). Depending on the type of ties, we use one of the following test:

- zero–differences test;
- compound–differences test; and
- zero–differences plus compound–differences test.

The following methods are comprehensively described in Lienert (1986, pp. 327–332).

1. Zero–Differences Test

(a) Sample reduction method of Wilcoxon and Hemelrijk (Hemelrijk, 1952):
This method is used if the sample size is large enough ($n \geq 10$) and the percentage of ties is less than 10% ($t_0/n \leq 1/10$, with t_0 denoting the number of zero–differences).

Zero–differences are excluded from the sample and the test is conducted using the remaining $n_0 = n - t_0$ pairs.

(b) Pratt's partial–rank randomization method (Pratt, 1959):
This method is used for small sample sizes with more than 10% of zero–differences.

The zero–differences are included during the association of ranks but are excluded from the test statistic. The exact distribution of W_0^+ under H_0 is calculated for the remaining n_0 signed ranks. The probabilities of rejection are given by:

- Test A (two–sided):

$$P_0' = \frac{2A_0' + a_0'}{2^{n_0}}\,.$$

– Test B (one–sided):

$$P_0' = \frac{A_0' + a_0'}{2^{n_0}} .$$

Here A_0' denotes the number of orderings which give $W_0^+ > w_0$ and a_0' denotes the number of orderings which give $W_0^+ = w_0$.

(c) Cureton's asymptotic version of the partial–rank randomization test (Cureton, 1967):
This test is used for large sample sizes and many zero–differences ($t_0/n > 0.1$). The test statistic is given by

$$Z_{W_0} = \frac{W_0^+ - \mathrm{E}(W_0^+)}{\sqrt{\mathrm{Var}(W_0^+)}}$$

with

$$\mathrm{E}(W_0^+) = \frac{n(n+1) - t_0(t_0+1)}{4} ,$$
$$\mathrm{Var}(W_0^+) = \frac{n(n+1)(2n+1) - t_0(t_0+1)(2t_0+1)}{24} .$$

Under H_0, the statistic Z_{W_0} follows asymptotically the standard normal distribution.

2. Compound–Differences Test

(a) Shared–ranks randomization method.
In small samples and for any percentage of compound–differences we assign averaged ranks to the compound–differences. The exact distributions as well as one– and two–sided critical values, are calculated as shown in Test 1(b).

(b) Approximated compound–differences test.
If we have a larger sample ($n > 10$) and a small percentage of compound–differences ($t/n \leq 1/5$ with $t =$ the number of compound–differences), then we assign averaged ranks to the compounded values. The test statistic is calculated and tested as usual.

(c) Asymptotic sign–rank test corrected for ties.
This method is useful for large samples with $t/n > 1/5$.

In equation (2.36) we replace $\mathrm{Var}(W^+)$ by a corrected variance (due to the association of ranks) $\mathrm{Var}(W^+_{\mathrm{corr.}})$ given by

$$\mathrm{Var}(W^+_{\mathrm{corr.}}) = \frac{n(n+1)(2n+1)}{24} - \sum_{j=1}^{r} \frac{t_j^3 - t_j}{48} ,$$

with r denoting the number of groups of ties and t_j denoting the number of ties in the jth group ($1 \leq j \leq r$). Unbounded observations are regarded as groups of size 1. If there are no ties, then $r = n$ and $t_j = 1$ for all j, e.g., the correction term becomes zero.

3. Zero–Differences Plus Compound–Differences Test

These tests are used if there are both zero–differences and compound–differences.

(a) Pratt's randomization method.
For small samples which are cleared up for zeros ($n_0 \leq 10$), we proceed as in Test 1(b) but additionally assign averaged ranks to the compound–differences.

(b) Cureton's approximation method.
In larger zero–cleared samples the test statistic is calculated analogously to Test 3(a). The expectation $\mathrm{E}(W_0^+)$ equals that in Test 1(c) and is given by

$$\mathrm{E}(W_0^+) = \frac{n(n+1) - t_0(t_0+1)}{4} .$$

The variance in Test 1(c) has to be corrected due to ties and is given by

$$\mathrm{Var}_{\mathrm{corr.}}(W_0^+) = \frac{n(n+1)(2n+1) - t_0(t_0+1)(2t_0+1)}{24} - \sum_{j=1}^{r} \frac{t_j^3 - t_j}{48} .$$

Finally, the test statistic is given by

$$Z_{W_0,\mathrm{corr.}} = \frac{W_0^+ - \mathrm{E}(W_0^+)}{\sqrt{\mathrm{Var}_{\mathrm{corr.}}(W_0^+)}} . \quad (2.36)$$

2.5 Rank Test for Homogeneity of Wilcoxon, Mann and Whitney

We consider two independent continuous random variables, X and Y, with unknown distribution or nonnormal distribution. We would like to test whether the samples of the two variables are samples of the same population (homogeneity). The so–called U–test of Wilcoxon, Mann, and Whitney is a rank test. As the Kruskal Wallis test (as the generalization of the Wilcoxon test) defines the null hypothesis that k populations are identical, i.e., testing for the homogeneity of these k populations, the Mann Whitney Wilcoxon test could also be seen as a test for homogeneity for the case $k = 2$ (cf. Gibbons, (1976), p. 173). This is the nonparametric analog of the t–test and is used if the assumptions for the use of the t–test are not justified or called into question. The relative efficiency of the U–test compared to the t–test is about 95% in the case of normally distributed variables. The U–test is often used as a quick test or as a control if the test statistic of the t–test gives values close to the critical values.

The hypothesis to be tested is H_0 : the probability P to observe a value from the first population X that is greater than any given value of the population Y is equal to 0.5. The two–sided alternative is $H_1 : P \neq 0.5$. The one–sided alternative $H_1 : P > 0.5$ means that X is *stochastically larger than* Y.

We combine the observations of the samples $(x_1, \ldots, x_m)$ and $(y_1, \ldots, y_n)$ in ascending order of ranks and note for each rank the sample it belongs to. Let R_1 and R_2 denote the sum of ranks of the X– and Y–samples, respectively. The test statistic U is the smaller of the values U_1 and U_2:

$$U_1 = m \cdot n + \frac{m(m+1)}{2} - R_1 , \tag{2.37}$$

$$U_2 = m \cdot n + \frac{n(n+1)}{2} - R_2 , \tag{2.38}$$

with $U_1 + U_2 = m \cdot n$ (control).

H_0 is rejected if $U \leq U(m, n; \alpha)$ (Table 2.3 contains some values for $\alpha = 0.05$ (one–sided) and $\alpha = 0.10$ (two–sided)).

	n								
m	2	3	4	5	6	7	8	9	10
4	–	0	1						
5	0	1	2	4					
6	0	2	3	5	7				
7	0	2	4	6	8	11			
8	1	3	5	8	10	13	15		
9	1	4	6	9	12	15	18	21	
10	1	4	7	11	14	17	20	24	27

TABLE 2.3. Critical values of the U–test ($\alpha = 0.05$ one–sided, $\alpha = 0.10$ two–sided).

In the case of m and $n \geq 8$, the excellent approximation

$$u = \frac{U - m \cdot n/2}{\sqrt{m \cdot n(m+n+1)/12}} \sim N(0,1) \tag{2.39}$$

is used. For $|u| > u_{1-\alpha/2}$ the hypothesis H_0 is rejected (type I error α two–sided and $\alpha/2$ one–sided).

Example 2.4. We test the equality of means of the two series of measurements given in Table 2.4 using the U–test. Let variable X be the flexibility of PMMA with silan and let variable Y be the flexibility of PMMA without silan. We put the $(16 + 15)$ values of both series in ascending order, apply ranks and calculate the sums of ranks $R_1 = 231$ and $R_2 = 265$ (Table 2.5).

PMMA 2.2 Vol% quartz without silan	PMMA 2.2 Vol% quartz with silan
98.47	106.75
106.20	111.75
100.47	96.67
98.72	98.70
91.42	118.61
108.17	111.03
98.36	90.92
92.36	104.62
80.00	94.63
114.43	110.91
104.99	104.62
101.11	108.77
102.94	98.97
103.95	98.78
99.00	102.65
106.05	
$\bar{x} = 100.42$	$\bar{y} = 103.91$
$s_x^2 = 7.9^2$	$s_y^2 = 7.6^2$
$n = 16$	$m = 15$

TABLE 2.4. Flexibility of PMMA with and without silan (cf. Toutenburg, Toutenburg and Walther, 1991, p. 100).

Rank	1	2	3	4	5	6	7	8	9
Observation	80.00	90.92	91.42	92.36	94.63	96.67	98.36	98.47	98.70
Variable	X	Y	X	X	Y	Y	X	X	Y
Sum of ranks X	1		+3	+4			+7	+8	
Sum of ranks Y		2			+5	+6			+9
Rank	10	11	12	13	14	15	16	17	
Observation	98.72	98.78	98.97	99.00	100.47	101.11	102.65	102.94	
Variable	X	Y	Y	X	X	X	Y	X	
Sum of ranks X	+10	+11		+13	+14	+15		+17	
Sum of ranks Y			+12				+16		
Rank	18	19	20	21	22	23	24		
Observation	103.95	104.62	104.75	104.99	106.05	106.20	106.75		
Variable	X	Y	Y	X	X	X	Y		
Sum of ranks X	+18			+21	+22	+23			
Sum of ranks Y		+19	+20				+24		
Rank	25	26	27	28	29	30	31		
Observation	108.17	108.77	110.91	111.03	111.75	114.43	118.61		
Variable	X	Y	Y	Y	Y	X	Y		
Sum of ranks X	+25					+30			
Sum of ranks Y		+26	+27	+28	+29		+31		

TABLE 2.5. Computing the sums of ranks (Example 2.3, cf. Table 2.4).

Then we get

$$U_1 = 16 \cdot 15 + \frac{16(16+1)}{2} - 231 = 145\,,$$

$$U_2 = 16 \cdot 15 + \frac{15(15+1)}{2} - 265 = 95\,,$$

$$U_1 + U_2 = 240 = 16 \cdot 15\,.$$

Since $m = 16$ and $n = 15$ (both sample sizes ≥ 8), we calculate the test statistic according to (2.39) with $U = U_2$ being the smaller of the two values of U:

$$u = \frac{95 - 120}{\sqrt{240(16 + 15 + 1)/12}} = -\frac{25}{\sqrt{640}} = -0.99\,,$$

and therefore $|u| = 0.99 < 1.96 = u_{1-0.05F/2} = u_{0.975}$.

The null hypothesis is not rejected (type I error 5% and 2.5% using two– and one–sided alternatives, respectively). The exact critical value of U is $U(16, 15, 0.05_{\text{two–sided}}) = 70$ (Tables in Sachs, 1974, p. 232), i.e., the decision is the same (H_0 is not rejected).

Correction of the U–Statistic in the Case of Equal Ranks

If observations occur more than once in the combined and ordered samples $(x_1, \ldots, x_m)$ and $(y_1, \ldots, y_n)$, we assign an averaged rank to each of them. The corrected U–test (with $m + n = S$) is given by

$$u = \frac{U - m \cdot n/2}{\sqrt{[m \cdot n/S(S-1)][(S^3 - S)/12 - \sum_{i=1}^{r}(t_i^3 - t_i)/12]}}\,. \tag{2.40}$$

The number of groups of equal observations (ties) is r, and t_i denotes the number of equal observations in each group.

Example 2.5. We compare the time that two dentists B and C need to manufacture an inlay (Table 4.1). First, we combine the two samples in ascending order (Table 2.6).

Observation	19.5	31.5	31.5	33.5	37.0	40.0	43.5	50.5	53.0	54.0
Dentist	C	C	C	B	B	C	B	C	C	B
Rank	1	2.5	2.5	4	5	6	7	8	9	10
Observation	56.0	57.0	59.5	60.0	62.5	62.5	65.5	67.0	75.0	
Dentist	B	B	B	B	C	C	B	B	B	
Rank	11	12	13	14	15.5	15.5	17	18	19	

TABLE 2.6. Association of ranks (cf. Table 4.1) .

We have $r = 2$ groups with equal data:

Group 1 : twice the value of 31.5; $t_1 = 2$,
Group 2 : twice the value of 62.5; $t_2 = 2$.

The correction term then is

$$\sum_{i=1}^{2} \frac{t_i^3 - t_i}{12} = \frac{2^3 - 2}{12} + \frac{2^3 - 2}{12} = 1\,.$$

The sums of ranks are given by

$$\begin{aligned} R_1 \text{ (dentist B)} &= 4 + 5 + \cdots + 19 = 130\,, \\ R_2 \text{ (dentist C)} &= 1 + 2.5 + \cdots + 15.5 = 60\,, \end{aligned}$$

and, according to (2.37), we get

$$U_1 = 11 \cdot 8 + \frac{11(11+1)}{2} - 130 = 24$$

and, according to (2.38),

$$\begin{aligned} U_2 &= 11 \cdot 8 + \frac{8(8+1)}{2} - 60 = 64\,, \\ U_1 + U_2 &= 88 = 11 \cdot 8 \quad \text{(control)}. \end{aligned}$$

With $S = m + n = 11 + 8 = 19$ and with $U = U_1$, the test statistic (2.40) becomes

$$u = \frac{24 - 44}{\sqrt{\left[\frac{88}{19 \cdot 18}\right]\left[\frac{19^3 - 19}{12} - 1\right]}} = -1.65\,,$$

and, therefore, $|u| = 1.65 < 1.96 = u_{1-0.05/2}$.

The null hypothesis H_0 : *Both dentists need the same time to make an inlay* is not rejected. Both samples can be regarded as homogeneous and may be combined in a single sample for further evaluation.

We now assume the working time to be normally distributed. Hence, we can apply the t–test and get

$$\begin{aligned} \text{dentist B} &: \bar{x} = 55.27, s_x^2 = 12.74^2, n_1 = 11\,, \\ \text{dentist C} &: \bar{y} = 43.88, s_y^2 = 15.75^2, n_2 = 8\,, \end{aligned}$$

(see Table 4.1).

The test statistic (2.15) is given by

$$F_{10,7} = \frac{15.75^2}{12.74^2} = 1.53 < 3.15 = F_{10,7;0.95}\,,$$

and the hypothesis of equal variance is not rejected. To test the hypothesis $H_0 : \mu_x = \mu_y$ the test statistic (1.5) is used. The pooled sample variance is calculated according to (1.6) and gives $s^2 = (10 \cdot 12.74^2 + 7 \cdot 15.75^2)/17 = 14.06^2$. We now can evaluate the test statistic (1.5) and get

$$t_{17} = \frac{55.27 - 43.88}{14.06}\sqrt{\frac{11 \cdot 8}{11 + 8}} = 1.74 < 2.11 = t_{17;0.95(\text{two-sided})}\,.$$

As before, the null hypothesis is not rejected.

2.6 Comparison of Two Groups with Categorical Response

In the previous sections the comparisons in the matched–pair designs and in designs with two independent groups were based on the assumption of continuous response. Now we want to compare two groups with categorical response. The distributions (binomial, multinomial, and Poisson distributions) and the maximum–likelihood–estimation are discussed in detail in Chapter 7.

To start with, we first focus on binary response, e.g., to recover/not to recover from an illness, success/no success in a game, scoring more/less than a given level.

2.6.1 McNemar's Test and Matched–Pair Design

In the case of binary response we use the codings 0 and 1, so that the pairs in a matched design are one of the tuples of response $(0,0)$, $(0,1)$, $(1,0)$, or $(1,1)$. The observations are summarized in a 2×2 table:

		Group 1		
		0	1	Sum
Group 2	0	a	c	$a+c$
	1	b	d	$b+d$
	Sum	$a+b$	$c+d$	$a+b+c+d=n$

The null hypothesis is $H_0 : p_1 = p_2$, where p_i is the probability $P(1 \mid \text{group } i)$ $(i = 1,2)$. The test is based on the relative frequencies $h_1 = (c+d)/n$ and $h_2 = (b+d)/n$ for response 1, which differ in b and c (these are the frequencies for the discordant results $(0,1)$ and $(1,0)$).

Under H_0, the values of b and c are expected to be equal or, analogously, the expression $b-(b+c)/2$ is expected to be zero. For a given value of $b+c$, the number of discordant pairs follows a binomial distribution with the parameter $p = 1/2$ (probability to observe a discordant pair $(0,1)$ or $(1,0)$). As a result, we get $E[(0,1)\text{–response}] = (b+c)/2$ and $\text{Var}[(0,1)\text{–response}] = (b+c)\cdot\frac{1}{2}\cdot\frac{1}{2}$ (analogously, this holds symmetrically for $[(1,0)\text{–response}]$).

The following ratio then has expectation 0 and variance 1:

$$\frac{b-(b+c)/2}{\sqrt{(b+c)\cdot 1/2 \cdot 1/2}} = \frac{b-c}{\sqrt{b+c}} \overset{H_0}{\sim} (0,1)$$

and follows the standard normal distribution for reasonably large $(b+c)$ due to the central limit theorem. This approximation can be used for $(b+c) \geq 20$. For the continuity correction, the absolute value of $|b-c|$ is decreased by 1. Finally, we get the following test statistic:

$$Z = \frac{(b-c)-1}{\sqrt{b+c}} \quad \text{if } b \geq c\,, \tag{2.41}$$

$$Z = \frac{(b-c)+1}{\sqrt{b+c}} \quad \text{if } b < c\,. \tag{2.42}$$

Critical values are the quantiles of the cumulated binomial distribution $B(b+c, \frac{1}{2})$ in the case of a small sample size. For larger samples (i.e., $b+c \geq 20$), we choose the quantiles of the standard normal distribution. The test statistic of McNemar is a certain combination of the two Z–statistics given above. This is used for a two–sided test problem in the case of $b+c \geq 20$ and follows a χ^2–distribution

$$Z^2 = \frac{(|b-c|-1)^2}{b+c} \sim \chi_1^2\,. \tag{2.43}$$

Example 2.6. A clinical experiment is used to examine two different teeth–cleaning techniques and their effect on oral hygiene. The response is coded binary: reduction of tartar yes/no. The patients are stratified into matched pairs according to sex, actual teeth–cleaning technique, and age. We assume the following outcome of the trial:

		Group 1		
		0	1	Sum
Group 2	0	10	50	60
	1	70	80	150
	Sum	80	130	210

We test $H_0: p_1 = p_2$ versus $H_1: p_1 \neq p_2$. Since $b+c = 70+50 > 20$, we choose the McNemar statistic

$$Z^2 = \frac{(|70-50|-1)^2}{70+50} = \frac{19^2}{120} = 3.01 < 3.84 = \chi^2_{1;0.95}$$

and do not reject H_0.

Remark. Modifications of the McNemar test can be constructed similarly to sign tests. Let n be the number of nonzero differences in the response of the pairs and let T_+ and T_- be the number of positive and negative differences, respectively. Then the test statistic, analogously to the Z–statistics (2.41) and (2.42), is given by

$$Z = \frac{(T_+/n - 1/2) \pm n/2}{1/\sqrt{4n}}\,, \tag{2.44}$$

in which we use $+n/2$ if $T_+/n < 1/2$ and $-n/2$ if $T_+/n \geq 1/2$. The null hypothesis is $H_0: \mu_d = 0$. Depending on the sample size ($n \geq 20$ or $n < 20$) we use the quantiles of the normal or binomial distributions.

2.6.2 Fisher's Exact Test for Two Independent Groups

Regarding two independent groups of size n_1 and n_2 with binary response, we get the following 2×2 table

	Group 1	Group 2	
1	a	c	$a+c$
0	b	d	$b+d$
	n_1	n_2	n

The relative frequencies of response 1 are $\hat{p}_1 = a/n_1$ and $\hat{p}_2 = c/n_2$. The null hypothesis is $\mathrm{H}_0 : p_1 = p_2 = p$. In this contingency table, we identify the cell with the smallest cell count and calculate the probability for this and all other tables with an even smaller cell count in the smallest cell. In doing so, we have to ensure that the marginal sums keep constant.

Assume $(1, 1)$ to be the weakest cell. Under H_0 we have, for response 1 in both groups (for given n, n_1, n_2, and p):

$$P((a+c)|n,p) = \binom{n}{a+c} p^{a+c}(1-p)^{n-(a+c)},$$

for Group 1 and response 1:

$$P(a|(a+b),p) = \binom{a+b}{a} p^{a}(1-p)^{b},$$

for Group 2 and response 1:

$$P(c|(c+d),p) = \binom{c+d}{c} p^{c}(1-p)^{d}.$$

Since the two groups are independent, the joint probability is given by

$$P(\text{Group 1} = a \wedge \text{ Group 2} = c) = \binom{a+b}{a} p^{a}(1-p)^{b} \binom{c+d}{c} p^{c}(1-p)^{d}$$

and the conditional probability of a and c (for the given marginal sum $a+c$) is

$$\begin{aligned} P(a, c \mid a+c) &= \binom{a+b}{a}\binom{c+d}{c} \Big/ \binom{n}{a+c} \\ &= \frac{(a+b)!\;(c+d)!\;(a+c)!\;(b+d)!}{n!} \cdot \frac{1}{a!\;b!\;c!\;d!}. \end{aligned}$$

Hence, the probability to observe the given table or a table with an even smaller count in the weakest cell is

$$P = \frac{(a+b)!\;(c+d)!\;(a+c)!\;(b+d)!}{n!} \cdot \sum_i \frac{1}{a_i!\;b_i!\;c_i!\;d_i!},$$

with summation over all cases i with $a_i \le a$. If $P < 0.05$ (one–sided) or $2P < 0.05$ (two–sided) hold, then hypothesis $\mathrm{H}_0 : p_1 = p_2$ is rejected.

Example 2.7. We compare two independent groups of subjects receiving either type A or type B of an implanted denture and observe whether it is lost during the healing process (8 weeks after implantation). The data are

		A	B	
Loss	Yes	2	8	10
	No	10	4	14
		12	12	24

The two tables with a smaller count in the (yes | A) cell are

$$\begin{array}{|cc|}\hline 1 & 9 \\ 11 & 3 \\ \hline\end{array} \quad \text{and} \quad \begin{array}{|cc|}\hline 0 & 10 \\ 12 & 2 \\ \hline\end{array}$$

and, therefore, we get

$$P = \frac{10!\ 14!\ 12!\ 12!}{24!}\left(\frac{1}{2!\ 8!\ 10!\ 4!} + \frac{1}{1!\ 9!\ 11!\ 3!} + \frac{1}{0!\ 10!\ 12!\ 2!}\right) = 0.018\,,$$

$$\left.\begin{array}{ll}\text{one–sided test:} & P = 0.018 \\ \text{two–sided test:} & 2P = 0.036\end{array}\right\} < 0.05\,.$$

Decision. $H_0 : p_1 = p_2$ is rejected in both cases. The risk of loss is significantly higher for type B than for type A.

Recurrence Relation

Instead of using tables, we can also use the following recurrence relation (cited by Sachs, 1974, p. 289):

$$P_{i+1} = \frac{a_i d_i}{b_{i+1} c_{i+1}} P_i\,.$$

In our example, we get

$$\begin{aligned} P &= P_1 + P_2 + P_3\,, \\ P_1 &= \frac{10!\ 14!\ 12!\ 12!}{24!}\frac{1}{2!\ 8!\ 10!\ 4!} \\ &= 0.0166\,, \\ P_2 &= \frac{2 \cdot 4}{11 \cdot 9} P_1 = 0.0013\,, \\ P_3 &= \frac{1 \cdot 3}{12 \cdot 10} P_2 = 0.0000\,, \end{aligned}$$

and, therefore, $P = 0.0179 \approx 0.0180$.

2.7 Exercises and Questions

2.7.1 What are the differences between the paired t–test and the two–sample t–test (degrees of freedom, power)?

2.7.2 Consider two samples with $n_1 = n_2$, $\alpha = 0.05$ and $\beta = 0.05$ in a matched–pair design and in a design of two independent groups. What is the minimum sample size needed to achieve a power of 0.95, assuming $\sigma^2 = 1$ and $\delta^2 = 4$.

2.7.3 Apply Wilcoxon's sign–rank test for a matched–pair design to the following table:

TABLE 2.7. Scorings of students who took a cup of coffee either before or after a lecture.

Student	Before	After
1	17	25
2	18	45
3	25	37
4	12	10
5	19	21
6	34	27
7	29	29

Does treatment B (coffee before) significantly influence the score?

2.7.4 For a comparison of two independent samples, X : leaf–length of strawberries with manuring A, and Y : manuring B, the normal distribution is put in question. Test $\text{H}_0 : \mu_X = \mu_Y$ using the homogeneity test of Wilcoxon, Mann, and Whitney.

A	B
37	45
49	51
51	62
62	73
74	87
89	45
44	33
53	
17	

Note that there are ties.

2.7.5 Recode the response in Table 2.4 into binary response with:
flexibility $< 100 : 0$,
flexibility $\geq 100 : 1$,
and apply Fisher's exact test for $\text{H}_0 : p_1 = p_2$ ($p_i = P(1 \mid \text{group } i)$).

2.7.6 Considering Exercise 2.7.3, we assume that the response has been binary recoded according to scoring higher/lower than average: 1/0. A sample of $n = 100$ shows the following outcome:

		Before		
		0	1	
After	0	20	25	45
	1	15	40	55
		35	65	100

Test for $H_0 : p_1 = p_2$ using McNemar's test.

3
The Linear Regression Model

3.1 Descriptive Linear Regression

The main focus of this chapter will be the linear regression model and its basic principle of estimation. We introduce the fundamental method of *least squares* by looking at the least squares geometry and discussing some of its algebraic properties.

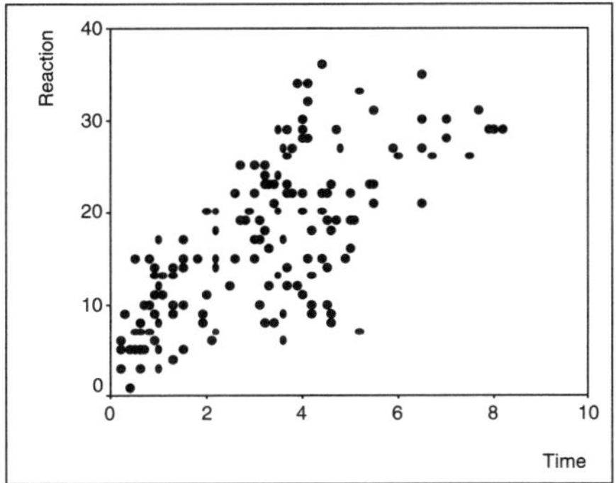

FIGURE 3.1. Scatterplot of advertising time and number of positive reactions.

In empirical work, it is quite often appropriate to specify the relationship between two sets of data by a simple linear function. For example, we model the influence of advertising time on the number of positive reactions from the public. From the scatterplot in Figure 3.1 one could suspect a linear function between advertising time (x–axis) and the number of positive reactions (y–axis). The study was done on 66 people in order to investigate the impact and cognition of advertising on TV.

Let Y denote the dependent variable which is related to a set of K independent variables $X_1, \ldots, X_K$ by a function f. As both sets comprise T observations on each variable, it is convenient to use the following notation:

$$(y, X) = \begin{pmatrix} y_1 & x_{11} & \cdots & x_{K1} \\ \vdots & \vdots & & \vdots \\ y_T & x_{1T} & \cdots & x_{KT} \end{pmatrix} = (y\, x_{(1)} \ldots x_{(K)}) = \begin{pmatrix} y_1\, x_1' \\ \vdots \\ y_T\, x_T' \end{pmatrix}, \tag{3.1}$$

where $x_{(t)}$ denotes a column vector and x_t' a row vector. We intend to obtain a good overall fit of the model and easy mathematical tractability. Choosing f to be linear seems to be realistic as almost every specification of f suffers from the exclusion of important variables or the inclusion of unimportant variables. Additionally, even a correct set of variables is often measured with at least some error such that a correct functional relationship between y and X will most unlikely be precise. On the other hand, the linear approach may serve as a suitable approximation to several nonlinear functional relationships.

If we assume Y to be *generated* additively by a linear combination of the independent variables, we may write

$$Y = X_1\beta_1 + \ldots + X_K\beta_K\,. \tag{3.2}$$

The β's in (3.2) are unknown (scalar–valued) coefficients explaining the direction and magnitude of their influence on Y. The magnitude of the β's indicates their importance in explaining Y. Therefore, an obvious goal of empirical regression analysis consists of finding those values for $\beta_1, \ldots, \beta_K$ which minimize the differences

$$e_t := y_t - x_t'\beta \quad (t = 1, \ldots, T)\,,$$

where $\beta' = (\beta_1, \ldots, \beta_K)$. The e_t's are called *residuals* and play an important role in regression analysis (e.g., in regression diagnostics, see, e.g., Rao and Toutenburg (1999, Chapter 7)). In general, we cannot expect that $e_t = 0$ will hold for all $t = 1, \ldots, T$, i.e., the scatterplot in Figure 3.1 would be a straight line. Accordingly, the residuals are incorporated into the linear approach upon setting

$$y_t = x_t'\beta + e_t \quad (t = 1, \ldots, T)\,. \tag{3.3}$$

This may be summarized in matrix notation by

$$y = X\beta + e\,. \tag{3.4}$$

Obviously, a successful choice for β is indicated by small values of all e_t. Thus, there are quite a few conceivable principles by which the quality of an actual choice for β may be evaluated.

Among others, the following measures have been proposed:

$$\begin{aligned} &\textstyle\sum_{t=1}^T \quad |e_t|\,, \quad \max_t |e_t|\,, \\ &\textstyle\sum_{t=1}^T \quad e_t^2 = e'e\,. \end{aligned} \tag{3.5}$$

Whereas the first two proposals are subject to either some complicated mathematics or poor statistical properties, the last principle has become widely accepted. This provides the basis for the famous method of least squares.

3.2 The Principle of Ordinary Least Squares

Let B be the set of all possible vectors β. If there is no further information, we have $B = \mathcal{R}^K$ (K–dimensional real Euclidean space). The idea is to find a vector $b' = (b_1, \ldots, b_K)$ from B that minimizes (3.5), the sum of squared residuals,

$$S(\beta) = \sum_{t=1}^{T} e_t^2 = e'e = (y - X\beta)'(y - X\beta)\,, \tag{3.6}$$

given y and X. Remembering the scatterplot in Figure 3.1 we can explain (3.6) by drawing the regression line and visualizing the individual difference ϵ_i between the original value (x_i, y_i) and the corresponding value $(x_i, \hat{y}_i)$ on the regression line. This can be seen in Figure 3.2 where these differences are shown for seven values.

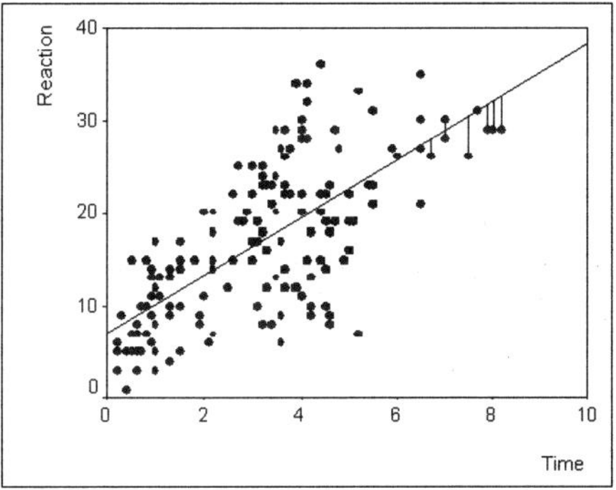

FIGURE 3.2. Scatterplot with regression line and some ϵ_i.

A minimum will always exist, as $S(\beta)$ is a real–valued convex differentiable function. If we rewrite $S(\beta)$ as

$$S(\beta) = y'y + \beta' X' X \beta - 2\beta' X' y' \tag{3.7}$$

and differentiate with respect to β (by help of A.63–A.67), we obtain

$$\frac{\partial S(\beta)}{\partial \beta} = 2X'X\beta - 2X'y\,, \tag{3.8}$$

$$\frac{\partial^2 S(\beta)}{\partial \beta^2} = 2X'X\,, \tag{3.9}$$

with $2X'X$ being nonnegative definite. Equating the first derivative to zero yields the *normal equations*

$$X'X\beta = X'y. \tag{3.10}$$

The solution of (3.10) is now straightforwardly obtainable by considering a system of linear equations

$$Ax = a\,, \tag{3.11}$$

where A is an $(n \times m)$–matrix and a is an $(n \times 1)$–vector. The $(m \times 1)$–vector x solves the equation. Let A^- be a generalized inverse of A (cf. Definition A.26). Then we have:

Theorem 3.1. The linear equation $Ax = a$ has a solution if and only if

$$AA^- a = a\,. \tag{3.12}$$

If (3.12) holds, then all solutions are given by

$$x = A^- a + (I - A^- A)w\,, \tag{3.13}$$

where w is an arbitrary $(m \times 1)$–vector. (Proof 1, Appendix B.)

Remark. $x = A^- a$ (i.e., (3.13) and $w = \mathbf{0}$) is a particular solution of $Ax = a$.

We apply this result to our problem, i.e., to (3.10), and check the solvability of the linear equation first.

X is a $(T \times K)$–matrix, thus $X'X$ is a symmetric $(K \times K)$–matrix of rank $(X'X) = p \le K$. Equation (3.10) has a solution if and only if (cf. (3.12))

$$(X'X)(X'X)^- X'y = X'y\,. \tag{3.14}$$

Following the definition of a g–inverse

$$(X'X)(X'X)^-(X'X) = (X'X)$$

we have with Theorem A.46

$$X'X(X'X)^- X' = X'\,,$$

such that (3.14) holds. Thus, the normal equation (3.10) always has a solution. The set of all solutions of (3.10) are, by (3.13), of the form

$$b = (X'X)^{-}X'y + (I - (X'X)^{-}X'X)w\,, \tag{3.15}$$

where w is an arbitrary $(K \times 1)$–vector. For the choice $w = 0$, we have with

$$b = (X'X)^{-}X'y \tag{3.16}$$

a particular solution, which is nonunique as the generalized inverse $(X'X)^{-}$ is nonunique.

An interesting algebraic property can be seen from the following theorem.

Theorem 3.2. The vector $\beta = b$ minimizes the sum of squared errors if and only if it is a solution of $X'Xb = X'y$. All solutions are located on the hyperplane Xb. (Proof 2, Appendix B.)

The solutions b of the normal equations are called *empirical regression coefficients* or empirical least squares estimates of β. $\hat{y} = Xb$ is called the *empirical regression hyperplane.* An important property of the sum of squared errors $S(b)$ is

$$y'y = \hat{y}'\hat{y} + \hat{e}'\hat{e}\ , \tag{3.17}$$

where $\hat{e}$ denotes the residuals $y - Xb$. This means that the sum of squared observations $y'y$ may be decomposed additively into the sum of squared values $\hat{y}'\hat{y}$, explained by regression and the sum of (unexplained) squared residuals $\hat{e}'\hat{e}$.

We derive (3.17) by premultiplication of (3.10) with b':

$$b'X'Xb = b'X'y$$

and

$$\hat{y}'\hat{y} = (Xb)'(Xb) = b'X'Xb = b'X'y \tag{3.18}$$

according to

$$\begin{aligned} S(b) = \hat{e}'\hat{e} &= (y - Xb)'(y - Xb) \\ &= y'y - 2b'X'y + b'X'Xb \\ &= y'y - b'X'y \\ &= y'y - \hat{y}'\hat{y}\,. \end{aligned} \tag{3.19}$$

Remark. In analysis of variance, $\hat{y}'\hat{y}$ will be decomposed further into orthogonal components which are related to the main and mixed effects of treatments.

3.3 Geometric Properties of Ordinary Least Squares Estimation

This section gives a short survey of some of the geometric properties of ordinary least squares (OLS) Estimation. Because of its geometric and algebraic characteristics it may be more theoretical than other sections and, therefore, the reader with practical interest may skip these pages.

Once again, we consider the linear model (3.4), i.e.,

$$y = X\beta + e\,,$$

where $X\beta \in \mathcal{R}(X) = \{\Theta : \Theta = X\tilde{\beta}\}$. $\mathcal{R}(X)$ is the *column space*, the set of all vectors Θ such that $\Theta = X\beta$ is fulfilled for all vectors β from $\mathcal{R}^p$. $\mathcal{R}(X) = \{\Theta : \Theta = Xb\}$ and the *null space* $\mathcal{N}(X) = \{\Phi : X\Phi = \mathbf{0}\}$ are vector spaces. The basic relation between the column space and the null space is given by

$$\mathcal{N}(X) = \mathcal{R}(X')^{\perp}\,. \tag{3.20}$$

If we assume that $\operatorname{rank}(X) = p$, then $\mathcal{R}(X)$ is of dimension p. Let $\mathcal{R}(X)^{\perp}$ denote the orthogonal complement of $\mathcal{R}(X)$ and let Xb be denoted by Θ_0 where b is the OLS estimation of β. Then we have:

Theorem 3.3. The OLS estimation Θ_0 of Xb minimizing

$$\begin{aligned} S(\beta) &= (y - X\beta)'(y - X\beta) \\ &= (y - \Theta)'(y - \Theta) = \tilde{S}(\Theta) \end{aligned} \tag{3.21}$$

for $\Theta \in \mathcal{R}(X)$, is given by the orthogonal projection of y on the space $\mathcal{R}(X)$. (Proof 3, Appendix B.)

As the context of theorem 3.3 is difficult to imagine Figure 3.3 may help to get a better impression.

The OLS estimator Xb of $X\beta$ may also be obtained in a more direct way by using idempotent projection matrices.

Theorem 3.4. Let P be a symmetric and idempotent matrix of rank p, representing the orthogonal projection of $\mathcal{R}^T$ on $\mathcal{R}(X)$.
Then $Xb = \Theta_0 = Py$. (Proof 4, Appendix B.)

The determination of P depends on the rank of X. Whereas for $\operatorname{rank}(X) = K$, i.e., X is of full rank, P is determined by $X(X'X)^{-1}X'$, it turns out to be more difficult when $\operatorname{rank}(X) = p < K$. As shown in Proof 4, Appendix B, unique solutions are derived, based on $(K-p)$ linear restrictions on β by $R\beta = r$, leading to the conditional Ordinary Least Squares Estimator (OLSE)

$$b(R, r) = (X'X + R'R)^{-1}(X'y + R'r)\,. \tag{3.22}$$

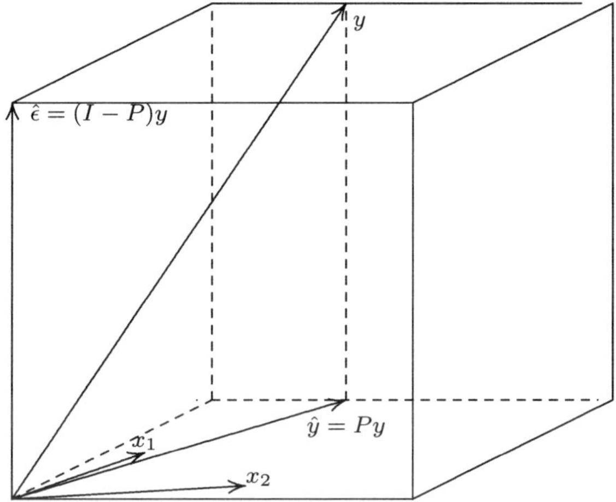

FIGURE 3.3. Orthogonal projection of y on $\mathcal{R}(X)$.

The conditional OLSE (in the sense of being restricted by $R\beta = r$) $b(R, r)$ will be most useful in tackling the problem of multicollinearity which is typical for design matrices in ANOVA models (see Section 3.5).

3.4 Best Linear Unbiased Estimation

After introducing the classical linear model, with its assumptions and measures for evaluating linear estimates, we want to show that b is the best linear unbiased estimator of β. As estimation of variance is always of practical interest we describe the estimation of σ^2 in general and for the special case $K = 2$.

In descriptive regression analysis, the regression coefficient β is allowed to vary and is then determined by the method of least squares in an algebraical way by using projection matrices. The classical linear regression model now interprets the vector β as a fixed but unknown model parameter. Then estimation is carried out by minimizing an appropriate risk function. The model and its main assumptions are given as follows:

$$\left.\begin{array}{ll} y = X\beta + \epsilon\,, & \\ \mathrm{E}(\epsilon) = 0\,, & \mathrm{E}(\epsilon\epsilon') = \sigma^2 I\,, \\ X \text{ nonstochastic}\,, & \mathrm{rank}(X) = K\,. \end{array}\right\} \tag{3.23}$$

As X is assumed to be nonstochastic, X and ϵ are independent, i.e.,

$$\mathrm{E}(\epsilon \mid X) = \mathrm{E}(\epsilon) = 0\,, \tag{3.24}$$

$$\mathrm{E}(X'\epsilon \mid X) = X'\,\mathrm{E}(\epsilon) = 0\,, \tag{3.25}$$

and

$$\mathrm{E}(\epsilon\epsilon' \mid X) = \mathrm{E}(\epsilon\epsilon') = \sigma^2 I\,. \tag{3.26}$$

The rank condition on X means that there are no linear relations between the K regressors $X_1, \ldots, X_K$; especially, the inverse matrix $(X'X)^{-1}$ exists. Using (3.23) and (3.24) we get the conditional expectation

$$\mathrm{E}(y|X) = X\beta + \mathrm{E}(\epsilon|X) = X\beta\,, \tag{3.27}$$

and by (3.26) the covariance matrix of y is of the form

$$\mathrm{E}[(y - \mathrm{E}(y))(y - \mathrm{E}(y))'|X] = \mathrm{E}(\epsilon\epsilon'|X) = \sigma^2 I\,. \tag{3.28}$$

In the following, all expected values should be understood as conditional on a fixed matrix X.

3.4.1 Linear Estimators

The statistician's task is now to estimate the true but unknown vector β of regression parameters in the model (3.23) on the basis of observations (y, X) and the assumptions already stated. This will be done by choosing a suitable estimator $\hat\beta$ which will then be used to calculate the conditional expectation $\mathrm{E}(y|X) = X\beta$, and an estimate for the error variance σ^2. It is common to choose an estimator $\hat\beta$ that is linear in y, i.e.,

$$\hat\beta = \underset{K\times T}{C}\; y + \underset{K\times 1}{d}\,. \tag{3.29}$$

C and d are nonstochastic matrices, which have been determined by minimizing a suitably chosen risk function in an optimal way.

At first, we have to introduce some definitions.

Definition 3.5. $\hat\beta$ is called a homogeneous estimator of β, if $d = 0$; otherwise $\hat\beta$ is called inhomogeneous.

In descriptive regression analysis, we measured the goodness of fit of the model by the sum of squared errors $S(\beta)$. Analogously, we define for the random variable $\hat\beta$ the quadratic loss function

$$L(\hat\beta, \beta, A) = (\hat\beta - \beta)'A(\hat\beta - \beta)\,, \tag{3.30}$$

where A is a symmetric and, at least, nonnegative–definite $(K \times K)$–matrix.

Remark. We say that $A \geq 0$ (A nonnegative definite) and $A > 0$ (A positive definite) in accordance with Theorems A.21–A.23.

Obviously, the loss (3.30) depends on the sample. Thus, we have to consider the average or expected loss over all possible samples. The expected loss of an estimator will be called risk.

Definition 3.6. The quadratic risk of an estimator $\hat{\beta}$ of β is defined as

$$R(\hat{\beta}, \beta, A) = \mathrm{E}(\hat{\beta} - \beta)' A (\hat{\beta} - \beta) . \tag{3.31}$$

The next step now consists of finding an estimator $\hat{\beta}$ that minimizes the quadratic risk function over a class of appropriate functions. Therefore, we have to define a criterion to compare estimators.

Definition 3.7 ($R(A)$–Superiority). An estimator $\hat{\beta}_2$ of β is called $R(A)$ superior or an $R(A)$ improvement over another estimator $\hat{\beta}_1$ of β, if

$$R(\hat{\beta}_1, \beta, A) - R(\hat{\beta}_2, \beta, A) \geq 0. \tag{3.32}$$

3.4.2 Mean Square Error

The quadratic risk is related closely to the matrix–valued criterion of the mean square error (MSE) of an estimator. The MSE is defined as

$$M(\hat{\beta}, \beta) = \mathrm{E}(\hat{\beta} - \beta)(\hat{\beta} - \beta)' . \tag{3.33}$$

We will denote the covariance matrix (see also Example A.1, Appendix A) of an estimator $\hat{\beta}$ by $\mathrm{V}(\hat{\beta})$:

$$\mathrm{V}(\hat{\beta}) = \mathrm{E}(\hat{\beta} - \mathrm{E}(\hat{\beta}))(\hat{\beta} - \mathrm{E}(\hat{\beta}))' . \tag{3.34}$$

If $\mathrm{E}(\hat{\beta}) = \beta$, then $\hat{\beta}$ will be called unbiased (for β). If $\mathrm{E}(\hat{\beta}) \neq \beta$, then $\hat{\beta}$ is called biased. The difference between $\mathrm{E}(\hat{\beta})$ and β is called

$$\mathrm{Bias}(\hat{\beta}, \beta) = \mathrm{E}(\hat{\beta}) - \beta . \tag{3.35}$$

If $\hat{\beta}$ is unbiased, then obviously $\mathrm{Bias}(\hat{\beta}, \beta) = 0$.

The following decomposition of the mean square error often proves to be useful

$$\begin{aligned} M(\hat{\beta}, \beta) &= \mathrm{E}[(\hat{\beta} - \mathrm{E}(\hat{\beta})) + (\mathrm{E}(\hat{\beta}) - \beta)][(\hat{\beta} - \mathrm{E}(\hat{\beta})) + (\mathrm{E}(\hat{\beta}) - \beta)]' \\ &= \mathrm{V}(\hat{\beta}) + (\mathrm{Bias}(\hat{\beta}, \beta))(\mathrm{Bias}(\hat{\beta}, \beta))' , \end{aligned} \tag{3.36}$$

i.e., the MSE of an estimator is the sum of the covariance matrix and the squared bias. In terms of statistical inference the MSE could be explained as the sum of stochastic and systematic errors made by estimating β through $\hat{\beta}$.

Mean Square Error Superiority

As the MSE contains all relevant information about the quality of an estimator, comparisons between different estimators may be made by comparing their MSE matrices.

Definition 3.8 (MSE–I Criterion). We consider two estimators $\hat{\beta}_1$ and $\hat{\beta}_2$ of β. Then $\hat{\beta}_2$ is called MSE superior to $\hat{\beta}_1$ (or $\hat{\beta}_2$ is called an MSE improvement to $\hat{\beta}_1$), if the difference of their MSE matrices is nonnegative definite, i.e., if

$$\Delta(\hat{\beta}_1, \hat{\beta}_2) = M(\hat{\beta}_1, \beta) - M(\hat{\beta}_2, \beta) \geq 0\,. \tag{3.37}$$

MSE superiority is a local property in the sense that it depends on the particular value of β. The quadratic risk function (3.30) is just a scalar–valued version of the MSE:

$$R(\hat{\beta}, \beta, A) = \operatorname{tr}\{AM(\hat{\beta}, \beta)\}\,. \tag{3.38}$$

One important connection between $R(A)$ and MSE superiority has been given by Theobald (1974) and Trenkler (1981):

Theorem 3.9. Consider two estimators $\hat{\beta}_1$ and $\hat{\beta}_2$ of β. The following two statements are equivalent:

$$\Delta(\hat{\beta}_1, \hat{\beta}_2) \geq 0\,, \tag{3.39}$$

$$R(\hat{\beta}_1, \beta, A) - R(\hat{\beta}_2, \beta, A) = \operatorname{tr}\{A\Delta(\hat{\beta}_1, \hat{\beta}_2)\} \geq 0\,, \tag{3.40}$$

for all matrices of the type $A = aa'$.

Proof. Using (3.37) and (3.38) we get

$$R(\hat{\beta}_1, \beta, A) - R(\hat{\beta}_2, \beta, A) = \operatorname{tr}\{A\Delta(\hat{\beta}_1, \hat{\beta}_2)\}. \tag{3.41}$$

Following Theorem A.20, it holds that $\operatorname{tr}\{A\Delta(\hat{\beta}_1, \hat{\beta}_2)\} \geq 0$ for all matrices $A = aa' \geq 0$ if and only if $\Delta(\hat{\beta}_1, \hat{\beta}_2) \geq 0$.

In practice, β is usually unknown, i.e., expressions like bias or MSE can not be determined. Within simulation experiments where β is determined, the value of these parameters can be estimated ("estimated" because of the individuality of the experiment).

3.4.3 Best Linear Unbiased Estimation

The previous definitions and theorems now enable us to evaluate the estimator β.

In (3.29), the matrix C and vector d are unknown and have to be estimated in an optimal way by minimizing the expectation of the sum of squared errors $S(\hat{\beta})$, namely, the risk function

$$r(\beta, \hat{\beta}) = \mathrm{E}(y - X\hat{\beta})'(y - X\hat{\beta})\,. \tag{3.42}$$

Direct calculus yields the following result:

$$\begin{aligned} y - X\hat{\beta} &= X\beta + \epsilon - X\hat{\beta} \\ &= \epsilon - X(\hat{\beta} - \beta)\,, \end{aligned} \tag{3.43}$$

such that

$$\begin{aligned} r(\beta, \hat{\beta}) &= \mathrm{tr}\{\mathrm{E}(\epsilon - X(\hat{\beta} - \beta))(\epsilon - X(\hat{\beta} - \beta))'\} \\ &= \mathrm{tr}\{\sigma^2 I_T + XM(\hat{\beta}, \beta)X' - 2X\,\mathrm{E}[(\hat{\beta} - \beta)\epsilon']\} \\ &= \sigma^2 T + \mathrm{tr}\{X'XM(\hat{\beta}, \beta)\} - 2\,\mathrm{tr}\{X\,\mathrm{E}[(\hat{\beta} - \beta)\epsilon']\}\,. \end{aligned} \tag{3.44}$$

Now we will specify the risk function $r(\hat{\beta}, \beta)$ for linear estimators, considering *unbiased* estimators only.

Unbiasedness of $\hat{\beta}$ requires that $\mathrm{E}(\hat{\beta} \mid \beta) = \beta$ holds independently of the true β in model (3.23). We will see that this imposes some new restrictions on the matrices to be estimated, i.e.,

$$\begin{aligned} \mathrm{E}(\hat{\beta} \mid \beta) &= C\,\mathrm{E}(y) + d \\ &= CX\beta + d = \beta \quad \text{for all } \beta\,. \end{aligned} \tag{3.45}$$

For the choice $\beta = \mathbf{0}$, we immediately have

$$d = 0 \tag{3.46}$$

and the condition, equivalent to (3.45), is

$$CX = I\,. \tag{3.47}$$

Inserting this into (3.43) yields

$$\begin{aligned} y - X\hat{\beta} &= X\beta + \epsilon - XCX\beta - XC\epsilon \\ &= \epsilon - XC\epsilon\,, \end{aligned} \tag{3.48}$$

and (cf. (3.44))

$$\begin{aligned} \mathrm{tr}\{X\,\mathrm{E}[(\hat{\beta} - \beta)\epsilon']\} &= \mathrm{tr}\{X\,\mathrm{E}(C\epsilon\epsilon')\} \\ &= \sigma^2\,\mathrm{tr}\{XC\} \\ &= \sigma^2\,\mathrm{tr}\{CX\} = \sigma^2\,\mathrm{tr}\{I_K\} = \sigma^2 K\,. \end{aligned} \tag{3.49}$$

Thus we can state the following:

Theorem 3.10. For linear unbiased estimators $\hat{\beta} = Cy$ with $CX = I$, it holds that $M(\hat{\beta}, \beta) = \mathrm{V}(\hat{\beta}) = \sigma^2 CC'$ and

$$r(\hat{\beta}, \beta) = \mathrm{tr}\{(X'X)\,\mathrm{V}(\hat{\beta})\} + \sigma^2(T - 2K)\,. \tag{3.50}$$

If we consider the risk functions $r(\hat{\beta}, \beta)$ and $R(\hat{\beta}, \beta, X'X)$, then we may state:

Theorem 3.11. Let $\hat{\beta}_1$ and $\hat{\beta}_2$ be two linear unbiased estimators. Then

$$\begin{aligned} r(\hat{\beta}_1, \beta) - r(\hat{\beta}_2, \beta) &= \mathrm{tr}\{(X'X) \bigtriangleup (\hat{\beta}_1, \hat{\beta}_2)\} \\ &= R(\hat{\beta}_1, \beta, X'X) - R(\hat{\beta}_2, \beta, X'X)\,, \end{aligned} \tag{3.51}$$

where $\bigtriangleup(\hat{\beta}_1, \hat{\beta}_2) = \mathrm{V}(\hat{\beta}_1) - \mathrm{V}(\hat{\beta}_2)$, i.e., the difference of the covariance matrices only.

Using Theorem 3.10 we get, with $CX = I$,

$$\begin{aligned} r(\hat{\beta}, \beta) &= \sigma^2(T - 2K) + \mathrm{tr}\{X'X\,\mathrm{V}(\hat{\beta})\} \\ &= \sigma^2(T - 2K) + \sigma^2\,\mathrm{tr}\{X'XCC'\}\,. \end{aligned}$$

Minimizing $r(\hat{\beta}, \beta)$ with respect to C leads to an optimum matrix $\hat{C} = (X'X)^{-1}X'$ (Proof 5, Appendix B). Therefore the actual linear unbiased estimator coincides with the descriptive or empirical OLS estimator b and is given by

$$\hat{\beta}_{\mathrm{opt}} = \hat{C}y = (X'X)^{-1}X'y\,, \tag{3.52}$$

being unbiased with the$(K \times K)$–covariance matrix

$$V_b = \sigma^2(X'X)^{-1}\,, \text{(see also Proof 5, Appendix B)}. \tag{3.53}$$

The main reason for the popularity of the OLS b in contrast to other estimators is obvious, as b possesses the minimum variance property among all members of the class of linear unbiased estimators $\tilde{\beta}$. More precisely:

Theorem 3.12. Let $\tilde{\beta}$ be an arbitrary linear unbiased estimator of β with covariance matrix $V_{\tilde{\beta}}$ and let a be an arbitrary $(K \times 1)$–vector.

Then the following two equivalent statements hold:

(a) The difference $V_{\tilde{\beta}} - V_b$ is always nonnegative definite (nnd).

(b) The variance of the linear form $a'b$ is always less than or equal to the variance of $a'b$:

$$a'V_b a \leq a'V_{\tilde{\beta}}a \quad \text{or} \quad a'(V_{\tilde{\beta}} - V_b)a \geq 0\,. \tag{3.54}$$

Proof. See Proof 6, Appendix B; note that Theorem 3.12 also holds for components, i.e., $\mathrm{Var}(\tilde{\beta}_i)$ and $\mathrm{Var}(\tilde{b}_i)$.

The minimum property of b is usually expressed by the fundamental Gauss–Markov theorem.

Theorem 3.13 (Gauss–Markov Theorem). Consider the classical linear regression model (3.23). The OLS estimator

$$b_0 = (X'X)^{-1}X'y\,, \tag{3.55}$$

with covariance matrix

$$V_{b_0} = \sigma^2(X'X)^{-1}\,, \tag{3.56}$$

is the best homogeneous linear unbiased estimator of β in the sense of the two properties of Theorem 3.12. b_0 will also be denoted as a Gauss–Markov estimator.

Estimation of a Linear Function of β

If we are interested in estimating a linear combination of the components of β, e.g., linear contrasts in ANOVA models, then we have to consider

$$d = a'\beta\,, \tag{3.57}$$

where a is a known $(K \times 1)$–vector. For now, it is sufficient to restrict consideration to the linear homogeneous estimators $\tilde{d} = c'y$. Then we have:

Theorem 3.14. In the classical linear regression model (3.23)

$$\hat{d} = a'b_0\,, \tag{3.58}$$

with the variance

$$\mathrm{Var}(\hat{d}) = \sigma^2 a'(X'X)^{-1}a = a'V_{b_0}a\,, \tag{3.59}$$

is the best linear unbiased estimator of $d = a'\beta$. (Proof 7, Appendix B.)

3.4.4 Estimation of σ^2

In this section we want to estimate σ^2, an important parameter characterizing the deviation between the actual and predicted response values. We decided not to put the derivation of $\hat{\sigma}^2$ in the appendix B because it is a simple proof supporting the exposure with the classical linear model.

We start the proof by rewriting $\hat{\epsilon}$ with the help of projection matrices to simplify the computation of $\mathrm{E}(\hat{\epsilon}'\hat{\epsilon})$. This leads to the estimation of σ^2 whose unbiasedness we subsequently prove. Finally, we demonstrate the special case $K = 2$.

The sum of squares $\hat{\epsilon}'\hat{\epsilon}$ of the estimated errors $\hat{\epsilon} = y - \hat{y}$ obviously provides a basis appropriate for estimating σ^2.

In detail, we get

$$\hat{\epsilon} \;=\; y - \hat{y} = X\beta + \epsilon - Xb_0$$

$$\begin{aligned} &= \epsilon - X(X'X)^{-1}X'\epsilon \\ &= (I - X(X'X)^{-1}X')\epsilon \\ &= M\epsilon\,. \end{aligned} \tag{3.60}$$

The matrix M is idempotent by Theorem A.36. As a consequence, the sum of squared errors

$$\hat{\epsilon}'\hat{\epsilon} = \epsilon' M M \epsilon = \epsilon' M \epsilon$$

has expectation

$$\begin{aligned} \mathrm{E}(\hat{\epsilon}'\hat{\epsilon}) &= \mathrm{E}(\epsilon' M \epsilon) \\ &= \mathrm{E}(\mathrm{tr}\{\epsilon' M \epsilon\}) \quad \text{[Theorem A.1(vi)]} \\ &= \mathrm{E}(\mathrm{tr}\{M \epsilon' \epsilon\}) \\ &= \mathrm{tr}\{M\,\mathrm{E}(\epsilon\epsilon')\} \\ &= \sigma^2\,\mathrm{tr}\{M\} \\ &= \sigma^2\,\mathrm{tr}\{I_T\} - \sigma^2\,\mathrm{tr}\{X(X'X)^{-1}X'\} \quad \text{[Theorem A.1(i)]} \\ &= \sigma^2\,\mathrm{tr}\{I_T\} - \sigma^2\,\mathrm{tr}\{(X'X)^{-1}X'X\} \\ &= \sigma^2\,\mathrm{tr}\{I_T\} - \sigma^2\,\mathrm{tr}\{I_K\} \\ &= \sigma^2(T-K)\,. \end{aligned} \tag{3.61}$$

An unbiased estimator for σ^2 is then given by

$$s^2 = \hat{\epsilon}'\hat{\epsilon}(T-K)^{-1} = (y - Xb_0)'(y - Xb_0)(T-K)^{-1}\,. \tag{3.62}$$

Hence, an unbiased estimator of V_{b_0} is given by

$$\hat{V}_{b_0} = s^2(X'X)^{-1}\,. \tag{3.63}$$

Bivariate Regression $K = 2$

The important special case $K = 2$ of the general linear model with K regressors $X_1, \ldots, X_K$ deserves attention. If there is only one true explanatory variable accompanied by a dummy regressor, i.e., a column of 1's, then we speak of the simple linear regression model

$$y_t = \alpha + \beta x_t + \epsilon_t \quad (t = 1, \ldots, T)\,. \tag{3.64}$$

It is often useful to transform the observations (x_t, y_t) in a way that $(\tilde{x}_t, \tilde{y}_t)$ represent deviations of the sample means $(\bar{x}_t, \bar{y}_t)$:

$$\tilde{y}_t = y_t - \bar{y}\,, \quad \tilde{x}_t = x_t - \bar{x}\,. \tag{3.65}$$

As

$$\mathrm{E}(\tilde{y}_t | x_1, \ldots, x_T) = \alpha + \beta x_t - (\alpha + \beta\bar{x}) = \beta\tilde{x}_t\,,$$

we are able to obtain an even simpler form of the model (3.64), while the parameter β remains unchanged, i.e.,

$$\tilde{y}_t = \beta\tilde{x}_t + \tilde{\epsilon}_t \quad (t = 1, \ldots, T)\,. \tag{3.66}$$

Assuming that $\bar{\epsilon} = 1/T \sum \epsilon_t = 0$, we have $\tilde{\epsilon}_t = \epsilon_t$ for all t. The OLS estimator of β and the unbiased estimator of σ^2 are obtained by (B.34) and (3.62) as

$$b = \frac{\sum \tilde{x}_t \tilde{y}_t}{\sum \tilde{x}_t^2} \quad \text{with} \quad \text{Var}(b) = \frac{\sigma^2}{\sum \tilde{x}_t^2} \,. \tag{3.67}$$

$$s^2 = (T-2)^{-1} \sum (\tilde{y}_t - \tilde{x}_t b)^2 \,. \tag{3.68}$$

From the right–hand side of (3.67) one can easily see what $\sigma^2(X'X)^{-1}$ looks like for $K = 2$.

It is easy to see that the OLS estimator for α is given by

$$\hat{\alpha} = \bar{y} - b\bar{x} \,. \tag{3.69}$$

Example 3.1. We are interested in modeling the dependence of advertising x, on sales increase y, of 10 department stores:

i	y_i	x_i	$y_i - \bar{y}$	$x_i - \bar{x}$	$(x_i - \bar{x})(y_i - \bar{y})$
1	2.0	1.5	−5.0	−2.5	12.5
2	3.0	2.0	−4.0	−2.0	8.0
3	6.0	3.5	−1.0	−0.5	0.5
4	5.0	2.5	−2.0	−1.5	3.0
5	1.0	0.5	−6.0	−3.5	21.0
6	6.0	4.5	−1.0	0.5	−0.5
7	5.0	4.0	−2.0	0.0	0.0
8	11.0	5.5	4.0	1.5	6.0
9	14.0	7.5	7.0	3.5	24.5
10	17.0	8.5	10.0	4.5	45.0
$\sum$	70	40	0.0	0.0	
	$\bar{y} = 7$	$\bar{x} = 4$	$S_{yy} = 252$	$S_{xx} = 60$	$S_{xy} = 120$

Using $\hat{\beta} = s_{xy}/s_{xx}$ and (3.69) leads to the model

$$y_t = -1 + 2x_t$$

which is easily calculated by $\hat{\beta} = \frac{120}{60}$ and $\hat{\alpha} = 7 - 2 * 4$. The coefficient of determination results from

$$R^2 = r^2 = s_{xy}^2/(s_{xx}s_{yy}) = 120^2/(60 * 252) \,.$$

Running the linear regression in SPLUS for the above data set produces the following output:

```
*** Linear Model ***

Call: lm(formula = Y ~ X, data = kaufhaus, na.action = na.omit)
Residuals:
 Min         1Q      Median 3Q Max
  -2 -9.384e-016 1.404e-015  1   1
```

```
Coefficients:
                Value Std. Error  t value Pr(>|t|)
(Intercept)  -1.0000   0.7416    -1.3484   0.2145
          X   2.0000   0.1581    12.6491   0.0000

Residual standard error: 1.225 on 8 degrees of freedom
Multiple R--Squared: 0.9524
F--statistic: 160 on 1 and 8 degrees of freedom, the $p$--value is 1.434e-006
```

Running the "linear regression" procedure with $\tilde{y}$ and $\tilde{x}$ leads to the results shown in (3.66).

3.5 Multicollinearity

3.5.1 *Extreme Multicollinearity and Estimability*

A typical problem in practical work is that there is almost always at least some correlation between the exogeneous variables in X. We speak of extreme multicollinearity if two or more columns in X are linearly dependent, i.e., if one is a linear combination of the others. As a consequence, we have $\text{rank}(X) < K$ such that one basic assumption of model (3.23) is violated. In this case, no unbiased linear estimators for β exist.

We recall that the condition for unbiasedness is equivalent to $d = 0$ and $CX = I$ (cf. (3.47)). If $\text{rank}(X) = p < K$, then CX is of rank p at most, cf. Theorem A.6(iv), whereas the identity matrix I_K is of rank K. Condition (3.47) is thus never fulfilled.

This result could be proven in an alternative way, as you will see in Proof 8, Appendix B.

The matrix $(X'X)$ is singular, since $\text{rank}(X) < K$ and solutions to the normal equation (3.10) are no longer unique.

We say that the parameter vector β is not estimable in the sense that no linear unbiased estimator exists.

Another problem occurring with extreme multicollinearity becomes apparent when considering, without loss of generality, for x_1, a linear combination consisting of all other columns, i.e.,

$$x_1 = \sum_{k=2}^{K} \alpha_k x_k \,.$$

For an arbitrary scalar $\lambda \neq 0$, we can derive the decomposition

$$X\beta = \sum_{k=1}^{K} x_k \beta_k = (1 - \lambda)\beta_1 x_1 + \sum_{k=2}^{K} (\beta_k + \lambda \alpha_k \beta_1) x_k$$

$$= \tilde{\beta}_1 x_1 + \sum_{k=2}^{K} \tilde{\beta}_k x_k = X\tilde{\beta}, \tag{3.70}$$

where $\tilde{\beta}_1 = (1-\lambda)\beta_1$, $\tilde{\beta}_k = (\beta_k + \lambda\alpha_k\beta_1), k = 2, \ldots, K$. This means, that the parameter vectors β and $\tilde{\beta}$ with $\beta \neq \tilde{\beta}$ yield the same systematical component $X\beta = X\tilde{\beta}$. Now the observations y do not depend directly, but over $X\beta$ on β.

This means that the information in y therefore does not allow us to distinguish between β and $\tilde{\beta}$. The regression coefficients are *not identifiable*, the related models are *observational equivalent.*

Example 3.2. We consider the model

$$y_t = \alpha + \beta x_t + \epsilon_t \quad (t = 1, \ldots, T). \tag{3.71}$$

Exact linear dependence between $X_1 \equiv 1$ and $X_2 = X$ means that $x_1 = \ldots = x_t = a$ (a constant), such that $\sum(x_t - \bar{x})^2 = 0$ and b (3.67) cannot be calculated.

Let $\binom{\hat{\alpha}}{\hat{\beta}} = Cy$ be a linear homogeneous estimator of $(\alpha, \beta)'$. Unbiasedness requires that (3.47) is fulfilled, such that

$$\begin{pmatrix} \sum c_{1t} & a\sum c_{1t} \\ \sum c_{2t} & a\sum c_{2t} \end{pmatrix} = \begin{pmatrix} 1 & 0 \\ 0 & 1 \end{pmatrix}. \tag{3.72}$$

There exists no matrix C and no real–valued $a \neq 0$; $(\alpha, \beta)'$ are not estimable. Since $x_t = a$ for all t, we have $y_t = (\alpha + \beta a) + \epsilon_t$, such that α and β are only jointly estimable as $\widehat{(\alpha + \beta a)} = \bar{y}$.

3.5.2 Estimation within Extreme Multicollinearity

We are mainly interested in making use of a prior restriction of the form (B.12) with $r = 0$, i.e.,

$$0 = R\beta. \tag{3.73}$$

Parameter values that are observational equivalent are thus excluded.

The identifiability of β is guaranteed if $RX = 0$ and the assumptions of Theorem B.1 are fulfilled. Following Theorem B.1, the OLS estimator of β is of the form

$$b(R, 0) = b(R) = (X'X + R'R)^{-1}X'y, \tag{3.74}$$

if $r = 0$. Summarizing, we may state: In the classical linear restrictive regression model

$$\left.\begin{array}{l} y = X\beta + \epsilon\,, \\ \mathrm{E}(\epsilon) = 0, \quad \mathrm{E}(\epsilon\epsilon') = \sigma^2 I\,, \\ X \text{ nonstochastic, } \operatorname{rank}(X) = p < K\,, \\ 0 = R\beta, \ \operatorname{rank}(R) = K - p, \ \operatorname{rank}(D) = K\,, \end{array}\right\} \tag{3.75}$$

with $D' = (X', R')$, the following fundamental theorem is valid.

Theorem 3.15. In model (3.75), the conditional OLS estimator

$$b(R) = (X'X + R'R)^{-1}X'y = (D'D)^{-1}X'y\,, \tag{3.76}$$

with covariance matrix

$$V_{b(R)} = \sigma^2(D'D)^{-1}X'X(D'D)^{-1}\,, \tag{3.77}$$

is the best linear unbiased estimator of β.

Definition 3.16. A linear estimator $\hat\beta$ is called conditionally unbiased under

$$\underset{K\times K}{A}\ \beta - \underset{K\times 1}{a} = 0\,,$$

if

$$\mathrm{E}(\hat\beta - \beta \mid A\beta - a = 0) = 0\,. \tag{3.78}$$

Proof of Theorem 3.15. See Proof 9, Appendix B.

Extreme multicollinearity is a problem usually not occurring in descriptive linear regression, i.e., when analyzing sample data, because an exact linear dependency between sampled data is unusual. In experimental designs, however, where factors are fixed, extreme multicollinearity is present. Assuming a simple case with one factor on $s = 2$ levels with n_s observations each, the linear model $y = X\beta + \epsilon$ could be written according to

$$\begin{pmatrix} y_{11} \\ \vdots \\ y_{1n_1} \\ y_{21} \\ \vdots \\ y_{2n_2} \end{pmatrix} = \begin{pmatrix} 1 & 1 & 0 \\ \vdots & \vdots & \vdots \\ 1 & 1 & 0 \\ 1 & 0 & 1 \\ \vdots & \vdots & \vdots \\ 1 & 0 & 1 \end{pmatrix} \begin{pmatrix} \mu \\ \alpha_1 \\ \alpha_2 \end{pmatrix} + \begin{pmatrix} \epsilon_{11} \\ \vdots \\ \epsilon_{1n_1} \\ \epsilon_{21} \\ \vdots \\ \epsilon_{2n_2} \end{pmatrix}. \tag{3.79}$$

As can easily be seen from (3.79) the $(n \times 3)$–matrix X has rank $s = 2$ because the first column representing the intercept is the sum of the last two columns, leading to a case of extreme multicollinearity. Using the conditional least squares by (3.73) with $r = 0$ and $R' = (0, n_1, n_2)$, i.e., $\sum \alpha_i n_i = 0$, guarantees the estimability of β because $\operatorname{rank}(X, R')' = s + 1 = 3$.

3.5.3 Weak Multicollinearity

When analyzing a data set by the linear model $y = X\beta + \epsilon$ with X not being a fixed factor (which would mean having the problem of extreme multicollinearity), a more common problem is weak multicollinearity. Weak multicollinearity means that there is no exact (but close) linear dependency between the exogenous variables, i.e., X is still of full rank. $X'X$ is regular and the results remain valid, especially, b still is the best linear unbiased estimator. The problem, however, occurs because one or more eigenvalues, which are nearly zero, lead to a determinant of $X'X$ used for computing $\sigma^2(X'X)^{-1}$ which is also going to be near zero. This means that $V_b = \sigma^2(X'X)^{-1}$ grows large and the estimates become unreliable.

In other words, there is not enough information to estimate the independent influences of some covariates on the response. The effect of each independent variable cannot be separated from the remaining variables. Ridge, shrinkage, or principal component regression are ad–hoc procedures which cope with multicollinearity in its weak form. However, they are controversial, and popular statistical software does not offer these methods; so we abandon a description of these.

Apart from considering the correlation between the exogenous variables, in order to find the source of the problem and possibly remove it in practice, some other alternatives might be:

- additional observations to reduce the correlation between some variables within a fixed model (experimental designs);
- linear transformations, e.g., building differences;
- eliminate trends (Schneeweiß (1990));
- use additional information such as a priori estimates $r = R\beta + d$, d being an error term; and
- exact linear restrictions.

Our main interest is the use of linear restrictions and external information. Using exact linear restrictions with $r = 0$, i.e.,

$$0 = R\beta\,, \tag{3.80}$$

means that the parameter β is subjected to limitations in the range of values in its components.

Finally, we want to illustrate the problem of weak multicollinearity with the help of a multiple regression, analyzing data from the demographic information of 122 countries (with the most data being from 1992). We decided to use SPSS within this framework because it provides some diagnostics for evaluating multicollinearity in a simple way.

Example 3.3. We are interested in predicting female life expectancy for a sample of 122 countries. Within a multiple regression model the variables shown in Table 3.1 specifying economic and health–care delivery characteristics are included in the analysis.

Variable Name	Description
Urban	Percentage of the population living in urban areas
lndocs	ln(number of doctors per 10,000 people)
lnbeds	ln(number of hospital beds per 10,000 people)
lngdp	ln(per capita gross domestic product in dollars)
lnradios	ln(radios per 100 people)

TABLE 3.1. Variable declaration.

When plotting each independent variable against the response it can be seen that only "urban" shows a linear relation to female life expectancy. In order to attain this relation for all other covariates also they should be transformed by the natural log leading to the variables described in Table 3.1.

First of all we consider the partial correlation coefficients. Each independent variable should correlate with the response because of the postulated linear relation. Between the independent variables correlation should not be present because of the possible problems already described theoretically.

	lifeexpf	urban	lndocs	lnbeds	lngdp	lnradio
lifeexpf	1.000	0.698**	0.870**	0.678**	0.806**	0.652**
urban	0.698**	1.000	0.765**	0.576**	0.751**	0.583**
lndocs	0.870**	0.765**	1.000	0.711**	0.824**	0.621**
lnbeds	0.678**	0.576**	0.711**	1.000	0.741**	0.616**
lngdp	0.806**	0.751**	0.824**	0.741**	1.000	0.709**
lnradios	0.652**	0.583**	0.621**	0.616**	0.709**	1.000

TABLE 3.2. ** Correlation (Pearson) is significant at the 0.01 level (two–tailed).

We abandon the p–values of the corresponding test for $H_0 : \rho = 0$ because they all indicate a significance at the 1% level. The first row shows the correlation between the response and the covariates. We see that a linear relation seems to be adequate. However, we also identify high correlation between the independent variables themselves, especially for "lndocs" and "lngdp". Whether this leads to a problem of multicollinearity has to be verified by further analysis. In the next step we run the linear regression by entering all variables.

R^2	R^2_{adj}	Standard error of the estimate	Change statistics		
			R Square Ch.	F Ch.	Sig. F Ch.
0.827	0.819	4.74	0.827	105.336	0.000

TABLE 3.3. Model summary.

From table 3.3 we should especially remember R^2 and R^2_{adj} for comparisons with other models. The ANOVA table was also abandoned because the focus here lies on coefficients and first collinearity diagnostics.

	Unstand. coefficients				Collinearity statistics	
Model	β	Std. error	t	Sig.	Tolerance	VIF
(Constant)	40.767	3.174	12.845	0.000		
lndocs	4.069	0.563	7.228	0.000	0.253	3.950
lnradios	1.542	0.686	2.247	0.027	0.467	2.140
lngdp	1.709	0.616	2.776	0.006	0.217	4.614
urban	-2.002E-02	0.029	0.686	0.494	0.371	2.699
lnbeds	1.147	0.749	1.532	0.128	0.406	2.461

TABLE 3.4. Coefficients (dependent variable: female life expectancy, 1992).

"lndocs", "lnradios" and "lngdp" have an influence on the female life expectancy within the saturated model (see Table 3.4). The last two columns give evidence to the existence of multicollinearity. The tolerance tells us whether linear relations upon the independent variables are present. This is the proportion of a variable's variance not accounted for by other independent variables. "VIF" is the reciprocal of tolerance and stands for the inflation factor. Its increase means an increase in the variance of $\hat{\beta}$ and thus an unstable estimate $\hat{\beta}$. A large "VIF" is therefore an indicator for multicollinearity.

Considering the variance inflation factor may cause doubt, in the independence between 'lngdp'" and the further covariates, because of its high value. Indicators for multicollinearity known from matrix theory are the eigenvalues of $X'X$, X denoting the independent variables. SPSS offers the eigenvalues within the "collinearity diagnostics" as well as the condition index which is the square root of the ratio between the largest eigenvalue and the actual eigenvalue. Condition indices larger than 15 indicate a problem with multicollinearity, values larger than 30 indicate a serious problem.

As we could not specify variables directly from the table containing the eigenvalues we remember the above results (especially the correlation and variance inflation factor), and we may conclude, that the variable describing

	Eigenvalue	Condition
1	5.510	1.000
2	0.360	3.911
3	6.608E-02	9.132
4	3.356E-02	12.813
5	2.360E-02	15.281
6	6.798E-03	28.469

TABLE 3.5. Collinearity diagnostics.

the per capita gross domestic product could be the reason for multicollinearity. A first way to check this may be the elimination of "lngdp" and then rerun the analysis leading to the following results.

R^2	R^2_{adj}	Standard error of the estimate	Change statistics		
			R Square Ch.	F Ch.	Sig. F Ch.
0.815	0.808	4.88	0.815	122.352	0.000

TABLE 3.6. Model summary.

	Unstand. coefficients				Collinearity statistics	
Model	β	Std. Error	t	Sig.	Tolerance	VIF
(Constant)	47.222	2.224	21.229	0.000		
lndocs	4.670	0.535	8.728	0.000	0.297	3.365
lnradios	2.177	0.666	3.268	0.001	0.526	1.902
urban	2.798E-03	0.006	0.097	0.923	0.402	2.485
lnbeds	1.786	0.148	2.434	0.017	0.449	2.229

TABLE 3.7. Coefficients (dependent variable: female life expectancy, 1992).

Comparing the primary model with the reduced model step by step (see Tables 3.6, 3.7, and 3.8) confirms the elimination of "lngdp". The elimination of "lngdp" leads to a decrease in the adjusted R^2 but the difference is just marginal. Analyzing the coefficients shows that the standard errors of all variables have decreased denoting more stable estimates. The parameter estimates changed more or less slightly to a larger value, especially that of "urban" where even the sign changed and whose values of the relative change (here not shown) are maximal. The two variables "lndocs" and "lnradios" are still significantly different from zero and, additionally, "lnbeds" is now a further covariate with an essential influence on female life expectancy. Last, but not least, we observe a decrease in the condition indices, especially a decrease in the maximum ratio which changed from 28.469 to 14.251.

	Eigenvalue	Condition
1	4.532	1.000
2	.347	3.615
3	6.579E-02	8.300
4	3.312E-02	11.697
5	2.232E-02	14.251

TABLE 3.8. Collinearity diagnostics.

There is no general guide as to when multicollinearity seems to be a problem even though indicators point to this more or less explicitly. We have demonstrated a possible solution which, in practice, should be arranged in terms of logical consistency concerning its context. This proceeding seems to be similar to a variable selection. But here we have just tried to overcome the problem of multicollinearity by eliminiating possible sources with the help of criteria concerning the constitution of X.

3.6 Classical Regression under Normal Errors

All the results obtained so far are valid, irrespective of the actual distribution of the random disturbances ϵ, provided that $\mathrm{E}(\epsilon) = \mathbf{0}$ and $\mathrm{E}(\epsilon\epsilon') = \sigma^2 I$. Now we shall specify the type of the distribution of ϵ by additionally imposing the following condition: The vector ϵ of the random disturbances ϵ_t is distributed according to a T–dimensional normal distribution $N(\mathbf{0}, \sigma^2\mathbf{I})$, i.e., $\epsilon \sim N(0, \sigma^2 I)$. The probability density of ϵ is given by

$$\begin{aligned} f(\epsilon; 0, \sigma^2 I) &= \prod_{t=1}^{T} (2\pi\sigma^2)^{-1/2} \exp\left(-\frac{1}{2\sigma^2}\epsilon_t^2\right) \\ &= (2\pi\sigma^2)^{-T/2} \exp\left\{-\frac{1}{2\sigma^2}\sum_{t=1}^{T}\epsilon_t^2\right\}, \end{aligned} \tag{3.81}$$

such that its components $\epsilon_t, t = 1, \ldots, T$, are independent and identically distributed (i.i.d.) as $N(0, \sigma^2)$. Equation (3.81) is a special case of the general T–dimensional normal distribution $N(\mu, \Sigma)$.

Let $\Xi \sim N_T(\mu, \Sigma)$, i.e., $\mathrm{E}(\Xi) = \mu$, $\mathrm{E}(\Xi - \mu)(\Xi - \mu)' = \Sigma$. Then Ξ is normally distributed with density

$$f(\Xi; \mu, \Sigma) = \{(2\pi)^T|\Sigma|\}^{-1/2} \exp\{-1/2(\Xi - \mu)'\Sigma^{-1}(\Xi - \mu)\}\,. \tag{3.82}$$

The classical linear regression model under normal errors is given by

$$\left.\begin{array}{l} y = X\beta + \epsilon\,, \\ \epsilon \sim N(0, \sigma^2 I)\,, \\ X \text{ nonstochastic, } \operatorname{rank}(X) = K\,. \end{array}\right\} \tag{3.83}$$

The Maximum Likelihood Principle

Definition 3.17. Let $\Xi = (\xi_1, \ldots, \xi_n)'$ be a random variable with density function $f(\Xi; \Theta)$, where the parameter vector $\Theta = (\Theta_1, \ldots, \Theta_m)'$ is a member of the parameter space Ω comprising all values that are a priori admissible.

The basic idea of the Maximum Likelihood (ML) principle is to interpret the density $f(\Xi; \Theta)$ for a specific realization of the sample Ξ_0 of Ξ as a function of Θ:

$$L(\Theta) = L(\Theta_1, \ldots, \Theta_m) = f(\Xi_0; \Theta) .$$

$L(\Theta)$ will be denoted as the likelihood function of Ξ_0.

The ML principle now postulates to choose a value $\hat{\Theta} \in \Omega$ which maximizes the likelihood function, i.e.,

$$L(\hat{\Theta}) \geq L(\Theta) \quad \text{for all} \quad \Theta \in \Omega .$$

Note that $\hat{\Theta}$ may not be unique. If we consider all possible samples, then $\hat{\Theta}$ is a function of Ξ and is thus a random variable itself. We will call it the maximum likelihood estimator (MLE) of Θ.

ML Estimation in Classical Normal Regression

Following Theorem A.55, we have for y, from (3.23),

$$y = X\beta + \epsilon \sim N(X\beta, \sigma^2 I) , \tag{3.84}$$

such that the Likelihood function of y is given by

$$L(\beta, \sigma^2) = (2\pi\sigma^2)^{-T/2} \exp\left\{ -\frac{1}{2\sigma^2}(y - X\beta)'(y - X\beta) \right\} . \tag{3.85}$$

The logarithmic transformation is monotonic. Hence, it is appropriate to maximize $\ln L(\beta, \sigma^2)$ instead of $L(\beta, \sigma^2)$, as the maximizing argument remains unchanged,

$$\ln L(\beta, \sigma^2) = -\frac{T}{2} \ln(2\pi\sigma^2) - \frac{1}{2\sigma^2}(y - X\beta)'(y - X\beta) . \tag{3.86}$$

If there are no a priori restrictions on the parameters, then the parameter space is given by $\Omega = \{\beta; \sigma^2 : \beta \in \mathcal{R}^K; \sigma^2 > 0\}$. We derive the ML estimators of β and σ^2 by equating the first derivatives to zero (Theorems A.63–A.67)

$$\partial \ln L / \partial \beta = 1/2\sigma^2 2X'(y - X\beta) = 0 , \tag{3.87}$$
$$\partial \ln L / \partial \sigma^2 = -T/2\sigma^2 + 1/2(\sigma^2)^2 (y - X\beta)'(y - X\beta) = 0 . \tag{3.88}$$

The *likelihood equations* are given by

$$\left.\begin{aligned} (I) \quad X'X\hat{\beta} &= X'y\,, \\ (II) \quad \hat{\sigma}^2 &= 1/T(y - X\hat{\beta})'(y - X\hat{\beta})\,. \end{aligned}\right\} \tag{3.89}$$

Equation (I) is identical to the well-known normal equation (3.10). Its solution is unique, as $\text{rank}(X) = K$, and we get the unique ML estimator

$$\hat{\beta} = b = (X'X)^{-1}X'y\,. \tag{3.90}$$

If we compare (II) with the unbiased estimator s^2 (3.62) for σ^2, we immediately see that

$$\hat{\sigma}^2 = \frac{T-K}{T}s^2\,, \tag{3.91}$$

such that $\hat{\sigma}^2$ is a biased estimator. The asymptotic expectation is given by (cf. A.71 (i))

$$\lim_{T\to\infty} \mathrm{E}(\hat{\sigma}^2) = \bar{\mathrm{E}}(\hat{\sigma}^2) = \mathrm{E}(s^2) = \sigma^2\,. \tag{3.92}$$

Thus we can state:

Theorem 3.18. The maximum likelihood estimator and the ordinary least squares estimator of β are identical in the model (3.84) of classical normal regression. The ML estimator $\hat{\sigma}^2$ of σ^2 is asymptotically unbiased.

Remark. The Cramér–Rao bound defines a lower bound (in the sense of the definiteness of matrices) for the covariance matrix of unbiased estimators. In the model of normal regression, the Cramér–Rao bound is given by (Amemiya, 1985, p. 19)

$$\mathrm{V}(\tilde{\beta}) \geq \sigma^2(X'X)^{-1},$$

where $\tilde{\beta}$ is an arbitrary estimator. The covariance matrix of the ML estimator is just identical to this lower bound, such that b is the best unbiased estimator in the linear regression model under normal errors.

3.7 Testing Linear Hypotheses

In this section, testing procedures, such as for $\mathrm{H}_0 : \beta_1 = \beta_2 = \beta_3$, for example, are being derived in order to test linear hypotheses in the model (3.83) of classical normal regression. The general linear hypothesis,

$$\mathrm{H}_0 : R\beta = r, \quad \sigma^2 > 0 \quad \text{arbitrary}\,, \tag{3.93}$$

is usually tested against the alternative

$$\mathrm{H}_1 : R\beta \neq r, \quad \sigma^2 > 0 \quad \text{arbitrary}\,, \tag{3.94}$$

where the following will be assumed:

$$\left.\begin{array}{l} \underset{(K-s)\times K}{R}, \\ \underset{(K-s)\times 1}{r}, \\ \\ R, r \text{ nonstochastic and known}, \\ \text{rank}(R) = K - s, \\ s \in \{0, 1, \ldots, K-1\}\,. \end{array}\right\} \tag{3.95}$$

The hypothesis H_0 expresses the fact that the parameter vector β obeys $(K - s)$ exact linear restrictions which are independent, as it is required that $\text{rank}(R) = K - s$. The general linear hypothesis (3.93) contains two main special cases:

Case 1: $s = 0$
The $(K \times K)$–matrix R is regular, by assumption (3.95), and we may express H_0 and H_1 in the following form:

$$\begin{array}{lll} H_0: & \beta = R^{-1}r = \beta^*, & \sigma^2 > 0 \text{ arbitrary}, \\ H_1: & \beta \neq \beta^*, & \sigma^2 > 0 \text{ arbitrary}. \end{array} \tag{3.96}$$

Case 2: $s > 0$
We choose an $(s \times K)$–matrix G complementary to R such that the $(K \times K)$–matrix $\begin{pmatrix} G \\ R \end{pmatrix}$ is regular of rank K. For exact notation, see Proof 10, Appendix B.

Then we may write

$$\begin{aligned} y = X\beta + \epsilon &= X \begin{pmatrix} G \\ R \end{pmatrix}^{-1} \begin{pmatrix} G \\ R \end{pmatrix} \beta + \epsilon \\ &= \tilde{X} \begin{pmatrix} \tilde{\beta}_1 \\ \tilde{\beta}_2 \end{pmatrix} + \epsilon \\ &= \tilde{X}_1 \tilde{\beta}_1 + \tilde{X}_2 \tilde{\beta}_2 + \epsilon\,. \end{aligned}$$

The latter model obeys all the assumptions (3.23). The hypotheses H_0 and H_1 are thus equivalent to

$$\begin{array}{llll} H_0: & \tilde{\beta}_2 = r, & \tilde{\beta}_1 & \text{and } \sigma^2 > 0 \text{ arbitrary}, \\ H_1: & \tilde{\beta}_2 \neq r, & \tilde{\beta}_1 & \text{and } \sigma^2 > 0 \text{ arbitrary}. \end{array} \tag{3.97}$$

Let Ω be the whole parameter space (either H_0 or H_1 are valid) and let $\omega \subset \Omega$ be the subspace in which only H_0 is true, i.e.,

$$\begin{aligned} \Omega &= \{\beta; \sigma^2 : \beta \in E^K, \sigma^2 > 0\}, \\ \omega &= \{\beta; \sigma^2 : \beta \in E^K \text{ and } R\beta = r, \sigma^2 > 0\}. \end{aligned} \tag{3.98}$$

As a genuine test statistic, we will use the likelihood ratio

$$\lambda(y) = \frac{\max_\omega L(\Theta)}{\max_\Omega L(\Theta)}, \tag{3.99}$$

which may be derived in terms of model (3.84) in the following way. $L(\Theta)$ attains its maximum at the ML estimator $\hat{\Theta}$. Let $\Theta = (\beta, \sigma^2)$, then it holds that

$$\begin{aligned}
\max_{\beta,\sigma^2} L(\beta, \sigma^2) &= L(\hat{\beta}, \hat{\sigma}^2) \\
&= (2\pi\hat{\sigma}^2)^{-T/2} \exp\left\{-1/2\hat{\sigma}^2 (y - X\hat{\beta})'(y - X\hat{\beta})\right\} \\
&= (2\pi\hat{\sigma}^2)^{-T/2} \exp\{-T/2\}
\end{aligned} \tag{3.100}$$

and, therefore,

$$\lambda(y) = \left(\frac{\hat{\sigma}^2_\omega}{\hat{\sigma}^2_\Omega}\right)^{-T/2}, \tag{3.101}$$

where $\hat{\sigma}^2_\omega$ and $\hat{\sigma}^2_\Omega$ are the ML estimators of σ^2 under H_0 and in Ω. The random variable $\lambda(y)$ can take values between 0 and 1, as is obvious from (3.99). If H_0 is true, the numerator of $\lambda(y)$ should be greater than the denominator, so that $\lambda(y)$ should be close to one in repeated samples. On the other hand, $\lambda(y)$ should be close to zero if H_1 is true. Consider the linear transform of $\lambda(y)$:

$$\begin{aligned}
F &= \{(\lambda(y))^{-2/T} - 1\}(T - K)(K - s)^{-1} \\
&= \frac{\hat{\sigma}^2_\omega - \hat{\sigma}^2_\Omega}{\hat{\sigma}^2_\Omega} \cdot \frac{T - K}{K - s}.
\end{aligned} \tag{3.102}$$

If $\lambda \to 0$, then $F \to \infty$ and if $\lambda \to 1$ we have $F \to 0$, such that "F is close to 0" if H_0 seems to be true and "F is sufficiently large" if H_1 is supposed to be true. The determination of F and its distribution for the two special cases $s = 0$ and $s > 0$ is shown in Proof 11, Appendix B. The resulting distribution of the test statistic F is $F_{K-s,T-K}(\sigma^{-2}(\beta_2 - r)'D(\beta_2 - r))$ under H_1, D being symmetric and regular, resulting from the inversion of the partitioned matrix, and central $F_{K-s,T-K}$ under H_0. The region of acceptance of H_0 at a level of significance α is then given by

$$0 \le F \le F_{K-s,T-K,1-\alpha}. \tag{3.103}$$

Accordingly, the critical area of H_0 is given by

$$F > F_{K-s,T-K,1-\alpha}. \tag{3.104}$$

Example 3.4. Assume that we want to test for $\mathrm{H}_0 : \beta_1 = \beta_2 = \beta_3$. One solution to this problem, with respect to $R\beta = r$ with its assumptions

(3.95), is based on the equations

$$(1) \quad \beta_1 - \beta_2 = 0\,, \tag{3.105}$$

and

$$(2) \quad \beta_2 - \beta_3 = 0\,, \tag{3.106}$$

leading to

$$R = \begin{pmatrix} 1 & -1 & 0 \\ 0 & 1 & -1 \end{pmatrix} \begin{pmatrix} \beta_1 \\ \beta_2 \\ \beta_3 \end{pmatrix} = \begin{pmatrix} 0 \\ 0 \end{pmatrix} . \tag{3.107}$$

R in (3.107) has rank 2 but is not the only solution. Its structure depends on the system of equations (3.105) and (3.106). A similar, but not the same case is the test for $H_0 : \beta_1 = \beta_2 = \beta_3 = 0$. One system of equations may be

$$(1) \quad \beta_1 = 0\,, \tag{3.108}$$

$$(2) \quad \beta_2 - \beta_1 = 0\,, \tag{3.109}$$

$$(3) \quad \beta_3 - \beta_2 = 0 \tag{3.110}$$

(3.111)

leading to

$$R = \begin{pmatrix} 1 & 0 & 0 \\ -1 & 1 & 0 \\ 0 & -1 & 1 \end{pmatrix} \tag{3.112}$$

or, in another way, simply to

$$R = \begin{pmatrix} 1 & 0 & 0 \\ 0 & 1 & 0 \\ 0 & 0 & 1 \end{pmatrix} , \tag{3.113}$$

i.e., $I\beta = 0$. Obviously, one has to be careful when handling linear hypotheses with its test situation and the corresponding estimation.

One simple example of testing a linear hypothesis is $H_0 : \beta_1 = 0$. This corresponds to the well–known t–test for testing if the parameter β differs from zero concerning its influence on y. Another example comes from analysis of variance where linear contrasts can be tested. Assuming a categorical covariate and a linear contrast, which tests if the means $\bar{y}_1, \bar{y}_2$ for different levels of factor A are the same, is the analog for testing $H_0 : \beta_1 = \beta_2$. Concerning the use of statistical software within testing linear hypotheses the user may hope to have a simple problem as above. A similar problem occurs when the aim is the estimation of a restrictive least squares estimator. One possibility is to compute R by the corresponding system of equations such as (3.105) and (3.106) and the well–known estimate $(X'X + R'R)^{-1}X'y$ by a software such as MAPLE used for analytical solutions.

3.8 Analysis of Variance and Goodness of Fit

Having only independent variables which are noncontinuous leads to the analysis of variance. One main aim is to test if factors have individual or joint influence on the response. The analysis of variance is also an instrument for reviewing the goodness of fit of the chosen model. The decomposition of the sum of squares is building the body for the analysis of variance which causes us to start with bivariate regression illustrating the derivation of this main context.

3.8.1 *Bivariate Regression*

To illustrate the basic ideas, we shall consider the model (3.64) with a constant dummy variable 1 and a regressor x:

$$y_t = \beta_0 + \beta_1 x_t + e_t \quad (t = 1, \dots, T). \tag{3.114}$$

Ordinary Least Squares estimators of $\beta = (\beta_0, \beta_1)'$ are given by

$$b_1 = \frac{\sum(x_t - \bar{x})(y_t - \bar{y})}{\sum(x_t - \bar{x})^2}, \tag{3.115}$$

$$b_0 = \bar{y} - b_1 \bar{x}. \tag{3.116}$$

The best predictor of y on the basis of a given x is

$$\hat{y} = b_0 + b_1 x, \tag{3.117}$$

Especially, we have, for $x = x_t$,

$$\begin{aligned} \hat{y}_t &= b_0 + b_1 x_t \\ &= \bar{y} + b_1(x_t - \bar{x}) \end{aligned} \tag{3.118}$$

(cf. (3.115)).

On the basis of the identity

$$y_t - \hat{y}_t = (y_t - \bar{y}) - (\hat{y}_t - \bar{y}) \tag{3.119}$$

we may express the sum of squared residuals (cf. (3.19)) as

$$\begin{aligned} S(b) = \sum(y_t - \hat{y}_t)^2 &= \sum(y_t - \bar{y})^2 + \sum(\hat{y}_t - \bar{y})^2 \\ &\quad -2\sum(y_t - \bar{y})(\hat{y}_t - \bar{y}). \end{aligned}$$

Further manipulation yields

$$\begin{aligned} \sum(y_t - \bar{y})(\hat{y}_t - \bar{y}) &= \sum(y_t - \bar{y}) b_1 (x_t - \bar{x}) && \text{[cf. (3.118)]} \\ &= b_1^2 \sum(x_t - \bar{x})^2 && \text{[cf. (3.115)]} \\ &= \sum(\hat{y}_t - \bar{y})^2 && \text{[cf. (3.118)].} \end{aligned}$$

Thus, we have

$$\sum(y_t - \bar{y})^2 = \sum(y_t - \hat{y}_t)^2 + \sum(\hat{y}_t - \bar{y})^2. \tag{3.120}$$

This relation has already been established in (3.17). The left–hand side of (3.120) is called the **sum of squares about the mean** or the **corrected sum of squares of** Y (i.e., SS(corrected)) or SYY.

The first term on the right–hand side describes the deviation: "observation – predicted value", i.e., the residual sum of squares

$$SS \text{ residual:} \quad RSS = \sum (y_t - \hat{y}_t)^2 , \tag{3.121}$$

whereas the second term describes the proportion of variability explained by regression

$$SS \text{ regression:} \quad SS_{\text{Reg}} = \sum (\hat{y}_t - \bar{y})^2 . \tag{3.122}$$

If all the observations y_t are located on a straight line, we obviously have $\sum(y_t - \hat{y}_t)^2 = 0$ and thus $SS(\text{corrected}) = SS_{\text{Reg}}$. Accordingly, the goodness of fit of a regression is measured by the ratio

$$R^2 = \frac{SS_{Reg}}{SS\ (\text{corrected})} . \tag{3.123}$$

We will discuss R^2 in some detail. The degrees of freedom (df) of the sum of squares are

$$\sum_{t=1}^{T} (y_t - \bar{y})^2 : df \ = T - 1 ,$$

and

$$\sum_{t=1}^{T} (\hat{y}_t - \bar{y})^2 = b_1^2 \sum (x_t - \bar{x})^2 : df \ = 1 ,$$

as *one* function in y_t – namely, b_1 – is sufficient to calculate SS_{Reg}. In view of (3.120), the degree of freedom for the sum of squares $\sum(y_t - \hat{y}_t)^2$ is just the difference of the other two df's, i.e., $df = T - 2$. This enables us to establish the following analysis of variance table:

Source of variation	SS	df	Mean Square ($= SS/df$)	F
Regression	SS regression	1	MS_{Reg}	MS_{Reg}/s^2
Residual	RSS	$T-2$	$s^2 = RSS/T-2$	
Total	SS (corrected) $= SYY$	$T-1$		

The following example illustrates the basics of the ANOVA table with a real data set from the 1993 General Social Survey. If the errors e_t are normally distributed, the sum of squares are distributed independently as χ^2_{df} and F follows an F–distribution.

Example 3.5. We are interested in the influence of the degree of education on the average hours worked per week. The degree of education is a categorical variable on five levels. Running Analysis of Variance in SPLUS produces the following output as an analog to the above table:

```
*** Analysis of Variance Model ***

Short Output:
Call:
   aov(formula = HRS1 ~ DEGREE, data = anova, na.action = na.omit)

Terms:
                 DEGREE Residuals
 Sum of Squares  1825.92  92148.28
Deg. of Freedom        4       736

Residual standard error: 11.18935
Estimated effects may be unbalanced

Analysis of Variance Table:
           Df Sum of Sq  Mean Sq  F Value       Pr(F)
   DEGREE   4   1825.92 456.4794 3.645958 0.005960708
Residuals 736  92148.28 125.2015
```

The overall hypothesis is significant and for further analysis one has to compute multiple comparisons for detecting local differences.

For goodness of fit and confidence intervals we need some tools and will use the following abbreviations for these essential quantities:

$$SXX = \sum (x_t - \bar{x})^2 , \tag{3.124}$$

$$SYY = \sum (y_t - \bar{y})^2 , \tag{3.125}$$

$$SXY = \sum (x_t - \bar{x})(y_t - \bar{y}) . \tag{3.126}$$

The sample correlation coefficient may then be written as

$$r_{XY} = \frac{SXY}{\sqrt{SXX}\sqrt{SYY}} . \tag{3.127}$$

Moreover, we have (cf. (3.115))

$$b_1 = \frac{SXY}{SXX} = r_{XY}\sqrt{\frac{SYY}{SXX}} . \tag{3.128}$$

The estimator of σ^2 may be expressed by using (3.127)as

$$s^2 = \frac{1}{T-2}\sum \hat{e}_t^2 = \frac{1}{T-2} RSS. \tag{3.129}$$

Various alternative formulations for RSS are in use as well

$$\begin{aligned} RSS &= \sum (y_t - (b_0 + b_1 x_t))^2 \\ &= \sum [(y_t - \bar{y}) - b_1(x_t - \bar{x})]^2 \end{aligned}$$

$$\begin{aligned} &= SYY + b_1^2 SXX - 2b_1 SXY \\ &= SYY - b_1^2 SXX \qquad (3.130) \end{aligned}$$

$$= SYY - \frac{(SXY)^2}{SXX}\,. \qquad (3.131)$$

Further relations immediately become apparent

$$SS\ (\text{corrected}) = SYY \qquad (3.132)$$

and

$$\begin{aligned} SS_{\text{Reg}} &= SYY - RSS \\ &= \frac{(SXY)^2}{SXX} = b_1^2\, SXX\,. \qquad (3.133) \end{aligned}$$

Testing the Model

If the model (3.114)

$$y_t = \beta_0 + \beta_1 x_t + \epsilon_t$$

is appropriate, the coefficient b_1 should be significantly different from zero. This is equivalent to the fact that X and Y are significantly correlated. Formally, we compare the models (cf. Weisberg, 1980, p. 17)

$$\begin{aligned} \text{H}_0 : y_t &= \beta_0 + \epsilon_t\,, \\ \text{H}_1 : y_t &= \beta_0 + \beta_1 x_t + \epsilon_t\,, \end{aligned}$$

by testing $\text{H}_0 : \beta_1 = 0$ against $\text{H}_1 : \beta_1 \neq 0$.

We assume normality of the errors $\epsilon \sim N(\mathbf{0}, \sigma^2 I)$. If we recall (B.65), i.e.,

$$\begin{aligned} D &= x'x - x'\mathbf{1}(\mathbf{1}'\mathbf{1})^{-1}\mathbf{1}'x \\ &= \sum x_t^2 - \frac{(\sum x_t)^2}{T} = \sum (x_t - \bar{x})^2 = SXX\,, \qquad (3.134) \end{aligned}$$

then the likelihood ratio test (B.78) is given by

$$\begin{aligned} F_{1,T-2} &= \frac{b_1^2 SXX}{s^2} \\ &= \frac{SS_{\text{Reg}}}{RSS} \cdot (T-2) \\ &= \frac{MS_{\text{Reg}}}{s^2}. \qquad (3.135) \end{aligned}$$

The Coefficient of Determination

In (3.123) R^2 has been introduced as a measure of goodness of fit. Using (3.133), we get

$$R^2 = \frac{SS_{\text{Reg}}}{SYY} = 1 - \frac{RSS}{SYY}. \tag{3.136}$$

The ratio SS_{Reg}/SYY describes the proportion of variability that is covered by regression in relation to the total variability of y. The right–hand side of the equation is 1 minus the proportion of variability that is not covered by regression.

Definition 3.19. R^2 is called the coefficient of determination.

By using (3.127) and (3.133), we get the basic relation between R^2 and the sample correlation coefficient

$$R^2 = r_{XY}^2. \tag{3.137}$$

As one can see from the model summary on page 65 the coefficient of determination could be computed when analyzing a linear model by software.

Confidence Intervals for b_0 and b_1

The covariance matrix of OLS is generally of the form $V_b = \sigma^2(X'X)^{-1} = \sigma^2 S^{-1}$. In model (3.114) we get

$$S = \begin{pmatrix} 1'1 & 1'x \\ 1'x & x'x \end{pmatrix} = \begin{pmatrix} T & T\bar{x} \\ T\bar{x} & \sum x_t^2 \end{pmatrix}, \tag{3.138}$$

$$S^{-1} = \frac{1}{SXX}\begin{pmatrix} 1/T\sum x_t^2 & -\bar{x} \\ -\bar{x} & 1 \end{pmatrix} \tag{3.139}$$

and, therefore,

$$\operatorname{Var}(b_1) = \sigma^2 \frac{1}{SXX}, \tag{3.140}$$

$$\begin{aligned} \operatorname{Var}(b_0) &= \frac{\sigma^2}{T}\cdot\frac{\sum x_t^2}{SXX} = \frac{\sigma^2}{T}\frac{\sum x_t^2 - T\bar{x}^2 + T\bar{x}^2}{SXX} \\ &= \sigma^2\left(\frac{1}{T} + \frac{\bar{x}^2}{SXX}\right). \end{aligned} \tag{3.141}$$

The estimated standard deviations are

$$SE(b_1) = s\sqrt{\frac{1}{SXX}} \tag{3.142}$$

and

$$SE(b_0) = s\sqrt{\frac{1}{T} + \frac{\bar{x}^2}{SXX}} \tag{3.143}$$

with s from (3.129).

Under normal errors $\epsilon \sim N(0, \sigma^2 I)$ in model (3.114), we have

$$b_1 \sim N\left(\beta_1, \sigma^2 \cdot \frac{1}{SXX}\right). \tag{3.144}$$

Thus it holds that

$$\frac{b_1 - \beta_1}{s}\sqrt{SXX} \sim t_{T-2}. \tag{3.145}$$

Analogously, we get

$$b_0 \sim N\left(\beta_0, \sigma^2\left(\frac{1}{T} + \frac{\bar{x}^2}{SXX}\right)\right), \tag{3.146}$$

$$\frac{b_0 - \beta_0}{s}\sqrt{\frac{1}{T} + \frac{\bar{x}^2}{SXX}} \sim t_{T-2}. \tag{3.147}$$

This enables us to calculate confidence intervals at level $1 - \alpha$:

$$b_0 - t_{T-2,1-\alpha/2} \cdot SE(b_0) \le \beta_0 \le b_0 + t_{T-2,1-\alpha/2} \cdot SE(b_0), \tag{3.148}$$

and

$$b_1 - t_{T-2,1-\alpha/2} \cdot SE(b_1) \le \beta_1 \le b_1 + t_{T-2,1-\alpha/2} \cdot SE(b_1). \tag{3.149}$$

		Unst. coefficients		95% Confidence interval for β	
Model		β	Std. error	Lower bound	Upper bound
1	(Constant)	6.019	1.104	3.838	8.199
	adv	3.079	0.300	2.486	3.672

TABLE 3.9. Dependent variable: reaction.

For the "advertise" model (see page 45) we computed the confidence intervals for the estimates using SPSS. It is not a standard output but one has to choose this option.

The above confidence intervals correspond to the region of acceptance of a two–sided test at the same level.

(i) Testing $H_0 : \beta_0 = \beta_0^*$

The test statistic is

$$t_{T-2} = \frac{b_0 - \beta_0^*}{SE(b_0)}. \tag{3.150}$$

H_0 is not rejected, if

$$|t_{T-2}| \le t_{T-2,1-\alpha/2}$$

or, equivalently, if (3.148) holds, with $\beta_0 = \beta_0^*$.

(ii) Testing $H_0 : \beta_1 = \beta_1^*$

The test statistic is

$$t_{T-2} = \frac{b_1 - \beta_1^*}{SE(b_1)} \tag{3.151}$$

or, equivalently,

$$t_{T-2}^2 = F_{1,T-2} = \frac{(b_1 - \beta_1^*)^2}{(SE(b_1))^2} \,. \tag{3.152}$$

This is identical to (3.135), if $H_0 : \beta_1 = 0$ is being tested.

H_0 will not be rejected, if

$$|t_{T-2}| \leq t_{T-2,1-\alpha/2}$$

or, equivalently, if (3.149) holds, with $\beta_1 = \beta_1^*$.

3.8.2 Multiple Regression

If we consider more than two regressors, still under the assumption of normality of the errors, we find the methods of analysis of variance to be most convenient in distinguishing the two models $y = 1\beta_0 + X\beta_* + \epsilon = \tilde{X}\beta + \epsilon$ and $y = 1\beta_0 + \epsilon$. In the latter model, we have $\hat{\beta}_0 = \bar{y}$ and the related residual sum of squares is

$$\sum (y_t - \hat{y}_t)^2 = \sum (y_t - \bar{y})^2 = SYY. \tag{3.153}$$

In the former model, the unknown parameter $\beta = (\beta_0, \beta_*)'$ will again be estimated by $b = (\tilde{X}'\tilde{X})^{-1} \tilde{X}' y$.

The two components of the parameter vector β in the full model may be estimated by

$$b = \begin{pmatrix} \hat{\beta}_0 \\ \hat{\beta}_* \end{pmatrix}, \hat{\beta}_* = (X'X)^{-1}X'y, \hat{\beta}_0 = \bar{y} - \hat{\beta}_*'\bar{x} \,. \tag{3.154}$$

Thus, we have (cf. Weisberg, 1980, p. 43)

$$\begin{aligned} RSS &= (y - \tilde{X}b)'(y - \tilde{X}b) \\ &= y'y - b'\tilde{X}'\tilde{X}b \\ &= (y - 1\bar{y})'(y - 1\bar{y}) - \hat{\beta}_*'(X'X)\hat{\beta}_* + T\bar{y}^2 \,. \end{aligned} \tag{3.155}$$

The proportion of variability explained by regression is (cf. (3.133))

$$SS_{\text{Reg}} = SYY - RSS \tag{3.156}$$

with RSS from (3.155) and SYY from (3.153). The ANOVA table is of the form

Source of variation	SS	df	MS
Regression on $X_1, \ldots, X_K$	SS_{Reg}	K	SS_{Reg}/K
Residual	RSS	$T-K-1$	$s^2 = RSS/(T-K-1)$
Total	SYY	$T-1$	

As before, the multiple coefficient of determination

$$R^2 = \frac{SS_{\text{Reg}}}{SYY} \tag{3.157}$$

is a measure of the proportion of variability explained by the regression of y on $X_1, \ldots, X_K$ in relation to the total variability SYY.

The F–test of

$$\text{H}_0 : \beta_* = 0$$

versus

$$\text{H}_1 : \beta_* \neq 0$$

(i.e., $\text{H}_0 : y = 1\beta_0 + \epsilon$ versus $\text{H}_1 : y = 1\beta_0 + X\beta_* + \epsilon$) is based on the test statistic

$$F_{K,T-K-1} = \frac{SS_{\text{Reg}}/K}{s^2}. \tag{3.158}$$

Often it is of interest to test for the significance of the single components of β. This type of problem arises, for example, in stepwise model selection, if an optimal subset is selected with respect to the coefficient of determination.

Criteria for Model Choice

Draper and Smith (1966) and Weisberg (1980) have established a variety of criteria to find the right model. We will follow the strategy proposed by Weisberg.

(i) Ad–Hoc Criteria

Denote by $X_1, \ldots, X_K$ all the available regressors and let $\{X_{i1}, \ldots, X_{ip}\}$ be a subset of $p \leq K$ regressors. We denote the residual sum of squares by RSS_K (resp. RSS_p). The parameter vectors are

$$\begin{array}{ll} \beta & \text{for } X_1, \ldots, X_K, \\ \beta_1 & \text{for } X_{i1}, \ldots, X_{ip}, \end{array}$$

and

$$\beta_2 \quad \text{for } (X_1, \ldots, X_K) \backslash (X_{i1}, \ldots, X_{ip}).$$

A choice between both models can be conducted by testing $H_0 : \beta_2 = \mathbf{0}$. We apply the F–test, since the hypotheses are nested,

$$F_{(K-p),T-K} = \frac{(RSS_p - RSS_K)/(K-p)}{RSS_K/(T-K)} . \tag{3.159}$$

We prefer the full model against the partial model if $H_0 : \beta_2 = \mathbf{0}$ is rejected, i.e., if $F > F_{1-\alpha}$ (with degrees of freedom $K-p$ and $T-K$).

Model Choice Based on an Adjusted Coefficient of Determination

The coefficient of determination (see (3.156) and (3.157))

$$R_p^2 = 1 - \frac{RSS_p}{SYY} \tag{3.160}$$

is inappropriate to compare a model with K and one with $p < K$, since R^2 always increases if an additional regressor is incorporated into the model, irrespective of its values. The full model always has the greatest value of R^2 (see Theorem 3.20). So we have to adjust R^2 with respect to the number of variables.

Example 3.6. Remembering our example from page 64, concerning the prediction of female life expectancy, we want to show the behavior of the coefficient of determination. Using a "Forward Selection" within the linear regression in SPSS leads to a model including "lndocs", "lngdp", and "lnradios" as predictors. Table 3.10 illustrates the varying coefficient of determination.

			Change statistics				
Model	R^2	R^2_{adj}	R Square change	F change	$df1$	$df2$	Sig. F change
1	0.775	0.773	0.775	391.724	1	114	0.000
2	0.813	0.809	0.038	23.055	1	113	0.000
3	0.823	0.818	0.010	6.161	1	112	0.015

TABLE 3.10. 1 (Constant), natural log of doctors per 10,000; 2 (Constant), natural log of doctors per 10,000, natural log of GDP; 3 (Constant), natural log of doctors per 10,000, natural log of GDP, natural log of radios per 100 people.

Beginning with Step 1 and $R^2 = 0.775$, $R^2_{\text{adj}} = 0.773$ the stepwise inclusion of two further variables leads to $R^2 = 0.823$ and an adjusted coefficient of determination of 0.818 – the coefficients of the model resulting from a "forward selection". In order to illustrate the possible effect of an increasing R^2 and a decreasing R^2_{adj} we first include additionally "lnbeds" into the above model (see Table 3.11). The result is shown in Table 3.11.

Model	R^2	R^2_{adj}	Change statistics				
			R Square change	F change	$df1$	$df2$	Sig. F change
4	0.826	0.820	0.826	132.183	4	111	0.000

TABLE 3.11. 4: (Constant), natural log of doctors per 10,000, natural log of GDP, natural log of radios per 100 people, natural log hospital beds/10,000.

Again, both R^2 and R^2_{adj} are increased. Including "urban" as a further variable (see Table 3.12), however illustrates the effect already described. The fact, that Models 4 and 5 have higher R^2_{adj}'s than the model resulting

Model	R^2	R^2_{adj}	Change statistics				
			R Square change	F change	$df1$	$df2$	Sig. F change
4	0.827	0.819	0.827	105.336	5	110	0.000

TABLE 3.12. 5: (Constant), natural log of doctors per 10,000, natural log of GDP, natural log of radios per 100 people, natural log hospital beds/10,000, percent urban, 1992.

from the "forward selection" has its reason in non significant parameter estimates of the variables "lnbeds" and "urban".

Theorem 3.20. Let $y = X_1\beta_1 + X_2\beta_2 + \epsilon = X\beta + \epsilon$ be a full model and let $y = X_1\beta_1 + \epsilon$ be a submodel. Then it holds that

$$R_X^2 - R_{X_1}^2 \geq 0. \tag{3.161}$$

(See Proof 12, Appendix B.)

On the basis of Theorem 3.20 we define the statistic

$$F\text{–change} = \frac{(RSS_{X_1} - RSS_X)/(K-p)}{RSS_X/(T-K)}, \tag{3.162}$$

which is distributed as $F_{K-p,T-K}$ under H_0 : "submodel is valid". In model choice procedures, F–change tests for the significance of the change of R^2 by adding further $K-p$ variables to the submodel.

In multiple regression, the appropriate adjustment of the ordinary coefficient of determination is provided by the coefficient of determination adjusted by the degrees of freedom of the multiple model

$$\bar{R}_p^2 = 1 - \left(\frac{T-1}{T-p}\right)(1 - R_p^2). \tag{3.163}$$

Remark. If there is no constant β_0 present in the model, then the numerator is T instead of $T-1$, such that $\bar{R}_p^2$ may possibly take negative values. This disadvantage cannot occur when using the ordinary R^2.

If we consider two models, the smaller of which is assumed to be completely included in the bigger one, and we find the relation

$$\bar{R}^2_{p+q} < \bar{R}^2_p \,,$$

then the smaller model obviously shows a better goodness of fit.

Further criteria are, for example, Mallows' C_p (cf. Weisberg, 1980, p. 88), or criteria based on the residual MSE $\hat{\sigma}^2_p = RSS_p/(T-p)$ which are closely related.

Confidence Regions

As in bivariate regression, there are close relations between the region of acceptance of the F–test and the confidence intervals for β in the multiple linear regression model as well.

Confidence Ellipsoids for the Whole Parameter Vector β

Considering (B.51) and (B.54), we get for $\beta^* = \beta$ a confidence ellipsoid at level $1 - \alpha$:

$$\frac{(b-\beta)'X'X(b-\beta)}{(y-Xb)'(y-Xb)} \cdot \frac{T-K}{K} \le F_{K,T-K,1-\alpha} \,. \tag{3.164}$$

Confidence Ellipsoids for Subvectors of β

From (B.78) and (3.103), we have that

$$\frac{(b_2-\beta_2)'D(b_2-\beta_2)}{(y-Xb)'(y-Xb)} \cdot \frac{T-K}{K-s} \le F_{K-s,T-K,1-\alpha} \tag{3.165}$$

is a $(1-\alpha)$–confidence ellipsoid for β_2.

Further results may be found in Judge, Griffiths, Hill and Lee (1980), Goldberger (1964), Pollock (1979), Weisberg (1980), and Kmenta (1971).

3.9 The General Linear Regression Model

3.9.1 *Introduction*

In many applications, it cannot be justified that the response values y_t $(t = 1, \ldots, T)$ are independent. Consider, for example, a time series with autocorrelated errors or processes typically arising in medicine or sociology, when measurements are being repeated several times on a single person or cluster analysis. We will discuss these types of models at a later stage. Here we present a first step to generalize the classical model assuming a less restrictive form of the error covariance matrix.

The general linear regression model is of the form

$$\left.\begin{array}{c} y = X\beta + \epsilon, \\ \mathrm{E}(\epsilon) = 0, \quad \mathrm{E}(\epsilon\epsilon') = \sigma^2 W, \\ W \text{ positive definite and known}, \\ X \text{ nonstochastic, } \operatorname{rank}(X) = K\,. \end{array}\right\} \tag{3.166}$$

The first problem is now, that in the case of an unknown matrix W, the number of additional parameters to be estimated may increase by $T(T+1)/2$, at the most, because $\sum_{i=1}^{T} i$ is the number of different parameters in W. This problem cannot be solved on the basis of T observations only. Therefore we assume, for the present, that W is known. Furthermore, it is useful to impose several restrictions on W in the sense that $\operatorname{tr}(W) = T$ or $w_{ii} = 1$ $(i = 1, \dots, T)$.

Aitken Estimator

In order to facilitate the estimation in a general linear regression model (3.166), we shall transform the model. For the exact transformation, see Proof 13, Appendix B.

This transformation leads to $b = (\tilde{X}'\tilde{X})^{-1}\tilde{X}'\tilde{y}$ which is, as we know, identical to the Gauss–Markov (GM) estimator in the transformed model. The Gauss–Markov property of b (with $S = (X'W^{-1}X)$) also remains valid in model (3.166):

$$\begin{aligned} b &= S^{-1}X'W^{-1}y \text{ is unbiased}, \\ \mathrm{E}(b) &= (X'W^{-1}X)^{-1}X'W^{-1}\,\mathrm{E}(y) \\ &= (X'W^{-1}X)^{-1}X'W^{-1}X\beta = \beta\,. \end{aligned} \tag{3.167}$$

Moreover, b possesses the smallest variance (in the sense of Theorem 3.12, see Proof 14, Appendix B).

These results are summarized in:

Theorem 3.21 (Gauss–Markov–Aitken Theorem). In the general linear regression model, the generalized OLS estimator

$$b = (X'W^{-1}X)^{-1}X'W^{-1}y\,, \tag{3.168}$$

with covariance matrix

$$V_b = \sigma^2(X'W^{-1}X)^{-1} = \sigma^2 S^{-1}\,, \tag{3.169}$$

is the best linear unbiased estimator of β.

(We denote b also as an Aitken estimator or a generalized least squares (GLS) estimator). Analogously to the classical model, we estimate σ^2 and V_b by

$$s^2 = (y - Xb)'W^{-1}(y - Xb)(T - K)^{-1} \tag{3.170}$$

and

$$\hat{V}_b = s^2 S^{-1} . \tag{3.171}$$

Both estimators are unbiased

$$\mathrm{E}(s^2) = \sigma^2 \quad \text{and} \quad \mathrm{E}(\hat{V}_b) = \sigma^2 S^{-1} . \tag{3.172}$$

Some statistical software packages offer a procedure for solving the problem of $\mathrm{E}(\epsilon\epsilon') \neq \sigma^2 I$. SPSS, for example, suggests using the "weight estimation procedure" where cases with less variability are given greater weights. The coefficients are computed by weighted least squares and a range of weight transformations is tested to get the best fit.

3.9.2 Misspecification of the Covariance Matrix

Assuming the general linear regression model (3.166) and W to be true, we want to examine the influence of a misspecification of the covariance matrix on the estimator of β and σ^2, compared to the GLS estimator b (3.168) and s^2 (3.170). Reasons for the misspecification could be:

- the use of the classical OLS estimator because the correlation between the errors ϵ_t was not recognized;
- that the correlation is generally described by a matrix $\tilde{W} \neq W$; and
- that the matrix W is unknown and is estimated independent of y from a presample through $\hat{W}$.

In any case, we get the estimator

$$\hat{\beta} = (X'AX)^{-1} X'Ay , \tag{3.173}$$

with $A \neq W^{-1}$ symmetric, nonstochastic, and with $(X'AX)$ regular. Then we have

$$\mathrm{E}(\hat{\beta}) = \beta , \tag{3.174}$$

where $\hat{\beta}$ [(3.173)] is unbiased for every misspecified matrix A (if $\operatorname{rank}(X'AX) = K$). For the covariance matrix of $\hat{\beta}$ we get

$$V_{\hat{\beta}} = \sigma^2 (X'AX)^{-1} X'AWAX(X'AX)^{-1} . \tag{3.175}$$

The loss of efficiency, due to the use of $\hat{\beta}$ instead of the GLS estimator $b = S^{-1}X'W^{-1}y$, becomes

$$\begin{aligned} V_{\hat{\beta}} - V_b &= \sigma^2[(X'AX)^{-1}X'A - S^{-1}X'W^{-1}] \\ &\quad \times W[(X'AX)^{-1}X'A - S^{-1}X'W^{-1}]' . \end{aligned} \tag{3.176}$$

Following Theorem A.18(iv), this matrix is nonnegative definite. There is no loss in efficiency if

$$(X'AX)^{-1}X'A = S^{-1}X'W^{-1} \quad \text{or} \quad \hat{\beta} = b . \tag{3.177}$$

Assume the first column of X being 1 and let $A = I$, i.e., implying the use of the classical OLS estimator $(X'X)^{-1}X'y$. Then the following theorem is valid (McElroy (1967)):

Theorem 3.22. The OLS estimator $b_0 = (X'X)^{-1}X'y$ is Gauss–Markov estimator in the generalized linear regression model if and only if $X = (1\tilde{X})$, and

$$W = (1-\rho)I + \rho 11' \tag{3.178}$$

with $0 \leq \rho < 1$ and $1' = (1, 1, \ldots, 1)$.

In other words, we have

$$(X'X)^{-1}X'y = (X'W^{-1}X)^{-1}X'W^{-1}y \tag{3.179}$$

for all y, if and only if the errors ϵ_t have the same variance σ^2 and equal nonnegative covariances $\sigma^2\rho$. A matrix of this form is called compound symmetric.

Moreover, a loss in efficiency occurs if σ^2 is estimated by an estimator $\hat{\sigma}^2$ that is based on $\hat{\beta}$. The average bias of the estimator $\hat{\sigma}^2$ which is based on OLS is given by $[\sigma^2/T - K](K - \text{tr}[(X'X)^{-1}X'WX])$ (see Proof 15, Appendix B). It is to be expected that the bias will tend to be negative, especially in processes with positive correlation. As a consequence, the variance will be underestimated, leading in turn to a better goodness of fit (cf. several examples in Goldberger, 1964, pp. 288, in cases of heteroscedasticity and first–order autoregression).

3.10 Diagnostic Tools

3.10.1 Introduction

This chapter discusses the influence of individual observations on the estimated values of parameters and the prediction of the dependent variable for given values of regressor variables. Methods for detecting the outliers, and deviation from normality of the distribution of errors, are given in some detail. The material of this chapter is drawn mainly from the excellent book by Chatterjee and Hadi (1988).

3.10.2 Prediction Matrix

We consider the classical linear model

$$y = X\beta + \epsilon, \quad \epsilon \sim (0, \sigma^2 I),$$

with the usual assumptions. In particular, we assume that the matrix X of order $T \times K$ has the full rank K. The quality of the classical ex–post predictor $\hat{p} = Xb_0 = \hat{y}$ of y with $b_0 = (X'X)^{-1}X'y$, the OLSE (ordinary least–squares estimator), is strongly determined by the $(T \times T)$–matrix

$$P = X(X'X)^{-1}X' = (p_{ij}), \tag{3.180}$$

which is symmetric and idempotent of $\text{rank}(P) = \text{tr}(P) = \text{tr}(I_K) = K$. The matrix $M = I - P$ is also symmetric and idempotent and has $\text{rank}(M) = T - K$. The estimated residuals are defined by

$$\begin{aligned} \hat{\epsilon} = (I-P)y &= y - Xb_0 \\ &= y - \hat{y} = (I-P)\epsilon . \end{aligned} \tag{3.181}$$

Definition 3.23 (Chatterjee and Hadi, 1988). The matrix P given in (3.180) is called the prediction matrix, and the matrix $I - P$ is called the residuals matrix.

Remark: The matrix P is sometimes called the *hat matrix* because it maps y onto $\hat{y}$.

The (i,j)th element of the matrix P is denoted by p_{ij} where

$$p_{ij} = p_{ji} = x_j'(X'X)^{-1}x_i \quad (i,j = 1, \ldots, T). \tag{3.182}$$

The ex–post predictor $\hat{y} = Xb_0 = Py$ has the dispersion matrix

$$V(\hat{y}) = \sigma^2 P . \tag{3.183}$$

Therefore, we obtain (denoting the ith component of $\hat{y}$ by $\hat{y}_i$ and the ith component of $\hat{\epsilon}$ by $\hat{\epsilon}_i$)

$$\text{var}(\hat{y}_i) = \sigma^2 p_{ii}, \tag{3.184}$$

$$\text{V}(\hat{\epsilon}) = \text{V}\left((I-P)y\right) = \sigma^2(I-P), \tag{3.185}$$

$$\text{var}(\hat{\epsilon}_i) = \sigma^2(1 - p_{ii}) \tag{3.186}$$

and, for $i \neq j$,

$$\text{cov}(\hat{\epsilon}_i, \hat{\epsilon}_j) = -\sigma^2 p_{ij} . \tag{3.187}$$

The correlation coefficient between $\hat{\epsilon}_i$ and $\hat{\epsilon}_j$ then becomes

$$\rho_{ij} = \text{corr}(\hat{\epsilon}_i, \hat{\epsilon}_j) = \frac{-p_{ij}}{\sqrt{1-p_{ii}}\sqrt{1-p_{jj}}} . \tag{3.188}$$

Thus the covariance matrices of the predictor Xb_0 and the estimator of error $\hat{\epsilon}$ are entirely determined by P. Although the disturbances ϵ_i of the model are independent and identically distributed, the estimated residuals $\hat{\epsilon}_i$ are not identically distributed and, moreover, they are correlated.

Observe that

$$\hat{y}_i = \sum_{j=1}^{T} p_{ij} y_i = p_{ii} y_i + \sum_{j \neq i} p_{ij} y_j \quad (i = 1, \ldots, T)\,, \tag{3.189}$$

implying that

$$\frac{\partial \hat{y}_i}{\partial y_i} = p_{ii} \quad \text{and} \quad \frac{\partial \hat{y}_i}{\partial y_j} = p_{ij}\,. \tag{3.190}$$

Therefore, p_{ii} can be interpreted as the amount of *leverage* each value y_i has in determining $\hat{y}_i$ regardless of the realized value y_i. The second relation of (3.190) may be interpreted, analogously, as the influence of y_j in determining $\hat{y}_i$.

Elements of P

The size and range of the elements of P are measures for the influence of data on the predicted values $\hat{y}_t$. Because of the symmetry of P, we have $p_{ij} = p_{ji}$, and the idempotence of P implies

$$p_{ii} = \sum_{j=1}^{n} p_{ij}^2 = p_{ii}^2 + \sum_{j \neq i} p_{ij}^2\,. \tag{3.191}$$

From this equation we obtain the important property

$$0 \leq p_{ii} \leq 1\,. \tag{3.192}$$

Reformulating (3.191)

$$p_{ii} = p_{ii}^2 + p_{ij}^2 + \sum_{k \neq i,j} p_{ik}^2 \quad (j \text{ fixed})\,, \tag{3.193}$$

which implies that $p_{ij}^2 \leq p_{ii}(1 - p_{ii})$ and, therefore, using (3.192), we obtain

$$-0.5 \leq p_{ij} \leq 0.5 \quad (i \neq j)\,. \tag{3.194}$$

If X contains a column of constants (1 or $c1$), then in addition to (3.192) we obtain

$$p_{ii} \geq T^{-1} \quad (\text{for all } i) \tag{3.195}$$

and

$$P1 = 1\,. \tag{3.196}$$

Relationship (3.195) is a direct consequence of (B.101) resulting from the decomposition of P shown in Proof 16, Appendix B.

The diagonal elements p_{ii} and the off–diagonal elements p_{ij} $(i \neq j)$ are interrelated according to properties (i)–(iii) as follows (Chatterjee and Hadi, 1988, p. 19):

(i) If $p_{ii} = 1$ or $p_{ii} = 0$, then $p_{ij} = 0$.

Proof. Use (3.191).

(ii) We have

$$(p_{ii}p_{jj} - p_{ij}^2) \geq 0\,. \text{ Proof17, AppendixB.} \tag{3.197}$$

(iii) We have

$$(1 - p_{ii})(1 - p_{jj}) - p_{ij}^2 \geq 0\,. \text{ Proof 18, Appendix B.} \tag{3.198}$$

Interpretation. If a diagonal element p_{ii} is close to either 1 or 0, then the elements p_{ij} (for all $j \neq i$) are close to 0.

The classical predictor of y is given by $\hat{y} = Xb_0 = Py$, and its first component is $\hat{y}_1 = \sum p_{1j}y_j$. If, for instance, $p_{11} = 1$, then $\hat{y}_1$ is fully determined by the observation y_1. On the other hand, if p_{11} is close to 0, then y_1 itself, and all the other observations $y_2, \ldots, y_T$, have low influence on $\hat{y}_1$. Relationship (B.105) indicates that if p_{ii} is large, then the standardized residual $\hat{\epsilon}_i/\hat{\epsilon}'\hat{\epsilon}$ becomes small.

Conditions for p_{ii} to Be Large

If we assume the simple linear model

$$y_t = \alpha + \beta x_t + \epsilon_t, \quad t = 1, \ldots, T\,,$$

then we obtain, from (B.101),

$$p_{ii} = \frac{1}{T} + \frac{(x_i - \bar{x})^2}{\sum_{t=1}^{T}(x_t - \bar{x})^2}\,. \tag{3.199}$$

The size of p_{ii} is dependent on the distance $|x_i - \bar{x}|$. Therefore, the influence of any observation (y_i, x_i) on $\hat{y}_i$ will be increasing with increasing distance $|x_i - \bar{x}|$.

In the case of multiple regression we have a similar relationship. Let λ_i denote the eigenvalues and let γ_i $(i = 1, \ldots, K)$ be the orthonormal eigenvectors of the matrix $X'X$. Furthermore, let θ_{ij} be the angle between the column vector x_i and the eigenvector γ_j $(i, j = 1, \ldots, K)$. Then we have

$$p_{ij} = \|x_i\|\,\|x_j\| \sum_{r=1}^{K} \lambda_r^{-1} \cos\theta_{ir} \cos\theta_{rj} \tag{3.200}$$

and

$$p_{ii} = x_i'x_i \sum_{r=1}^{K} \lambda_r^{-1} (\cos\theta_{ir})^2\,. \tag{3.201}$$

See Proof 19, Appendix B.

Therefore, p_{ii} tends to be large if:

(i) $x_i'x_i$ is large in relation to the square of the vector norm $x_j'x_j$ of the other vectors x_j (i.e., x_i is far from the other vectors x_j); or

(ii) x_i is parallel (or almost parallel) to the eigenvector corresponding to the smallest eigenvalue. For instance, let λ_K be the smallest eigenvalue of $X'X$, and assume x_i to be parallel to the corresponding eigenvector γ_K. Then we have $\cos\theta_{iK} = 1$, and this is multiplied by λ_K^{-1}, resulting in a large value of p_{ii} (cf. Cook and Weisberg, 1982, p. 13).

Multiple X Rows

In the statistical analysis of linear models there are designs (as, e.g., in the analysis of variance of factorial experiments) that allow a repeated response y_t for the same fixed x–vector. Let us assume that the ith row $(x_{i1}, \ldots, x_{iK})$ occurs a times in X. Then it holds that

$$p_{ii} \leq a^{-1}. \tag{3.202}$$

This property is a direct consequence of (3.193). Let $J = \{j : x_i = x_j\}$ denote the set of indices of rows identical to the ith row. This implies $p_{ij} = p_{ii}$ for $j \in J$ and, hence, (3.193) becomes

$$p_{ii} = ap_{ii}^2 + \sum_{j \notin J} p_{ij}^2 \geq ap_{ii}^2 \,,$$

including (3.202).

Example 3.7. We consider the matrix

$$X = \begin{pmatrix} 1 & 2 \\ 1 & 2 \\ 1 & 1 \end{pmatrix}$$

with $K = 2$ and $T = 3$, and calculate

$$\begin{aligned} X'X &= \begin{pmatrix} 3 & 5 \\ 5 & 9 \end{pmatrix}, \quad |X'X| = 2, \quad (X'X)^{-1} = \frac{1}{2}\begin{pmatrix} 9 & -5 \\ -5 & 3 \end{pmatrix}, \\ P &= X(X'X)^{-1}X' = \begin{pmatrix} 0.5 & 0.5 & 0 \\ 0.5 & 0.5 & 0 \\ 0 & 0 & 1 \end{pmatrix}. \end{aligned}$$

The first and second rows of P coincide. Therefore we have $p_{11} \leq \frac{1}{2}$. Inserting $\bar{x} = \frac{5}{3}$ and $\sum_{t=1}^{3}(x_t - \bar{x})^2 = \frac{6}{9}$ in (3.199) results in

$$p_{ii} = \frac{1}{3} + \frac{(x_i - \bar{x})^2}{\sum(x_t - \bar{x}^2)},$$

that is, $p_{11} = p_{22} = \frac{1}{3} + \frac{1/9}{6/9} = \frac{1}{2}$ and $p_{33} = \frac{1}{3} + \frac{4/9}{6/9} = 1$.

3.10.3 The Effect of a Single Observation on the Estimation of Parameters

In Section 3.8 we investigated the effect of one variable X_i (or sets of variables) on the fit of the model. The effect of including or excluding columns of X is measured and tested by the statistic F.

In this section we wish to investigate the effect of rows (y_t, x_t') instead of columns x_t on the estimation of β. Usually, not all observations (y_t, x_t') have equal influence in a least squares fit or on the estimator $(X'X)^{-1}X'y$. It is important for the data analyst to be able to identify observations that individually or collectively have excessive influence compared to other observations. Such rows of the data matrix (y, X) will be called *influential observations*.

The measures for the goodness of fit of a model are mainly based on the residual sum of squares

$$\begin{aligned} \hat{\epsilon}'\hat{\epsilon} &= (y - Xb)'(y - Xb) \\ &= y'(I - P)y = \epsilon'(I - P)\epsilon\,. \end{aligned} \tag{3.203}$$

This quadratic form and the residual vector $\hat{\epsilon} = (I - P)\epsilon$ itself may change considerably if an observation is excluded or added. Depending on the change in $\hat{\epsilon}$ or $\hat{\epsilon}'\hat{\epsilon}$, an observation may be identified as influential or not. In the literature, a large number of statistical measures have been proposed for diagnosing influential observations. We describe some of them and focus attention on the detection of a single influential observation. A more detailed presentation is given by Chatterjee and Hadi (1988, Chapter 4).

Measures Based on Residuals

Residuals play an important role in regression diagnostics, since the ith residual $\hat{\epsilon}_i$ may be regarded as an appropriate guess for the unknown random error ϵ_i.

The relationship $\hat{\epsilon} = (I - P)\epsilon$ implies that $\hat{\epsilon}$ would even be a good estimator for ϵ if $(I - P) \approx I$, that is, if all p_{ij} are sufficiently small and if the diagonal elements p_{ii} are of the same size. Furthermore, even if the random errors ϵ_i are independent and identically distributed. (i.e., $\mathrm{E}\,\epsilon\epsilon' = \sigma^2 I$), the identity $\hat{\epsilon} = (I - P)\epsilon$ indicates that the residuals are not independent (unless P is diagonal) and do not have the same variance (unless the diagonal elements of P are equal). Consequently, the residuals can be expected to be reasonable substitutes for the random errors if:

(i) the diagonal elements p_{ii} of the matrix P are almost equal, that is, the rows of X are almost homogeneous, implying homogeneity of variances of the $\hat{\epsilon}_t$; and

(ii) the off–diagonal elements p_{ij} $(i \neq j)$ are sufficiently small, implying uncorrelated residuals.

Hence it is preferable to use transformed residuals for diagnostic purposes. That is, instead of $\hat{\epsilon}_i$, we may use a transformed standardized residual $\tilde{\epsilon}_i = \hat{\epsilon}_i/\sigma_i$, where σ_i is the standard deviation of the ith residual. Several standardized residuals with specific diagnostic power are obtained by different choices of $\hat{\sigma}_i$ (Chatterjee and Hadi, 1988, p. 73).

(i) *Normalized Residual.* Replacing σ_i by $(\hat{\epsilon}'\hat{\epsilon})^{1/2}$ gives

$$a_i = \frac{\hat{\epsilon}_i}{(\hat{\epsilon}'\hat{\epsilon})^{1/2}} \quad (i = 1, \ldots, T). \tag{3.204}$$

(ii) *Standardized Residual.* Replacing σ_i by $s = \sqrt{\hat{\epsilon}'\hat{\epsilon}/(T-K)}$, we obtain

$$b_i = \frac{\hat{\epsilon}_i}{s} \quad (i = 1, \ldots, T). \tag{3.205}$$

(iii) *Internally Studentized Residual.* With $\hat{\sigma}_i = s\sqrt{1 - p_{ii}}$ we obtain

$$r_i = \frac{\hat{\epsilon}_i}{s\sqrt{1 - p_{ii}}} \quad (i = 1, \ldots, T). \tag{3.206}$$

(iv) *Externally Studentized Residual.* Let us assume that the ith observation is omitted. This fact is indicated by writing the index (i) in parantheses. Using this indicator, we may define the estimator of σ_i^2 when the ith row (y_i, x_i') is omitted as

$$s_{(i)}^2 = \frac{y_{(i)}'(I - P_{(i)})y_{(i)}}{T - K - 1} \quad (i = 1, \ldots, T). \tag{3.207}$$

If we take $\hat{\sigma}_i = s_{(i)}\sqrt{1 - p_{ii}}$, the ith externally Studentized residual is defined as

$$r_i^* = \frac{\hat{\epsilon}_i}{s_{(i)}\sqrt{1 - p_{ii}}} \quad (i = 1, \ldots, T). \tag{3.208}$$

Detection of Outliers

To find the relationships between the ith internally and externally Studentized residuals, we need to write $(T-K)s^2 = y'(I-P)y$ as a function of $s_{(i)}^2$, that is, as $(T - K - 1)s_{(i)}^2 = y_{(i)}'(I - P_{(i)})y_{(i)}$. This is done by noting that omitting the ith observation is equivalent to fitting the *mean–shift outlier model*

$$y = X\beta + e_i\delta + \epsilon\,, \tag{3.209}$$

where e_i is the ith unit vector; that is, $e_i' = (0, \ldots, 0, 1, 0, \ldots, 0)$. The argument is as follows. Suppose that either y_i or $x_i'\beta$ deviates systematically by δ from the model $y_i = x_i'\beta + \epsilon_i$. Then the ith observation $(y_i, x_i'\beta)$ would have a different intercept than the remaining observations and $(y_i, x_i'\beta)$ would hence be an outlier. To check this fact, we test the hypothesis

$$H_0 : \delta = 0 \quad (\text{i.e., } \mathrm{E}(y) = X\beta)$$

against the alternative

$$H_1 : \delta \neq 0 \quad (\text{i.e., } \mathrm{E}(y) = X\beta + e_i\delta)$$

using the likelihood–ratio test (LRT) statistic

$$F_i = \frac{\big(SSE(H_0) - SSE(H_1)\big)/1}{SSE(H_1)/(T-K-1)}, \tag{3.210}$$

where $SSE(H_0)$ is the residual sum of squares in the model $y = X\beta + \epsilon$ containing all the T observations

$$SSE(H_0) = y'(I-P)y = (T-K)s^2$$

and $SSE(H_1)$ is the residual sum of squares in the model $y = X\beta + e_i\delta + \epsilon$.

The test statistic (3.210) may be written as

$$F_i = \frac{\hat{\epsilon}_i^2}{(1-p_{ii})s_{(i)}^2} = (r_i^*)^2, \tag{3.211}$$

where r_i^* is the ith externally Studentized residual (see Proof 20, Appendix B).

Theorem 3.24 (Beckman and Trussel, 1974). Assume the design matrix X is of full column rank K.

(i) If $\operatorname{rank}(X_{(i)}) = K$ and $\epsilon \sim N_T(0, \sigma^2 I)$, then the externally Studentized residuals r_i^* $(i = 1, \ldots, T)$ are t_{T-K-1}-distributed.

(ii) If $\operatorname{rank}(X_{(i)}) = K - 1$, then the residual r_i^* is not defined.

Assume $\operatorname{rank}(X_{(i)}) = K$. Then Theorem 3.24(i) implies that the test statistic $(r_i^*)^2 = F_i$ from (3.211) is distributed as central $F_{1,T-K-1}$ under H_0 and noncentral $F_{1,T-K-1}(\delta^2(1-p_{ii})\sigma^2)$ under H_1, respectively. The noncentrality parameter decreases (tending to zero) as p_{ii} increases. That is, the detection of outliers becomes difficult when p_{ii} is large.

Relationships Between r_i^* and r_i

Equations (B.108) and (3.206) imply that

$$\begin{aligned} s_{(i)}^2 &= \frac{(T-K)s^2}{T-K-1} - \frac{\hat{\epsilon}_i^2}{(T-K-1)(1-p_{ii})} \\ &= s^2\left(\frac{T-K-r_i^2}{T-K-1}\right) \end{aligned} \tag{3.212}$$

and, hence,

$$r_i^* = r_i\sqrt{\frac{T-K-1}{T-K-r_i^2}}. \tag{3.213}$$

Inspecting the Four Types of Residuals

The normalized, standardized, and internally and externally Studentized residuals are transformations of the OLS residuals $\hat{\epsilon}_i$ according to $\hat{\epsilon}_i/\sigma_i$, where σ_i is estimated by the corresponding statistics defined in (3.204)–(3.207), respectively. The normalized, as well as the standardized, residuals a_i and b_i, respectively, are easy to calculate but they do not measure the variability of the variances of the $\hat{\epsilon}_i$. Therefore, in the case of large differences in the diagonal elements p_{ii} of P or, equivalently (cf. (3.186)), of the variances of $\hat{\epsilon}_i$, application of the Studentized residuals r_i or r_i^* is well recommended. The externally Studentized residuals r_i^* are advantageous in the following sense:

(i) $(r_i^*)^2$ may be interpreted as the F–statistic for testing the significance of the unit vector e_i in the mean–shift outlier model (3.209).

(ii) The internally Studentized residual r_i follows a beta distribution (cf. Chatterjee and Hadi, 1988, p. 76) whose quantiles are not included in standard textbooks.

(iii) If $r_i^2 \to T - K$ then $r_i^{*2} \to \infty$ (cf. (3.213)). Hence, compared to r_i, the residual r_i^* is more sensitive to outliers.

Example 3.8. We consider the following data set including the response vector y and the variable X_4 (which was already detected to be the most important variable compared to X_1, X_2, and X_3):

$$\begin{pmatrix} y \\ X_4 \end{pmatrix}' = \begin{pmatrix} 18 & 47 & 125 & 40 & 37 & 20 & 24 & 35 & 59 & 50 \\ -10 & 19 & 100 & 17 & 13 & 10 & 5 & 22 & 35 & 20 \end{pmatrix}.$$

Including the dummy variable 1, the matrix $X = (1, X_4)$ gives

$$\begin{aligned} X'X &= \begin{pmatrix} 10 & 231 \\ 231 & 13153 \end{pmatrix}, \quad |X'X| = 78169, \\ (X'X)^{-1} &= \frac{1}{78169}\begin{pmatrix} 13153 & -231 \\ -231 & 10 \end{pmatrix}. \end{aligned}$$

The diagonal elements of $P = X(X'X)^{-1}X'$ are

$$\begin{aligned} p_{11} &= 0.24, & p_{66} &= 0.12, \\ p_{22} &= 0.10, & p_{77} &= 0.14, \\ p_{33} &= 0.86, & p_{88} &= 0.10, \\ p_{44} &= 0.10, & p_{99} &= 0.12, \\ p_{55} &= 0.11, & p_{1010} &= 0.11, \end{aligned}$$

where $\sum p_{ii} = 2 = K = \operatorname{tr} P$ and $p_{ii} \geq \frac{1}{10}$ (cf. (3.195)). The value p_{33} differs considerably from the other p_{ii}. To calculate the test statistic F_i (3.211), we have to find the residuals $\hat{\epsilon}_i = y_i - \hat{y}_i = y_i - x_i' b_0$, where $\hat{\beta}$ was (21.80, 1.03). The results are summarized in Table 3.13.

i	$1 - p_{ii}$	$\hat{y}_i$	$\hat{\epsilon}_i$	r_i^2	$r_i^{*2} = F_i$
1	0.76	11.55	6.45	1.15	1.18
2	0.90	41.29	5.71	0.76	0.74
3	0.14	124.38	0.62	0.06	0.05
4	0.90	39.24	0.76	0.01	0.01
5	0.89	35.14	1.86	0.08	0.07
6	0.88	32.06	-12.06	3.48	5.38
7	0.86	26.93	-2.93	0.21	0.19
8	0.90	44.37	-9.37	2.05	2.41
9	0.88	57.71	1.29	0.04	0.03
10	0.90	42.32	7.68	1.38	1.46

TABLE 3.13. Internally and externally Studentized residuals.

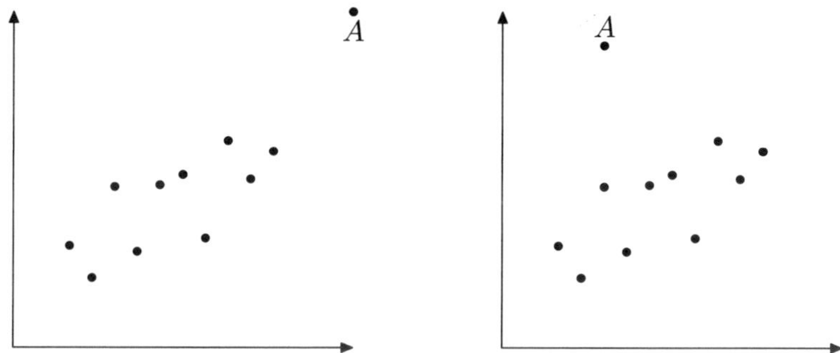

FIGURE 3.4. High–leverage point A.

FIGURE 3.5. Outlier A.

The residuals r_i^2 and r_i^{*2} are calculated according to (3.206) and (3.213), respectively. The standard deviation was found to be $s = 6.9$.

From Table C.6 (Appendix C) we have the quantile $F_{1,7,0.95} = 5.59$, implying that the null hypothesis H_0 : "ith observation $(y_i, 1, x_{4i})$ is not an outlier" is not rejected for all $i = 1, \ldots, 10$. The third observation may be identified as a high–leverage point having remarkable influence on the regression line. Taking $\bar{x}_4 = 23.1$ and $s^2(x_4) = 868.544$ and applying formula (3.199), we obtain

$$\begin{aligned} p_{33} = \frac{1}{10} + \frac{(100 - 23.1)^2}{\sum_{t=1}^{10}(x_t - \bar{x})^2} &= \frac{1}{10} + \frac{76.9^2}{9 \cdot 868.544} \\ &= 0.10 + 0.76 = 0.86. \end{aligned}$$

Therefore, the large value of $p_{33} = 0.86$ is mainly caused by the large distance between x_{43} and the mean value $\bar{x}_4 = 23.1$.

Figures 3.4 and 3.5 show typical situations for points that are very far from the others. Outliers correspond to extremely large residuals, but high–leverage points correspond to extremely small residuals in each case when compared with other residuals.

3.10.4 Diagnostic Plots for Testing the Model Assumptions

Many graphical methods make use of the residuals to detect deviations from the stated assumptions. From experience one may prefer graphical methods over numerical tests based on residuals. The most common residual plots are:

(i) empirical distribution of the residuals, stem–and–leaf diagrams, Box–Whisker plots;

(ii) normal probability plots; and

(iii) residuals versus fitted values or residuals versus x_i plots (see Figures 3.6 and 3.7).

These plots are useful in detecting deviations from assumptions made on the linear model.

The externally Studentized residuals may also be used to detect a violation of normality. If normality is present, then approximately 68% of the residuals r_i^* will be in the interval $[-1, 1]$. As a rule of thumb, one may identify the ith observation as an outlier if $|r_i^*| > 3$.

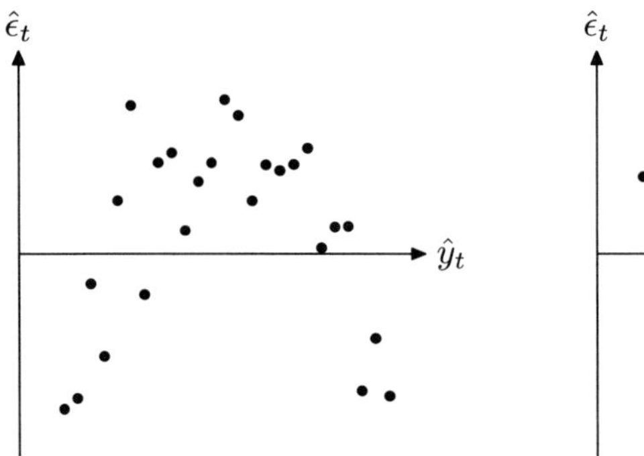

FIGURE 3.6. Plot of the residuals $\hat{\epsilon}_t$ versus the fitted values $\hat{y}_t$ (suggests deviation from linearity).

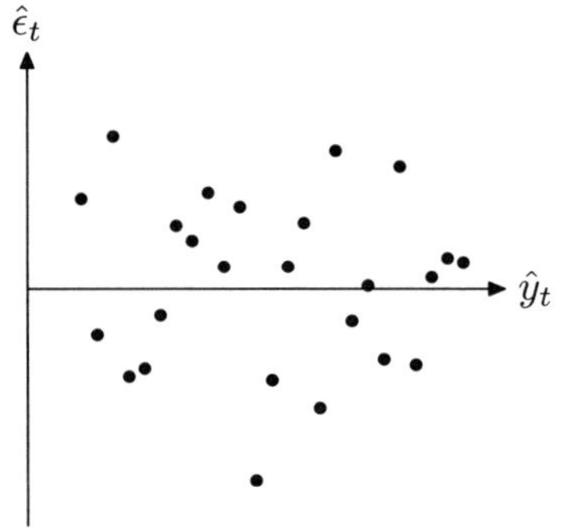

FIGURE 3.7. No violation of linearity.

If the assumptions of the model are correctly specified, then we have

$$\operatorname{cov}(\hat{\epsilon}, \hat{y}') = \mathrm{E}\left((I - P)\epsilon\epsilon' P\right) = 0 . \tag{3.214}$$

Therefore, plotting $\hat{\epsilon}_t$ versus $\hat{y}_t$ (Figures 3.6 and 3.7) exhibits a random scatter of points. Such a situation, as in Figure 3.7, is called a null plot. A plot, as in Figure 3.8, indicates heteroscedasticity of the covariance matrix.

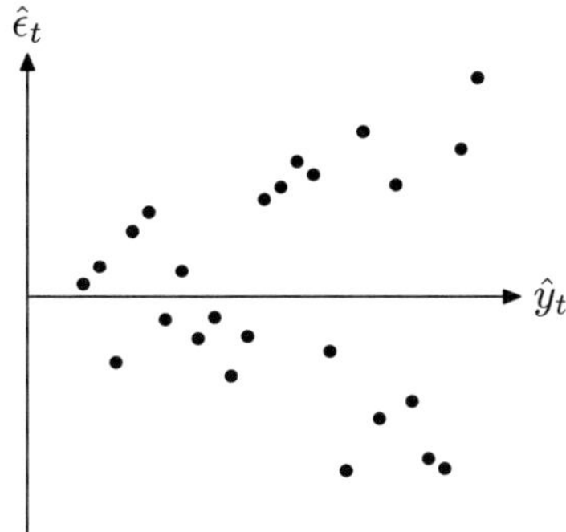

FIGURE 3.8. Signals for heteroscedasticity.

3.10.5 Measures Based on the Confidence Ellipsoid

Under the assumption of normally distributed disturbances, that is, $\epsilon \sim N(0, \sigma^2 I)$, we have $b_0 = (X'X)^{-1}X'y \sim N(\beta, \sigma^2(X'X)^{-1})$ and

$$\frac{(\beta - b_0)'(X'X)(\beta - b_0)}{Ks^2} \sim F_{K,T-K} \,. \tag{3.215}$$

Then the inequality

$$\frac{(\beta - b_0)'(X'X)(\beta - b_0)}{Ks^2} \leq F_{K,T-K,1-\alpha} \tag{3.216}$$

defines a $100(1-\alpha)\%$ confidence ellipsoid for β centered at b_0. The influence of the ith observation (y_i, x_i') can be measured by the change of various parameters of the ellipsoid when the ith observation is omitted. Strong influence of the ith observation would be equivalent to a significant change of the corresponding measure.

Cook's Distance

Cook (1977) suggested the index

$$C_i = \frac{(b - \hat{\beta}_{(i)})'X'X(b - \hat{\beta}_{(i)})}{Ks^2} \tag{3.217}$$

$$= \frac{(\hat{y} - \hat{y}_{(i)})'(\hat{y} - \hat{y}_{(i)})}{Ks^2} \quad (i = 1, \ldots, T)\,, \tag{3.218}$$

to measure the influence of the ith observation on the center of the confidence ellipsoid or, equivalently, on the estimated coefficients $\hat{\beta}_{(i)}$ or the predictors $\hat{y}_{(i)} = X\hat{\beta}_{(i)}$. The measure C_i can be thought of as the scaled distance between b and $\hat{\beta}_{(i)}$ or $\hat{y}$ and $\hat{y}_{(i)}$, respectively. Using

$$b - \beta_{(i)} = \frac{(X'X)^{-1}x_i\hat{\epsilon}_i}{1 - p_{ii}}\,, \tag{3.219}$$

the difference between the OLSEs in the full model and the reduced data sets, we immediately obtain the following relationship:

$$C_i = \frac{1}{K} \frac{p_{ii}}{1 - p_{ii}} r_i^2 , \tag{3.220}$$

where r_i is the ith internally Studentized residual. C_i becomes large if p_{ii} and/or r_i^2 are large. Furthermore, C_i is proportional to r_i^2. Applying (3.211) and (3.213), we get

$$\frac{r_i^2(T - K - 1)}{T - K - r_i^2} \sim F_{1,T-K-1} ,$$

indicating that C_i is not exactly F–distributed. To inspect the relative size of C_i for all the observations, Cook (1977), by analogy of (3.216) and (3.217), suggests comparing C_i with the $F_{K,T-K}$–percentiles. The greater the percentile corresponding to C_i, the more influential is the ith observation.

Let, for example, $K = 2$ and $T = 32$, that is, $(T - K) = 30$. The 95% and 99% quantiles of $F_{2,30}$ are 3.32 and 5.59, respectively. When $C_i = 3.32$, $\hat{\beta}_{(i)}$ lies on the surface of the 95% confidence ellipsoid. If $C_j = 5.59$ for $j \neq i$, then $\hat{\beta}_{(j)}$ lies on the surface of the 99% confidence ellipsoid and, hence, the jth observation would be more influential than the ith observation.

Welsch–Kuh's Distance

The influence of the ith observation on the predicted value $\hat{y}_i$ can be measured by the scaled difference $(\hat{y}_i - \hat{y}_{i(i)})$ – by the change in predicting y_i when the ith observation is omitted. The scaling factor is the standard deviation of $\hat{y}_i$ (cf. (3.184)):

$$\frac{|\hat{y}_i - \hat{y}_{i(i)}|}{\sigma\sqrt{p_{ii}}} = \frac{|x_i'(b - \hat{\beta}_{(i)})|}{\sigma\sqrt{p_{ii}}} . \tag{3.221}$$

suggesting the use of $s_{(i)}$ [(3.207)] as an estimate of σ in (3.221). Using (3.219) and (3.208), (3.221) can be written as

$$\begin{aligned} WK_i &= \frac{|\hat{\epsilon}_i/(1 - p_{ii})x_i'(X'X)^{-1}x_i|}{s_{(i)}\sqrt{p_{ii}}} \\ &= |r_i^*|\sqrt{\frac{p_{ii}}{1 - p_{ii}}} . \end{aligned} \tag{3.222}$$

WK_i is called the Welsch–Kuh statistic. When $r_i^* \sim t_{T-K-1}$ (see Theorem 3.24), we can judge the size of WK_i by comparing it to the quantiles of the t_{T-K-1}–distribution. For sufficiently large sample sizes, one may use $2\sqrt{K/(T-K)}$ as a cutoff point for WK_i, signaling an influential ith observation.

Remark: The literature contains various modifications of Cook's distance (cf. Chatterjee and Hadi, 1988, pp. 122–135).

Measures Based on the Volume of Confidence Ellipsoids

Let $x'Ax \leq 1$ define an ellipsoid and assume A to be a symmetric (positive–definite or nonnegative–definite) matrix. From spectral decomposition (Theorem A.30), we have $A = \Gamma\Lambda\Gamma'$, $\Gamma\Gamma' = I$. The volume of the ellipsoid $x'Ax = (x'\,\Gamma)\Lambda(\Gamma' x) = 1$ is then seen to be

$$V = c_K \prod_{i=1}^{K} \lambda_i^{-1/2} = c_K \sqrt{|\Lambda^{-1}|}\,,$$

that is, inversely proportional to the root of $|A|$. Applying these arguments to (3.216), we may conclude that the volume of the confidence ellipsoid (3.216) is inversely proportional to $|X'X|$. Large values of $|X'X|$ indicate an informative design. If we take the confidence ellipsoid when the ith observation is omitted, namely,

$$\frac{(\beta - \hat{\beta}_{(i)})'(X'_{(i)}X_{(i)})(\beta - \hat{\beta}_{(i)})}{K s^2_{(i)}} \leq F_{K,T-K-1,1-\alpha}\,, \tag{3.223}$$

then its volume is inversely proportional to $|X'_{(i)}X_{(i)}|$. Therefore, omitting an influential (informative) observation would decrease $|X'_{(i)}X_{(i)}|$ relative to $|X'X|$. On the other hand, omitting an observation having a large residual will decrease the residual sum of squares $s^2_{(i)}$ relative to s^2. These two ideas can be combined in one measure.

Andrews–Pregibon Statistic

Andrews and Pregibon (1978) have compared the volume of the ellipsoids (3.216) and (3.223) according to the ratio

$$\frac{(T-K-1)s^2_{(i)}|X'_{(i)}X_{(i)}|}{(T-K)s^2|X'X|}\,. \tag{3.224}$$

An equivalent representation, proved in Proof 21, Appendix B, is

$$\frac{|Z'_{(i)}Z_{(i)}|}{|Z'Z|}. \tag{3.225}$$

Omitting an observation that is far from the center of data will result in a large reduction in the determinant and, consequently, a large increase in volume. Hence, small values of (3.225) correspond to this fact. For the sake of convenience, we define

$$AP_i = 1 - \frac{|Z'_{(i)}Z_{(i)}|}{|Z'Z|}\,, \tag{3.226}$$

so that large values will indicate influential observations. AP_i is called the Andrews–Pregibon statistic and could be rewritten to

$$AP_i = p_{zii}, \text{ Proof 22, Appendix B,} \tag{3.227}$$

where p_{zii} is the ith diagonal element of the prediction matrix $P_Z = Z(Z'Z)^{-1}Z'$. From (B.106) we get

$$p_{zii} = p_{ii} + \frac{\hat{\epsilon}_i^2}{\hat{\epsilon}'\hat{\epsilon}} \,. \tag{3.228}$$

Thus AP_i does not distinguish between high–leverage points in the X–space and outliers in the Z–space. Since $0 \le p_{zii} \le 1$ (cf. (3.192)), we get

$$0 \le AP_i \le 1 \,. \tag{3.229}$$

If we apply the definition (3.206) of the internally Studentized residuals r_i and use $s^2 = \hat{\epsilon}'\hat{\epsilon}/(T-K)$, (3.229) implies

$$AP_i = p_{ii} + (1 - p_{ii})\frac{r_i^2}{T-K} \tag{3.230}$$

or

$$(1 - AP_i) = (1 - p_{ii})\left(1 - \frac{r_i^2}{T-K}\right) . \tag{3.231}$$

The first quantity of (3.231) identifies high–leverage points and the second identifies outliers. Small values of $(1-AP_i)$ indicate influential points (high–leverage points or outliers), whereas independent examination of the single factors in (3.231) is necessary to identify the nature of influence.

Variance Ratio

As an alternative to the Andrews–Pregibon statistic and the other measures, one can identify the influence of the ith observation by comparing the estimated dispersion matrices of b_0 and $\hat{\beta}_{(i)}$:

$$V(b_0) = s^2(X'X)^{-1} \quad \text{and} \quad V(\hat{\beta}_{(i)}) = s^2_{(i)}(X'_{(i)}X_{(i)})^{-1}$$

by using measures based on the determinant or the trace of these matrices. If $(X'_{(i)}X_{(i)})$ and $(X'X)$ are positive definite, one may apply the following variance ratio suggested by Belsley, Kuh and Welsch (1980):

$$VR_i = \frac{|s^2_{(i)}(X'_{(i)}X_{(i)})^{-1}|}{|s^2(X'X)^{-1}|} \tag{3.232}$$

$$= \left(\frac{s^2_{(i)}}{s^2}\right)^K \frac{|X'X|}{|X'_{(i)}X_{(i)}|} \,. \tag{3.233}$$

Applying Theorem A.2(x), we obtain

$$\begin{aligned} |X'_{(i)}X_{(i)}| &= |X'X - x_i x_i'| \\ &= |X'X|(1 - x_i'(X'X)^{-1}x_i) \\ &= |X'X|(1 - p_{ii}) \,. \end{aligned}$$

i	C_i	WK_i	AP_i	VR_i
1	0.182	0.610	0.349	1.260
2	0.043	0.289	0.188	1.191
3	0.166	0.541	0.858	8.967
4	0.001	0.037	0.106	1.455
5	0.005	0.096	0.122	1.443
6	0.241	0.864	0.504	0.475
7	0.017	0.177	0.164	1.443
8	0.114	0.518	0.331	0.803
9	0.003	0.068	0.123	1.466
10	0.078	0.405	0.256	0.995

TABLE 3.14. Cook's C_i; Welsch–Kuh, WK_i; Andrews–Pregibon, AP_i; variance ratio VR_i, for the data set of Table 3.13.

With this relationship, and using (3.212), we may conclude that

$$VR_i = \left(\frac{T-K-r_i^2}{T-K-1}\right)^K \frac{1}{1-p_{ii}} . \tag{3.234}$$

Therefore, VR_i will exceed 1 when r_i^2 is small (no outliers) and p_{ii} is large (high–leverage point), and it will be smaller than 1 whenever r_i^2 is large and p_{ii} is small. But if both r_i^2 and p_{ii} are large (or small), then VR_i tends toward 1. When all observations have equal influence on the dispersion matrix, VR_i is approximately equal to 1. Deviation from unity then will signal that the ith observation has more influence than the others. Belsley et al. (1980) propose the approximate cut–off "quantile"

$$|VR_i - 1| \geq \frac{3K}{T} . \tag{3.235}$$

Example 3.9 (Example 3.8, continued). We calculate the measures defined before for the data of Example 3.8 (cf. Table 3.13). Examining Table 3.14, we see that Cook's C_i has identified the sixth data point to be the most influential one. The cutoff quantile $2\sqrt{K/T-K} = 1$ for the Welsch–Kuh distance is not exceeded, but the sixth data point has the largest indication, again.

In calculating the Andrews–Pregibon statistic AP_i (cf. (3.227) and (3.228)), we insert $\hat{\epsilon}'\hat{\epsilon} = (T-K)s^2 = 8 \cdot (6.9)^2 = 380.88$. The smallest value $(1 - AP_i) = 0.14$ corresponds to the third observation, and we obtain

$$\begin{aligned}(1 - AP_3) = 0.14 &= (1-p_{33})\left(1 - \frac{r_3^2}{8}\right) \\ &= 0.14 \cdot (1 - 0.000387),\end{aligned}$$

indicating that (y_3, x_3) is a high–leverage point, as we have noted already. The sixth observation has an AP_i value next to that of the third observation. An inspection of the factors of $(1 - AP_6)$ indicates that (y_6, x_6) tends to be an outlier

$$(1 - AP_6) = 0.496 = 0.88 \cdot (1 - 0.437).$$

These conclusions also hold for the variance ratio. Condition (3.235), namely, $|VR_i - 1| \geq \frac{6}{10}$, is fulfilled for the third observation, indicating significance, in the sense of (3.235).

Remark: In the literature one may find many variants and generalizations of the measures discussed here. A suitable recommendation is the monograph by Chatterjee and Hadi (1988).

3.10.6 Partial Regression Plots

Plotting the residuals against a fixed independent variable can be used to check the assumption that this regression has a linear effect on Y. If the residual plot shows the inadequacy of a linear relation between Y and some fixed X_i, it does not display the true (nonlinear) relation between Y and X_i. *Partial regression plots* are refined residual plots to represent the correct relation for a regressor in a multiple model under consideration. Suppose that we want to investigate the nature of the marginal effect of a variable X_k, say, on Y in case the other independent variables under consideration are already included in the model. Thus partial regression plots may provide information about the marginal importance of the variable X_k that may be added to the regression model.

Let us assume that one variable X_1 is included and that we wish to add a second variable X_2 to the model (cf. Neter, Wassermann and Kutner, 1990, p. 387). Regressing Y on X_1, we obtain the fitted values

$$\hat{y}_i(X_1) = \hat{\beta}_0 + x_{1i}\hat{\beta}_1 = \tilde{x}'_{1i}\tilde{\beta}_1 , \tag{3.236}$$

where

$$\tilde{\beta}_1 = (\hat{\beta}_0, \hat{\beta}_1)' = (\tilde{X}_1'\tilde{X}_1)^{-1}\tilde{X}_1'y \tag{3.237}$$

and $\tilde{X}_1 = (1, x_1)$.

Hence, we may define the residuals

$$e_i(Y|X_1) = y_i - \hat{y}_i(X_1) . \tag{3.238}$$

Regressing X_2 on $\tilde{X}_1$, we obtain the fitted values

$$\hat{x}_{2i}(X_1) = \tilde{x}'_{1i}b_1^* \tag{3.239}$$

with $b_1^* = (\tilde{X}_1'\tilde{X}_1)^{-1}\tilde{X}_1'x_2$ and the residuals

$$e_i(X_2|X_1) = x_{2i} - \hat{x}_{2i}(X_1) . \tag{3.240}$$

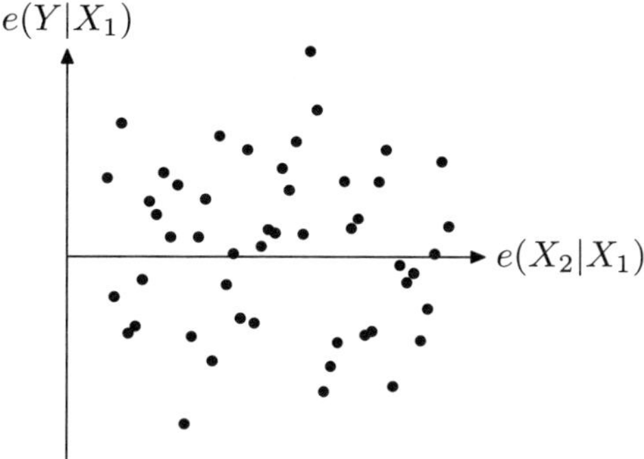

FIGURE 3.9. Partial regression plot (of $e(X_2 \mid X_1)$ versus $e(Y \mid X_1)$) indicating no additional influence of X_2 compared to the model $y = \beta_0 + X_1\beta_1 + \epsilon$.

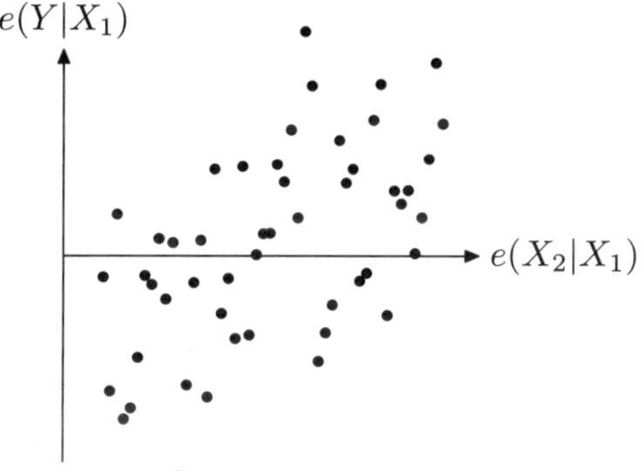

FIGURE 3.10. Partial regression plot (of $e(X_2 \mid X_1)$ versus $e(Y \mid X_1)$) indicating additional linear influence of X_2.

Analogously, in the full model $y = \beta_0 + X_1\beta_1 + X_2\beta_2 + \epsilon$, we have

$$e_i(Y|X_1, X_2) = y_i - \hat{y}_i(X_1, X_2), \tag{3.241}$$

where

$$\hat{y}_i(X_1, X_2) = \tilde{X}_1 b_1 + X_2 b_2 \tag{3.242}$$

and b_1 and b_2 are the two components resulting from the separation of b (replace X_1 by $\tilde{X}_1$), for example, see Rao and Toutenburg (1999). Then we have

$$e(Y \mid X_1, X_2) = e(Y \mid X_1) - b_2 e(X_2 \mid X_1). \tag{3.243}$$

The partial regression plot is obtained by plotting the residuals $e_i(Y \mid X_1)$ against the residuals $e_i(X_2 \mid X_1)$. Figures 3.9 and 3.10 present some standard partial regression plots. If the vertical deviations of the plotted points around the line $e(Y \mid X_1) = 0$ are squared and summed, we obtain the residual sum of squares

$$\begin{aligned} RSS_{\tilde{X}_1} &= (y - \tilde{X}_1(\tilde{X}_1'\tilde{X}_1)^{-1}\tilde{X}_1'y)'(y - \tilde{X}_1(\tilde{X}_1'\tilde{X}_1)^{-1}\tilde{X}_1'y) \\ &= y'\tilde{M}_1 y \\ &= \left[e(y \mid X_1)\right]'\left[e(Y \mid X_1)\right]. \end{aligned} \tag{3.244}$$

The vertical deviations of the plotted points in Figure 3.9, taken with respect to the line through the origin with slope b_1 are the estimated residuals $e(Y \mid X_1, X_2)$.

The extra sum of squares relationship is

$$SS_{\text{Reg}}(X_2 \mid \tilde{X}_1) = RSS_{\tilde{X}_1} - RSS_{\tilde{X}_1, X_2}. \tag{3.245}$$

This relation is the basis for the interpretation of the partial regression plot: If the scatter of the points around the line with slope b_2 is much less than the scatter around the horizontal line, then adding an additional independent variable X_2 to the regression model will lead to a substantial reduction of the error sum of squares and, hence, will substantially increase the fit of the model.

3.10.7 Regression Diagnostics by Animating Graphics

Graphical techniques are an essential part of statistical methodology. One of the important graphics in regression analysis is the residual plot. In regression analysis the plotting of residuals versus the independent variable or predicted values has been recommended by Draper and Smith (1966) and Cox and Snell (1968). These plots help to detect outliers, to assess the presence of the inhomogeneity of variance, and to check model adequacy. Larsen and McCleary (1972) introduced partial residual plots, which can detect the importance of each independent variable and assess some nonlinearity or necessary transformation of variables.

For the purpose of regression diagnostics, Cook and Weisberg (1989) introduced dynamic statistical graphics. They considered the interpretation of two proposed types of dynamic displays, rotation and animation, in regression diagnostics. Some of the issues that they addressed by using dynamic graphics include adding predictors to a model, assessing the need to transform, and checking for interactions and normality. They used animation to show the dynamic effects of adding a variable to a model and provided methods for simultaneously adding variables to a model.

Assume the classical linear, normal model

$$\begin{aligned} y &= X\beta + \epsilon \\ &= X_1\beta_1 + X_2\beta_2 + \epsilon, \quad \epsilon \sim N(0, \sigma^2 I). \end{aligned} \tag{3.246}$$

X consists of X_1 and X_2 where X_1 is a $[T \times (K-1)]$–matrix, and X_2 is a $(T \times 1)$–matrix, that is, $X = (X_1, X_2)$. The basic idea of Cook and Weisberg (1989) is to begin with the model $y = X_1\beta_1 + \epsilon$ and then smoothly add

X_2, ending with a fit of the full model $y = X_1\beta_1 + X_2\beta_2 + \epsilon$, where β_1 is a $[(K-1)\times 1]$–vector and β_2 is an unknown scalar. Since the animated plot that they proposed involves only fitted values and residuals, they worked in terms of a modified version of the full model (3.246) given by

$$\begin{aligned} y &= Z\beta^* + \epsilon \\ &= X_1\beta_1^* + \tilde{X}_2\beta_2^* + \epsilon\,, \end{aligned} \tag{3.247}$$

where $\tilde{X}_2 = Q_1X_2/||Q_1X_2||$ is the part of X_2 orthogonal to X_1, normalized to unit length, $Q_1 = I - P_1$, $P_1 = X_1(X_1'X_1)^{-1}X_1'$, $Z = (X_1, \tilde{X}_2)$, and $\beta^* = (\beta_1^{*\prime}, \beta_2^{*\prime})'$.

Next, for each $0 < \lambda \le 1$, they estimated β^* by

$$\hat{\beta}_\lambda = \left(Z'Z + \frac{1-\lambda}{\lambda}ee'\right)^{-1} Z'y\,, \tag{3.248}$$

where e is a $(K\times 1)$–vector of zeros except for a single 1 corresponding to X_2. Since

$$\begin{aligned} \left(Z'Z + \frac{1-\lambda}{\lambda}ee'\right)^{-1} &= \begin{pmatrix} X_1'X_1 & 0 \\ 0' & \tilde{X}_2'\tilde{X}_2 + (1-\lambda)/\lambda \end{pmatrix}^{-1} \\ &= \begin{pmatrix} X_1'X_1 & 0 \\ 0' & 1/\lambda \end{pmatrix}^{-1}, \end{aligned}$$

we obtain

$$\hat{\beta}_\lambda = \begin{pmatrix} (X_1'X_1)^{-1}X_1'y \\ \lambda\tilde{X}_2'y \end{pmatrix}.$$

So as λ tends to 0, (3.248) corresponds to the regression of y on X_1 alone. And if $\lambda = 1$, then (3.248) corresponds to the ordinary least squares regression of y on X_1 and X_2. Thus as λ increases from 0 to 1, $\hat{\beta}_\lambda$ represents a continuous change of estimators that add X_2 to the model, and an animated plot of $\hat{\epsilon}(\lambda)$ versus $\hat{y}(\lambda)$, where $\hat{\epsilon}(\lambda) = y - \hat{y}(\lambda)$ and $\hat{y}(\lambda) = Z\hat{\beta}_\lambda$, gives a dynamic view of the effects of adding X_2 to the model that already includes X_1. This idea corresponds to the weighted mixed regression estimator, see Rao and Toutenburg (1999), for example.

Using Cook and Weisberg's idea of animation, Park, Kim and Toutenburg (1992) proposed an animating graphical method to display the effects of removing an outlier from a model for regression diagnostic purposes.

We want to view the dynamic effects of removing the ith observation from the model (3.246). First, we consider the mean shift model $y = X\beta + \gamma_i e_i + \epsilon$ (see (3.209)) where e_i is the vector of zeros except for a single 1 corresponding to the ith observation. We can work in terms of a modified version of the mean shift model given by

$$\begin{aligned} y &= Z\beta^* + \epsilon \\ &= X\tilde{\beta} + \gamma_i^*\tilde{e} + \epsilon\,, \end{aligned} \tag{3.249}$$

where $\tilde{e}_i = Q_x e_i / ||Q_x e_i||$ is the orthogonal part of e_i to X normalized to unit length, $Q = I - P$, $P = X(X'X)^{-1}X'$, $Z = (X, \tilde{e}_i)$, and $\beta^* = (\tilde{\beta}\gamma_i^*)'$. And then, for each $0 < \lambda \leq 1$, we estimate β^* by

$$\hat{\beta}_\lambda = \left(Z'Z + \frac{1-\lambda}{\lambda} ee' \right)^{-1} Z'y\,, \tag{3.250}$$

where e is the $[(K+1) \times 1]$–vector of zeros except for a single 1 for the $(K+1)$th element. Now we can think of some properties of $\hat{\beta}_\lambda$. First, without loss of generality, we take X and y of the forms $X = (X_{(i)} x_i')'$ and $y = (y_{(i)} y_i)'$, where x_i' is the ith row vector of X, $X_{(i)}$ is the matrix X without the ith row, and $y_{(i)}$ is the vector y without y_i. That is, place the ith observation to the bottom and so e_i and e become vectors of zeros except for the last 1. Then, since

$$\left(Z'Z + \frac{1-\lambda}{\lambda} ee' \right)^{-1} = \begin{pmatrix} X'X & 0 \\ 0' & 1/\lambda \end{pmatrix}^{-1} = \begin{pmatrix} (X'X)^{-1} & 0 \\ 0' & \lambda \end{pmatrix}$$

and

$$Z'y = \begin{pmatrix} X'y \\ \tilde{e}_i' y \end{pmatrix}$$

we obtain

$$\hat{\beta}_\lambda = \begin{pmatrix} \hat{\tilde{\beta}} \\ \hat{\gamma}_i^* \end{pmatrix} = \begin{pmatrix} (X'X)^{-1}X'y \\ \lambda \tilde{e}_i^* y \end{pmatrix}$$

and

$$\hat{y}(\lambda) = Z\hat{\beta}_\lambda = X(X'X)^{-1}X'y + \lambda \tilde{e}\tilde{e}'y\,.$$

Hence at $\lambda = 0$, $\hat{y}(\lambda) = (X'X)^{-1}X'y$ is the predicted vector of observed values for the full model by the method of ordinary least squares. And at $\lambda = 1$, we can get the following lemma, where $\hat{\beta}_{(i)} = (X_{(i)}'X_{(i)})^{-1}X_{(i)}y_{(i)}$.

Lemma 3.25.

$$\hat{y}(1) = \begin{pmatrix} X_{(i)}\hat{\beta}_{(i)} \\ y_{(i)} \end{pmatrix}.$$

Proof. See Proof 23, Appendix B.

Thus as λ increases from 0 to 1, an animated plot of $\hat{\epsilon}(\lambda)$ versus $\hat{\lambda}$ gives a dynamic view of the effects of removing the ith observation from model (3.246).

The following lemma shows that the residuals $\hat{\epsilon}(\lambda)$ and fitted values $\hat{y}(\lambda)$ can be computed from the residuals $\hat{\epsilon}$, fitted values $\hat{y} = \hat{y}(0)$ from the full model, and the fitted values $\hat{y}(1)$ from the model that does not contain the ith observation.

Lemma 3.26.

(i) $\hat{y}(\lambda) = \lambda\hat{y}(1) + (1-\lambda)\hat{y}(0)$; and

(ii) $\hat{\epsilon}(\lambda) = \hat{\epsilon} - \lambda(\hat{y}(1) - \hat{y}(0))$.

Proof. See Proof 24, Appendix B.

Because of the simplicity of Lemma 3.26, an animated plot of $\hat{\epsilon}(\lambda)$ versus $\hat{y}(\lambda)$ as λ is varied between 0 and 1 can easily be computed.

The appropriate number of frames (values of λ) for an animated residual plot depends on the speed with which the computer screen can be refreshed and, thus, on the hardware being used. With too many frames, changes often become too small to be noticed and, as a consequence, the overall trend can be missed. With too few frames, smoothness and the behavior of individual points cannot be detected.

When there are too many observations, and it is difficult to check all the animated plots, it is advisable to select several suspicious observations based on nonanimated diagnostic measures, such as Studentized residuals, Cook's distance, and so on.

From animated residual plots for individual observations, $i = 1, 2, \ldots, n$, it would be possible to diagnose which observation is most influential in changing the residuals $\hat{\epsilon}$, and the fitted values of y, $\hat{y}(\lambda)$, as λ changes from 0 to 1. Thus, it may be possible to formulate a measure to reflect which observation is most influential, and which kind of influential points can be diagnosed in addition to those that can already be diagnosed by well–known diagnostics. However, our primary intent is only to provide a graphical tool to display and see the effects of continuously removing a single observation from a model. For this reason, we do not develop a new diagnostic measure that could give a criterion when an animated plot of removing an observation is significant or not. Hence, development of a new measure based on such animated plots remains open to further research.

Example 3.10 (Phosphorus Data). In this example, we illustrate the use of $\hat{\epsilon}(\lambda)$ versus $\hat{y}(\lambda)$ as an aid to understanding the dynamic effects of removing an observation from a model. Our illustration is based on the phosphorus data reported in Snedecor and Cochran (1967, p. 384). An investigation of the source from which corn plants obtain their phosphorus was carried out. Concentrations of phosphorus, in parts per million, in each of 18 soils was measured. The variables are

$$\begin{aligned} X_1 &= \text{concentrations of inorganic phosphorus in the soil,} \\ X_2 &= \text{concentrations of organic phosphorus in the soil,} \\ \text{and} & \\ y &= \text{phosphorus content of corn grown in the soil at 20 °C.} \end{aligned}$$

Soil	X_1	X_2	y	e_i	h_{ii}	r_i	C_i
1	0.4	53	64	2.44	0.26	0.14	0.002243
2	0.4	23	60	1.04	0.19	0.06	0.000243
3	3.1	19	71	7.55	0.23	0.42	0.016711
4	0.6	34	61	0.73	0.13	0.04	0.000071
5	4.7	24	54	-12.74	0.16	-0.67	0.028762
6	1.7	65	77	12.07	0.46	0.79	0.178790
7	9.4	44	81	4.11	0.06	0.21	0.000965
8	10.1	31	93	15.99	0.10	0.81	0.023851
9	11.6	29	93	13.47	0.12	0.70	0.022543
10	12.6	58	51	-32.83	0.15	-1.72	0.178095
11	10.9	37	76	-2.97	0.06	-0.15	0.000503
12	23.1	46	96	-5.58	0.13	-0.29	0.004179
13	23.1	50	77	-24.93	0.13	-1.29	0.080664
14	21.6	44	93	-5.72	0.12	-0.29	0.003768
15	23.1	56	95	-7.45	0.15	-0.39	0.008668
16	1.9	36	54	-8.77	0.11	-0.45	0.008624
17	26.8	58	168	58.76	0.20	3.18	0.837675
18	29.9	51	99	-15.18	0.24	-0.84	0.075463

TABLE 3.15. Data, ordinary residuals e_i, diagonal terms h_{ii} of hat matrix $H = X(X'X)^{-1}X'$, Studentized residuals r_i, and Cook's distances C_i from Example 3.10.

The data set, together with the ordinary residuals e_i, the diagonal terms h_{ii} of the hat matrix $H = X(X'X)^{-1}X'$, the Studentized residuals r_i, and Cook's distances C_i are shown in Table 3.15 under the linear model assumption. We developed computer software that plots the animated residuals and some related regression results. The plot for the seventeenth observation shows the most significant changes in residuals among eighteen plots. In fact, the seventeenth observation has the largest target residual e_i, Studentized residuals r_{ii}, and Cook's distances C_i, as shown in Table 3.15.

Figure 3.10 shows four frames of an animated plot of $\hat{\epsilon}(\lambda)$ versus $\hat{y}(\lambda)$ for removing the seventeenth observation. The first frame (a) is for $\lambda = 0$ and thus corresponds to the usual plot of residuals versus fitted values from the regression of y on $X = (X_1, X_2)$, and we can see that in (a) the seventeenth observation is located in the upper–right corner. The second (b), third (c), and fourth (d) frames correspond to $\lambda = \frac{1}{2}, \frac{2}{3}$, and 1, respectively. So the fourth frame (d) is the usual plot of the residuals versus the fitted values from the regression of $y_{(17)}$ on $X_{(17)}$ where the subscript represents omission of the corresponding observation. We can see that as λ increases from 0 to 1, the seventeenth observation moves to the right and down, becoming the rightmost point in (b), (c), and (d). Considering the plotting form, the residual plot in (a) has an undesirable form because it does not have

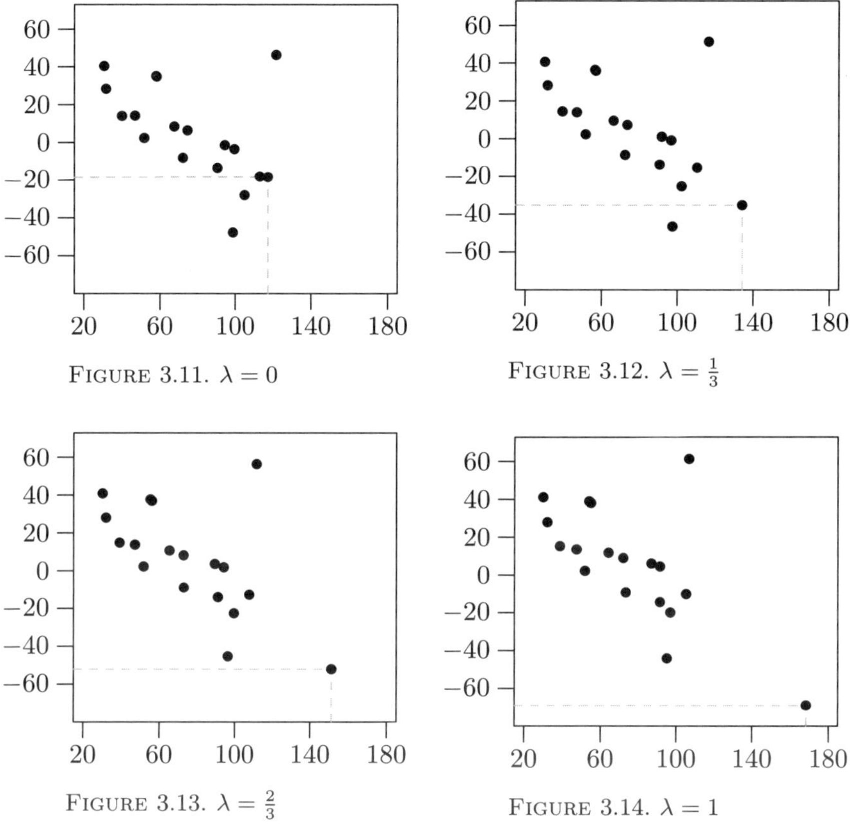

FIGURE 3.11. $\lambda = 0$

FIGURE 3.12. $\lambda = \frac{1}{3}$

FIGURE 3.13. $\lambda = \frac{2}{3}$

FIGURE 3.14. $\lambda = 1$

a random form in a band between -60 and $+60$, but in (d) its form has randomness in a band between -20 and $+20$.

Figure 3.11–3.14 show animated plots of $\hat{\epsilon}(\lambda)$ versus $\hat{y}(\lambda)$ for data in Example 3.10 when removing the seventeenth observation (marked by dotted lines).

Apart from the problems we described within this section there exist many other problems which the user may be confronted with in practical work. Based on the usual notation of the linear model, problems may arise by its components, i.e., ϵ (heteroscedasticity, autocorrelation), X (exclusion of relevant variables, inclusion of irrelevant variables, correlation between X and ϵ), or with the parameter β. Especially, the constancy of β as an important assumption may be violated. Several testing procedures, e.g., the Chow or Hansen tests, are described in Johnston (1984). Also helpful is the description of tests of slope coefficients or of an intercept (see also Johnston (1984)).

3.11 Exercises and Questions

3.11.1 Define the principle of least squares.

3.11.2 Given the normal equation $X'X\beta = X'y$, what are the conditions for a unique solution?

3.11.3 Assume $\text{rank}(X) = p < K$. What are the linear restrictions to ensure estimability of β? Give the definition of the restricted least squares estimator.

3.11.4 Define the matrix–valued mean square error of a linear estimator and the MSE–I superiority.

3.11.5 Let $\hat{\beta} = Cy + d$ be a linear estimator. Give the condition of unbiasedness of $\hat{\beta}$. What is the best linear unbiased estimator?

3.11.6 What is the relation of the covariance matrices of the best linear unbiased estimator $\hat{\beta}$ and any linear estimator $\tilde{\beta}$?

3.11.7 How can you get an unbiased estimate of σ^2?

3.11.8 Characterize weak and extreme multicollinearity in terms of the rank of $X'X$, unbiasedness of the least squares estimator and identifiability.

3.11.9 Assume $\epsilon \sim N(\mathbf{0}, \sigma^2 I)$ and give the ML estimators of β and σ^2.

4
Single–Factor Experiments with Fixed and Random Effects

4.1 Models I and II in the Analysis of Variance

The analysis of variance, which was originally developed by R.A. Fisher for field experiments, is one of the most widely used and one of the most general statistical procedures for testing and analyzing data. These procedures require a large amount of computation, especially in the case of complicated classifications. For this reason, these procedures are available as software.

We distinguish between two fundamental problems.

Model I with fixed effects is used for the *multiple comparison of means* of quantitative normally distributed factors that are observed on fixed selected experimental units. We test the null hypothesis $H_0 : \mu_1 = \mu_2 = \ldots = \mu_s$ against the general alternative H_1 : *at least two means are different*, i.e., we compare s normally distributed populations with respect to their means. The corresponding F–test is a generalization of the t–test, that compares two normal distributions. In general, this comparison is called *comparison of the effects of treatments*. If *specific* treatments are to be compared, then it is wise not to choose them at random, but to assume them as *fixed.*

Example 4.1. Comparison of the average manufacturing time for an inlay by three different *prespecified* dentists (Table 4.1).

Dentist A	Dentist B	Dentist C
55.5	67.0	62.5
40.0	57.0	31.5
38.5	33.5	31.5
31.5	37.0	53.0
45.5	75.0	50.5
70.0	60.0	62.5
78.0	43.5	40.0
80.0	56.0	19.5
74.5	65.5	
57.5	54.0	
72.0	59.5	
70.0		
48.0		
59.0		
$n_1 = 14$	$n_2 = 11$	$n_3 = 8$
$\bar{x}_1 = 58.57$	$\bar{x}_2 = 55.27$	$\bar{x}_3 = 43.88$
$n = n_1 + n_2 + n_3$		

TABLE 4.1. Manufacturing time (in minutes) for the making of inlays, measured for three dentists (cf. Toutenburg, 1977).

Model II with random effects is used for the *decomposition* of the *total variability* produced by the effect of several factors. This total variability (variance) is decomposed into components that reflect the effect of each factor and into a component that cannot be explained by the factors, i.e., the error variance. The experimental units are chosen at *random*, as opposed to Model I. The treatments are then to be regarded as a random sample from an assumed infinite population. Hence, we have no interest in the treatments chosen at random, but only in the respective proportion of the total variability.

Example 4.2. From a total population, the manufacturing times of (e.g., three) dentists *chosen at random* are to be analyzed with respect to their proportion of the total variability.

4.2 One–Way Classification for the Multiple Comparison of Means

Assume we have s samples from s normally distributed populations $N(\mu_i, \sigma^2)$. Furthermore, assume the sample sizes to be n_i and the total sample size to be n with

$$\sum_{i=1}^{s} n_i = n. \tag{4.1}$$

The variances σ^2 are *unknown, but equal* in all populations.

Definition 4.1. If all n_i are equal, then the sampling design (experimental design) is called *balanced.* Otherwise, it is called *unbalanced.*

The s different levels of a Factor A are called *treatments.* Since only one factor is investigated, we call this type of experimental design *one–way classification.*

Examples:

1. Factor A: plastic PMMA:
   s levels: s different concentrations of quartz in PMMA;
   s effects: flexibility of the different PMMA materials.
2. Factor A: fertilization:
   s levels: s different fertilizers (or one fertilizer with s different concentrations of phosphate);
   s effects: output per acre.

	Single experiments per level of Factor A				Sum of the observations per sample	Sample mean
	1	2	...	n_i		
1	y_{11}	y_{12}	...	y_{1n_1}	$\sum y_{1j} = Y_{1.}$	$Y_{1.}/n_1 = y_{1.}$
2	y_{21}	y_{22}	...	y_{2n_2}	$\sum y_{2j} = Y_{2.}$	$Y_{2.}/n_2 = y_{2.}$
⋮						
s	y_{s1}	y_{s2}	...	y_{sn_s}	$\sum y_{sj} = Y_{s.}$	$Y_{s.}/n_s = y_{s.}$
				$n = \sum n_i$	$\sum Y_{i.} = Y_{..}$	$Y_{..}/n = y_{..}$

TABLE 4.2. Sample design (one–way classification).

The observations of the s samples are arranged according to Table 4.2. A period in the subscript indicates that we summed over this subscript. For example, $y_{1.}$ is the sum of the first row, $y_{..}$ is the total sum. For the observations y_{ij} we assume the following model:

$$y_{ij} = \mu + \alpha_i + \epsilon_{ij} \quad (i = 1, \ldots, s, j = 1, \ldots, n_i), \tag{4.2}$$

in which μ is the overall mean, α_i is the effect of the ith level of Factor A (i.e., the deviation (treatment effect) from the overall mean μ caused by the ith level), and ϵ_{ij} is a random error (i.e., random deviation from μ and α_i).
μ and α_i are fixed parameters, the ϵ_{ij} are random. The following assumptions have to hold:

- the errors ϵ_{ij} are independent and identically distributed with mean 0 and variance σ^2;

- the errors are normal, i.e., we have $\epsilon_{ij} \sim N(0, \sigma^2)$; and
- the following constraint holds

$$\sum \alpha_i n_i = 0. \tag{4.3}$$

In experimental designs, it is important to have equal sample sizes n_i in the groups (balanced case), otherwise the analysis of variance is not *robust* against deviations from the assumptions (normal distribution, equal variances).

Remark. Model I (with fixed effects) assumes that the s treatments are given in advance, i.e., they are *fixed* before the experiment. Hence, the α_i are nonstochastic factors. If the s treatments were selected by a random mechanism from a set of possible treatments, then the α_i would be stochastic, i.e., random variables with a certain distribution. For the analysis of linear models with stochastic parameters the methods of linear models have to be modified. For now, we restrict ourselves to the case with fixed effects. Models with random effects are discussed in Section 4.6.

Completely Randomized Experimental Design

The simplest and least restrictive design (CRD: completely randomized design) consists of assigning the s treatments to the n experimental units in the following manner. We choose n_1 experimental units at random and assign them to treatment $i = 1$. After that, n_2 experimental units are selected from the remaining $n - n_1$ units, once again at random, and are assigned to treatment $i = 2$, and so on. The remaining $n - \sum_{i=1}^{s-1} n_i = n_s$ units receive the sth treatment. This experimental design has the following *advantages* (cf., e.g., Petersen, 1985, p. 7):

- Flexibility: The number s of treatments and the amounts n_i are not restricted; in particular, unbalanced designs are allowed. However, balanced design should be preferred, since for these designs the power of the tests is the highest.
- Degrees of freedom: The design provides a maximum number of degrees of freedom for the error variance.
- Statistical analysis: The employment of standard procedures is possible in the unbalanced case as well (e.g., in the case of missing values due to nonresponse).

A *disadvantage* of this design arises in case of inhomogeneous experimental units: a decrease in the precision of the results. Often, however, the experimental units can be grouped into homogeneous subgroups (blocking) with a resulting increase in precision.

4.2.1 Representation as a Restrictive Model

The linear model (4.2) can be formulated in matrix notation

$$\begin{pmatrix} y_{11} \\ \vdots \\ y_{1n_1} \\ \vdots \\ y_{s1} \\ \vdots \\ y_{sn_s} \end{pmatrix} = \begin{pmatrix} 1 & 1 & 0 & \dots & 0 \\ \vdots & \vdots & \vdots & \vdots & \vdots \\ 1 & 1 & 0 & \dots & 0 \\ \vdots & \vdots & \vdots & \vdots & \vdots \\ 1 & 0 & \cdots & 0 & 1 \\ \vdots & \vdots & \vdots & \vdots & \vdots \\ 1 & 0 & \cdots & 0 & 1 \end{pmatrix} \begin{pmatrix} \mu \\ \alpha_1 \\ \vdots \\ \alpha_s \end{pmatrix} + \begin{pmatrix} \epsilon_{11} \\ \vdots \\ \epsilon_{1n_1} \\ \vdots \\ \epsilon_{s1} \\ \vdots \\ \epsilon_{sn_s} \end{pmatrix},$$

i.e.,

$$y = X\beta + \epsilon, \quad \epsilon \sim N(\mathbf{0}, \sigma^2 I), \tag{4.4}$$

with X of type $n \times (s+1)$ and $\operatorname{rank}(X) = s$. Hence, we have exact multicollinearity. $X'X$ is now singular, and a linear restriction $r = R'\beta$ with $\operatorname{rank}(R) = J = 1$ and $\operatorname{rank}(XR')' = s+1$ has to be introduced for the estimation of the $[(s+1) \times 1]$–vector $\beta' = (\mu, \alpha_1, \dots, \alpha_s)$ (cf. Theorem B.1). We choose

$$r = 0, \quad R' = (0, n_1, \dots, n_s), \tag{4.5}$$

and, hence,

$$\sum \alpha_i n_i = 0 \tag{4.6}$$

(cf. (4.3)).

Remark. The estimability of β is ensured according to Theorem B.1 for *every* restriction $r = R'\beta$ with $\operatorname{rank}(R') = J = 1$ and $\operatorname{rank}(XR')' = s+1$. However, the *selected* restriction (4.6) has the advantage of an interpretation, justified by the subject matter, that follows the effect coding of a loglinear model. The parameters α_i are then the deviations from the overall mean μ and hence standardized with respect to μ. Thus, the α_i determine the relative (positive or negative) factors, with which the ith treatment leads to deviations from the overall mean, by their magnitude and sign.

According to (B.16), the conditional OLS estimate of $\beta' = (\mu, \alpha_1, \dots, \alpha_s)$ is of the following form:

$$b(R', 0) = (X'X + RR')^{-1} X'y. \tag{4.7}$$

As we can easily check, the matrix $(XR')'$ with X from (4.4), and R' from (4.5), is of full column rank $s+1$.

Case $s = 2$

We demonstrate the computation of the estimate $b(R', 0)$ for $s = 2$. With the notation $\mathbf{1}'_{n_i} = (1, \dots, 1)$ for the $(n_i \times 1)$–vector of ones, we obtain the following representation:

$$\underset{n,3}{X} = \begin{pmatrix} \mathbf{1}_{n_1} & \mathbf{1}_{n_1} & \mathbf{0} \\ \mathbf{1}_{n_2} & \mathbf{0} & \mathbf{1}_{n_2} \end{pmatrix}, \tag{4.8}$$

$$X'X = \begin{pmatrix} \mathbf{1}'_{n_1} & \mathbf{1}'_{n_2} \\ \mathbf{1}'_{n_1} & \mathbf{0}' \\ \mathbf{0}' & \mathbf{1}'_{n_2} \end{pmatrix} \begin{pmatrix} \mathbf{1}_{n_1} & \mathbf{1}_{n_1} & \mathbf{0} \\ \mathbf{1}_{n_2} & \mathbf{0} & \mathbf{1}_{n_2} \end{pmatrix}$$

$$= \begin{pmatrix} n_1+n_2 & n_1 & n_2 \\ n_1 & n_1 & 0 \\ n_2 & 0 & n_2 \end{pmatrix},$$

$$RR' = \begin{pmatrix} 0 \\ n_1 \\ n_2 \end{pmatrix} (0 \quad n_1 \quad n_2) \tag{4.9}$$

$$= \begin{pmatrix} 0 & 0 & 0 \\ 0 & n_1^2 & n_1 n_2 \\ 0 & n_1 n_2 & n_2^2 \end{pmatrix}.$$

With $n = n_1 + n_2$ we have

$$(X'X + RR') = \begin{pmatrix} n & n_1 & n_2 \\ n_1 & n_1 + n_1^2 & n_1 n_2 \\ n_2 & n_1 n_2 & n_2 + n_2^2 \end{pmatrix},$$

$$|X'X + RR'| = n_1 n_2 n^2, \tag{4.10}$$

following that $(X'X + RR')^{-1}$ equals

$$\frac{1}{n_1 n_2 n^2} \cdot \begin{pmatrix} n_1 n_2(1+n) & -n_1 n_2 & -n_1 n_2 \\ -n_1 n_2 & n_2(n(1+n_2)-n_2) & -n_1 n_2(n-1) \\ -n_1 n_2 & -n_1 n_2(n-1) & n_1(n(1+n_1)-n_1) \end{pmatrix}, \tag{4.11}$$

$$X'y = \begin{pmatrix} \mathbf{1}'_{n_1} & \mathbf{1}'_{n_2} \\ \mathbf{1}'_{n_1} & \mathbf{0}' \\ \mathbf{0}' & \mathbf{1}'_{n_2} \end{pmatrix} \begin{pmatrix} y_1 \\ y_2 \end{pmatrix}$$

$$= \begin{pmatrix} Y_{\cdot\cdot} \\ Y_{1\cdot} \\ Y_{2\cdot} \end{pmatrix}. \tag{4.12}$$

Here we have

$$y_1 = \begin{pmatrix} y_{11} \\ \vdots \\ y_{1n_1} \end{pmatrix}, \quad y_2 = \begin{pmatrix} y_{21} \\ \vdots \\ y_{2n_2} \end{pmatrix},$$

$$Y_{1\cdot} = \sum_{j=1}^{n_1} y_{1j}, \quad Y_{2\cdot} = \sum_{j=1}^{n_2} y_{2j},$$

$$Y_{\cdot\cdot} = Y_{1\cdot} + Y_{2\cdot}\,.$$

Finally, we receive the conditional OLS estimate (4.7) for the case $s = 2$ according to

$$b((0, n_1, n_2), 0) = (X'X + RR')^{-1} X'y = \begin{pmatrix} \hat{\mu} \\ \hat{\alpha}_1 \\ \hat{\alpha}_2 \end{pmatrix} = \begin{pmatrix} y_{\cdot\cdot} \\ y_{1\cdot} - y_{\cdot\cdot} \\ y_{2\cdot} - y_{\cdot\cdot} \end{pmatrix}. \quad (4.13)$$

Proof. See Proof 25, Appendix B.2.

4.2.2 Decomposition of the Error Sum of Squares

With $b(R', 0)$ from (4.13) we receive

$$\hat{y} = Xb(R', 0) = \begin{pmatrix} y_{1\cdot}\mathbf{1}_{n_1} \\ y_{2\cdot}\mathbf{1}_{n_2} \end{pmatrix}. \quad (4.14)$$

The decomposition (3.120), i.e.,

$$\sum (y_t - \bar{y})^2 = \sum (y_t - \hat{y}_t)^2 + \sum (\hat{y}_t - \bar{y})^2,$$

is of the following form in the model (4.4) with the new notation

$$\sum_{i=1}^{s}\sum_{j=1}^{n_i}(y_{ij} - y_{\cdot\cdot})^2 = \sum_{i=1}^{s}\sum_{j=1}^{n_i}(y_{ij} - y_{i\cdot})^2 + \sum_{i=1}^{s} n_i (y_{i\cdot} - y_{\cdot\cdot})^2 \quad (4.15)$$

or, written according to (3.121) and (3.122),

$$SS_{\text{Corr}} = RSS + SS_{\text{Reg}} \quad (4.16)$$

or, in the notation of the analysis of variance,

$$SS_{\text{Total}} = SS_{\text{Within}} + SS_{\text{Between}}. \quad (4.17)$$

The sum of squares

$$SS_{\text{Within}} = \sum\sum (y_{ij} - y_{i\cdot})^2$$

measures the variability within each treatment. On the other hand, the sum of squares

$$SS_{\text{Between}} = \sum_{i=1}^{s} n_i (y_{i\cdot} - y_{\cdot\cdot})^2$$

measures the differences in variability between the treatments, i.e., the actual treatment effects.

Testing the Regression

We consider the linear model

$$y_{ij} = \mu + \alpha_i + \epsilon_{ij} \ (i = 1, \ldots, s, j = 1, \ldots, n_i) \quad (4.18)$$

with

$$\sum n_i \alpha_i = 0 . \tag{4.19}$$

Testing the hypothesis

$$\mathrm{H}_0 : \alpha_1 = \cdots = \alpha_s = 0 \tag{4.20}$$

is equivalent to comparing the models

$$\mathrm{H}_0 : y_{ij} = \mu + \epsilon_{ij} \tag{4.21}$$

and

$$\mathrm{H}_1 : y_{ij} = \mu + \alpha_i + \epsilon_{ij} \quad \text{with} \quad \sum n_i \alpha_i = 0 , \tag{4.22}$$

i.e., is equivalent to testing

$$\mathrm{H}_0 : \alpha_1 = \cdots = \alpha_s = 0 \text{ (parameter space } \omega) \tag{4.23}$$

against

$$\mathrm{H}_1 : \alpha_i \neq 0 \text{ for at least two } i \text{ (parameter space } \Omega) . \tag{4.24}$$

In the case of an assumed normal distribution $\epsilon_{ij} \sim N(0, \sigma^2)$ for all i, j the corresponding likelihood ratio test statistic (3.102)

$$F = \frac{\hat{\sigma}^2_\omega - \hat{\sigma}^2_\Omega}{\hat{\sigma}^2_\Omega} \frac{T-K}{K-s}$$

changes to

$$F = \frac{SS_{\text{Total}} - SS_{\text{Within}}}{SS_{\text{Within}}} \frac{n-s}{s-1} \tag{4.25}$$

$$= \frac{SS_{\text{Between}}}{SS_{\text{Within}}} \frac{n-s}{s-1} \tag{4.26}$$

$$= \frac{MS_{\text{Between}}}{MS_{\text{Within}}} . \tag{4.27}$$

Remark. The sum of squares

$$SS_{\text{Between}} = \sum_{i=1}^{s} n_i (y_{i\cdot} - y_{\cdot\cdot})^2$$

is named according to the factor, e.g., SS_A, if Factor A represents a treatment in s different levels. Analogously, we also denote

$$SS_{\text{Within}} = \sum_{i=1}^{s} \sum_{j=1}^{n_i} (y_{ij} - y_{i\cdot})^2$$

as SS_{Error} (SSE, error sum of squares).

The sums of squares with respect to $SS_{\text{Between}} = SS_A$ can also be written in detail as follows:

$$SS_{\text{Total}} = \sum_i \sum_j (y_{ij} - y_{..})^2 = \sum_i \sum_j y_{ij}^2 - ny_{..}^2 , \qquad (4.28)$$

$$SS_A = \sum_i \sum_j (y_{i.} - y_{..})^2 = \sum_i n_i y_{i.}^2 - ny_{..}^2 , \qquad (4.29)$$

$$SS_{\text{Error}} = \sum_i \sum_j (y_{ij} - y_{i.})^2 = \sum_i \sum_j y_{ij}^2 - \sum_i n_i y_{i.}^2 . \qquad (4.30)$$

These formulas make the computation a lot easier (i.e., if calculators are used).

Under the assumption of a normal distribution, the sums of squares have a χ^2–distribution with the corresponding degrees of freedom. The ratios SS/df are called MS (Mean Square). As we will show further on,

$$MS_E = \frac{SS_{\text{Error}}}{n-s} \qquad (4.31)$$

is an unbiased estimate of σ^2. For the test of hypothesis (4.23), the test statistic (4.27) is used, i.e.,

$$F = \frac{MS_A}{MS_E} = \frac{n-s}{s-1} \frac{SS_A}{SS_{\text{Error}}} . \qquad (4.32)$$

Under H_0, F has an $F_{s-1,n-s}$–distribution. If

$$F > F_{s-1,n-s;1-\alpha} , \qquad (4.33)$$

then H_0 is rejected. For the realization of the analysis of variance we use Table 4.3.

Source of variation	SS	Degrees of freedom	MS	Test statistics F
Between the levels of Factor A	$SS_A = \sum_{i=1}^{s} n_i y_{i.}^2 - ny_{..}^2$	$df_A = s-1$	$MS_A = \frac{SS_A}{df_A}$	MS_A/MS_E
Within the levels of Factor A	$SS_{\text{Error}} = \sum_i \sum_j y_{ij}^2 - \sum_i n_i y_{i.}^2$	$df_E = n-s$	$MS_E = \frac{SS_E}{df_E}$	
	$SS_{\text{Total}} = \sum_i \sum_j y_{ij}^2 - ny_{..}^2$	$df_T = n-1$		

TABLE 4.3. Layout for the analysis of variance; one–way classification.

Remark. For the derivation of the test statistic (4.32) we used the results of Chapter 3 and those of Section 3.7 in particular. Hence, we did not again prove the independence of the χ^2–distributions in the numerator and denominator of F (4.32).

Theorem 4.2 (Theorem by Cochran). Let $z_i \sim N(0,1), i = 1,\ldots,v$, be independent random variables and assume the following disjunctive decomposition

$$\sum_{i=1}^{v} z_i^2 = Q_1 + Q_2 + \cdots + Q_s \tag{4.34}$$

with $s \leq v$. Hence, the $Q_1, \ldots, Q_s$ are independent $\chi^2_{v_1}, \ldots, \chi^2_{v_s}$–distributed random variables if and only if

$$v = v_1 + \cdots + v_s \tag{4.35}$$

holds.

Employing this theorem yields the following:

$$\text{(i)} \qquad SS_{\text{Total}} = \sum_{i=1}^{s}\sum_{j=1}^{n_i}(y_{ij} - y_{\cdot\cdot})^2 \tag{4.36}$$

has $n = \sum_{i=1}^{s} n_i$ summands, that have to satisfy one linear restriction $(\sum\sum y_{ij} = ny_{\cdot\cdot})$. Hence, SS_{Total} has $n-1$ degrees of freedom:

$$\text{(ii)} \qquad SS_{\text{Within}} = SS_{\text{Error}} = \sum\sum(y_{ij} - y_{i\cdot})^2 \tag{4.37}$$

has s linear restrictions $\sum_{j=1}^{n_i} y_{ij} = n_i y_{i\cdot} \quad (i = 1,\ldots,s)$ in the case of n summands. Hence, SS_{Within} has $n-s$ degrees of freedom:

$$\text{(iii)} \qquad SS_{\text{Between}} = SS_A = \sum_{i=1}^{s} n_i(y_{i\cdot} - y_{\cdot\cdot})^2 \tag{4.38}$$

has s summands, that have to satisfy one linear restriction $(\sum_{i=1}^{s} n_i y_{i\cdot} = ny_{\cdot\cdot})$, and thus SS_{Between} has $s-1$ degrees of freedom. Hence, for the decomposition (4.34), according to

$$SS_{\text{Total}} = SS_{\text{Error}} + SS_A$$

we have the decomposition (4.35) of the degrees of freedom, i.e.,

$$n - 1 = (n-s) + (s-1)\,,$$

such that according to Theorem 4.2, SS_{Error} and SS_A have independent χ^2–distributions, i.e., their ratio F [(4.32)] has an F–distribution.

4.2.3 Estimation of σ^2 by MS_{Error}

In (3.62) we derived the statistic

$$s^2 = \frac{1}{T-K}(y - Xb_0)'(y - Xb_0)$$

as an unbiased estimate for σ^2 in the linear model. In our special case of model (4.4) and using

$$\hat{y} = Xb_0 = \begin{pmatrix} y_{1\cdot}\mathbf{1}_{n_1} \\ y_{2\cdot}\mathbf{1}_{n_2} \\ \vdots \\ y_{s\cdot}\mathbf{1}_{n_s} \end{pmatrix} \tag{4.39}$$

according to (4.14) for $s > 2$, we receive (equating $K = s$, $T = n$):

$$\begin{aligned} s^2 &= \frac{1}{n-s}((y_1 - y_{1\cdot}\mathbf{1}_{n_1})', \ldots, (y_s - y_{s.}\mathbf{1}_{n_s})') \begin{pmatrix} y_1 - y_{1\cdot}\mathbf{1}_{n_1} \\ \vdots \\ y_s - y_{s\cdot}\mathbf{1}_{n_s} \end{pmatrix} \\ &= \frac{1}{n-s}\sum_{i=1}^{s}\sum_{j=1}^{n_i}(y_{ij} - y_{i\cdot})^2 \qquad (4.40) \\ &= MS_{\text{Error}}. \qquad (4.41) \end{aligned}$$

Model (4.2) yields

$$y_{i\cdot} = \mu + \alpha_i + \epsilon_{i\cdot}\,, \epsilon_{i\cdot} \sim N\left(0, \frac{\sigma^2}{n_i}\right), \tag{4.42}$$

and, hence, in analogy to (3.61),

$$\begin{aligned} \mathrm{E}(MS_{\text{Error}}) &= \frac{1}{n-s}\,\mathrm{E}\left[\sum\sum(y_{ij} - y_{i\cdot})^2\right] \\ &= \frac{1}{n-s}\,\mathrm{E}\left[\sum\sum(\epsilon_{ij}^2 + \epsilon_{i\cdot}^2 - 2\epsilon_{ij}\epsilon_{i\cdot})\right] \\ &= \frac{1}{n-s}\sum_i\sum_j\left(\sigma^2 + \frac{\sigma^2}{n_i} - 2\frac{\sigma^2}{n_i}\right) \\ &= \sigma^2. \qquad (4.43) \end{aligned}$$

Furthermore, it follows, from (4.42) with (4.6), that

$$\begin{aligned} y_{\cdot\cdot} &= \mu + \frac{1}{n}\sum_{i=1}^{s} n_i\alpha_i + \epsilon_{\cdot\cdot} \\ &= \mu + \epsilon_{\cdot\cdot}\,, \qquad \epsilon_{\cdot\cdot} \sim N\left(0, \frac{\sigma^2}{n}\right), \qquad (4.44) \\ \mathrm{E}(\epsilon_{i\cdot}\epsilon_{\cdot\cdot}) &= \frac{1}{n_i n}\,\mathrm{E}\left[\sum_{j=1}^{n_i}\epsilon_{ij}\sum_{i=1}^{s}\sum_{j=1}^{n_i}\epsilon_{ij}\right] \\ &= \frac{\sigma^2}{n}. \qquad (4.45) \end{aligned}$$

Hence

$$y_{i\cdot} - y_{\cdot\cdot} = \alpha_i + \epsilon_{i\cdot} - \epsilon_{\cdot\cdot}, \tag{4.46}$$

$$\mathrm{E}(y_{i\cdot} - y_{\cdot\cdot})^2 = \alpha_i^2 + \frac{\sigma^2}{n_i} - \frac{\sigma^2}{n}, \tag{4.47}$$

holds and, thus,

$$\begin{aligned} \mathrm{E}(MS_A) &= \frac{1}{s-1}\sum\sum \mathrm{E}(y_{i\cdot} - y_{\cdot\cdot})^2 \\ &= \sigma^2 + \frac{\sum n_i \alpha_i^2}{s-1}. \end{aligned} \tag{4.48}$$

Hence, under $\mathrm{H}_0 : \alpha_1 = \cdots = \alpha_s = 0$, MS_A is an unbiased estimate for σ^2 as well. Thus, if H_0 does not hold, the test statistic F [(4.32)] has an expectation larger than one.

Example 4.3. The measured manufacturing times for the making of inlays (Table 4.1) represent one–way classified data material. Here, Factor A represents the effect of a dentist on the manufacturing times, it has $s = 3$ levels (dentists A, B, C).

We may assume that the assumptions for a normal distribution hold, if we replace the manufacturing times in Table 4.1 by their natural logarithm (the reason for this transformation is that time values usually have a skewed distribution).

	i	1	2	3	4	5	6	7	8	9	10
(A)	1	4.02	3.69	3.65	3.45	3.82	4.25	4.36	4.38	4.31	4.05
(B)	2	4.20	4.04	3.51	3.61	4.32	4.09	3.77	4.03	4.18	3.99
(C)	3	4.14	3.45	3.45	3.97	3.92	4.14	3.69	2.97		

	i	11	12	13	14	$Y_{i\cdot}$	$y_{i\cdot}$
(A)	1	4.28	4.25	3.87	4.08	$56.46 = Y_{1\cdot}$	$4.03 = y_{1\cdot}$
(B)	2	4.09				$43.83 = Y_{2\cdot}$	$3.98 = y_{2\cdot}$
(C)	3					$29.73 = Y_{3\cdot}$	$3.72 = y_{3\cdot}$
			$n = 33$			$130.02 = Y_{\cdot\cdot}$	$3.94 = y_{\cdot\cdot}$

TABLE 4.4. Logarithms of the manufacturing times from Table 4.1.

The arrangement in Table 4.4 of the measured values is done according to Table 4.1, the analysis is done in Table 4.5. The analysis yields the test statistic $F = 2.70 < 3.32 = F_{2,30;0.95}$ (Table C.6). Hence, the null hypothesis *The mean manufacturing times per inlay are equal for all three dentists* is *not* rejected.

Once again we want to point out the difference between Models I and II: The above result indicates that the three selected dentists do not differ with respect to their average manufacturing times per inlay. If, however, we want to test the effect that the factor *dentist* has on the manufacturing time, then the manufacturing times would have to be measured in a sample

		SS	df	MS	F
SS_A	=	512.82 - 512.28	2	$MS_A = 0.27$	$F = 2.70$
	=	0.54			
SS_{Error}	=	515.76 - 512.82	30	$MS_E = 0.10$	
	=	2.94			
SS_{Total}	=	515.76 - 512.28	32		
	=	3.48			

TABLE 4.5. Analysis of variance table for Example 4.1.

of s dentists *selected at random*, and the proportion of the variability due to dentists compared to the total variation would have to be tested. Hence, the comparison of means is not the point of interest, but the decomposition of the total variation into components (Model II).

Remark.

(i) The above analysis was done on a PC with maximum precision. If calculators are used, and in the case of two–digital precision, deviations in the $SS's$ arise, but not in the test decision.

(ii) The model (4.4) assumes identical variances of ϵ_{ij} in the s populations. ANOVA under unequal error variances is a Behrens–Fisher problem which is discussed in Weerahandi (1995), which gives an exact test for comparing more than two variances.

4.3 Comparison of Single Means

4.3.1 Linear Contrasts

The multiple comparison of means, i.e., the test of H_0 [(4.23)] against H_1 [(4.24)], has two possible outcomes–acceptance of H_0 (no treatment effect) and rejection of H_0 (treatment effect). In the case of the first decision the analysis is finished, although a second run for the proof of an effect with a larger sample size could be done after appropriate power calculations.

If, however, $\mathrm{H}_1 : \alpha_i \neq 0$ for at least one i (or, equivalently, $\mu_i = \mu + \alpha_i \neq \mu + \alpha_j = \mu_j$ for at least one pair $(i, j), i \neq j$) is accepted, i.e., an overall treatment effect is proven, then the main interest lies in finding those populations that caused this overall effect. Hence, in this situation comparisons of pairs or of linear combinations are appropriate, that is, we test, for example,

$$\mathrm{H}_0 : \mu_1 = \mu_2$$

against

$$\mathrm{H}_1 : \mu_1 \neq \mu_2$$

with the two–sample t–test by comparing $y_{1\cdot}$ and $y_{2\cdot}$ according to (1.5). Another possible hypothesis would be, for example, $\mu_1 + \mu_2 = \mu_3 + \mu_4$.

These hypotheses stand for one linear constraint $r = R'\beta$ each, with $\mathrm{rank}(R') = 1$. In the analysis of variance, a linear combination of means (in the population or in the sample) is called a linear contrast, as long as the following assumption is fulfilled.

Definition 4.3. A linear combination

$$\sum_{i=1}^{a} c_i y_{i\cdot} = c'y$$

of means is called a linear contrast if

$$c'c \neq 0 \quad \text{and} \quad \sum_{i=1}^{a} c_i = 0 \tag{4.49}$$

holds.

Suppose we want to compare s populations with respect to their means, i.e., if we assume

$$y_{ij} \sim N(\mu_i, \sigma^2), \quad i = 1, \ldots, s, j = 1, \ldots, n_i, \tag{4.50}$$

with y_{ij} and $y_{i'j}$ independent for $i \neq i'$, then

$$y_{i\cdot} \sim N\left(\mu_i, \frac{\sigma^2}{n_i}\right). \tag{4.51}$$

Denote by

$$\mu = (\mu_1, \ldots, \mu_s)' \tag{4.52}$$

the vector of the s expectations. Then every linear contrast in the expectations can be written as

$$c'\mu \quad \text{with} \quad \sum c_i = 0 \quad \text{and} \quad c'c \neq 0. \tag{4.53}$$

The vector μ is not to be mistaken for the overall mean μ from (4.4). Hence, the test statistic for testing $\mathrm{H}_0 : c'\mu = 0$ has the typical form

$$\frac{(c'y)^2}{\mathrm{Var}(c'y)} \tag{4.54}$$

with the vector

$$y' = (y_{1\cdot}, \ldots, y_{s\cdot}) \tag{4.55}$$

of the sample means. Thus, because of the independence of the s populations, we have (cf. (4.4))

$$c'y \sim N\left(c'\mu, \sigma^2 \sum \frac{c_i^2}{n_i}\right) \tag{4.56}$$

and, hence, under H_0:

$$\frac{(c'y)^2}{\sigma^2 \sum c_i^2/n_i} \sim \chi_1^2 . \tag{4.57}$$

As always, the MS_{Error} [(4.41)] is an unbiased estimate of the variance σ^2, hence the test statistic is of the following form:

$$t_{n-s}^2 = F_{1,n-s} = \frac{(c'y)^2}{\text{MS}_{\text{Error}} \sum c_i^2/n_i} \tag{4.58}$$

if the χ^2–distributions of the numerator and denominator are independent which could be proven by Cochran's Theorem 4.2. For the exact proof, see Proof 26, Appendix B.

Since, under $\text{H}_0 : c'\mu = 0$, a linear contrast is invariant to a multiplication with a constant $a \neq 0$:

$$ac'\mu = 0, \quad a \sum c_i = 0, \tag{4.59}$$

it is advisable to eliminate the ambiguity by the standardization

$$c'c = 1. \tag{4.60}$$

Definition 4.4. A linear contrast $c'\mu$ is normed if $c'c = 1$.

Definition 4.5. Two linear contrasts $c_1'\mu$ and $c_2'\mu$ are orthogonal if

$$c_1'c_2 = 0. \tag{4.61}$$

Analogously, a system $(c_1'\mu, \ldots, c_v'\mu)$ of orthogonal contrasts is called an *orthonormal system* if

$$c_i'c_j = \delta_{ij} \ (i, j = 1, \ldots, v) \tag{4.62}$$

holds, where δ_{ij} is the Kronecker symbol.

The orthogonal contrasts are an essential aid in reducing the number of possible pairwise comparisons to the maximum number of independent hypotheses, and hence in ensuring the testability.

Example 4.4. Assume we have $s = 3$ samples (3 levels of Factor A) and let the design be balanced ($n_i = r$). The overall null hypothesis

$$\text{H}_0 : \mu_1 = \mu_2 = \mu_3 \quad (\text{i.e., } \text{H}_0 : \alpha_i = 0 \text{ for } i = 1, 2, 3) \tag{4.63}$$

can be written, for example, as

$$\text{H}_0 : \mu_1 = \mu_2 \quad \text{and} \quad \mu_2 = \mu_3 , \tag{4.64}$$

or with linear contrasts as

$$\mathrm{H}_0 : \begin{pmatrix} c_1' \\ c_2' \end{pmatrix} \mu = \begin{pmatrix} 0 \\ 0 \end{pmatrix} \tag{4.65}$$

with

$$\mu' = (\mu_1, \mu_2, \mu_3)$$

and

$$c_1' = (1, -1, 0)\,, \tag{4.66}$$
$$c_2' = (0, 1, -1)\,. \tag{4.67}$$

We have $c_1'c_2 = -1$, hence $c_1'\mu$ and $c_2'\mu$ are not orthogonal and the quadratic forms $(c_1'y)^2$ and $(c_2'y)^2$ are not stochastically independent. If, however, we choose

$$c_1' = (1, -1, 0), \quad c_1'c_1 = 2, \tag{4.68}$$

as before, and

$$c_2' = (1, 1, -2), \quad c_2'c_2 = 6\,, \tag{4.69}$$

then $c_1'c_2 = 0$. $c_1'\mu = 0$ means $\mu_1 = \mu_2$ and $c_2'\mu = 0$ means $(\mu_1+\mu_2)/2 = \mu_3$, so that both contrasts represent $\mathrm{H}_0 : \mu_1 = \mu_2 = \mu_3$ simultaneously. The test statistic for H_0 [(4.65)] is then of the form

$$F_{2,n-2} = \left(\frac{r(c_1'y)^2}{c_1'c_1} + \frac{r(c_2'y)^2}{c_2'c_2} \right) / MS_{\text{Error}}. \tag{4.70}$$

With the contrasts (4.68) and (4.69), we thus have, for the hypothesis H_0 [(4.63)],

$$F_{2,n-2} = \left(\frac{r(y_{1\cdot} - y_{2\cdot})^2}{2} + \frac{r(y_{1\cdot} + y_{2\cdot} - 2y_{3\cdot})^2}{6} \right) / MS_{\text{Error}}\,. \tag{4.71}$$

4.3.2 Contrasts of the Total Response Values in the Balanced Case

We want to derive an interesting decomposition of the sum of squares SS_A. We assume:

- s levels of Factor A (treatments);
- $n_i = r$ repetitions per treatment (balanced design);
- $n = rs$ the total number of response values;
- $Y_{i\cdot} = \sum_{j=1}^{r} y_{ij}$ the total response of treatment i;
- $Y' = (Y_{1\cdot}, \ldots, Y_{s\cdot})$ the vector of the total response values; and

- $$SS_A = \frac{1}{r}\sum_{i=1}^{s} Y_{i\cdot}^2 - \frac{1}{rs}\left(\sum_{i=1}^{s} Y_{i\cdot}\right)^2 \tag{4.72}$$

  (cf. (4.29) for the balanced case).

Under these assumptions the following rules apply (cf., e.g., Petersen, 1985, p. 92):

(i) Let $c_1'Y$ be a linear contrast of the total response values. Then

$$S_1^2 = \frac{\left(\sum_{i=1}^{s} c_{1i}Y_{i\cdot}\right)^2}{\left(r\sum c_{1i}^2\right)} = \frac{(c_1'Y)^2}{(rc_1'c_1)} \tag{4.73}$$

is a component of SS_A with one degree of freedom. Hence, with

$$\begin{aligned} c_{1i}Y_{i\cdot} &\sim N(0, r\sigma^2 c_{1i}^2), \\ c_1'Y &\sim N(0, r\sigma^2 \sum c_{1i}^2) \\ &= N(0, r\sigma^2 c_1'c_1), \end{aligned}$$

we have under H_0:

$$\frac{(c_1'Y)^2}{rc_1'c_1} = S_1^2 \sim \sigma^2\chi_1^2 . \tag{4.74}$$

(ii) If $c_2'Y$ and $c_1'Y$ are orthogonal contrasts, then

$$S_2^2 = \frac{(c_2'Y)^2}{(rc_2'c_2)} \tag{4.75}$$

is a component of $SS_A - S_1^2$.

(iii) If $c_1'Y, \ldots, c_{s-1}'Y$ is a complete system of orthogonal contrasts, then

$$S_1^2 + \ldots + S_{s-1}^2 = SS_A \tag{4.76}$$

holds.

We now have a decomposition of SS_A into $s-1$ independent sums of squares. In the case of a normal distribution, these components have independent χ^2–distributions. This decomposition corresponds to the decomposition of the G^2–statistic in $(I \times 2)$–contingency tables into $(I-1)$ independent, χ^2–distributed G^2–statistics for the analysis of the subeffects. In the case of a significant overall treatment effect the main subeffects that contributed to the significance can thus be discovered. The significance of the subeffects, i.e., $H_0 : c_i'Y = 0$ against H_1: $c_i'Y \neq 0$, is tested with

$$t_{n-s}^2 = F_{1,n-s} = F_{1,s(r-1)} = \frac{S_i^2}{MS_{\text{Error}}} . \tag{4.77}$$

	Repetitions								
i	1	2	3	4	5	6	$Y_{i\cdot}$	$y_{i\cdot}$	s_i
1	4.5	5.0	3.5	3.7	4.8	4.0	25.5	4.25	0.6091
2	3.8	4.0	3.9	4.2	3.6	4.4	23.9	3.98	0.2858
3	3.5	4.5	3.2	2.1	3.5	4.0	20.8	3.47	0.8116
4	3.0	2.8	2.2	3.4	4.0	3.9	19.3	3.22	0.6882
							$Y_{\cdot\cdot} = 89.5$	$y_{\cdot\cdot} = 3.73$	

TABLE 4.6. Flexibility in dependency of four levels of Factor A (additives).

Source	df	Sum of squares	Mean squares	F ratio	F prob.
Between groups	3	4.0046	1.3349	3.3687	0.0389
Within groups	20	7.9250	0.3962		
Total	23	11.9296			

TABLE 4.7. Analysis of variance table for Table 4.6 in SPSS format.

Variance of Linear Contrasts

If the s samples are independent, then the variance of a linear contrast is computed as follows:

(i) **Contrast of the means**
Let $c'y = c_1 y_{1\cdot} + \ldots + c_s y_{s\cdot}$, then

$$\mathrm{Var}(c'y) = \left(\frac{c_1^2}{n_1} + \ldots + \frac{c_s^2}{n_s} \right) \sigma^2 \tag{4.78}$$

holds in general. In the balanced case ($n_i = r, i = 1, \ldots, s$) this expression simplifies to

$$\mathrm{Var}(c'y) = \frac{c'c}{r} \sigma^2 . \tag{4.79}$$

(ii) **Contrast of the totals**
Let $c'Y = c_1 Y_{1\cdot} + \ldots + c_s Y_{s\cdot}$, then

$$\mathrm{Var}(c'Y) = (n_1 c_1^2 + \ldots + n_s c_s^2)\sigma^2 \tag{4.80}$$

holds in general, and in the balanced design

$$\mathrm{Var}(c'Y) = r c' c \sigma^2 . \tag{4.81}$$

The variance σ^2 of the population is estimated by $\mathrm{MS}_{\mathrm{Error}} = s^2$, hence

$$\widehat{\mathrm{Var}}(c'y) = s^2 \sum \frac{c_i^2}{n_i} \tag{4.82}$$

and

$$\widehat{\mathrm{Var}}(c'Y) = s^2 \sum n_i c_i^2 \tag{4.83}$$

are unbiased estimates of $\text{Var}(c'y)$ and $\text{Var}(c'Y)$.

Example 4.5. Consider the following balanced experimental design with $r = 6$ repetitions:

Factor A:	Level 1:	control group (neither A_1 nor A_2);
	Level 2:	additive A_1;
	Level 3:	additive A_2;
	Level 4:	additives A_1 and A_2 (combination).

Suppose response Y is the flexibility of a plastic material, and that we are interested in the most favorable mixture in the sense of a reduction of the flexibility. The data are shown in Table 4.6.

We receive the analysis of variance table (Table 4.7) according to the layout of Table 4.3 in the SPSS format. The F–test rejects the hypothesis $H_0 : \mu_1 = \mu_2 = \mu_3$ with the statistic $F_{3,20} = 3.3687$ (p–value, 0.0389). Hence, we can now compare pairs or combinations of treatments. For $s = 4$ levels, systems exist with $s - 1 = 3$ orthogonal contrasts. We consider the two systems in Tables 4.8 and 4.9.

Contrast	Treatment response $Y_{i\cdot}$	1 25.5	2 23.9	3 20.8	4 19.3	$c'Y$	S^2
A_1 against A_2		0	+1	−1	0	3.1	0.8008
A_1 or A_2 against A_1 and A_2		0	−1	−1	2	−6.1	1.0336
A_1 or A_2 or A_1 and A_2 against control group		−3	+1	+1	+1	−12.5	2.1702
							$\sum = 4.0046$

TABLE 4.8. Orthogonal contrasts and test statistics S^2.

Contrast	Treatment response $Y_{i\cdot}$	1 25.5	2 23.9	3 20.8	4 19.3	$c'Y$	S^2
A_1		−1	+1	−1	+1	−3.1	0.4004
A_2		−1	−1	+1	+1	−9.3	3.6038
$A_1 \times A_2$		+1	−1	−1	+1	0.1	0.0004
							$\sum = 4.0046$

TABLE 4.9. Orthogonal contrasts and test statistics S^2.

In both systems the sums of squares S^2 of the contrasts add up to SS_A (SS Between Groups in Table 4.7) according to (4.76). With $MS_{\text{Error}} = 0.3962$, the test statistics (4.77) are

Table 4.8	Table 4.9
2.02	1.01
2.61	9.10 *
5.48 *	0.00

The 95%–quantile of the $F_{1,23}$–distribution is 4.15, so that:

- the employment of at least one additive, compared to the control group, is significant (i.e., reduces the flexibility significantly); and
- the employment of A_2 (alone or in combination with A_1) reduces the flexibility significantly.

The orthogonal contrasts of the response sums $Y_{i\cdot}$ make a decomposition of the variability SS_A possible, i.e., of the treatment effect, and hence enable the determination of significant subeffects. With F from (4.58), the orthogonal contrast of means, on the other hand, yields a test statistic for testing differences of treatments according to the linear function of the means given by the contrast.

We demonstrate this with the same systems of orthogonal contrasts as in Tables 4.8 and 4.9. The results are shown in Tables 4.10 and 4.11. We have, for example (Table 4.11, first row),

$$\begin{aligned} c'y &= (y_{2\cdot} + y_{4\cdot}) - (y_{1\cdot} + y_{3\cdot}) \\ &= 3.98 + 3.22 - (4.25 + 3.47) = -0.52\,, \\ \widehat{\text{Var}}(c'y) &= \frac{c'c}{r} s^2 \\ &= 4/6 \cdot 0.3962 = 0.2641 \\ &= 0.5140^2 \end{aligned}$$

with $s^2 = MS_{\text{Error}} = 0.3962$ from Table 4.7. The test statistic from (4.58), for

$$\text{H}_0 : c'\mu = (\mu_2 + \mu_4) - (\mu_1 + \mu_3) = 0\,,$$

i.e., for $\text{H}_0 : (\alpha_2 + \alpha_4) = (\alpha_1 + \alpha_3)$, is now

$$t_{24-4} = t_{20} = \frac{-0.520}{0.514} = -1.002\,.$$

The critical value is (Table C.5)

$$t_{20;0.95,\text{one–sided}} = -1.73$$

and

$$t_{20;0.95,\text{two–sided}} = \pm 2.09\,,$$

so that H_0 is not rejected. We can see from Tables 4.10 and 4.11 that the following contrasts are significant:

$$\frac{\mu_2 + \mu_3 + \mu_4}{3} - \mu_1 < 0$$

(the control group has a higher flexibility than the mean of the three treatments),

$$\mu_3 + \mu_4 - (\mu_1 + \mu_2) < 0$$

(A_2 plus (A_1 and A_2) have a lower mean flexibility than the control group plus A_1). Commands and output in SPSS: The contrasts from Table 4.11 are called, with the command,

```
/contrast = -1 1 -1 1
/contrast = -1 -1 1 1
/contrast = 1 -1 -1 1
```

which is inserted into the SPSS procedure.

Contrast	Treatment mean $y_{i\cdot}$	1 4.25	2 3.98	3 3.47	4 3.22	$c'y$	$\mathrm{Var}(c'y)$	t_{20}	
A_1 against A_2		0	+1	−1	0	0.52	0.363^2	1.42	
A_1 or A_2 against A_1 and A_2		0	−1	−1	2	−1.02	0.629^2	−1.61	
A_1 or A_2 or A_1 and A_2 against control group		−3	+1	+1	+1	−2.08	0.890^2	−2.33	*

TABLE 4.10. Orthogonal contrasts of the means.

Contrast	Treatment mean $y_{i\cdot}$	1 4.25	2 3.98	3 3.47	4 3.22	$c'y$	$\mathrm{Var}(c'y)$	t_{20}	
A_1		−1	+1	−1	+1	−0.52	0.514^2	−1.002	
A_2		−1	−1	+1	+1	−1.54	0.514^2	−2.996	*
$A_1 \times A_2$		+1	−1	−1	+1	0.02	0.514^2	0.039	

TABLE 4.11. Orthogonal contrasts of the means.

The obvious question, as whether A_2 should be employed alone or in combination with A_1, could be tested with the two–sample t–test according to (1.5). We compute with $s_{A_2} = 0.8116$, $s_{A_1 \text{ and } A_2} = 0.6882$ (Table 4.6) the pooled variance (1.6)

$$s^2 = \frac{5(0.8116^2 + 0.6882^2)}{6+6-2} = 0.7524^2$$

and

$$t_{10} = \frac{20.8/6 - 19.3/6}{0.7524}\sqrt{6 \cdot 6/(6+6)} = 0.5755\,,$$

so that $H_0 : \mu_{A_2} = \mu_{(A_1 \text{ and } A_2)}$ is not rejected ($t_{10,0.95,\text{one–sided}} = 1.81$). Hence, the two treatments A_2 and (A_1 and A_2) show no significant difference.

In the next section, however, we will integrate this problem of pairwise comparisons in the case of s treatments into the multiple test problem. As we will see, this shows that an adjustment of the degrees of freedom, or of the applied quantile, respectively, has to be made.

4.4 Multiple Comparisons

4.4.1 Introduction

With the linear and, especially, with the orthogonal contrasts we have the possibility of testing selected linear combinations for significance and thus structure the treatments. The starting point is a rejection of the overall equality $\mu_1 = \ldots = \mu_s$ of the means of the response.

A number of statistical procedures exist for the comparison of single means or of groups of means. These procedures have the following different objectives:

- Comparison of all possible pairs of means (for s levels of A we have $s(s-1)/2$ different pairs).
- Comparison of all $s-1$ means with a control group selected in advance.
- Comparison of all pairs of treatments that were selected in advance.
- Comparison of any linear combinations of the means.

These procedures differ, next to their aims, especially with respect to the way in which they control for the type I error. In one case, the error is controlled on a *per comparison basis*, in the other case the error is controlled simultaneously for all comparisons.

A multiple test procedure, that conducts every pairwise comparison at a significance level α, i.e., that works per comparison basis, is possible if the group comparisons are already planned at the beginning of the experiment. This is based mainly on the t–statistic. If we want to ensure the significance level α simultaneously for all group comparisons of interest, the appropriate multiple test procedure is one that controls the error rate *per experiment basis*.

The decision for one of the two procedures is to be made ahead of the experiment.

4.4.2 Experimentwise Comparisons

The most popular multiple procedures that control the error simultaneously are those of Dunnett (1955) for the comparison of $s-1$ groups with a control group, of Tukey (1953) for all $s(s-1)/2 = \binom{s}{2}$ pairwise comparisons, and those of Scheffé (1953) for any linear combinations. The procedures of Tukey and Scheffé should be applied in the explorative phase of an experiment, in order to avoid comparisons that are suggested by the data. The main condition for all multiple procedures is the rejection of $H_0 : \mu_1 = \cdots = \mu_s$.

Hint. A detailed representation and rating of the multiple test procedures can be found in Miller, Jr. (1981).

Procedure by Scheffé

Let $c'\mu$ be any linear contrast of μ and $c'y$, with $\sum_{i=1}^{s} c_i = 0$ and $y' = (y_{1\cdot}, \ldots, y_{s\cdot})$ the corresponding contrast of the vector of means. We then have, for all c,

$$P(c'y - \sqrt{S_{1-\alpha}} \leq c'\mu \leq c'y + \sqrt{S_{1-\alpha}}) = 1 - \alpha \tag{4.84}$$

with (cf. (4.78))

$$S_{1-\alpha} = MS_{\text{Error}}(s-1)\left(\frac{c_1^2}{n_1} + \cdots + \frac{c_s^2}{n_s}\right) F_{s-1,n-s;1-\alpha} \,. \tag{4.85}$$

The null hypothesis $\text{H}_0 : c'\mu = 0$ is rejected if zero is not within the confidence interval. The multiple level is α.

Procedure by Dunnett

Let group $i = 1$ be selected as the control group that is to be compared with the treatments (groups) $i = 2, \ldots, s$. The $[(1-\alpha) \cdot 100\%]$–confidence intervals for the $s - 1$ pairwise comparisons "control – treatment" are of the form

$$(y_{1\cdot} - y_{i\cdot}) \pm C_{1-\alpha}(s-1, n-s) s_{\bar{d}_i} \tag{4.86}$$

with

$$s_{\bar{d}_i} = \sqrt{MS_{\text{Error}}\left(\frac{1}{n_1} + \frac{1}{n_i}\right)} \,. \tag{4.87}$$

The quantiles $C_{1-\alpha}(s-1, n-s)$ are given in special tables (one– and two–sided, cf. Woolson, 1987, Tables 13a and 13b, p. 502–503; or Dunnett (1955; 1964)). We show an excerpt for $C_{0.95}(s-1, n-s)$ in Table 4.12 and 4.13.

	$s-1$				
$n-s$	1	2	3	4	5
5	2.57	3.03	3.39	3.66	3.88
10	2.23	2.57	2.81	2.97	3.11
15	2.13	2.44	2.64	2.79	2.90
20	2.09	2.38	2.57	2.70	2.81

TABLE 4.12. $[C_{0.95}(s-1, n-s)]$–quantiles (two–sided).

The hypothesis $\text{H}_0 : \mu_1 = \mu_i$ $(i = 2, \ldots, s)$ is rejected:

	$s-1$				
$n-s$	1	2	3	4	5
5	2.02	2.44	2.68	2.85	2.98
10	1.81	2.15	2.34	2.47	2.56
15	1.75	2.07	2.24	2.36	2.44
20	1.72	2.03	2.19	2.30	2.39

TABLE 4.13. $[\tilde{C}_{0.95}(s-1, n-s)]$–quantiles (one–sided).

- two–sided in favor of $\mathrm{H}_1 : \mu_1 \neq \mu_i$, if

$$|y_{1\cdot} - y_{i\cdot}| > C_{1-\alpha}(s-1, n-s) \cdot s_{\bar{d}_i}\,; \tag{4.88}$$

- one–sided in favor of $\mathrm{H}_1 : \mu_1 > \mu_i$, if

$$y_{1\cdot} - y_{i\cdot} > \tilde{C}_{1-\alpha}(s-1, n-s) \cdot s_{\bar{d}_i}\,; \tag{4.89}$$

- one–sided in favor of $\mathrm{H}_1 : \mu_1 < \mu_i$, if

$$y_{1\cdot} - y_{i\cdot} < -\tilde{C}_{1-\alpha}(s-1, n-s) \cdot s_{\bar{d}_i} \tag{4.90}$$

holds. For all $s-1$ comparisons the multiple level α is ensured.

Procedure by Tukey

In the case of experiments in the explorative phase it is often not possible to fix the set of planned comparisons in advance. Hence, all $s(s-1)/2$ possible pairwise comparisons are done. The two–sided test procedure by Tukey assumes the balanced case $n_i = r$ and controls for the error experimentwise, i.e., for all $s(s-1)/2$ comparisons the multiple level α holds. We compute the confidence intervals

$$(y_{i\cdot} - y_{j\cdot}) \pm T_\alpha \quad (i > j) \tag{4.91}$$

with

$$T_\alpha = Q_\alpha(s, n-s)\, s_{\bar{d}}\,, \tag{4.92}$$

$$s_{\bar{d}} = \sqrt{MS_{\text{Error}}/r}\,. \tag{4.93}$$

The quantiles $Q_{1-\alpha}(s, n-s)$ are so–called Studentized rank–values, that are given in special tables (cf., e.g., Woolson, 1987, Table 14, pp. 504–505). The set of null hypotheses $\mathrm{H}_0(i,j) : \mu_i = \mu_j$ $(i > j)$ is rejected in favor of $\mathrm{H}_1 : \mathrm{H}_0$ incorrect (i.e., $\mu_i \neq \mu_j$ for at least one pair $i > j$), if

$$|y_{i\cdot} - y_{j\cdot}| > T_\alpha \tag{4.94}$$

holds. For all pairs (i,j), $i > j$ with $|y_{i\cdot} - y_{j\cdot}| > T_\alpha$, we have a statistically significant treatment difference.

Bonferroni Method

Suppose, we want to conduct $k \leq s$ comparisons with a multiple level of α at the most. In this situation the Bonferroni method can be applied. This method splits up the risk α into equal parts α/k for the k comparisons. The basis is Bonferroni's inequality.

Let $H_1, \ldots, H_k$ be the confidence intervals for the k comparisons. Denote by $P(H_i)$ the probability that H_i is true (i.e., H_i covers the respective parameter of the ith comparison). Then $P(H_1 \cap \cdots \cap H_k)$ is the probability that all k confidence intervals cover the respective parameters. According to Bonferroni's inequality, we have

$$P(H_1 \cap \cdots \cap H_k) \geq 1 - \sum_{i=1}^{k} P(\bar{H}_i)\,, \tag{4.95}$$

where $\bar{H}_i$ is the complementary event to H_i. If $P(\bar{H}_i) = \alpha/k$ is chosen, then the following holds for the simultaneous probability

$$P(H_1 \cap \cdots \cap H_k) \geq 1 - \alpha\,. \tag{4.96}$$

Assume, for example, $k \leq s$ contrasts $c_i'\mu$ are to be tested simultaneously. The confidence intervals for $c_i'\mu$, according to the Bonferroni method, are then of the following form:

$$c_i'y \pm t_{n-s;1-\alpha/2k}\sqrt{MS_{\text{Error}}}\sqrt{\frac{c_1^2}{n_1} + \cdots + \frac{c_s^2}{n_s}}\,. \tag{4.97}$$

The test runs analogously to the procedure by Scheffé, i.e., if (4.97) does not contain the zero, then H_0 is rejected and the respective comparison is significant.

4.4.3 Select Pairwise Comparisons

The "Least Significant Difference" (LSD)

Suppose we want to compare the means of two selected treatments, i.e., suppose we want to test $\text{H}_0 : \mu_1 = \mu_2$ against $\text{H}_1 : \mu_1 \neq \mu_2$. The appropriate test statistic is

$$t_{df} = \frac{y_{1\cdot} - y_{2\cdot}}{\sqrt{\widehat{\text{Var}}(y_{1\cdot} - y_{2\cdot})}}\,, \tag{4.98}$$

where df is the number of degrees of freedom. For $|t| > t_{df;1-\alpha/2}$ we reject H_0, where $t_{df;1-\alpha/2}$ is the two–sided quantile at the α probability level. If H_0 is rejected, then μ_1 is significantly different from μ_2 at the α level.

$|t| > t_{df;1-\alpha/2}$ is equivalent with

$$t_{df;1-\alpha/2}\sqrt{\widehat{\text{Var}}(y_{1\cdot} - y_{2\cdot})} < |y_{1\cdot} - y_{2\cdot}|\,. \tag{4.99}$$

Hence, every sample with a difference $|y_{1\cdot} - y_{2\cdot}|$ that exceeds $t_{df;1-\alpha/2}\sqrt{\text{Var}(y_{1\cdot} - y_{2\cdot})}$, indicates a significant difference between μ_1 and μ_2. According to (4.99), the left side would be the smallest difference of $y_{1\cdot}$ and $y_{2\cdot}$ for which significance would be declared. Thus, we define (*df* is the number of degrees of freedom of s^2, the pooled variance of the two samples)

$$\begin{aligned} LSD &= t_{df;1-\alpha/2}\sqrt{\widehat{\text{Var}}(y_{1\cdot} - y_{2\cdot})} \\ &= t_{df;1-\alpha/2}\sqrt{s^2\left(\frac{1}{n_1} + \frac{1}{n_2}\right)}. \end{aligned} \tag{4.100}$$

In the balanced case ($n_1 = n_2 = r$) we receive

$$LSD = t_{df;1-\alpha/2}\sqrt{\frac{2s^2}{r}}. \tag{4.101}$$

Using the *LSD* is controversial, especially if it is used for comparisons suggested by the data (largest/smallest sample mean) or if all pairwise comparisons are done without correction of the test level. If the *LSD* is used for all pairwise comparisons (i.e., for $s(s-1)/2$ comparisons in the case of s treatments), then these tests are not independent. Procedures based on the *LSD*, that ensure the test level due to corrections of the quantiles, exist (*HSD*, Duncan test). *FPLSD* and *SNK* on the other hand, only ensure the global level.

Fisher's Protected LSD (FPLSD)

This procedure starts out with the analysis of variance and tests the global hypothesis $H_0: \mu_1 = \cdots = \mu_s$ with the statistic $F = MS_A/MS_{\text{Error}}$ from (4.32). If F is not significant the procedure stops. If $F > F_{s-1,n-s;1-\alpha}$, i.e., differences of the means are significant, then all pairs of means $y_{i\cdot}$ and $y_{j\cdot}$ $(i \neq j)$ are tested for differences with

$$FPLSD = t_{n-s;1-\alpha/2}\sqrt{MS_{\text{Error}}\left(\frac{1}{n_i} + \frac{1}{n_j}\right)}. \tag{4.102}$$

For $|y_{i\cdot} - y_{j\cdot}| > FPLSD$ we have a significant difference of means. Note that in (4.102) σ^2 is estimated by MS_{Error}. Hence, t now has $n-s$ degrees of freedom (instead of $n_1 + n_2 - 2$ degrees of freedom as in the two–sample case).

Tukey's Honestly Significant Difference (HSD)

This procedure uses the Studentized rank values $Q_{\alpha,(s,n-s)}$ (cf. (4.92)) instead of the t–quantiles and replaces the standard error of the mean by the standard error of the difference (pooled sample). We compute

$$HSD = Q_{\alpha,(s,n-s)}\sqrt{MS_{\text{Error}}/r}. \tag{4.103}$$

All differences of pairs $|y_{i\cdot} - y_{j\cdot}|\,(i < j)$ are compared with HSD. For $|y_{i\cdot} - y_{j\cdot}| > HSD$ we have a significant difference between μ_i and μ_j.

Student–Newman–Keuls Test (SNK)

The SNK test is a test in which the difference needed for significance varies with the degree of separation. Suppose we want to compare k means. The sample means are sorted in descending order

$$y_{(1)\cdot}, \ldots, y_{(k)\cdot}\,,$$

where $y_{(i)\cdot}$ is the mean with the ith rank (i.e., $y_{(1)\cdot}$ is the largest mean, $y_{(k)\cdot}$ the smallest mean). We compute the SNK differences

$$SNK_i = Q_{\alpha,(i,df)}\sqrt{MS_{\text{Error}}/r} \quad (i = 2, \ldots, k), \tag{4.104}$$

with $Q_{\alpha,(i,df)}$ for df degrees of freedom of SS_{Error} and (in succession) $i = 2, 3, \ldots, k$ means.

If $|y_{(1)\cdot} - y_{(k)\cdot}| < SNK_k$, then none of the differences of means are significant and the procedure stops.
If $|y_{(1)\cdot} - y_{(k)\cdot}| > SNK_k$, then this (largest) difference is significant. We proceed by testing whether

$$|y_{(2)\cdot} - y_{(k)\cdot}| > SNK_{k-1} \tag{4.105}$$

and

$$|y_{(1)\cdot} - y_{(k-1)\cdot}| > SNK_{k-1} \tag{4.106}$$

holds. If both conditions hold, then those differences of the rank–ordered means are tested, where the ranks differ by $k - 3$. This procedure is continued up to the comparison of rank–neighbored means.

Duncan Test

Duncan (1975) modified the procedure FPLSD by computing alternative quantiles. The least significant difference is Bayes adjusted and reads as follows:

$$BLSD = t_B\sqrt{2MS_{\text{Error}}/r}\,. \tag{4.107}$$

The values t_B are given in special tables (Waller and Duncan, 1972) and are printed in the SPSS procedure.

Hint. A number of multiple test procedures exist that work with other rank values. These are implemented in the standard software.

Example 4.6. (Continuation of Example 4.5)
Table 4.6 yields:

Treatment	1	2	3	4
Rank	1	2	3	4
Mean	4.25	3.98	3.47	3.22

We had $s = 4$, $r = 6$, and $n = 4 \cdot 6 = 24$, as well as $MS_{\text{Error}} = 0.3962$ for $n - s = 20$ degrees of freedom (Table 4.7). The hypothesis $\text{H}_0: \mu_1 = \cdots = \mu_4$ was rejected.

Experimentwise Procedures

Procedure by Scheffé The critical value (4.85) of the confidence interval (4.84) for any contrast $c'\mu$ is, with $F_{3,20;0.95} = 3.10$,

$$\begin{aligned} S_{1-\alpha} &= 0.3962 \cdot 3 \cdot 3.10 \cdot \frac{c'c}{6} \\ &= 0.61 \cdot c'c\,. \end{aligned}$$

We test the complete system of orthogonal contrasts of the means from Table 4.11 and receive:

	$c'y$	$c'c$	$\sqrt{S_{1-\alpha}}$	$c'y \pm \sqrt{S_{1-\alpha}}$
A_1	-0.52	4	1.57	$[-2.09\,,\,1.05]$
A_2	-1.54	4	1.57	$[-3.11\,,\,0.03]$
$A_1 \times A_2$	0.02	4	1.57	$[-1.55\,,\,1.59]$

The zero lies in all three intervals, hence $\text{H}_0 : c'\mu = 0$ is never rejected.

Procedure by Dunnett In Example 4.5 Level 1 was designed as control group. We conduct the multiple comparison (according to Dunnett) of the control group with the Groups 2, 3, and4. The critical limits (4.86) are $(n_i = n_j = 6)$ (cf. Tables 4.12 and 4.13) two–sided:

$$C_{1-\alpha}(3, 20)\sqrt{0.3962 \cdot 2/6} = 2.57 \cdot 0.3634 = 0.9340$$

and one–sided:

$$\tilde{C}_{1-\alpha}(3, 20) \cdot 0.3634 = 2.19 \cdot 0.3634 = 0.7958\,.$$

For the one–sided tests we receive

$$\begin{aligned} y_{1\cdot} - y_{2\cdot} &= 0.27, \\ y_{1\cdot} - y_{3\cdot} &= 0.78, \\ y_{1\cdot} - y_{4\cdot} &= 1.03\ *\,, \end{aligned}$$

and, hence, a significant difference between the control group and Group 4.

Procedure by Tukey Here all $4 \cdot 3/2 = 6$ possible comparisons are conducted. With $Q_{0.05}(4, 20) = 3.95$ and $s_{\bar{d}} = \sqrt{MS_{\text{Error}}/r} = \sqrt{0.3962/6} = 0.2570$ the critical value (cf. (4.92)) is $T_{0.05} = 3.95 \cdot 0.2570 = 1.02$.

(i, j)	$\lvert y_{i\cdot} - y_{j\cdot}\rvert$	
(1, 2)	0.27	
(1, 3)	0.78	
(1, 4)	1.03	*
(2, 3)	0.51	
(2, 4)	0.76	
(3, 4)	0.25	

Again, the difference between treatments 1 and 4 is significant.

Bonferroni Method We conduct the $k = 3$ comparisons from Table 4.10 according to the Bonferroni method. The critical limit from (4.97) for the chosen contrast $c'\mu$ is

$$\begin{aligned} t_{20;1-0.05/2\cdot 3} \cdot \sqrt{0.3962} \cdot \sqrt{\frac{c'c}{6}} &= 2.95 \cdot \frac{0.6294}{2.4495} \cdot \sqrt{c'c} \\ &= 0.7580 \cdot \sqrt{c'c}\,. \end{aligned}$$

Contrast	$c'y$	$c'c$	$0.7580 \cdot \sqrt{c'c}$	Interval (4.97)
1/2	0.52	2	1.0720	[−0.5520, 1.5920]
1 or 2/4	−1.02	6	1.8567	[−2.8767, 0.8367]
1/2 or 3 or 4	−2.08	12	2.6258	[−4.7058, 0.6058]

In the multiple comparison according to Bonferroni no contrast is statistically significant.

Selected Pairwise Comparisons

SNK Test The Studentized ranges, $Q_{0.05,(i,df)}$ for $df = 20$ degrees of freedom, are

	2	3	4
$Q_{0.05,(i,20)}$	2.95	3.57	3.95
SNK_i	0.76	0.92	1.02

This yields the following comparisons

$$\begin{aligned} \lvert y_{(1)\cdot} - y_{(4)\cdot}\rvert &= \lvert 4.25 - 3.22\rvert \\ &= 1.03 > SNK_4 = 1.02\,. \end{aligned}$$

Hence the largest difference is significant. Thus, we can proceed with the procedure

$$\begin{aligned} \lvert y_{(1)\cdot} - y_{(3)\cdot}\rvert &= \lvert 4.25 - 3.47\rvert \\ &= 0.78 < SNK_3 = 0.92\,, \\ \lvert y_{(2)\cdot} - y_{(4)\cdot}\rvert &= \lvert 3.98 - 3.22\rvert \\ &= 0.76 < SNK_3 = 0.92\,. \end{aligned}$$

Here, the SNK test stops. Therefore, the only significant difference is that between treatment 1 (control group) and treatment 4 (A_1 and A_2). The treatments (1, 2, 3), or (2, 3, 4), respectively, may be regarded as homogeneous.

SNK in SPSS

The procedure is started with `/Ranges = snk`

Note. SPSS computes the SNK statistic according to

$$SNK = \sqrt{\frac{MS_{\text{Error}}}{2}}\, Q_{\alpha,(i,df)} \sqrt{\frac{1}{n_i} + \frac{1}{n_j}}\,, \tag{4.108}$$

for $n_i = n_j = r$ this yields the expression (4.104).

The SPSS printout is of the following form:

```
Multiple Range Test
Student--Newman--Keuls Procedure
Ranges for the .050 level
        2.95   3.57  3.95
The ranges above are table ranges.

The value actually compared with
Mean(J)-Mean(I) is
        .4451 * Range * Sqrt(1/N(I) + 1/N(J))

(*) Denotes pairs of groups significantly
    different at the .050 level


             G G G G
             r r r r
             p p p p
             4 3 2 1
Mean   Group
3.22   Grp 4
3.47   Grp 3
3.98   Grp 2
4.25   Grp 1 *

Homogeneous Subsets

Subset 1
Group  Grp 4  Grp 3  Grp 2
Mean    3.22   3.47   3.98
```

```
Subset 2
Group  Grp 3  Grp 2  Grp 1
Mean    3.47   3.98   4.25
```

Tukey's HSD Test We compute the HSD (4.103) according to

$$\begin{aligned} HSD &= Q_{\alpha,(4,20)}\sqrt{MS_{\text{Error}}/6} \\ &= 3.95 \cdot 0.2569 = 1.01 \,. \end{aligned}$$

The differences of pairs $y_{i\cdot} - y_{j\cdot}$ $(i < j)$ are

$$\begin{aligned} y_{1\cdot} - y_{2\cdot} &= 4.25 - 3.98 = 0.27, \\ y_{1\cdot} - y_{3\cdot} &= 0.78, \\ y_{1\cdot} - y_{4\cdot} &= 1.03, \ * \\ y_{2\cdot} - y_{3\cdot} &= 0.51, \\ y_{2\cdot} - y_{4\cdot} &= 0.76, \\ y_{3\cdot} - y_{4\cdot} &= 0.25\,, \end{aligned}$$

hence only $|y_{1\cdot} - y_{4\cdot}| > HSD$ holds.
SPSS call and printout:

```
/Ranges = tukey

Tukey--HSD Procedure
Ranges for the .050 level
      3.95  3.95  3.95


             G G G G
             r r r r
             p p p p
             4 3 2 1
Mean   Group
3.22   Grp 4
3.47   Grp 3
3.98   Grp 2
4.25   Grp 1 *
```

Fisher's Protected LSD (FPLSD)

The FPLSD (4.102) at the 5% level is

$$t_{20;0.975}\sqrt{0.3962 \cdot 2/6} = 2.09 \cdot 0.3634 = 0.76\,.$$

With the differences of means calculated above, we receive

```
               G G G G
               r r r r
               p p p p
               4 3 2 1
Mean   Group
3.22   Grp 4
3.47   Grp 3
3.98   Grp 2 *
4.25   Grp 1 * *
```

The means μ_1 and μ_4 and μ_1 and μ_3, as well as the means μ_2 and μ_4, are significantly different according to this test.

4.5 Regression Analysis of Variance

For the description of the dependence of a variable Y on another (fixed) variable X by a regression model of the form

$$Y = \alpha + \beta X + \epsilon$$

we need pairs of observations (x_i, y_i), $i = 1, \ldots, n$, i.e., for every x–value one y–value is observed.

Consider the following experimental design. For *every* x–value *several* observations of Y are realized

$$x_i, y_{i1}, \ldots, y_{in_i}\,.$$

This corresponds to the idea that a population of y–values belongs to a fixed x–value. The question of interest is whether a dependence exists between the y–samples, represented by their means $y_{i\cdot}$, and the factor X. First, we test whether the populations Y_i have equal means (analysis of variance – multiple comparison of means).

If this hypothesis is rejected, we have reason for assuming a simple linear relationship

$$y_{i\cdot} = \alpha + \beta x_i + \epsilon_i \quad (i = 1, \ldots, s)\,. \tag{4.109}$$

The estimates of α and β are determined, under consideration of the sample sizes n_i, according to the method of weighted least squares, i.e.,

$$\sum_{i=1}^{s} n_i(y_{i\cdot} - \alpha - \beta x_i)^2 \tag{4.110}$$

is minimized with respect to α and β. Let $n = \sum n_i$ be the sum of all observations. The *weighted least squares estimates* are then of the following form

$$\hat{\beta} = \frac{\sum n_i x_i y_{i\cdot} - 1/n \sum n_i x_i \sum n_i y_{i\cdot}}{\sum n_i x_i^2 - 1/n \left[\sum n_i x_i\right]^2}, \tag{4.111}$$

$$\hat{\alpha} = y_{\cdot\cdot} - b\bar{x}, \tag{4.112}$$

where $y_{i\cdot} = 1/n_i \sum_j y_{ij}$ is the ith sample mean and $y_{\cdot\cdot} = 1/n \sum_i \sum_j y_{ij}$ is the overall mean of all y–values. We receive the estimated means according to

$$\hat{y}_{i\cdot} = \hat{\alpha} + \hat{\beta} x_i\,. \tag{4.113}$$

We partition the sum of squares SS_A as follows:

$$\begin{aligned} SS_A &= \sum_{i=1}^{s} n_i(y_{i\cdot} - y_{\cdot\cdot})^2 \\ &= \sum_{i=1}^{s} n_i(\hat{y}_{i\cdot} - y_{\cdot\cdot})^2 + \sum_{i=1}^{s} n_i(y_{i\cdot} - \hat{y}_{i\cdot})^2 \\ &= SS_{\text{Model}} + SS_{\text{Deviation}}\,. \end{aligned} \tag{4.114}$$

For the degrees of freedom we have

$$df_A = df_M + df_{\text{Deviation}}, \tag{4.115}$$

i.e.,

$$(s-1) = 1 + s - 2\,. \tag{4.116}$$

If not only $K = 2$ parameters are to be estimated, but K parameters in general, then

$$df_A = s - 1,\ df_M = K - 1,\ df_{\text{Deviation}} = s - K\,. \tag{4.117}$$

The complete *table of the regression analysis of variance* is shown in Table 4.14. As a test value for the fit of the model we compute

$$F = \frac{MS_{\text{Model}}}{MS_{\text{Deviation}}}\,. \tag{4.118}$$

If $F > F_{s-1,n-s;1-\alpha}$ the fit of the model is significant at the α level.

Example 4.7. In a study the rate of abrasion of silanized plastic material PMMA was determined for various levels of the proportion of quartz (Table 4.15).

Source of variation	SS	df	$MS = SS/df$	Test value
Model	SS_M	$K-1$	MS_M	$MS_{\text{Model}}/MS_{\text{Dev}}$
Model deviation	SS_{Dev}	$s-K$	MS_{Dev}	
Between the y–groups	SS_A	$s-1$	MS_A	$F = MS_{\text{A}}/MS_{\text{Error}}$
Within the y–groups	SS_{Error}	$n-s$	MS_{Error}	
Total	SS_{Total}	$n-1$		

TABLE 4.14. Table of the regression analysis of variance.

x [in volume % quartz]			
$x_1 = 2.2$	$x_2 = 4.5$	$x_3 = 9.3$	$x_4 = 25.6$
0.1420	0.0964	0.0471	0.0451
0.1113	0.0680	0.0585	0.0311
0.1092	0.0964	0.0544	0.0458
0.1298	0.0764	0.0444	0.0534
0.0962	0.0749	0.0575	0.0488
0.0917	0.0813	0.0406	0.0508
0.0800	0.0813	0.0522	0.0440
0.0996	0.0813	0.0525	0.0549
0.1123		0.0570	0.0539
		0.0559	0.0526
$y_{1\cdot} = 0.1080$	$y_{2\cdot} = 0.0820$	$y_{3\cdot} = 0.0520$	$y_{4\cdot} = 0.0480$
$n_1 = 9$	$n_2 = 8$	$n_3 = 10$	$n_4 = 10$
$y_{\cdot\cdot} = 0.0710$		$n = 37$	
$\hat{y}_{1\cdot} = 0.0878$	$\hat{y}_{2\cdot} = 0.0831$	$\hat{y}_{3\cdot} = 0.0733$	$\hat{y}_{4\cdot} = 0.0400$

TABLE 4.15. Data of the rate of abrasion.

The null hypothesis H_0 : *All means are equal, i.e., the proportion of quartz has no effect on the rate of abrasion* is rejected, since the analysis of variance yields the test value (see Table 4.16)

$$F = \frac{MS_A}{MS_{\text{Error}}} = 55.80 > 2.74 = F_{3,33;0.95} \,. \tag{4.119}$$

Hence, we fit a linear regression (4.113) to the means $y_{i\cdot}$ of the $s = 4$ samples. The parameters are computed according to (4.111) and (4.112):

$$\hat{y}_{i\cdot} = 0.0923 - 0.0020\, x_i \quad (i = 1, \ldots, 4)\,.$$

SS	df	MS	Test value
$SS_M = 0.01340$	1	$MS_M = 0.01340$	$F = 3.02$
$SS_{\text{Dev.}} = 0.00886$	2	$MS_{\text{Dev.}} = 0.00443$	
$SS_A = 0.02226$	3	$MS_A = 0.00742$	$F = 55.80$
$SS_E = 0.00440$	33	$MS_E = 0.00013$	
$SS_T = 0.02667$	36		

TABLE 4.16. Table of the regression analysis of variance of the rate of abrasion.

These estimated values are shown in Table 4.15. We can now calculate the partition (4.114) of SS_A (Table 4.16), the test value is

$$F = \frac{MS_{\text{Model}}}{MS_{\text{Dev.}}} = 3.02 < 18.51 = F_{1,2;0.95} .$$

Hence, the null hypothesis $\text{H}_0 : \beta = 0$ cannot be rejected.

4.6 One–Factorial Models with Random Effects

So far, in this chapter, we have discussed models with fixed effects. In the Introduction, however, we have already referred to the difference to models with random effects.

Models with fixed effects for the analysis of treatment effects are the standard in designed experiments. Models with random effects, however, occur in sample surveys where the grouping categories are random effects.

Examples: Quality control:

(i) Fixed effects: The daily production of five particular machines from an assembly line.

(ii) Random effects: The daily production of five machines, chosen at random, that represent the machines as a class.

The model with random effects is of the same structure as the model (4.2) with fixed effects

$$y_{ij} = \mu + \alpha_i + \epsilon_{ij} \ (i = 1, \ldots, s; \ j = 1, \ldots, n_i) . \tag{4.120}$$

The meaning of the parameter α_i however has now changed. The α_i are now the random effects of the ith treatment (ith machine). Hence, the α_i are the random variables whose distributions we have to specify. We assume

$$\text{E}(\alpha_i) = 0, \text{Var}(\alpha_i) = \sigma_\alpha^2 , \tag{4.121}$$

and

$$\text{E}(\epsilon_{ij}\alpha_i) = 0, E(\alpha_i\alpha_j) = 0 \ (i \neq j) . \tag{4.122}$$

Then

$$y_{ij} \sim (\mu, \sigma_\alpha^2 + \sigma^2) \tag{4.123}$$

holds.

In the model with fixed effects, the treatment effect A was represented by the parameter estimates $\hat{\alpha}_i$, or $\hat{\mu}_i = \hat{\mu} + \hat{\alpha}_i$, respectively. In the model with random effects, a treatment effect can be expressed by the so–called variance components. The variance σ_α^2 is estimated as a component of the entire variance. The absolute or relative size of this component then makes conclusions about the treatment effect possible.

The estimation of the variances σ_α^2 and σ^2 requires no assumptions about the distribution. For the test procedure and the computation of confidence intervals, however, we assume the normal distribution, i.e.,

$$\begin{aligned} \epsilon_{ij} &\sim N(0, \sigma^2), \epsilon_{ij} \text{ independent}, \\ \alpha_i &\sim N(0, \sigma_\alpha^2), \alpha_i \text{ independent}, \end{aligned}$$

and, hence,

$$y_{ij} \sim N(\mu, \sigma_\alpha^2 + \sigma^2)\,. \tag{4.124}$$

Unlike the model with fixed effects, the response values y_{ij} of a level i of the treatment (i.e., of the ith sample) are no longer uncorrelated

$$\begin{aligned} \mathrm{E}(y_{ij} - \mu)(y_{ij'} - \mu) &= \mathrm{E}(\alpha_i + \epsilon_{ij})(\alpha_i + \epsilon_{ij'}) \\ &= \mathrm{E}(\alpha_i^2) = \sigma_\alpha^2\,. \end{aligned} \tag{4.125}$$

On the other hand, the response values of different samples are still uncorrelated ($i \neq i'$, for any j, j'):

$$\mathrm{E}(y_{ij} - \mu)(y_{i'j'} - \mu) = \mathrm{E}(\alpha_i \alpha_{i'}) + \mathrm{E}(\epsilon_{ij}\epsilon_{i'j'}) + \mathrm{E}(\alpha_i \epsilon_{i'j'}) + \mathrm{E}(\alpha_{i'}\epsilon_{ij}) = 0\,. \tag{4.126}$$

In the case of a normal distribution, uncorrelated can be replaced by independent.

Test of the Null Hypothesis $\mathrm{H}_0:\ \sigma_\alpha^2 = 0$ Against $\mathrm{H}_1:\ \sigma_\alpha^2 > 0$

The hypothesis H_0 : "no treatment effect" for the two models is:

– fixed effects: $\mathrm{H}_0:\ \alpha_i = 0 \quad \forall i$;
– random effects: $\mathrm{H}_0:\ \sigma_\alpha^2 = 0$.

With the results of Section 4.2.3, which we can partly adopt, we have, for the model with random effects,

$$\mathrm{E}(MS_{\text{Error}}) = \sigma^2\,,$$

i.e., $MS_{\text{Error}} = \hat{\sigma}^2$ is an unbiased estimate of σ^2. We compute $\mathrm{E}(MS_A)$ as follows:

$$\begin{aligned}
SS_A &= \sum_{i=1}^{s}\sum_{j=1}^{n_i}(y_{i\cdot} - y_{\cdot\cdot})^2 ,\\
y_{i\cdot} &= \mu + \alpha_i + \epsilon_{i\cdot} ,\\
y_{\cdot\cdot} &= \mu + \alpha + \epsilon_{\cdot\cdot} ,\\
\alpha &= \sum n_i\alpha_i/n ,\\
(y_{i\cdot} - y_{\cdot\cdot}) &= (\alpha_i - \alpha) + (\epsilon_{i\cdot} - \epsilon_{\cdot\cdot}) .
\end{aligned}$$

With (4.121) and (4.122) we have

$$\mathrm{E}(y_{i\cdot} - y_{\cdot\cdot})^2 = \mathrm{E}(\alpha_i - \alpha)^2 + \mathrm{E}(\epsilon_{i\cdot} - \epsilon_{\cdot\cdot})^2 , \tag{4.127}$$

$$\begin{aligned}
\mathrm{E}(\alpha_i - \alpha)^2 &= \mathrm{E}(\alpha_i^2) + \mathrm{E}(\alpha^2) - 2\mathrm{E}(\alpha_i\alpha)\\
&= \sigma_\alpha^2\left[1 + \frac{\sum n_i^2}{n^2} - 2\frac{n_i}{n}\right] ,
\end{aligned} \tag{4.128}$$

$$\begin{aligned}
\mathrm{E}(\epsilon_{i\cdot}^2 - \epsilon_{\cdot\cdot})^2 &= \mathrm{E}(\epsilon_{i\cdot}^2) + \mathrm{E}(\epsilon_{\cdot\cdot}^2) - 2\mathrm{E}(\epsilon_{i\cdot}\epsilon_{\cdot\cdot})\\
&= \frac{\sigma^2}{n_i} + \frac{\sigma^2}{n} - 2\frac{\sigma^2}{n}\\
&= \sigma^2\left(\frac{1}{n_i} - \frac{1}{n}\right) .
\end{aligned} \tag{4.129}$$

Hence

$$\begin{aligned}
\sum_{j=1}^{n_i}\mathrm{E}(y_{i\cdot} - y_{\cdot\cdot})^2 &= n_i\mathrm{E}(y_{i\cdot} - y_{\cdot\cdot})^2\\
&= \sigma_\alpha^2\left[n_i + \frac{n_i}{n}\frac{\sum n_i^2}{n} - 2\frac{n_i^2}{n}\right] + \sigma^2\left(1 - \frac{n_i}{n}\right)
\end{aligned}$$

and

$$\sum_{i=1}^{s} n_i\mathrm{E}(y_{i\cdot} - y_{\cdot\cdot})^2 = \sigma_\alpha^2\left[n - \frac{\sum n_i^2}{n}\right] + \sigma^2(s-1) .$$

We receive:

(i) in the unbalanced case

$$\mathrm{E}(MS_A) = \frac{1}{s-1}\mathrm{E}(SS_A) = \sigma^2 + k\sigma_\alpha^2 \tag{4.130}$$

with

$$k = \frac{1}{s-1}\left(n - \frac{1}{n}\sum n_i^2\right) ; \tag{4.131}$$

(ii) in the balanced case ($n_i = r$ for all i, $n = r \cdot s$)

$$k = \frac{1}{s-1}\left(r \cdot s - \frac{1}{r \cdot s} s \cdot r^2\right) = r, \tag{4.132}$$

$$\mathrm{E}(MS_A) = \sigma^2 + r\sigma_\alpha^2 \,. \tag{4.133}$$

This yields the unbiased estimate $\hat{\sigma}_\alpha^2$ of σ_α^2:

(i) in the unbalanced case

$$\hat{\sigma}_\alpha^2 = \frac{MS_A - MS_{\text{Error}}}{k}\,; \tag{4.134}$$

(ii) in the balanced case

$$\hat{\sigma}_\alpha^2 = \frac{MS_A - MS_{\text{Error}}}{r}, . \tag{4.135}$$

In the case of an assumed normal distribution we have

$$MS_{\text{Error}} \sim \sigma^2 \chi^2_{n-s}$$

and

$$MS_A \sim (\sigma^2 + k\sigma_\alpha^2)\chi^2_{s-1} \,.$$

The two distributions are independent, hence the ratio

$$\frac{MS_A}{MS_{\text{Error}}} \cdot \frac{\sigma^2}{\sigma^2 + k\sigma_\alpha^2}$$

has a central F–distribution under the assumption of equal variances, i.e., under $\mathrm{H}_0 : \sigma_\alpha^2 = 0$. Under $\mathrm{H}_0 : \sigma_\alpha^2 = 0$ we thus have

$$\frac{MS_A}{MS_{\text{Error}}} \sim F_{s-1,n-s} \,. \tag{4.136}$$

Hence, $\mathrm{H}_0 : \sigma_\alpha^2 = 0$ is tested with the same test statistic as $\mathrm{H}_0 : \alpha_i = 0$ (all i) in the model with fixed effects. The table of the analysis of variance remains unchanged.

			E(MS) Effects	
Source	SS	*df*	Fixed	Random
Treatment	SS_A	$s-1$	$\sigma^2 + \frac{\sum n_i\alpha_i^2}{s-1}$	$\sigma^2 + k\sigma_\alpha^2$
Error	SS_{Error}	$n-s$	σ^2	σ^2

TABLE 4.17. Expectations of MS_A and MS_{Error}.

Example 4.8. (Continuation of Example 4.5)
We now regard the design from Table 4.6 as a model with random effects.

The null hypothesis $H_0 : \sigma_\alpha^2 = 0$ is tested with the statistic from (4.136). Table 4.7 yields

$$F_{3,20} = \frac{1.3349}{0.3962} = 3.3687 \quad (p\text{–value: } 0.0389)\,,$$

hence $H_0 : \sigma_\alpha^2 = 0$ is rejected. The estimated components of variance are

$$\hat{\sigma}^2 = MS_{\text{Error}} = 0.3962$$

and (cf. (4.135))

$$\hat{\sigma}_\alpha^2 = \frac{1.3349 - 0.3962}{6} = 0.1564\,.$$

4.7 Rank Analysis of Variance in the Completely Randomized Design

4.7.1 Kruskal–Wallis Test

The previous models were designed for the case that the response values follow a normal distribution. We now consider the situation that the response is either continuous but not normal or that we have a categorical response. For this data situation, which is often found in practice, we want to conduct the one–factorial comparison of groups. We first discuss the completely randomized design.

The response values are y_{ij} with the two subscripts $i = 1, \ldots, s$ (groups) and $j = 1, \ldots, n_i$ (subscript within the ith group). The data are collected according to the completely randomized design: n_1 units are chosen at random from $n = \sum n_i$ units and are assigned to the treatment (group) 1, etc. The data structure is shown in Table 4.18.

Group			
1	2	$\cdots$	s
y_{11}	y_{21}	$\cdots$	y_{s1}
$\vdots$	$\vdots$		$\vdots$
y_{1n_1}	y_{2n_2}	$\cdots$	y_{sn_s}

TABLE 4.18. Data matrix in the completely randomized design.

To begin with, we choose the following linear additive model

$$y_{ij} = \mu_i + \epsilon_{ij} \tag{4.137}$$

and assume that

$$\epsilon_{ij} \sim F(0, \sigma^2) \tag{4.138}$$

holds (where F is any continuous distribution). Additionally, we assume that the observations are independent within and between the groups.

The major statistical task is the comparison of the group means μ_i according to

$$\mathrm{H}_0 : \mu_1 = \cdots = \mu_s \quad \text{against} \quad \mathrm{H}_1 : \mu_i \neq \mu_j \quad (\text{at least one pair } i, j,\ i \neq j).$$

The tests are based on the comparison of the rank sums of the groups, in analogy to the Wilcoxon test in the two–sample case. The ranking procedure assigns the rank 1 to the smallest value of all s groups, ..., the rank $n = \sum n_i$ to the largest value of all s groups. These ranks R_{ij} replace the original values y_{ij} of the response Table 4.18 according to Table 4.19.

	Group				
	1	2	$\cdots$	s	
	R_{11}	R_{21}		R_{s1}	
	$\vdots$	$\vdots$		$\vdots$	
	R_{1n_1}	R_{2n_2}		R_{sn_s}	
$\sum$	$R_{1\cdot}$	$R_{2\cdot}$	$\cdots$	$R_{s\cdot}$	$R_{\cdot\cdot}$
Mean	$r_{1\cdot}$	$r_{2\cdot}$	$\cdots$	$r_{s\cdot}$	$r_{\cdot\cdot}$

TABLE 4.19. Rank values for Table 4.18.

The rank sums and rank means are

$$\begin{aligned} R_{i\cdot} &= \textstyle\sum_{j=1}^{n_i} R_{ij}, & R_{\cdot\cdot} &= \textstyle\sum_{i=1}^{s} R_{i\cdot} = \frac{n(n+1)}{2}, \\ r_{i\cdot} &= \frac{R_{i\cdot}}{n_i}, & r_{\cdot\cdot} &= \frac{R_{\cdot\cdot}}{n} = \frac{n+1}{2}. \end{aligned}$$

Under the null hypothesis all $n!/n_1! \cdots n_s!$ possible arrangements of the ranks have equal possibility. Hence, for each of these arrangements we can compute a measure for the difference between the groups. One possible measure for the group difference is based on the comparison of the rank means $r_{i\cdot}$.

In analogy to the error sum of squares $SS_A = \sum_{i=1}^{s} n_i(y_{i\cdot} - y_{\cdot\cdot})^2$ (cf. (4.29)) Kruskal and Wallis constructed the following test statistic (Kruskal and Wallis, 1952):

$$\begin{aligned} H &= \frac{12}{n(n+1)} \sum_{i=1}^{s} n_i (r_{i\cdot} - r_{\cdot\cdot})^2 \\ &= \frac{12}{n(n+1)} \sum_{i=1}^{s} \frac{R_{i\cdot}^2}{n_i} - 3(n+1). \end{aligned} \tag{4.139}$$

The test statistic H is a measure for the variance of the sample rank means. For the case of $n_i \leq 5$, tables exist for the exact critical values (cf., e.g., Hollander and Wolfe, 1973, p. 294). For $n_i > 5$ $(i = 1, \ldots, s)$, H is approximatively χ^2_{s-1}–distributed.

Correction in the Case of Ties

If equal response values y_{ij} arise and mean ranks are assigned, then the following corrected test statistic is used

$$H_{\text{Corr}} = H\left(1 - \frac{\sum_{k=1}^{r}(t_k^3 - t_k)}{n^3 - n}\right)^{-1} . \tag{4.140}$$

Here r is the number of groups with equal ranks and t_k is the number of equal response values within a group. If $H > \chi^2_{s-1;1-\alpha}$, the hypothesis $H_0 : \mu_1 = \cdots = \mu_s$ is rejected in favor of H_1. If H_{Corr} has to be used, the corrected value does not have to be calculated in the case of significance of H, due to $H_{\text{Corr}} > H$.

Example 4.9. We now compare the manufacturing times from Table 4.1 according to the Kruskal–Wallis test. Hint: In Example 4.1 the analysis of variance was done with the logarithms of the response values, since a normal distribution of the original values was doubtful. The null hypothesis was not rejected, cf. Table 4.5. The test statistic based on Table 4.20 is

Dentist A		Dentist B		Dentist C	
Manufacturing time	Rank	Manufacturing time	Rank	Manufacturing time	Rank
31.5	3.0	33.5	5.0	19.5	1.0
38.5	7.0	37.0	6.0	31.5	3.0
40.0	8.5	43.5	10.0	31.5	3.0
45.5	11.0	54.0	15.0	40.0	8.5
48.0	12.0	56.0	17.0	50.5	13.0
55.5	16.0	57.0	18.0	53.0	14.0
57.5	19.0	59.5	21.0	62.5	23.5
59.0	20.0	60.0	22.0	62.5	23.5
70.0	27.5	65.5	25.0		
70.0	27.5	67.0	26.0		
72.0	29.0	75.0	31.0		
74.5	30.0				
78.0	32.0				
80.0	33.0				
$n_1 = 14$		$n_2 = 11$		$n_3 = 8$	
$R_{1\cdot} = 275.5$		$R_{2\cdot} = 196.0$		$R_{3\cdot} = 89.5$	
$r_{1\cdot} = 19.68$		$r_{2\cdot} = 17.82$		$r_{3\cdot} = 11.19$	

TABLE 4.20. Computation of the ranks and rank sums for Table 4.1.

$$\begin{aligned} H &= \frac{12}{33 \cdot 34}\left[\frac{275.5^2}{14} + \frac{196.0^2}{11} + \frac{89.5^2}{8}\right] - 3 \cdot 34 \\ &= 4.04 < 5.99 = \chi^2_{2;0.95} \,. \end{aligned}$$

Since H is not significant we have to compute H_{Corr}. Table 4.20 yields:

$$\begin{aligned} r = 4: \quad & t_1 = 3 \quad (3 \text{ ranks of } 3), \\ & t_2 = 2 \quad (2 \text{ ranks of } 8.5), \\ & t_3 = 2 \quad (2 \text{ ranks of } 23.5), \\ & t_4 = 2 \quad (2 \text{ ranks of } 27.5). \end{aligned}$$

Correction term:

$1 - [3 \cdot (2^3 - 2) + (3^3 - 3)]/(33^3 - 33) = 1 - 42/35904 = 0.9988,$

$H_{\text{Corr}} = 4.045$.

The decision is: the null hypothesis $H_0 : \mu_1 = \mu_2 = \mu_3$ is not rejected, the effect "dentist" cannot be proven.

4.7.2 Multiple Comparisons

In analogy to the reasoning in Section 4.4, we want to discuss the procedure in case of a rejection of the null hypothesis $H_0 : \mu_1 = \cdots = \mu_s$ for ranked data.

Planned Single Comparisons

If we plan a comparison of two particular groups before the data is collected, then the Wilcoxon rank–sum test is the appropriate test procedure (cf. Section 2.5). The type I error, however, only holds for this particular comparison.

Comparison of All Pairwise Differences

The procedure for comparing all $s(s-1)/2$ possible pairs (i, j) of differences with $i > j$ dates back to Dunn (1964). It is based on the Bonferroni method and assumes large sample sizes. The following statistics are computed from the differences $r_{i\cdot} - r_{j\cdot}$ of the rank means $(i \neq j\,, \, i > j)$:

$$z_{ij} = \frac{r_{i\cdot} - r_{j\cdot}}{\sqrt{(n(n+1)/12) \cdot (1/n_i + 1/n_j)}} \,. \tag{4.141}$$

Let $u_{1-\alpha/s(s-1)}$ be the $[1-\alpha/s(s-1)]$–quantile of the $N(0,1)$–distribution. The multiple testing rule that ensures the α–level overall for all $s(s-1)$ pairwise comparisons is

$$\text{H}_0\colon \mu_i = \mu_j \quad \text{for all } (i,j),\ i > j, \tag{4.142}$$

is rejected in favor of

$$\text{H}_1 : \mu_i \neq \mu_j \quad \text{for at least one pair } (i,j),$$

if

$$|z_{ij}| > z_{1-\alpha/s(s-1)} \quad \text{for at least one pair } (i,j), i > j\,. \tag{4.143}$$

Example 4.10. Table 4.6 shows the response values of the four treatments (i.e., control group, A_1, A_2, $A_1 \cup A_2$) in the balanced randomized design.

The analysis of variance, under the assumption of a normal distribution, rejected the null hypothesis $H_0 : \mu_1 = \cdots = \mu_4$. In the following, we conduct the analysis based on ranked data, i.e., we no longer assume a normal distribution. From Table 4.6 we compute the Rank Table 4.21

Control group		A_1		A_2		$A_1 \cup A_2$	
Value	Rank	Value	Rank	Value	Rank	Value	Rank
4.5	21.5	3.8	12.0	3.5	8.0	3.0	4.0
5.0	24.0	4.0	16.5	4.5	21.5	2.8	3.0
3.5	8.0	3.9	13.5	3.2	5.0	2.2	2.0
3.7	11.0	4.2	19.0	2.1	1.0	3.4	6.0
4.8	23.0	3.6	10.0	3.5	8.0	4.0	16.5
4.0	16.5	4.4	20.0	4.0	16.5	3.9	13.5
$R_{1\cdot} = 104$		$R_{2\cdot} = 91$		$R_{3\cdot} = 60$		$R_{4\cdot} = 45$	
$r_{1\cdot} = 17.33$		$r_{2\cdot} = 15.17$		$r_{3\cdot} = 10.00$		$r_{4\cdot} = 7.50$	

TABLE 4.21. Rank table for Table 4.6.

and receive the Kruskal–Wallis statistic

$$\begin{aligned} H &= \frac{12}{24 \cdot 25 \cdot 6} \sum R_{i\cdot}^2 - 3 \cdot 25 \\ &= \frac{1}{300} \sum (104^2 + 91^2 + 60^2 + 45^2) - 75 \\ &= 7.41\,. \end{aligned}$$

H_0 is not rejected on the 5% level, due to $7.41 < 7.81 = \chi^2_{3;0.95}$. Hence, the nonparametric analysis stops.

For the demonstration of nonparametric multiple comparisons we now change to the 10% level. This yields $H = 7.41 > 6.25 = \chi^2_{3;0.90}$. Since H already is significant, H_{Corr} does not have to be calculated. Hence, $H_0 : \mu_1 = \cdots = \mu_4$ can be rejected on the 10% level.

We can now conduct the multiple comparisons of the pairwise differences. The denominator of the test statistic z_{ij} (4.141) is

$$\sqrt{((24 \cdot 25)/12)(2/6)} = \sqrt{50/3} = 4.08.$$

Comparison	$r_{i\cdot} - r_{j\cdot}$	z_{ij}	
1/2	2.16	0.53	
1/3	7.33	1.80	
1/4	10.83	2.65	*
2/3	5.17	1.27	
2/4	8.67	2.13	
3/4	3.50	0.86	

For $\alpha = 0.10$ we receive $\alpha/s(s-1) = 0.10/12 = 0.0083$, $1 - \alpha/s(s-1) = 0.9917$, $u_{0.9917} = 2.39$. Hence, the comparison 1/4 is significant.

Comparison Control Group – All Other Treatments

If one treatment out of the s treatments is chosen as the control group and compared to the other $s-1$ treatments, then the test procedure is the same, but with the $[u_{1-\alpha/2(s-1)}]$–quantile.

Example 4.11. (Continuation of Example 4.10)
The control group is treatment 1 (no additives). The comparison with the treatments 2 (A_1), 3 (A_2), and 4 ($A_1 \cup A_2$) is done with the test statistics z_{12}, z_{13}, z_{14}. Here we have to use the $[u_{1-\alpha/2(s-1)}]$–quantile. We receive $1 - 0.10/6 = 0.9833$, $u_{1-0.10/6} = 2.126 \Rightarrow$ the comparisons 1/4 and 2/4 are significant.

4.8 Exercises and Questions

4.8.1 Formulate the one–factorial design with $s = 2$ fixed effects for the balanced case as a linear model in the usual coding and in effect coding.

4.8.2 What does the table of the analysis of variance look like in a two–factorial design with fixed effects?

4.8.3 What meaning does the theorem of Cochran have? What effects can be tested with it?

4.8.4 In a field experiment three fertilizers are to be tested. The table of the analysis of variance is:

	df	*MS*	*F*
$SS_A = 50$			
$SS_{\text{Error}} =$			
$SS_{\text{Total}} = 350$	32		

Name the hypothesis to be tested and the test decision.

4.8.5 Let $c'y.$ be a linear contrast of the means $y_{1\cdot}, \ldots, y_{s\cdot}$. Complete the following:

$$c'y. \sim N(?, ?).$$

The test statistic for testing $\text{H}_0 : c'\mu = 0$ is

$$? \sim \chi^2_{df}, \quad df = ?\,.$$

4.8.6 How many independent linear contrasts exist for s means? What is a complete system of linear contrasts? Is this system unique?

4.8.7 Let $c_1'Y., \ldots, c_{s-1}'Y.$ be a complete system of linear contrasts of the total response values $Y. = (Y_1., \ldots, Y_s.)'$. Assume that each contrast has the distribution

$$c_i'Y. \sim N(?, ?).$$

Then

$$\frac{(c_i'Y)^2}{?} \sim ?$$

and, if the contrasts are ..., then

$$SS_A = ?$$

holds.

4.8.8 Let A_1 be a control group and assume that A_2 and A_3 are two treatments. Name the contrasts for the comparison of:
A_1 against A_2 or A_3;
A_2 against A_1;
A_3 against A_1?

4.8.9 Describe the main concern of multiple comparisons and the two methods of comparison.

4.8.10 Assign the experimentwise designed multiple comparisons correctly into the following matrix:

	Scheffé	Dunnett	Tukey	Bonferroni
(i)				
(ii)				
(iii)				
(iv)				

(i) $k \leq s$ comparisons planned in advance;

(ii) set of any linear contrasts;

(iii) $(s-1)$ comparisons with a control group; and

(iv) all $s(s-1)/2$ comparisons of means.

4.8.11 In the case of the two–sample t–test (balanced) the critical value is $t_{n-1;1-\alpha}$. In the case of the Bonferroni procedure with three comparisons the critical value for each single comparison is $t_{?;?}$.

4.8.12 Name the assumptions in the model $y_{ij} = \mu + \alpha_i + \epsilon_{ij}$ with mixed effects. We have $y_{ij} \sim N(?, ?)$. Formulate the hypothesis H_0 : no treatment effect!

4.8.13 Conduct the rank analysis of variance according to Kruskal–Wallis for the following table:

Student A		Student B		Student C	
Points	Rank	Points	Rank	Points	Rank
32		34		38	
39		37		40	
45		42		43	
47		54		48	
53		60		52	
59		75		61	
71				80	
85				95	

Hint: Completely randomized design.

5

More Restrictive Designs

5.1 Randomized Block Design

In statistical practice, the experimental units are often not completely homogeneous. Usually, a grouping according to a stratification factor can be observed (clinical population: stratified according to patient's age, degree of disease, etc.). If we have such prior information then a gain in efficiency compared to the completely randomized experiment is possible by grouping into blocks. The experimental units are grouped together in homogeneous groups (blocks) and the treatments are assigned to the experimental units within each block by random. Hence the block effect (differences between the blocks) can now be separated from the experimental error. This leads to a higher precision. The strategy of building blocks should yield a variability within each block that is as small as possible and a variability between blocks that is as high as possible.

The most widely used block design is the randomized block design (RBD). Here s treatments with r repetitions each (i.e., balanced) are assigned to a total of $n = r \cdot s$ experimental units. First, the experimental units are divided into r blocks with s units each in such a way that the units within each block are as homogeneous as possible. The s treatments are then assigned to the s units at random, so that each treatment occurs only once per block.

Example 5.1. We want to test $s = 3$ treatments A, B, C with $r = 4$ repetitions each in the randomized block design with respect to their effect. Assume the blocking factor to be ordinal scaled (e.g., $r = 4$ levels of intensity of a disease or $r = 4$ age groups).

The block design of the $n = r \cdot s = 12$ experimental units is then of the structure displayed in Table 5.1. The assignment of the $s = 3$ treatments

Block								
I	II	III	IV		I	II	III	IV
1	1	1	1	$\longrightarrow$	A	B	C	B
2	2	2	2	Randomization	B	A	A	C
3	3	3	3		C	C	B	A

TABLE 5.1. Randomized assignment of treatments per block.

per block to the three units of the $r = 4$ blocks can be done via random numbers. Ranks 1, 2, or 3 are assigned to these random numbers and the assignment to the treatments is then done according to a previously specified coding (rank 1: treatment A, rank 2: treatment B, rank 3: treatment C).

Example 5.2. Block II in Table 5.1:

Unit	Random number	Rank	Treatment
1	182	2	B
2	037	1	A
3	217	3	C

The structure of the data is shown in Table 5.2, with

Sums			Means			
$Y_{i\cdot}$	=	$\sum_j y_{ij}$	$y_{i\cdot}$	=	$Y_{i\cdot}/s$	Block i
$Y_{\cdot j}$	=	$\sum_i y_{ij}$	$y_{\cdot j}$	=	$Y_{\cdot j}/r$	Treatment j
$Y_{\cdot\cdot}$	=	$\sum_i Y_{i\cdot} = \sum_j Y_{\cdot j}$	$y_{\cdot\cdot}$	=	$Y_{\cdot\cdot}/rs$	Total

	Treatment j					
Block i	1	2	$\cdots$	s	Sum	Mean
1	y_{11}	y_{12}	$\cdots$	y_{1s}	$Y_{1\cdot}$	$y_{1\cdot}$
2	y_{21}	y_{22}	$\cdots$	y_{2s}	$Y_{2\cdot}$	$y_{2\cdot}$
$\vdots$	$\vdots$	$\vdots$		$\vdots$	$\vdots$	$\vdots$
r	y_{r1}	y_{r2}	$\cdots$	y_{rs}	$Y_{r\cdot}$	$y_{r\cdot}$
Sum	$Y_{\cdot 1}$	$Y_{\cdot 2}$	$\cdots$	$Y_{\cdot s}$	$Y_{\cdot\cdot}$	
Mean	$y_{\cdot 1}$	$y_{\cdot 2}$	$\cdots$	$y_{\cdot s}$	$y_{\cdot\cdot}$	

TABLE 5.2. Data table for the randomized block design.

Source	SS	df	MS	F
Block	SS_{Block}	$r-1$	MS_{Block}	F_{Block}
Treatment	SS_{Treat}	$s-1$	MS_{Treat}	F_{Treat}
Error	SS_{Error}	$(r-1)(s-1)$	MS_{Error}	
Total	SS_{Total}	$sr-1$		

TABLE 5.3. Analysis of variance table for the randomized block design.

The linear model for the randomized block design (without interaction) is

$$y_{ij} = \mu + \beta_i + \tau_j + \epsilon_{ij} \tag{5.1}$$

where

y_{ij}	is the	response of the jth treatment in the ith block;
μ	is the	average response of all experimental units (overall mean);
β_i	is the	additive effect of the ith block;
τ_j	is the	additive effect of the jth treatment; and
ϵ_{ij}	is the	random error of the experimental unit that receives the jth treatment in the ith block.

The following assumptions are made:

(i) The blocks are used for error control, hence the β_i are random effects with

$$\beta_i \sim N(0, \sigma_\beta^2) \,. \tag{5.2}$$

(ii) Assume the treatments to be fixed factors. The τ_j are then fixed effects that represent the deviation from the overall mean μ. Hence the following constraint holds

$$\sum_{j=1}^{s} \tau_j = 0 \,. \tag{5.3}$$

Remark. If, however, the treatment effects are to be regarded as random effects, then we assume

$$\tau_j \sim N(0, \sigma_\tau^2) \tag{5.4}$$

and

$$\mathrm{E}(\beta_i \tau_j) = 0 \text{ (for all } i, j) \tag{5.5}$$

instead of (5.3).

(iii) The ϵ_{ij} are the random errors. Assume

$$\epsilon_{ij} \overset{\text{i.i.d.}}{\sim} N(0, \sigma^2) \tag{5.6}$$

and

$$\mathrm{E}(\epsilon_{ij}\beta_i) = 0 \tag{5.7}$$

as well as

$$\mathrm{E}(\epsilon_{ij}\tau_j) = 0\,. \tag{5.8}$$

Then

$$\mu_i = \mu + \beta_i \quad \text{is the mean of the } i\text{th block}$$

and

$$\mu_j = \mu + \tau_j \quad \text{is the mean of the } j\text{th treatment.}$$

Decomposition of the Error Sum of Squares

Using the identity

$$y_{ij} - y_{\cdot\cdot} = (y_{ij} - y_{i\cdot} - y_{\cdot j} + y_{\cdot\cdot}) + (y_{i\cdot} - y_{\cdot\cdot}) + (y_{\cdot j} - y_{\cdot\cdot})\,, \tag{5.9}$$

it can be shown that the following decomposition holds:

$$\begin{aligned} \sum_i \sum_j (y_{ij} - y_{\cdot\cdot})^2 &= \sum_i \sum_j (y_{ij} - y_{i\cdot} - y_{\cdot j} + y_{\cdot\cdot})^2 \\ &\quad + \sum_{i=1}^{r} s(y_{i\cdot} - y_{\cdot\cdot})^2 \\ &\quad + \sum_{j=1}^{s} r(y_{\cdot j} - y_{\cdot\cdot})^2\,. \end{aligned} \tag{5.10}$$

If the correction term is computed by

$$C = Y_{\cdot\cdot}^2 / rs\,, \tag{5.11}$$

then the above sums of squares can be expressed as

$$SS_{\text{Total}} = \sum_i \sum_j (y_{ij} - y_{\cdot\cdot})^2 = \sum_i \sum_j y_{ij}^2 - C, \tag{5.12}$$

$$SS_{\text{Block}} = s \sum_i (y_{i\cdot} - y_{\cdot\cdot})^2 = \frac{1}{s} \sum_i Y_{i\cdot}^2 - C, \tag{5.13}$$

$$SS_{\text{Treat}} = r \sum_j (y_{\cdot j} - y_{\cdot\cdot})^2 = \frac{1}{r} \sum_j Y_{\cdot j}^2 - C, \tag{5.14}$$

$$SS_{\text{Error}} = SS_{\text{Total}} - SS_{\text{Block}} - SS_{\text{Treat}}\,. \tag{5.15}$$

The F–ratios (cf. Table 5.3) are

$$\begin{aligned} F_{\text{Block}} &= \frac{SS_{\text{Block}}}{SS_{\text{Error}}} \cdot \frac{(r-1)(s-1)}{(r-1)} \\ &= \frac{MS_{\text{Block}}}{MS_{\text{Error}}} \end{aligned} \tag{5.16}$$

and

$$\begin{aligned} F_{\text{Treat}} &= \frac{SS_{\text{Treat}}}{SS_{\text{Error}}} \cdot \frac{(s-1)(r-1)}{(s-1)} \\ &= \frac{MS_{\text{Treat}}}{MS_{\text{Error}}} . \end{aligned} \tag{5.17}$$

The significance of the treatment effect, i.e., $H_0 : \tau_j = 0$ $(j = 1, \ldots, s)$ for fixed effects and $H_0 : \sigma_\tau^2 = 0$ for random effects, is tested with F_{Treat}.

Testing for Block Effects

Consider the completely randomized design of the model (4.2) for the balanced case ($n_i = r$ for all i) and exchange the rows and columns (i.e., the meaning of i and j) in Table 4.2. If we additionally assume $\alpha_i = \tau_j$, then the following model corresponds with the completely randomized design

$$y_{ij} = \mu + \tau_j + \epsilon_{ij} \tag{5.18}$$

with the constraint $\sum \tau_j = 0$. The subscript $i = 1, \ldots, r$ represents the repetitions of the jth treatment $(j = 1, \ldots, s)$. Hence the completely randomized design (5.18) is a nested submodel of the randomized block design (5.1). Testing for significance of the block effect is therefore equivalent to model choice between the complete model (here (5.1)) and a submodel restricted by constraints ($H_0 : \beta_i = 0$).

The appropriate test statistic for this problem was already derived in Section 3.8.2 with F_{Change} (cf. (3.162)). F_{Change} is of the following form:

$$\frac{\text{error variance (small model) - error variance (large model)}}{\text{error variance (large model)}} . \tag{5.19}$$

Applied to our problem we receive for the "large" model (5.1), according to (5.15),

$$SS_{\text{Error(large)}} = SS_{\text{Total}} - SS_{\text{Block}} - SS_{\text{Treat}} . \tag{5.20}$$

In the "small" model (5.18) we have

$$SS_{\text{Error(small)}} = SS_{\text{Total}} - SS_{\text{Treat}} , \tag{5.21}$$

hence F_{Change} is now

$$\frac{SS_{\text{Block}}/(r-1)}{SS_{\text{Error(large)}}/(r-1)(s-1)} = F_{\text{Block}} . \tag{5.22}$$

This statistic tests the significance of the transition from the smaller model (completely randomized design) to the larger model (randomized block design) and hence the significance of the block effects.

Estimates and Variances

The unbiased estimate of the jth treatment mean $\mu_j = \mu + \tau_j$ is given by

$$\hat{\mu}_j = \frac{Y_{\cdot j}}{r} = y_{\cdot j}\,. \tag{5.23}$$

The variance of this estimate is

$$\mathrm{Var}(y_{\cdot j}) = \frac{1}{r^2} r\,\mathrm{Var}(y_{ij}) = \frac{\sigma^2}{r} \quad \text{(for all } j). \tag{5.24}$$

The unbiased estimate of the standard deviation of the estimates $y_{\cdot j}$ is then

$$s_{y_{\cdot j}} = \sqrt{MS_{\text{Error}}/r} \quad (j = 1, \ldots, s)\,. \tag{5.25}$$

Hence, the $(1-\alpha)$-confidence intervals of the jth treatment means are given by

$$y_{\cdot j} \pm t_{(s-1)(r-1),1-\alpha/2}\sqrt{MS_{\text{Error}}/r}\,. \tag{5.26}$$

For the simple comparison of two treatment means we receive an unbiased estimate of their difference by

$$y_{\cdot j_1} - y_{\cdot j_2}$$

with the standard deviation

$$s_{(y_{\cdot j_1} - y_{\cdot j_2})} = \sqrt{2MS_{\text{Error}}/r}\,. \tag{5.27}$$

Hence the $(1-\alpha)$-confidence intervals for the differences of means are of the form

$$(y_{\cdot j_1} - y_{\cdot j_2}) \pm t_{(s-1)(r-1),1-\alpha/2}\sqrt{2MS_{Error}/r}\,. \tag{5.28}$$

Hint. Note the admissibility of simple comparisons.

Example 5.3. A physician wants to test the effect of three blood pressure lowering drugs (drug A, drug B, a combination of A and B) and of a placebo as control group. The 12 patients are assigned into three groups according to their weight. The "difference of the diastolic blood pressure from taking the drug at 6 o'clock am until 6 o'clock pm" is the measured response. The assignment to the treatments is done at random in each block. Table 5.4 shows the measured values from which the table of variance is calculated.

We now receive

$$\begin{aligned}
C &= Y_{\cdot\cdot}^2/rs = 108^2/12 = 972, \\
SS_{\text{Total}} &= 5^2 + \cdots + 18^2 - C \\
&= 1158 - 972 = 186, \\
SS_{\text{Block}} &= 1/4(28^2 + 36^2 + 44^2) - C \\
&= 1004 - 972 = 32, \\
SS_{\text{Treat}} &= 1/3(21^2 + 24^2 + 18^2 + 45^2) - C
\end{aligned}$$

Block	Placebo 1	A 2	B 3	A and B 4	$\sum$	$y_{i\cdot}$
1	5	7	4	12	28	7
2	7	8	6	15	36	9
3	9	9	8	18	44	11
$\sum$	21	24	18	45	108	
$y_{\cdot j}$	7	8	6	15	9	

TABLE 5.4. Blood pressure differences.

$$\begin{aligned} &= 1122 - 972 = 150, \\ SS_{\text{Error}} &= 186 - 32 - 150 = 4. \end{aligned}$$

	SS	df	MS	F
Block	32	2	16	24.00
Treat	150	3	50	75.00
Error	4	6	0.67	
Total	186	11		

The testing of $\text{H}_0 : \tau_j = 0$ $(j = 1, \ldots, 4)$ (no treatment effect) with $F_{\text{Treat}} = F_{3,6} = 75.00$ leads to a rejection of H_0 $(F_{3,6;0.95} = 4.76)$, hence the treatment effect is significant. The test of the block effect yields significance with $F_{\text{Block}} = F_{2,6} = 24.00$ $(F_{2,6;0.95} = 5.14)$, hence the randomized block design is significant compared to the completely randomized design.

Consider the analysis of variance table in the completely randomized design with the same response values as in Table 5.4:

	SS	df	MS	F
Treat	150	3	50	11.11
Error	36	8	4.5	
Total	186	11		

Due to

$$F = 11.11 > F_{3,8;0.95} = 4.07$$

the treatment effect here is significant as well:

Treatment means				Standard error
1	2	3	4	$\sqrt{MS_{\text{Error}}/r}$
7	8	6	15	$\sqrt{0.67/3} = 0.47$
Confidence intervals				
7 ± 1.15	8 ± 1.15	6 ± 1.15	15 ± 1.15	

(Hint. $t_{6,0.975} = 2.45$, $2.45\sqrt{MS_{\text{Error}}/r} = 1.15$.)
Confidence intervals for differences of means.
(Hint. $t_{6,0.975}\ \sqrt{2MS_{\text{Error}}/r} = 1.63$.)

Treatments					
1/2	−1	±	1.63	⟹	[−2.63, 0.63]
1/3	1	±	1.63	⟹	[−0.63, 2.63]
1/4	−8	±	1.63	⟹	[−9.63, −6.37] *
2/3	2	±	1.63	⟹	[0.37, 3.63] *
2/4	7	±	1.63	⟹	[5.37, 8.63] *
3/4	9	±	1.63	⟹	[7.37, 10.63] *

TABLE 5.5. Simple comparisons.

In the simple comparison of means the treatments 1 and 4, 2 and 3, 2 and 4, as well as 3 and 4, differ significantly. Using Scheffé (see Table 5.6) we get that treatments 1, 2, and 3 define a homogeneous subset which is separated from treatment 4, i.e., the means of treatments 2 and 3 do not differ significantly using the multiple tests.

Treatments					
1/2	−1	±	1.7321	⟹	[−7.0494, 5.0494]
1/3	1	±	1.7321	⟹	[−5.0494, 7.0494]
1/4	−8	±	1.7321	⟹	[−14.0494, −1.9506] *
2/3	2	±	1.7321	⟹	[−4.0494, 8.0494]
2/4	7	±	1.7321	⟹	[−13.0494, −0.9506] *
3/4	9	±	1.7321	⟹	[−15.0494, −2.9506] *

TABLE 5.6. Multiple comparisons according to Scheffé.

Example 5.4. $n = 16$ students are tested for $s = 4$ training methods. The students are divided into $r = 4$ blocks according to their previous level of performance and the training methods are then assigned at random within each block. The response is measured as the level of performance on a scale of 1 to 100 points. The results are shown in Table 5.7.

Again, we calculate the sums of squares and test for treatment effect and block effect

$$\begin{aligned}
C &= \frac{(918)^2}{16} = 52670.25 \\
SS_{\text{Total}} &= 41^2 + \cdots + 61^2 - \frac{(918)^2}{16} = 54524.00 - 52670.25, \\
&= 1853.75, \\
SS_{\text{Block}} &= \frac{190^2 + \cdots + 270^2}{4} - \frac{(918)^2}{16} = 53691.00 - 52670.25 \\
&= 1020.75,
\end{aligned}$$

Block	Training method 1	2	3	4	$\sum$	Means
1	41	53	54	42	190	47.5
2	47	62	58	41	208	52.0
3	55	71	66	58	250	62.5
4	59	78	72	61	270	67.5
$\sum$	202	264	250	202	918	
Means	50.5	66.0	62.5	50.5	57.375	

TABLE 5.7. Points.

$$\begin{aligned} SS_{\text{Treat}} &= \frac{202^2 + \cdots + 202^2}{4} - \frac{(918)^2}{16} = 53451.00 - 52670.25 \\ &= 780.75, \\ SS_{\text{Error}} &= 1853.75 - 1020.75 - 780.75 \\ &= 52.25. \end{aligned}$$

	SS	df	MS	F	
Block	1020.75	3	340.25	58.61	*
Treat	780.75	3	260.25	44.83	*
Error	52.25	9	5.81		
Total	1853.75	15			

Both effects are significant

$$\begin{aligned} F_{\text{Treat}} &= F_{3,9} = 44.83 > 3.86 = F_{3,9;0.95}, \\ F_{\text{Block}} &= F_{3,9} = 58.61 > 3.86 = F_{3,9;0.95}. \end{aligned}$$

5.2 Latin Squares

In the randomized block design we divided the experimental units into homogeneous blocks according to a blocking factor and hence eliminated the differences among the blocks from the experimental error, i.e., increased the part of the variability explained by a model.

We now consider the case that the experimental units can be grouped with respect to two factors, as in a contingency table. Hence two block effects can be removed from the experimental error. This design is called a Latin square.

If s treatments are to be compared, s^2 experimental units are required. These units are first classified into s blocks with s units each, based on one of the factors (row classification). The units are then classified into s groups with s units each, based on the other factor (column classification). The s treatments are then assigned to the units in such a way that each treatment occurs once, and only once, in each row and column.

Table 5.8 shows a Latin square for the $s = 4$ treatments A, B, C, D, which were assigned to the $n = 16$ experimental units by permutation.

A	B	C	D
B	C	D	A
C	D	A	B
D	A	B	C

TABLE 5.8. Latin square for $s = 4$ treatments.

This arrangement can be varied by randomization, e.g., by first defining the order of the rows by random numbers. We replace the lexicographical order A, B, C, D of the treatments by the numerical order 1, 2, 3, 4.

Row	Random number	Rank
1	131	2
2	079	1
3	284	3
4	521	4

This yields the following row randomization:

B	C	D	A
A	B	C	D
C	D	A	B
D	A	B	C

Assume the randomization by columns leads to:

Column	Random number	Rank
1	003	1
2	762	4
3	319	3
4	199	2

The final arrangement of the treatments would then be:

B	A	D	C
A	D	C	B
C	B	A	D
D	C	B	A

If a time trend is present, then the Latin square can be applied to separate these effects.

I	II	III	IV
A B C D	B C D A	C D A B	D A B C

———————> time axis

FIGURE 5.1. Latin square for the elimination of a time trend.

5.2.1 *Analysis of Variance*

The linear model of the Latin square (without interaction) is of the following form:

$$y_{ij(k)} = \mu + \rho_i + \gamma_j + \tau_{(k)} + \epsilon_{ij} \ (i, j, k = 1, \ldots, s) . \qquad (5.29)$$

Here $y_{ij(k)}$ is the response of the experimental unit in the ith row and the jth column, subjected to the kth treatment. The parameters are:

μ	is the average response (overall mean);
ρ_i	is the ith row effect;
γ_j	is the jth column effect;
$\tau_{(k)}$	is the kth treatment effect; and
ϵ_{ij}	is the experimental error.

We make the following assumptions:

$$\epsilon_{ij} \sim N(0, \sigma^2) , \qquad (5.30)$$

$$\rho_i \sim N(0, \sigma_\rho^2) , \qquad (5.31)$$

$$\gamma_j \sim N(0, \sigma_\gamma^2) . \qquad (5.32)$$

Additionally, we assume all random variables to be mutually independent. For the treatment effects we assume

(i) fixed: $\sum_{k=1}^{s} \tau_{(k)} = 0,$
or
(ii) random: $\tau_{(k)} \sim N(0, \sigma_\tau^2)$,

respectively. The treatments are distributed over all s^2 experimental units according to the randomization, such that each unit, or rather its response, has to have the subscript (k) in order to identify the treatment. From the data table of the Latin square we obtain the marginal sums

$Y_{i\cdot} = \sum_{j=1}^{s} y_{ij}$		is the sum of the ith row;
$Y_{\cdot j} = \sum_{i=1}^{s} y_{ij}$		is the sum of the jth column; and
$Y_{\cdot\cdot} = \sum_i Y_{i\cdot} = \sum_j Y_{\cdot j}$		is the total response.

For the treatments we calculate that

T_k	is the sum of the response values of the kth treatment; and
$m_k = T_k/s$	is the average response of the kth treatment.

The decomposition of the error sum of squares is as follows.

	Treatment				
	1	2	$\cdots$	s	
Sum	T_1	T_2	$\ldots$	T_s	$\sum_{k=1}^{s} T_k = Y_{..}$
Mean	m_1	m_2	$\ldots$	m_s	$Y_{..}/s^2 = y_{..}$

TABLE 5.9. Sums and means of the treatments.

Source	SS	*df*	MS	F
Rows	SS_{Row}	$s-1$	MS_{Row}	F_{Row}
Columns	SS_{Column}	$s-1$	MS_{Column}	F_{Column}
Treatment	SS_{Treat}	$s-1$	MS_{Treat}	F_{Treat}
Error	SS_{Error}	$(s-1)(s-2)$	MS_{Error}	
Total	SS_{Total}	s^2-1		

TABLE 5.10. Analysis of variance table for the Latin square.

Assume the correction term defined according to

$$C = Y_{..}^2/s^2\,. \tag{5.33}$$

Then we have

$$SS_{\text{Total}} = \sum_i \sum_j y_{ij}^2 - C, \tag{5.34}$$

$$SS_{\text{Row}} = \frac{1}{s}\sum_i Y_{i\cdot}^2 - C, \tag{5.35}$$

$$SS_{\text{Column}} = \frac{1}{s}\sum_j Y_{\cdot j}^2 - C, \tag{5.36}$$

$$SS_{\text{Treat}} = \frac{1}{s}\sum_k T_k^2 - C, \tag{5.37}$$

$$SS_{\text{Error}} = SS_{\text{Total}} - SS_{\text{Row}} - SS_{\text{Column}} - SS_{\text{Treat.}} \tag{5.38}$$

The MS–values are obtained by dividing the SS–values by their degrees of freedom. The F–ratios are MS/MS_{Error} (cf. Table 5.10). The expectations of the MS are shown in Table 5.11.

Source	MS	$\mathrm{E}(MS)$
Rows	MS_{Row}	$\sigma^2 + s\sigma_\rho^2$
Columns	MS_{Column}	$\sigma^2 + s\sigma_\gamma^2$
Treatment	MS_{Treat}	$\sigma^2 + s/(s-1)\sum_k \tau_{(k)}^2$
Error	MS_{Error}	σ^2

TABLE 5.11. $\mathrm{E}(MS)$.

The null hypothesis, H_0 : "no treatment effect", i.e., $H_0: \tau_1 = \cdots = \tau_s = 0$ against $H_1 : \tau_i \neq 0$ for at least one i, is tested with

$$F_{\text{Treat}} = \frac{MS_{\text{Treat}}}{MS_{\text{Error}}} . \tag{5.39}$$

Due to the design of the Latin square, the s treatments are repeated s–times each. Hence, treatment effects can be tested for. On the other hand, we cannot always speak of a repetition of rows and columns in the sense of blocks. Hence, F_{Row} and F_{Column} can only serve as indicators for additional effects which yield a reduction of MS_{Error} and thus an increase in precision. Row and column effects would be statistically detectable if repetitions were realized for each cell.

Point and Confidence Estimates of the Treatment Effects

The OLS estimate of the kth treatment mean $\mu_k = \mu + \tau_{(k)}$ is

$$m_k = T_k/s \tag{5.40}$$

with the variance

$$\text{Var}(m_k) = \sigma^2/s \tag{5.41}$$

and the estimated variance

$$\widehat{\text{Var}}(m_k) = MS_{\text{Error}}/s . \tag{5.42}$$

Hence the confidence interval is of the following form:

$$m_k \pm t_{(s-1)(s-2);1-\alpha/2} \sqrt{MS_{\text{Error}}/s} . \tag{5.43}$$

In the case of a simple comparison of two treatments the difference is estimated by the confidence interval

$$(m_{k_1} - m_{k_2}) \pm t_{(s-1)(s-2);1-\alpha/2} \sqrt{2MS_{\text{Error}}/s} . \tag{5.44}$$

Example 5.5. The effect of $s = 4$ sleeping pills is tested on $s^2 = 16$ persons, who are stratified according to the design of the Latin square, based on the ordinally classified factor's body weight and blood pressure. The response to be measured is the prolongation of sleep (in minutes) compared to an average value (without sleeping pills).

	Weight				
Blood pressure	1	2	3	4	$Y_{i\cdot}$
1	43	57	61	74	235
2	59	63	75	46	243
3	65	79	48	64	256
4	83	55	67	72	277
$Y_{\cdot j}$	250	254	251	256	1011

	Weight →			
Blood pressure ↓	^{A} 43	^{B} 57	^{C} 61	^{D} 74
	^{B} 59	^{C} 63	^{D} 75	^{A} 46
	^{C} 65	^{D} 79	^{A} 48	^{B} 64
	^{D} 83	^{A} 55	^{B} 67	^{C} 72

TABLE 5.12. Latin square (prolongation of sleep).

Medicament	A	B	C	D	Total
Total (T_k)	192	247	261	311	1011
Mean	48.00	61.75	65.25	77.75	63.19

We calculate the sums of squares

$$\begin{aligned} C &= 1011^2/16 = 63882.56, \\ SS_{\text{Total}} &= 65939 - C = 2056.44, \\ SS_{\text{Row}} &= 1/4 \cdot 256539 - C = 252.19, \\ SS_{\text{Column}} &= 1/4 \cdot 255553 - C = 5.69, \\ SS_{\text{Treat}} &= 1/4 \cdot 262715 - C = 1796.19, \\ SS_{\text{Error}} &= 2056.44 - (252.19 + 5.69 + 1796.19) \\ &= 2056.44 - 2054.07 \\ &= 2.37. \end{aligned}$$

Source	SS	df	MS	F	
Rows	252.19	3	84.06	212.8	*
Columns	5.69	3	1.90	4.802	
Treatment	1796.19	3	598.73	1496.83	*
Error	2.37	6	0.40		
Total	2056.44	15			

The critical value is $F_{3,6;0.95} = 4.757$. Hence the row effect (stratification according to blood pressure groups) is significant, the column effect (weight) however, is not significant. The treatment effect is significant as well. The final conclusion should be, that in further clinical tests of the four different sleeping pills the experiment should be conducted according to the randomized block design with the blocking factor "blood pressure groups".

The simple and multiple tests require SS_{Error} from the model with the main effect treatment:

Source	SS	df	MS	F
Treatment	1796.19	3	598.73	27.60 *
Error	260.25	12	21.69	
Total	2056.44	15		

For the simple mean comparisons we obtain ($t_{6;0.975}\sqrt{2MS_{\text{Error}}/4} = 8.058$):

Treatments	Difference	Confidence interval
2/1	13.75	[5.68, 21.82]
3/1	17.25	[9.18, 25.32]
4/1	29.75	[21.68, 37.82]
3/2	3.50	[−4.57, 11.57]
4/2	16.00	[7.93, 24.07]
4/3	12.50	[4.43, 20.57]

Result: In the case of the simple test all pairwise mean comparisons, except for 3/2, are significant. These tests however are not independent. Hence, we conduct the multiple tests.

Multiple Tests

The multiple test statistics (cf. (4.102)–(4.104)) with the degrees of freedom of the Latin square are

$$FPLSD = t_{s(s-1);1-\alpha/2}\sqrt{2MS_{\text{Error}}/s}\,, \quad (5.45)$$

$$HSD = Q_{\alpha,(s,s(s-1))}\sqrt{MS_{\text{Error}}/s}\,, \quad (5.46)$$

$$SNK_i = Q_{\alpha,(i,(s-1)(s-2))}\sqrt{MS_{\text{Error}}/s}\,. \quad (5.47)$$

Results of the Multiple Tests

Fisher's protected LSD test:

$$\begin{aligned} FPLSD &= t_{12,0.975}\sqrt{2MS_{\text{Error}}/4} \\ &= 2.18\sqrt{21.69/2} \\ &= 7.18. \end{aligned}$$

Hence, the means are different except for μ_2 and μ_3.

HSD test:

We have $Q_{0.05,(4,12)} = 4.20$, hence

$$HSD = 4.20\sqrt{21.69/4} = 9.78\,.$$

All the means except 2/3 differ significantly.

SNK test

The means ordered according to their size are

$$48.00(A),\ 61.75(B),\ 65.25(C),\ 77.75(D).$$

The Studentized rank values and the SNK_i values calculated from them are

i	2	3	4
$Q_{0.05,(i,6)}$	3.46	4.34	4.90
SNK_i	8.06	10.11	11.41

For the largest difference (D minus A) we have

$$77.75 - 48 = 29.75 > 11.41\,,$$

for the next differences (D minus B) and (C minus A) we receive

$$\begin{aligned} 77.75 - 61.75 &= 16.00 > 10.11\,, \\ 65.25 - 48.00 &= 17.25 > 10.11\,, \end{aligned}$$

and, finally, we have

$$\begin{array}{llll} (D \text{ minus } C): & 77.75 - 65.25 = & 12.50 & > 8.06\,, \\ (C \text{ minus } B): & & 3.50 & < 8.06\,, \\ (B \text{ minus } A): & & 13.75 & > 8.06\,. \end{array}$$

Hence all means except for 2/3 differ significantly.

5.3 Rank Variance Analysis in the Randomized Block Design

5.3.1 *Friedman Test*

In the randomized block design, the individuals are grouped into blocks and are assigned one of the s treatments, randomized within each block. The essential demand is that each treatment occurs once, and only once, within each block. The layout of the response values is shown in Table 5.2. Once again we assume the linear additive model (5.1). Furthermore, we assume

$$\epsilon_{ij} \overset{\text{i.i.d.}}{\sim} F(0, \sigma^2), \tag{5.48}$$

where F is any continuous distribution and does not have to be equal to the normal distribution. The randomization leads to independence of the ϵ_{ij}. Hence, the actual assumption in (5.48) refers to the homogeneity of variance.

The hypothesis of interest is H_0 : no treatment effect, i.e., we test

$$H_0 : \tau_1 = \cdots = \tau_s$$

against

$$H_1 : \tau_i \neq \tau_j \quad \text{for at least one } (i, j),\ i \neq j\,.$$

The test procedure is based on the rank assignment (ranks 1 to s) for the response values, which is to be done separately for each block. Under the null hypothesis each of the $s!$ possible orders per block have the same probability. Analogously, the $(s!)^r$ possible orders of the intra block ranks have equal possibilities.

If we take the sums of ranks per treatment $j = 1, \ldots, s$ over the r blocks, then they should be almost equal if H_0 holds. The test statistic by Friedman (1937) for testing H_0 compares these rank sums.

	Treatment		
Block	1	$\cdots$	s
1	R_{11}	$\cdots$	R_{s1}
$\vdots$	$\vdots$		$\vdots$
r	R_{1r}	$\cdots$	R_{sr}
Sum	$R_{1\cdot}$	$\cdots$	$R_{s\cdot}$
Mean	$r_{1\cdot}$	$\cdots$	$r_{s\cdot}$

TABLE 5.13. Rank sums and rank means in the randomized block design.

The test statistic by Friedman is

$$Q = \frac{12r}{s(s+1)} \sum_{j=1}^{s} (r_{j\cdot} - r_{\cdot\cdot})^2 \tag{5.49}$$

$$= \frac{12}{rs(s+1)} \sum_{j=1}^{s} R_{j\cdot}^2 - 3r(s+1)\,. \tag{5.50}$$

Here we have

$$R_{j\cdot} = \sum_{i=1}^{r} R_{ji} \quad \text{rank sum of the } j\text{th treatment},$$

$$r_{j\cdot} = R_{j\cdot}/r \quad \text{rank mean of the } j\text{th treatment},$$

$$r_{\cdot\cdot} = (s+1)/2\,.$$

If H_0 holds, then the differences $r_{i\cdot} - r_{\cdot\cdot}$ are almost equal and Q is sufficiently small. If, however, H_0 does not hold, then Q becomes large.

The test statistic Q is approximately (for r sufficiently large) χ^2_{s-1}-distributed. Hence, $H_0 : \tau_1 = \cdots = \tau_s$ is rejected for

$$Q > \chi^2_{s-1;1-\alpha}\,.$$

For small values of r ($r < 15$), this approximation is insufficient. In this case exact quantiles are used (cf. tables in Hollander and Wolfe (1973); Michaelis (1971); and Sachs (1974), p. 424). If ties are present, then the correction term

$$C_{\text{corr}} = 1 - \sum_{i=1}^{r} \sum_{k=1}^{s_i} (t_{ik}^3 - t_{ik})/(rs(s^2 - 1)) \tag{5.51}$$

is calculated. Here t_{i1} is the size of the first group of equally large response values, t_{i2} is the size of the second group of equally large response values, etc., in the ith block.

The corrected Friedman statistic is

$$Q_{\text{corr}} = \frac{Q}{C_{\text{corr}}} . \tag{5.52}$$

The Friedman test is a test of homogeneity. It tests whether the treatment samples could possibly come from the same population.

Example 5.6. (Continuation of Example 5.3) We conduct the comparison of the $s = 4$ treatments, that are arranged in $r = 3$ blocks, according to Table 5.4 with the Friedman test. From Table 5.4 we calculate the ranks in Table 5.14.

	Placebo	A	B	A and B
Block	1	2	3	4
1	2	3	1	4
2	2	3	1	4
3	2.5	2.5	1	4
Sum	6.5	8.5	3	12
$r_{j\cdot}$	2.17	2.83	1	4

TABLE 5.14. Rank table for Table 5.4.

The test statistic Q is

$$\begin{aligned} Q &= \frac{12}{3 \cdot 4 \cdot 5}(6.5^2 + 8.5^2 + 3^2 + 12^2) - 3 \cdot 3 \cdot 5 \\ &= \frac{267.5}{5} - 45 = 8.5 . \end{aligned}$$

Since we have ties in the third block, we compute

$$\begin{aligned} C_{\text{corr}} &= 1 - (2^3 - 2)/(3 \cdot 4 \cdot (4^2 - 1)) \\ &= 1 - 1/30 = 0.97 \end{aligned}$$

and

$$Q_{\text{corr}} = \frac{Q}{C_{\text{corr}}} = 8.76 .$$

The exact test yields the 95%–quantile as 7.4. Hence, H_0 : "homogeneity of the four treatments" is rejected.

5.3.2 *Multiple Comparisons*

We assume that the null hypothesis $H_0 : \tau_1 = \cdots = \tau_s$ is rejected by the Friedman test. Analogously to Section 4.7.2, we distinguish between the planned single comparisons, all pairwise comparisons, and the comparison of a control group with all other treatments.

Planned Single Comparisons

If the comparison of two selected treatments is planned before the data collection, then the Wilcoxon test (cf. Chapter 2) is applied.

Comparison of all Pairwise Differences According to Friedman

The comparison of all $s(s-1)/2$ possible pairs is based on a modification of the Friedman test (cf. Woolson, 1987, p. 387).

For each combination (j_1, j_2), $j_1 > j_2$, of treatments we compute the test statistic

$$Z_{j_1,j_2} = \frac{|r_{j_1\cdot} - r_{j_2\cdot}|}{\sqrt{s(s+1)/12r}} \tag{5.53}$$

for testing $H_0 : \tau_{j_1} = \tau_{j_2}$ against $H_1 : \tau_{j_1} \neq \tau_{j_2}$. All null hypotheses with $Z_{j_1,j_2} > QP_{1-\alpha}(r)$ are rejected and the multiple level is α. Tables for the critical values $QP_{1-\alpha}(r)$ exist (cf., e.g., Woolson 1987, Table 15, p. 506; Hollander and Wolfe, 1973). For $\alpha = 0.05$ some selected values are:

r	2	3	4	5	6	7	8	9	10
$QP_{0.95}(r)$	2.77	3.31	3.63	3.86	4.03	4.17	4.29	4.39	4.47

Example 5.7. (Continuation of Example 5.3) For the differences of the rank means we obtain from Table 5.14 the following table ($\sqrt{4(4+1)/12 \cdot 3} = \sqrt{20/36} = 0.745$):

Comparisons	$\lvert r_{j_1\cdot} - r_{j_2\cdot}\rvert$	Test statistic
1/2	$\lvert 2.17 - 2.83\rvert = 0.66$	0.86
1/3	$\lvert 2.17 - 1.0\rvert = 1.17$	1.57
1/4	$\lvert 2.17 - 4.0\rvert = 1.83$	2.46
2/3	$\lvert 2.83 - 1.0\rvert = 1.83$	2.46
2/4	$\lvert 2.83 - 4.0\rvert = 1.17$	1.57
3/4	$\lvert 1.0 - 4.0\rvert = 3.00$	4.03 *

Result: The treatment B and the combination (A and B) show differences in effect.

Remark: A well–known problem from screening trials is that of a large number s of treatments with limited replication r ($r \leq 4$ blocks). Brownie and Boos 1994, demonstrate the validity of standard ANOVA and of rank-based ANOVA under nonnormality with respect to type I error rates when s becomes large.

Comparison Control Group versus All Other Treatments

Let $j = 1$ be the subscript of the control group. The test statistic for the multiple comparison of treatment 1 with the $(s-1)$ other treatments is

$$Z_{1j} = \frac{|r_{1\cdot} - r_{j\cdot}|}{\sqrt{s(s+1)/6r}}\,, j = 2, \ldots, s\,.$$

The two–stage quantiles $QC_{1-\alpha}(s-1)$ are given in special tables (Woolson, 1987, p. 507; Hollander and Wolfe, 1973). For $Z_{1j} > QC_{1-\alpha}(s-1)$ the corresponding null hypothesis H_0 : "homogeneity of the treatments 1 and j" is rejected. The multiple level α is ensured. In the following table we give a few selected critical values $QC_{0.95}(s-1)$:

$s-1$	1	2	3	4	5
$QC_{0.95}(s-1)$	1.96	2.21	2.35	2.44	2.51

Example 5.8. (Continuation of Example 5.3) The above table of the $|r_{j1\cdot} - r_{j2\cdot}|$ yields the following results for the comparison "placebo against A, B, and combination":

$$\left.\begin{array}{lll} 1/2: & Z_{12} = 0.66/\sqrt{4 \cdot 5/6 \cdot 3} & = 0.63, \\ 1/3: & Z_{13} = 1.17/\sqrt{20/18} & = 1.11, \\ 1/4: & Z_{14} = 1.83/\sqrt{20/18} & = 1.74, \end{array}\right\} < 2.35\,.$$

Hence, no comparison is significant.

5.4 Exercises and Questions

5.4.1 Describe the strategy of building blocks (homogeneity/heterogeneity). Does the experimental error diminish or increase in the case of blocking?

5.4.2 How can it be shown that the completely randomized design is a submodel of the randomized block design? How can the block effect be tested? Name the correct F–test for the treatment effect in the following table:

	SS		MS	F
Block	20	3		
Treatment	60	3		
Error	10	9		
Total	90	15		

5.4.3 Conduct a comparison of means according to Scheffé and Bonferroni for Example 5.3 (Table 5.4). Compare the results with those from Example 5.3 for the simple comparisons.

5.4.4 A Latin square is to test the effect of the $s = 3$ eating habits of decathletes, who are classified according to the ordinally classified factors, sprinting speed and strength. Test for block effects and for the treatment effect (measured in points).

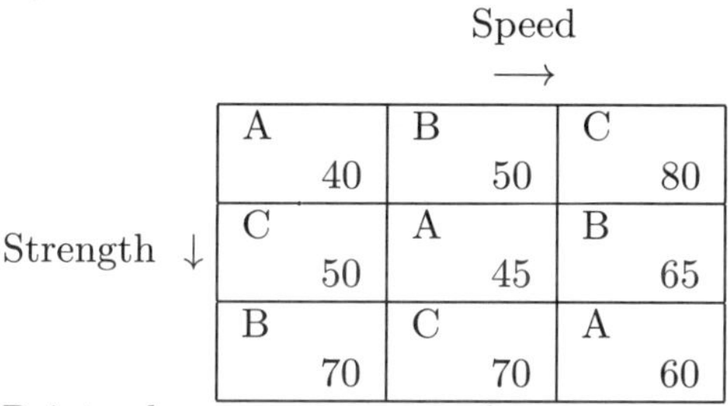
Speed →

Strength ↓

A 40	B 50	C 80
C 50	A 45	B 65
B 70	C 70	A 60

Points above an average value.

5.4.5 Conduct the Friedman test for Table 5.7. Define training method 1 as the control group and conduct a multiple comparison with the three other training methods.

6
Multifactor Experiments

6.1 Elementary Definitions and Principles

In practice, for most designed experiments it can be assumed that the response Y is not only dependent on a single variable but on a whole group of prognostic factors. If these variables are continuous, their influence on the response is taken into account by so–called factor levels. These are ranges (e.g., low, medium, high) that classify the continuous variables as ordinal variables. In Sections 1.7 and 1.8, we have already cited examples for designed experiments where the dependence of a response on two factors was to be examined.

Designs of experiments that analyze the response for all possible combinations of two or more factors are called *factorial experiments* or *cross–classification*. Suppose that we have s factors $A_1, \ldots, A_s$ with $r_1, \ldots, r_s$ factor levels. The complete factorial design then requires $r = \prod r_i$ observations for one trial. This shows that it is important to restrict the number of factors as well as the number of their levels.

For factorial experiments, two elementary models are distinguished—models with and without interaction. Assume the situation of two factors A and B with two factor levels each, i.e., A_1, A_2 and B_1, B_2.

The change in response produced by a change in the level of a factor is called the main effect of this factor. Considering Table 6.1, the main effect of Factor A can be interpreted as the difference between the average response of the two factor levels A_1 and A_2:

$$\lambda_A = \frac{60}{2} - \frac{40}{2} = 10 \,.$$

Similarly, the main effect of Factor B is

$$\lambda_B = \frac{70}{2} - \frac{30}{2} = 20 \,.$$

		Factor B		
		B_1	B_2	$\sum$
Factor A	A_1	10	30	40
	A_2	20	40	60
	$\sum$	30	70	100

TABLE 6.1. Two–factorial experiment without interaction.

The effects of Factor A at the two levels of Factor B are

$$\text{for } B_1\text{:}\quad 20 - 10 = 10; \quad \text{for } B_2\text{:}\quad 40 - 30 = 10,$$

and hence identical for both levels of Factor B. For the effect of Factor B we have

$$\text{for } A_1\text{:}\quad 30 - 10 = 20; \quad \text{for } A_2\text{:}\quad 40 - 20 = 20,$$

so that no effect dependent on Factor A can be seen. The response lines are parallel.
The analysis of Table 6.2, however, leads to the following effects:

$$\begin{aligned} \text{main effect } \lambda_A &= \frac{80 - 40}{2} = 20, \\ \text{main effect } \lambda_B &= \frac{90 - 30}{2} = 30, \end{aligned}$$

		Factor B		
		B_1	B_2	$\sum$
Factor A	A_1	10	30	40
	A_2	20	60	80
	$\sum$	30	90	120

TABLE 6.2. Two–factorial experiment with interaction.

effects of Factor A:

$$\text{for } B_1\text{:}\quad 20 - 10 = 10; \quad \text{for } B_2\text{:}\quad 60 - 30 = 30,$$

effects of Factor B:

$$\text{for } A_1\text{:}\quad 30 - 10 = 20; \quad \text{for } A_2\text{:}\quad 60 - 20 = 40.$$

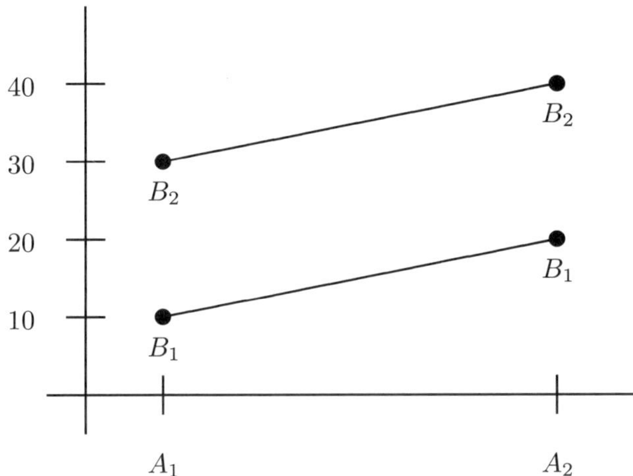

FIGURE 6.1. Two–factorial experiment without interaction

Here the effects depend on the levels of the other factor, the interaction effect amounts to 20. The response lines are no longer parallel (Figure 6.2).

Remark. The term factorial experiment describes the completely crossed combination of the factors (treatments) and not the design of experiment. Factorial experiments may be realized as completely randomized designs of experiments, as Latin squares, etc.

The factorial experiment should be used:

- in pilot studies that analyze the statistical relevance of possible covariates;
- for the determination of bivariate interaction; and
- for the determination of possible rank orders of the factors related to their influence on the response.

Compared to experiments with a single factor, the factorial experiment has the advantage that the main effects may be estimated with the same precision, but with a smaller sample size.

Assume that we want to estimate the main effects A and B as in the above examples. The following one–factor experiment with two repetitions would be appropriate (cf. Montgomery, 1976, p. 124):

$A_1B_1^{(1)}$	$A_1B_2^{(1)}$
$A_2B_1^{(1)}$	

$A_1B_1^{(2)}$	$A_1B_2^{(2)}$
$A_2B_1^{(2)}$	

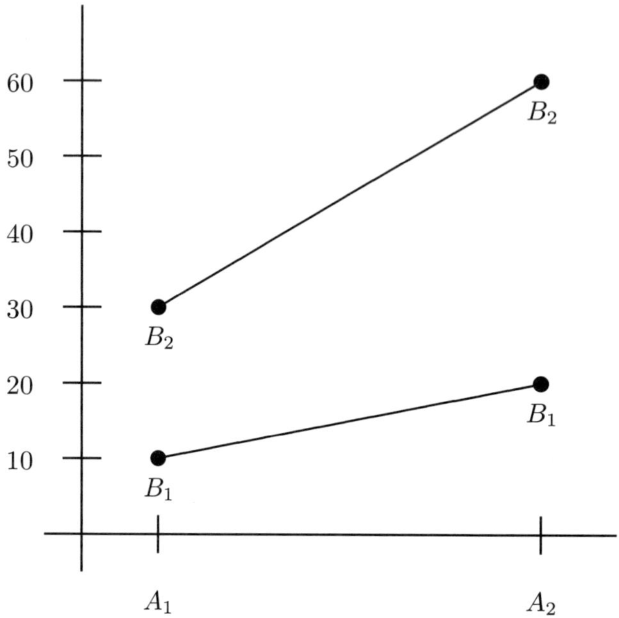

FIGURE 6.2. Two–factorial experiment with interaction.

$$n = 3 + 3 = 6 \text{ observations}$$

$$\text{estimation of } \lambda_A: \; 1/2\left[(A_2B_1^{(1)} - A_1B_1^{(1)}) + (A_2B_1^{(2)} - A_1B_1^{(2)})\right],$$

$$\text{estimation of } \lambda_B: \; 1/2\left[(A_1B_1^{(1)} - A_1B_2^{(1)}) + (A_1B_1^{(2)} - A_1B_2^{(2)})\right].$$

Estimation of the effects with the same precision is achieved by the factorial experiment

A_1B_1	A_1B_2
A_2B_1	A_2B_2

with only $n = 4$ observations according to

$$\lambda_A = 1/2\left[(A_2B_1 - A_1B_1) + (A_2B_2 - A_1B_2)\right]$$

and

$$\lambda_B = 1/2\left[(A_1B_2 - A_1B_1) + (A_2B_2 - A_2B_1)\right].$$

Additionally, the factorial experiment reveals existing interaction and hence leads to an adequate model.

If a present interaction is neglected or not revealed, a serious misinterpretation of the main effects may be the consequence. In principle, if significant interaction is present, then the main effects are of secondary importance since the effect of one factor on the response can no longer be segregated from the other factor.

6.2 Two–Factor Experiments (Fixed Effects)

Suppose that there are a levels of Factor A and b levels of Factor B. For each combination (i, j), r replicates are realized and the design is a completely randomized design. Hence the number of observations equals $N = rab$. The response is described by the linear model

$$\begin{aligned} y_{ijk} = \mu + \alpha_i + \beta_j + (\alpha\beta)_{ij} + \epsilon_{ijk}\,, \\ (i = 1, \ldots, a,\ j = 1, \ldots, b,\ k = 1, \ldots, r)\,. \end{aligned} \tag{6.1}$$

where we have:

y_{ijk}	is the response to the ith level of Factor A and the jth level of Factor B in the kth replicate;
μ	is the overall mean;
α_i	is the effect of the ith level of Factor A;
β_j	is the effect of the jth level of Factor B;
$(\alpha\beta)_{ij}$	is the effect of the interaction of the combination (i, j); and
ϵ_{ijk}	is the random error.

The following assumption is made for $\epsilon' = (\epsilon_{111}, \ldots, \epsilon_{\mathrm{abr}})$:

$$\epsilon \sim N(\mathbf{0}, \sigma^2 I)\,. \tag{6.2}$$

For the fixed effects, we have the following constraints:

$$\sum_{i=1}^{a} \alpha_i = 0\,, \tag{6.3}$$

$$\sum_{j=1}^{b} \beta_j = 0\,, \tag{6.4}$$

$$\sum_{i=1}^{a} (\alpha\beta)_{ij} = \sum_{j=1}^{b} (\alpha\beta)_{ij} = 0\,. \tag{6.5}$$

Remark. If the randomized block design is chosen as the design of experiment, the model (6.1) additionally contains the (additive) block effects ρ_k as random effects with $\rho_k \sim N(0, \sigma_\rho^2)$.

	B					
A	1	2	$\cdots$	b	$\sum$	Means
1	$Y_{11\cdot}$	$Y_{12\cdot}$	$\cdots$	$Y_{1b\cdot}$	$Y_{1\cdot\cdot}$	$y_{1\cdot\cdot}$
2	$Y_{21\cdot}$	$Y_{22\cdot}$	$\cdots$	$Y_{2b\cdot}$	$Y_{2\cdot\cdot}$	$y_{2\cdot\cdot}$
$\vdots$	$\vdots$	$\vdots$		$\vdots$	$\vdots$	$\vdots$
a	$Y_{a1\cdot}$	$Y_{a2\cdot}$	$\cdots$	$Y_{ab\cdot}$	$Y_{a\cdot\cdot}$	$y_{a\cdot\cdot}$
$\sum$	$Y_{\cdot1\cdot}$	$Y_{\cdot2\cdot}$	$\cdots$	$Y_{\cdot b\cdot}$	$Y_{\cdot\cdot\cdot}$	$y_{\cdot\cdot\cdot}$
Means	$y_{\cdot1\cdot}$	$y_{\cdot2\cdot}$	$\cdots$	$y_{\cdot b\cdot}$		

TABLE 6.3. Table of the total response values in the $(A \times B)$–design.

Source	SS	df	MS	F
Factor A	SS_A	$a-1$	MS_A	F_A
Factor B	SS_B	$b-1$	MS_B	F_B
Interaction $A \times B$	$SS_{A\times B}$	$(a-1)(b-1)$	$MS_{A\times B}$	$F_{A\times B}$
Error	SS_{Error}	$N-ab$ $=ab(r-1)$	MS_{Error}	
Total	SS_{Total}	$N-1$		

TABLE 6.4. Analysis of variance table in the $(A \times B)$–design with interaction.

Ordinary Least Squares Estimation of the Parameters

The score function (3.6) in model (6.1) is as follows:

$$S(\theta) = \sum_i \sum_j \sum_k (y_{ijk} - \mu - \alpha_i - \beta_j - (\alpha\beta)_{ij})^2 \tag{6.6}$$

under the constraints (6.3)–(6.5).

Here

$$\theta' = (\mu, \alpha_1, \ldots, \alpha_a, \beta_1, \ldots, \beta_b, (\alpha\beta)_{11}, \ldots, (\alpha\beta)_{ab}) \tag{6.7}$$

is the vector of the unknown parameters. The normal equations, taking the restrictions (6.3)–(6.5) into consideration, can easily be derived

$$\begin{aligned}
-\frac{1}{2}\frac{\partial S(\theta)}{\partial \mu} &= \sum\sum\sum (y_{ijk} - \mu - \alpha_i - \beta_j - (\alpha\beta)_{ij}) \\
&= Y_{\cdot\cdot\cdot} - N\mu = 0, & (6.8) \\
-\frac{1}{2}\frac{\partial S(\theta)}{\partial \alpha_i} &= Y_{i\cdot\cdot} - br\alpha_i - br\mu = 0 \quad (i \text{ fixed}), & (6.9) \\
-\frac{1}{2}\frac{\partial S(\theta)}{\partial \beta_j} &= Y_{\cdot j\cdot} - ar\beta_j - ar\mu = 0 \quad (j \text{ fixed}), & (6.10) \\
-\frac{1}{2}\frac{\partial S(\theta)}{\partial (\alpha\beta)_{ij}} &= Y_{ij\cdot} - r\mu - r\alpha_i - r\beta_j - (\alpha\beta)_{ij}
\end{aligned}$$

$$= 0 \quad (i, j \text{ fixed}) . \tag{6.11}$$

We now obtain the OLS estimates under the constraints (6.3)–(6.5), that is, the conditional OLS estimates

$$\hat{\mu} = Y_{\cdots}/N = y_{\cdots}, \tag{6.12}$$

$$\hat{\alpha}_i = \frac{Y_{i\cdot\cdot}}{br} - \hat{\mu} = y_{i\cdot\cdot} - y_{\cdots}, \tag{6.13}$$

$$\hat{\beta}_j = \frac{Y_{\cdot j\cdot}}{ar} - \hat{\mu} = y_{\cdot j\cdot} - y_{\cdots}, \tag{6.14}$$

$$(\hat{\alpha\beta})_{ij} = \frac{Y_{ij\cdot}}{r} - \hat{\mu} - \hat{\alpha}_i - \hat{\beta}_j = y_{ij\cdot} - y_{i\cdot\cdot} - y_{\cdot j\cdot} + y_{\cdots} . \tag{6.15}$$

The correction term is defined as

$$C = Y_{\cdots}^2/N \tag{6.16}$$

with $N = a\, b\, r$. The sums of squares can now be expressed as follows:

$$\begin{aligned} SS_{\text{Total}} &= \sum\sum\sum (y_{ijk} - y_{\cdots})^2 \\ &= \sum\sum\sum y_{ijk}^2 - C, \end{aligned} \tag{6.17}$$

$$SS_A = \frac{1}{br} \sum_i Y_{i\cdot\cdot}^2 - C, \tag{6.18}$$

$$SS_B = \frac{1}{ar} \sum_j Y_{\cdot j\cdot}^2 - C, \tag{6.19}$$

$$\begin{aligned} SS_{A\times B} &= \frac{1}{r} \sum_i \sum_j Y_{ij\cdot}^2 - \frac{1}{br} \sum_i Y_{i\cdot\cdot}^2 - \frac{1}{ar} \sum_j Y_{\cdot j\cdot}^2 + C \\ &= \left[\frac{1}{r} \sum_i \sum_j Y_{ij\cdot}^2 - C \right] - SS_A - SS_B, \end{aligned} \tag{6.20}$$

$$\begin{aligned} SS_{\text{Error}} &= SS_{\text{Total}} - SS_A - SS_B - SS_{A\times B} \\ &= SS_{\text{Total}} - \left[\frac{1}{r} \sum_i \sum_j Y_{ij\cdot}^2 - C \right] . \end{aligned} \tag{6.21}$$

Remark. The sum of squares between the $a \cdot b$ sums of response $Y_{ij\cdot}$ is also called SS_{Subtotal}, i.e.,

$$SS_{\text{Subtotal}} = \frac{1}{r} \sum_i \sum_j Y_{ij\cdot}^2 - C . \tag{6.22}$$

Hint. In order to ensure that the interaction effect is detectable (and hence $(\alpha\beta)_{ij}$ can be estimated), in the balanced design at least $r = 2$ replicates have to be realized for each combination (i, j). Otherwise, the interaction effect is included in the error and cannot be separated.

Test Procedure

The model (6.1) with interaction is called a *saturated model.* The model without interaction,

$$y_{ijk} = \mu + \alpha_i + \beta_j + \epsilon_{ijk}\,, \tag{6.23}$$

is called the *independence model.*

First, the hypothesis $H_0 : (\alpha\beta)_{ij} = 0$ (for all (i,j)) against $H_1 : (\alpha\beta)_{ij} \neq 0$ (for at least one pair (i,j)) is tested. This corresponds to the model choice *submodel (6.23) compared to the complete model (6.1)* according to our likelihood–ratio test strategy in Chapter 3. The interpretation of inferences obtained from the factorial experiment depends on the result of this test.

H_0 is rejected if

$$F_{A\times B} = \frac{MS_{A\times B}}{MS_{\text{Error}}} > F_{(a-1)(b-1),ab(r-1);1-\alpha}\,. \tag{6.24}$$

The interaction effects are significant in the case of a rejection of H_0. The main effects are of no importance, no matter whether they are significant or not.

Remark: This test procedure is a kind of philosophy representing one school. One could also consider a less dogmatic idea. If the main effect—being, for example, the average over the levels of another factor—is sensible within an application the test could also be interpretable and meaningful even in the presence of an interaction.

If, however, H_0 is not rejected, then the test results for $H_0 : \alpha_i = 0$ against $H_1 : \alpha_i \neq 0$ (for at least one i) with $F_A = MS_A/MS_{\text{Error}}$ and for $H_0 : \beta_j = 0$ against $H_1 : \beta_j \neq 0$ (for at least one j) with $F_B = MS_B/MS_{\text{Error}}$ are of importance for the interpretation in model (6.23). If only one factor effect is significant (e.g., Factor A), then the model is reduced further to a balanced one–factor model with a factor levels and br replicates each

$$y_{ijk} = \mu + \alpha_i + \epsilon_{ijk}\,. \tag{6.25}$$

Example 6.1. The influence of two factors A (fertilizer) and B (irrigation) on the yield of a type of grain is to be analyzed in a pilot study. The Factors A and B are applied at two levels (low, high) and $r = 2$ replicates each. Hence, we have $a = b = r = 2$ and $N = abr = 8$. The experimental units (plants) are assigned to the treatments at random. From Tables 6.5 and 6.6, we calculate

$$\begin{aligned} C &= 77.6^2/8 = 752.72, \\ SS_{\text{Total}} &= 866.92 - C = 114.20, \\ SS_A &= 1/4(39.6^2 + 38.0^2) - C \end{aligned}$$

$$\begin{aligned}
&= 753.04 - 752.72 = 0.32, \\
SS_B &= 1/4(26.4^2 + 51.2^2) - C \\
&= 892.60 - 752.72 = 76.88, \\
SS_{\text{Subtotal}} &= 1/2(17.8^2 + 21.8^2 + 8.6^2 + 29.4^2) - C \\
&= 865.20 - 752.72 = 112.48, \\
SS_{A\times B} &= SS_{\text{Subtotal}} - SS_A - SS_B = 35.28, \\
SS_{\text{Error}} &= 114.20 - 35.28 - 0.32 - 76.88 \\
&= 1.72\,.
\end{aligned}$$

		Factor B			
		1		2	
Factor A	1	8.6	9.2	10.4	11.4
	2	4.7	3.9	14.1	15.3

TABLE 6.5. Response values.

		Factor B		
		1	2	$\sum$
Factor A	1	17.8	21.8	39.6
	2	8.6	29.4	38.0
	$\sum$	26.4	51.2	77.6

TABLE 6.6. Total response.

Source	SS	df	MS	F	
A	0.32	1	0.32	0.74	
B	76.88	1	76.88	178.79	*
$A \times B$	35.28	1	35.28	82.05	*
Error	1.72	4	0.43		
Total	114.20	7			

TABLE 6.7. Analysis of variance table for Example 6.1.

Result: The test for interaction leads to a rejection of H_0 : *no interaction* with $F_{1,4} = 82.05$ ($F_{1,4;0.95} = 7.71$). A reduction to an experiment with a single factor is not possible, in spite of the nonsignificant main effect A.

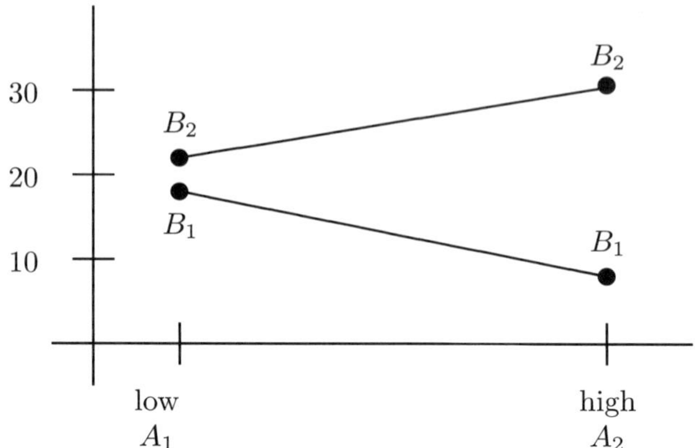

FIGURE 6.3. Interaction in Example 6.1.

6.3 Two–Factor Experiments in Effect Coding

In the above section, we have derived the parameter estimates of the components of θ (6.7) by minimizing the error sum of squares under the linear restrictions $\sum_i \alpha_i = 0$, $\sum_j \beta_j = 0$, and $\sum_i (\alpha\beta)_{ij} = \sum_j (\alpha\beta)_{ij} = 0$. This corresponds to the conditional OLS estimate $b(R)$ from (3.76).

We now want to achieve a reduction in the number of parameters. This is done by an alternative parametrization that includes the restrictions already in the model. The result is a set of parameters that corresponds to a design matrix of full column rank. The parameter estimation is now achieved by the OLS estimate b_0. For this purpose we use the so–called *effect coding* of categories. The effect coding for Factor A at $a = 3$ categories (levels) is as follows:

$$x_i^A = \begin{cases} 1 & \text{for category } i \quad (i = 1, \ldots, a-1), \\ -1 & \text{for category } a, \\ 0 & \text{else,} \end{cases}$$

so that

$$\alpha_a = -\sum_{i=1}^{a-1} \alpha_i \,, \tag{6.26}$$

or, expressed differently,

$$\sum_{i=1}^{a} \alpha_i = 0 \,. \tag{6.27}$$

Example: Assume Factor A has $a = 3$ levels, A_1: low, A_2: medium, A_3: high. The original link of design and parameters is as follows:

$$\begin{array}{l} \text{low:} \\ \text{medium:} \\ \text{high:} \end{array} \begin{pmatrix} 1 & 0 & 0 \\ 0 & 1 & 0 \\ 0 & 0 & 1 \end{pmatrix} \begin{pmatrix} \alpha_1 \\ \alpha_2 \\ \alpha_3 \end{pmatrix} \quad \text{and } \alpha_1 + \alpha_2 + \alpha_3 = 0.$$

If effect coding is applied, we obtain

$$\begin{array}{l} \text{low:} \\ \text{medium:} \\ \text{high:} \end{array} \begin{pmatrix} 1 & 0 \\ 0 & 1 \\ -1 & -1 \end{pmatrix} \begin{pmatrix} \alpha_1 \\ \alpha_2 \end{pmatrix}.$$

Case $a = b = 2$

In the case of a linear model with two two–level prognostic Factors A and B, we have, for fixed k ($k = 1, \ldots, r$), the following parametrization (cf. Toutenburg, 1992a, p. 255):

$$\begin{pmatrix} y_{11k} \\ y_{12k} \\ y_{21k} \\ y_{22k} \end{pmatrix} = \begin{pmatrix} 1 & 1 & 1 & 1 \\ 1 & 1 & -1 & -1 \\ 1 & -1 & 1 & -1 \\ 1 & -1 & -1 & 1 \end{pmatrix} \begin{pmatrix} \mu \\ \alpha_1 \\ \beta_1 \\ (\alpha\beta)_{11} \end{pmatrix} + \begin{pmatrix} \epsilon_{11k} \\ \epsilon_{12k} \\ \epsilon_{21k} \\ \epsilon_{22k} \end{pmatrix}. \qquad (6.28)$$

Here we get the constraints immediately

$$\begin{aligned} \alpha_1 + \alpha_2 = 0 &\;\Rightarrow\; \alpha_2 = -\alpha_1, \\ \beta_1 + \beta_2 = 0 &\;\Rightarrow\; \beta_2 = -\beta_1, \\ (\alpha\beta)_{11} + (\alpha\beta)_{12} = 0 &\;\Rightarrow\; (\alpha\beta)_{12} = -(\alpha\beta)_{11}, \\ (\alpha\beta)_{11} + (\alpha\beta)_{21} = 0 &\;\Rightarrow\; (\alpha\beta)_{21} = -(\alpha\beta)_{11}, \\ (\alpha\beta)_{21} + (\alpha\beta)_{22} = 0 &\;\Rightarrow\; (\alpha\beta)_{22} = -(\alpha\beta)_{21} = (\alpha\beta)_{11}\,. \end{aligned}$$

Of the original nine parameters, only four remain in the model. The others are calculated from these equations. The following notation is used:

$$\begin{aligned} \underset{r,4}{X_{11}} &= (\mathbf{1}_r \;\; \mathbf{1}_r \;\; \mathbf{1}_r \;\; \mathbf{1}_r), \\ \underset{r,4}{X_{12}} &= (\mathbf{1}_r \;\; \mathbf{1}_r \; -\mathbf{1}_r \; -\mathbf{1}_r), \\ \underset{r,4}{X_{21}} &= (\mathbf{1}_r \; -\mathbf{1}_r \;\; \mathbf{1}_r \; -\mathbf{1}_r), \\ \underset{r,4}{X_{22}} &= (\mathbf{1}_r \; -\mathbf{1}_r \; -\mathbf{1}_r \;\; \mathbf{1}_r), \\ \underset{4,4r}{X'} &= (X'_{11} \;\; X'_{12} \;\; X'_{21} \;\; X'_{22}), \end{aligned}$$

$$\theta_0' = (\mu,\ \alpha_1,\ \beta_1,\ (\alpha\beta)_{11}),$$

$$y_{ij} = \begin{pmatrix} y_{ij1} \\ \vdots \\ y_{ijr} \end{pmatrix}, \quad \epsilon_{ij} = \begin{pmatrix} \epsilon_{ij1} \\ \vdots \\ \epsilon_{ijr} \end{pmatrix},$$

$$y = \begin{pmatrix} y_{11} \\ y_{12} \\ y_{21} \\ y_{22} \end{pmatrix}, \quad \epsilon = \begin{pmatrix} \epsilon_{11} \\ \epsilon_{12} \\ \epsilon_{21} \\ \epsilon_{22} \end{pmatrix}.$$

In the case of $a = b = 2$ and r replicates, and considering the restrictions (6.3), (6.4), (6.5), the two–factorial model (6.1) can alternatively be expressed in effect coding:

$$y = X\theta_0 + \epsilon\,. \tag{6.29}$$

The OLS estimate of θ_0 is

$$\hat{\theta}_0 = (X'X)^{-1}X'y\,.$$

We now calculate $\hat{\theta}_0$:

$$\begin{aligned} \underset{4,4}{X'X} &= X_{11}'X_{11} + X_{12}'X_{12} + X_{21}'X_{21} + X_{22}'X_{22} \\ &= 4rI_4\,, \end{aligned}$$

$$\begin{aligned} X'y &= \begin{pmatrix} Y_{\cdot\cdot\cdot} \\ Y_{1\cdot\cdot} - Y_{2\cdot\cdot} \\ Y_{\cdot 1\cdot} - Y_{\cdot 2\cdot} \\ (Y_{11\cdot} + Y_{22\cdot}) - (Y_{12\cdot} + Y_{21\cdot}) \end{pmatrix} \\ &= \begin{pmatrix} Y_{\cdot\cdot\cdot} \\ 2Y_{1\cdot\cdot} - Y_{\cdot\cdot\cdot} \\ 2Y_{\cdot 1\cdot} - Y_{\cdot\cdot\cdot} \\ (Y_{11\cdot} + Y_{22\cdot}) - (Y_{12\cdot} + Y_{21\cdot}) \end{pmatrix}. \end{aligned} \tag{6.30}$$

With $(X'X)^{-1} = 1/4rI$, the OLS estimate $\hat{\theta}_0 = (X'X)^{-1}X'y$ can be written in detail as (cf. (6.12)–(6.15))

$$\begin{pmatrix} \hat{\mu} \\ \hat{\alpha}_1 \\ \hat{\beta}_1 \\ (\widehat{\alpha\beta})_{11} \end{pmatrix} = \begin{pmatrix} y_{\cdot\cdot\cdot} \\ y_{1\cdot\cdot} - y_{\cdot\cdot\cdot} \\ y_{\cdot 1\cdot} - y_{\cdot\cdot\cdot} \\ y_{11\cdot} - y_{1\cdot\cdot} - y_{\cdot 1\cdot} + y_{\cdot\cdot\cdot} \end{pmatrix}. \tag{6.31}$$

The first three relations in (6.31) can easily be detected. The transition from the fourth row in (6.30) to the fourth row in (6.31), however, has to be proven in detail.

With $a = b = 2$, we have

$$
\begin{aligned}
y_{11\cdot} - y_{1\cdot\cdot} - y_{\cdot 1\cdot} + y_{\cdot\cdot\cdot}
&= \frac{Y_{11\cdot}}{r} - \left[\frac{Y_{11\cdot}}{br} + \frac{Y_{12\cdot}}{br}\right] - \left[\frac{Y_{11\cdot}}{ar} + \frac{Y_{21\cdot}}{ar}\right] + \frac{Y_{11\cdot} + Y_{12\cdot} + Y_{21\cdot} + Y_{22\cdot}}{abr} \\
&= \frac{Y_{11\cdot}}{r}\left(1 - \frac{1}{b} - \frac{1}{a} + \frac{1}{ab}\right) - \frac{Y_{12\cdot}}{br}\left(1 - \frac{1}{a}\right) - \frac{Y_{21\cdot}}{ar}\left(1 - \frac{1}{b}\right) + \frac{Y_{22\cdot}}{abr} \\
&= \frac{Y_{11\cdot}}{r}\left(\frac{ab - a - b + 1}{ab}\right) + \frac{Y_{22\cdot}}{abr} - \frac{Y_{12\cdot}}{abr}(a-1) - \frac{Y_{21\cdot}}{abr}(b-1) \\
&= \frac{1}{4r}\left[(Y_{11\cdot} + Y_{22\cdot}) - (Y_{12\cdot} + Y_{21\cdot})\right].
\end{aligned}
$$

Remark. Here we wish to point out an important characteristic of the effect coding in the case of equal numbers r of replications. First, we write the matrix X in a different form

$$
X = \begin{pmatrix} X_{11} \\ X_{12} \\ X_{21} \\ X_{22} \end{pmatrix} = \begin{pmatrix} \mathbf{1}_r & \mathbf{1}_r & \mathbf{1}_r & \mathbf{1}_r \\ \mathbf{1}_r & \mathbf{1}_r & -\mathbf{1}_r & -\mathbf{1}_r \\ \mathbf{1}_r & -\mathbf{1}_r & \mathbf{1}_r & -\mathbf{1}_r \\ \mathbf{1}_r & -\mathbf{1}_r & -\mathbf{1}_r & \mathbf{1}_r \end{pmatrix}
= \left(\underset{4r,1}{x_\mu} \quad \underset{4r,1}{x_{\alpha_1}} \quad \underset{4r,1}{x_{\beta_1}} \quad \underset{4r,1}{x_{(\alpha\beta)_{11}}} \right)
$$

so that

$$
\begin{aligned}
x'_\mu x_\mu &= x'_{\alpha_1} x_{\alpha_1} = x'_{\beta_1} x_{\beta_1} = x'_{(\alpha\beta)_{11}} x_{(\alpha\beta)_{11}} = 4r, \\
x'_\mu x_{\alpha_1} &= x'_\mu x_{\beta_1} = x'_\mu x_{(\alpha\beta)_{11}} = 0, \\
x'_{\alpha_1} x_{\beta_1} &= x'_{\alpha_1} x_{(\alpha\beta)_{11}} = 0, \\
x'_{\beta_1} x_{(\alpha\beta)_{11}} &= 0.
\end{aligned}
$$

Hence, as we mentioned before, the following holds

$$
X'X = \begin{pmatrix} x'_\mu \\ x'_{\alpha_1} \\ x'_{\beta_1} \\ x_{(\alpha\beta)_{11}} \end{pmatrix} \left(x_\mu \; x_{\alpha_1} \; x_{\beta_1} \; x_{(\alpha\beta)_{11}} \right) = 4rI_4 .
$$

The vectors that belong to different effect groups $(\mu, \alpha, \beta, (\alpha\beta))$ are orthogonal. This property remains true in general for effect coding.

General Cases: $a > 2, b > 2$

In the general case of a two–factorial model with interaction with:

Factor A: a levels; and

Factor B: b levels;

the parameter vector (after taking the constraints into account, i.e., in effect coding) is as follows

$$\theta_0' = (\mu, \alpha_1, \ldots, \alpha_{a-1}, \beta_1, \ldots, \beta_{b-1}, (\alpha\beta)_{1,1}, \ldots, (\alpha\beta)_{a-1,b-1}) \tag{6.32}$$

and the design matrix is

$$X = \left(x_\mu \; X_\alpha \; X_\beta \; X_{(\alpha\beta)} \right) . \tag{6.33}$$

Here the column vectors of a submatrix are orthogonal to the column vectors of every other submatrix, e.g.,

$$X_\alpha' X_\beta = \mathbf{0} .$$

The matrix $X'X$ is now block-diagonal

$$X'X = \operatorname{diag}\Big(x_\mu' x_\mu,\; X_\alpha' X_\alpha,\; X_\beta' X_\beta,\; X_{(\alpha\beta)}' X_{(\alpha\beta)}\Big)$$

so that

$$(X'X)^{-1} = \operatorname{diag}\Big((x_\mu' x_\mu)^{-1},\; (X_\alpha' X_\alpha)^{-1},\; (X_\beta' X_\beta)^{-1},\; (X_{(\alpha\beta)}' X_{(\alpha\beta)})^{-1}\Big) \tag{6.34}$$

and the OLS estimate $\hat{\theta}_0$ can be written as

$$\hat{\theta}_0 = \begin{pmatrix} \hat{\mu} \\ \hat{\alpha} \\ \hat{\beta} \\ \widehat{(\alpha\beta)} \end{pmatrix} = \begin{pmatrix} (x_\mu' x_\mu)^{-1} x_\mu' y \\ (X_\alpha' X_\alpha)^{-1} X_\alpha' y \\ (X_\beta' X_\beta)^{-1} X_\beta' y \\ (X_{(\alpha\beta)}' X_{(\alpha\beta)})^{-1} X_{(\alpha\beta)}' y \end{pmatrix} . \tag{6.35}$$

For the covariance matrix of $\hat{\theta}$, we get a block-diagonal structure as well:

$$\mathrm{V}(\hat{\theta}) = \sigma^2 \begin{pmatrix} (x_\mu' x_\mu)^{-1} & \mathbf{0} & \mathbf{0} & \mathbf{0} \\ \mathbf{0} & (X_\alpha' X_\alpha)^{-1} & \mathbf{0} & \mathbf{0} \\ \mathbf{0} & \mathbf{0} & (X_\beta' X_\beta)^{-1} & \mathbf{0} \\ \mathbf{0} & \mathbf{0} & \mathbf{0} & (X_{(\alpha\beta)}' X_{(\alpha\beta)})^{-1} \end{pmatrix} . \tag{6.36}$$

This shows that the estimation vectors $\hat{\mu}, \hat{\alpha}, \hat{\beta}, \widehat{(\alpha\beta)}$ are uncorrelated and independent in the case of normal errors. From this it follows that the estimates $\hat{\mu}, \hat{\alpha}$ and $\hat{\beta}$ in model (6.1), with interaction and the estimates in the independence model (6.23), are identical. Hence, the estimates for one parameter group—e.g., the main effects of Factor B—are always the same, no matter whether the other parameters are contained in the model or not. Again, this holds only for balanced data.

In the case of rejection of $\mathrm{H}_0 : (\alpha\beta)_{ij} = 0$, σ^2 is estimated by

$$MS_{\text{Error}} = \frac{SS_{\text{Error}}}{N - ab} = \frac{1}{N - ab}(SS_{\text{Total}} - SS_A - SS_B - SS_{A\times B})$$

(cf. Table 6.4 and (6.21)). If H_0 is not rejected, then the independence model (6.23) holds and we have

$$SS_{\text{Error}} = SS_{\text{Total}} - SS_A - SS_B$$

for $N-1-(a-1)-(b-1)=N-a-b+1$ degrees of freedom.

The model (6.1) with interaction corresponds to the parameter space Ω, according to our notation in Chapter 3. The independence model is the submodel of the parameter space $\omega \subset \Omega$. With (B.77) we have

$$\hat{\sigma}_\omega - \hat{\sigma}^2_\Omega \geq 0\,. \tag{6.37}$$

Applied to our problem, we find

$$\hat{\sigma}^2{}_\Omega = \frac{SS_{\text{Total}} - SS_A - SS_B - SS_{A\times B}}{N-ab} \tag{6.38}$$

and

$$\hat{\sigma}^2_\omega = \frac{SS_{\text{Total}} - SS_A - SS_B}{N-ab+(a-1)(b-1)}\,. \tag{6.39}$$

Interpretation. In the independence model σ^2 is estimated by (6.39). Hence, the confidence intervals of the parameter estimates $\hat{\mu}, \hat{\alpha}$, and $\hat{\beta}$ are larger when compared with those obtained from the model with interaction. On the other hand, the parameter estimates themselves (which correspond to the center points of the confidence intervals) stay unchanged. Thus, the precision of the estimates $\hat{\mu}, \hat{\alpha}$, and $\hat{\beta}$ decreases. Simultaneously the test statistics change so that in the case of a rejection of the saturated model (6.1), tests of significance for μ, α, and β, based on the analysis of variance table for the independence model, are to be carried out.

Cases $a=2, b=3$

Considering the constraints (6.3)–(6.5), the model in effect coding is as follows:

$$\begin{pmatrix} y_{11} \\ y_{12} \\ y_{13} \\ y_{21} \\ y_{22} \\ y_{23} \end{pmatrix} = \begin{pmatrix} \mathbf{1}_r & \mathbf{1}_r & \mathbf{1}_r & \mathbf{0} & \mathbf{1}_r & \mathbf{0} \\ \mathbf{1}_r & \mathbf{1}_r & \mathbf{0} & \mathbf{1}_r & \mathbf{0} & \mathbf{1}_r \\ \mathbf{1}_r & \mathbf{1}_r & -\mathbf{1}_r & -\mathbf{1}_r & -\mathbf{1}_r & -\mathbf{1}_r \\ \mathbf{1}_r & -\mathbf{1}_r & \mathbf{1}_r & \mathbf{0} & -\mathbf{1}_r & \mathbf{0} \\ \mathbf{1}_r & -\mathbf{1}_r & \mathbf{0} & \mathbf{1}_r & \mathbf{0} & -\mathbf{1}_r \\ \mathbf{1}_r & -\mathbf{1}_r & -\mathbf{1}_r & -\mathbf{1}_r & \mathbf{1}_r & \mathbf{1}_r \end{pmatrix} \begin{pmatrix} \mu \\ \alpha_1 \\ \beta_1 \\ \beta_2 \\ (\alpha\beta)_{11} \\ (\alpha\beta)_{12} \end{pmatrix} + \begin{pmatrix} \epsilon_{11} \\ \epsilon_{12} \\ \epsilon_{13} \\ \epsilon_{21} \\ \epsilon_{22} \\ \epsilon_{23} \end{pmatrix}. \tag{6.40}$$

Here we once again find the constraints immediately:

$$\begin{aligned} \alpha_1 + \alpha_2 = 0 \quad &\Rightarrow \quad \alpha_2 = -\alpha_1, \\ \beta_1 + \beta_2 + \beta_3 = 0 \quad &\Rightarrow \quad \beta_3 = -\beta_1 - \beta_2, \\ (\alpha\beta)_{11} + (\alpha\beta)_{21} = 0 \quad &\Rightarrow \quad (\alpha\beta)_{21} = -(\alpha\beta)_{11}, \end{aligned}$$

$$\begin{aligned}
(\alpha\beta)_{12} + (\alpha\beta)_{22} = 0 \quad &\Rightarrow \quad (\alpha\beta)_{22} = -(\alpha\beta)_{12}, \\
(\alpha\beta)_{13} + (\alpha\beta)_{23} = 0 \quad &\Rightarrow \quad (\alpha\beta)_{23} = -(\alpha\beta)_{13}, \\
(\alpha\beta)_{11} + (\alpha\beta)_{12} + (\alpha\beta)_{13} = 0 \quad &\Rightarrow \quad (\alpha\beta)_{13} = -(\alpha\beta)_{11} - (\alpha\beta)_{12}, \\
(\alpha\beta)_{21} + (\alpha\beta)_{22} + (\alpha\beta)_{23} = 0 \quad &\Rightarrow \quad (\alpha\beta)_{23} = -(\alpha\beta)_{21} - (\alpha\beta)_{22}, \\
&\qquad\qquad = (\alpha\beta)_{11} + (\alpha\beta)_{12},
\end{aligned}$$

so that, of the original 12 parameters, only six remain in the model

$$\theta_0' = (\mu, \alpha_1, \beta_1, \beta_2, (\alpha\beta)_{11}, (\alpha\beta)_{12}) \ . \tag{6.41}$$

We now take advantage of the orthogonality of the submatrices and apply (6.35) for the determination of the OLS estimates. We thus have

$$\begin{aligned}
\hat{\mu} = (x_\mu' x_\mu)^{-1} x_\mu' y &= \frac{1}{6r} Y_{\cdots} = y_{\cdots}\,, \\
\hat{\alpha}_1 = (x_\alpha' x_\alpha)^{-1} x_\alpha' y &= \frac{1}{6r}(Y_{1\cdot\cdot} - Y_{2\cdot\cdot}) \\
&= \frac{1}{6r}(2Y_{1\cdot\cdot} - Y_{\cdots}) \\
&= y_{1\cdot\cdot} - y_{\cdots}\,, \\
\begin{pmatrix} \hat{\beta}_1 \\ \hat{\beta}_2 \end{pmatrix} &= \left(X_\beta' X_\beta\right)^{-1} X_\beta' y \\
&= \begin{pmatrix} 4r & 2r \\ 2r & 4r \end{pmatrix}^{-1} \begin{pmatrix} Y_{11\cdot} - Y_{13\cdot} + Y_{21\cdot} - Y_{23\cdot} \\ Y_{12\cdot} - Y_{13\cdot} + Y_{22\cdot} - Y_{23\cdot} \end{pmatrix} \\
&= \frac{1}{6r} \begin{pmatrix} 2 & -1 \\ -1 & 2 \end{pmatrix} \begin{pmatrix} Y_{\cdot1\cdot} - Y_{\cdot3\cdot} \\ Y_{\cdot2\cdot} - Y_{\cdot3\cdot} \end{pmatrix} \\
&= \frac{1}{6r} \begin{pmatrix} 2Y_{\cdot1\cdot} - Y_{\cdot2\cdot} - Y_{\cdot3\cdot} \\ 2Y_{\cdot2\cdot} - Y_{\cdot1\cdot} - Y_{\cdot3\cdot} \end{pmatrix} \\
&= \begin{pmatrix} y_{\cdot1\cdot} - y_{\cdots} \\ y_{\cdot2\cdot} - y_{\cdots} \end{pmatrix},
\end{aligned}$$

since, for instance,

$$\begin{aligned}
\frac{1}{6r}(2Y_{\cdot1\cdot} - Y_{\cdot2\cdot} - Y_{\cdot3\cdot}) &= \frac{3Y_{\cdot1\cdot} - Y_{\cdots}}{6r} \\
&= y_{\cdot1\cdot} - y_{\cdots}\,,
\end{aligned}$$

$$\begin{aligned}
\begin{pmatrix} \widehat{(\alpha\beta)}_{11} \\ \widehat{(\alpha\beta)}_{12} \end{pmatrix} &= \frac{1}{6r} \begin{pmatrix} 2 & -1 \\ -1 & 2 \end{pmatrix} \begin{pmatrix} Y_{11\cdot} - Y_{13\cdot} - Y_{21\cdot} + Y_{23\cdot} \\ Y_{12\cdot} - Y_{13\cdot} - Y_{22\cdot} + Y_{23\cdot} \end{pmatrix} \\
&= \frac{1}{6r} \begin{pmatrix} 2Y_{11\cdot} - Y_{13\cdot} - 2Y_{21\cdot} + Y_{23\cdot} - Y_{12\cdot} + Y_{22\cdot} \\ -Y_{11\cdot} - Y_{13\cdot} + Y_{21\cdot} + Y_{23\cdot} + 2Y_{12\cdot} - 2Y_{22\cdot} \end{pmatrix} \\
&= \begin{pmatrix} y_{11\cdot} - y_{1\cdot\cdot} - y_{\cdot1\cdot} + y_{\cdots} \\ y_{12\cdot} - y_{1\cdot\cdot} - y_{\cdot2\cdot} + y_{\cdots} \end{pmatrix}.
\end{aligned}$$

Example 6.2. A designed experiment is to analyze the effect of different concentrations of phosphate in a combination fertilizer (Factor B) on the yield of two types of beans (Factor A). A factorial experiment with two factors and fixed effects is chosen:

Factor A: A_1: type of beans I,
A_2: type of beans II;
Factor B: B_1: no phosphate,
B_2: 10% per unit,
B_3: 30% per unit.

Hence, in the case of the two–factor approach we have the six treatments A_1B_1, A_1B_2, A_1B_3, A_2B_1, A_2B_2, and A_2B_3. In order to be able to estimate the error variance, the treatments have to be repeated. Here we choose the completely randomized design of experiment with four replicates each. The response values are summarized in Table 6.8.

	B_1	B_2	B_3	Sum
A_1	15	18	22	
	17	19	29	
	14	20	31	
	16	21	35	
Sum	62	78	117	257
A_2	13	17	18	
	9	19	22	
	8	18	24	
	12	18	23	
Sum	42	72	87	201
Sum	104	150	204	458

TABLE 6.8. Response in the $(A \times B)$–design (Example 6.2).

We calculate the sums of squares ($a = 2$, $b = 3$, $r = 4$, $N = 3 \cdot 3 \cdot 4 = 24$):

$$\begin{aligned}
C &= Y_{\cdots}^2/N = 458^2/24 = 8740.17, \\
SS_{\text{Total}} &= (15^2 + 17^2 + \cdots + 23^2) - C \\
&= 9672 - C = 931.83, \\
SS_A &= \frac{1}{3 \cdot 4}(257^2 + 201^2) - C \\
&= 8870.83 - C = 130.66, \\
SS_B &= \frac{1}{2 \cdot 4}(104^2 + 150^2 + 204^2) - C \\
&= 9366.50 - C = 626.33, \\
SS_{\text{Subtotal}} &= 1/4(62^2 + 78^2 + \cdots + 87^2) - C \\
&= 9533.50 - C = 793.33,
\end{aligned}$$

$$\begin{aligned} SS_{A\times B} &= SS_{\text{Subtotal}} - SS_A - SS_B \\ &= 36.34 \\ SS_{\text{Error}} &= SS_{\text{Total}} - SS_{\text{Subtotal}} = 138.50\,. \end{aligned}$$

	SS	df	MS	F
Factor A	130.66	1	130.66	16.99 *
Factor B	626.33	2	313.17	40.72 *
$A \times B$	36.34	2	18.17	2.36
Error	138.50	18	7.69	
Total	931.83	23		

TABLE 6.9. Analysis of variance table for Table 6.8.

The test strategy starts by testing H_0 : *no interaction.* The test statistic is

$$F_{A\times B} = F_{2,18} = \frac{18.17}{7.69} = 2.36\,.$$

The critical value is

$$F_{2,18;0.95} = 3.55\,.$$

Hence, the interaction is not significant at the 5% level.

	SS	df	MS	F
Factor A	130.66	1	130.66	14.95 *
Factor B	626.33	2	313.17	35.83 *
Error	174.84	20	8.74	
Total	931.83	23		

TABLE 6.10. Analysis of variance table for Table 6.8 after omitting the interaction (independence model).

The test for significance of the main effects and the interaction effect in Table 6.9 is based on model (6.1) with interaction. The test statistics for $\text{H}_0 : \alpha_i = 0$, $\text{H}_0 : \beta_i = 0$, and $\text{H}_0 : (\alpha\beta)_{ij} = 0$ are independent. We did not reject $\text{H}_0 : (\alpha\beta)_{ij} = 0$ (cf. Figure 6.4). This leads us back to the independence model (6.23) and we test the significance of the main effects according to Table 6.10. Here both effects are significant as well.

6.4 Two–Factorial Experiment with Block Effects

We now realize the factorial design with Factors A (at a levels) and B (at b levels) as a randomized block design with ab observations for each block

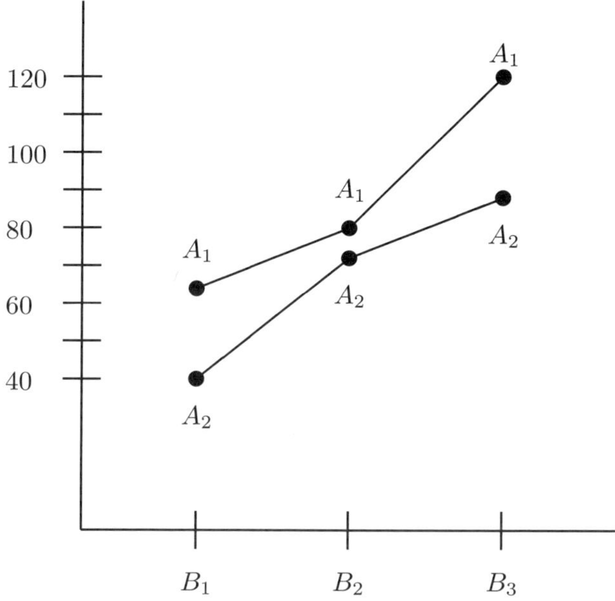

FIGURE 6.4. Interaction type × fertilization (not significant).

(Table 6.11). The appropriate linear model with interaction is then of the following form:

$$y_{ijk} = \mu + \alpha_i + \beta_j + \rho_k + (\alpha\beta)_{ij} + \epsilon_{ijk} \qquad (6.42)$$
$$(i = 1, \ldots, a,\ j = 1, \ldots, b,\ k = 1, \ldots, r).$$

Here ρ_k $(k = 1, \ldots, r)$ is the kth block effect and the constraints $\sum_{k=1}^{r} \rho_k = 0$ for fixed effects hold. The other parameters are the same as in model (6.1). In the case of random block effects we assume $\rho' = (\rho_1, \ldots, \rho_r) \sim N(\mathbf{0}, \sigma_\rho^2\, I)$ and $\mathrm{E}(\epsilon\rho') = \mathbf{0}$. Let

$$Y_{ij\cdot} = \sum_{k=1}^{r} y_{ijk} \qquad (6.43)$$

be the total response of the factor combination over all r blocks. The error sum of squares SS_{Total} (6.17), SS_A (6.18), SS_B (6.19), and $SS_{A\times B}$ (6.20) remain unchanged. For the additional block effect, we calculate

$$SS_{\text{Block}} = \frac{1}{ab} \sum_{k=1}^{r} Y_{\cdot\cdot k}^2 - C\,. \qquad (6.44)$$

The sum of squares SS_{Error} is now

$$SS_{\text{Error}} = SS_{\text{Total}} - SS_A - SS_B - SS_{A\times B} - SS_{\text{Block}}\,. \qquad (6.45)$$

The analysis of variance is shown in Table 6.12.

	Factor B				
Factor A	1	2	$\cdots$	b	Sum
1	$Y_{11\cdot}$	$Y_{12\cdot}$	$\cdots$	$Y_{1b\cdot}$	$Y_{1\cdot\cdot}$
2	$Y_{21\cdot}$	$Y_{22\cdot}$	$\cdots$	$Y_{2b\cdot}$	$Y_{2\cdot\cdot}$
$\vdots$	$\vdots$	$\vdots$		$\vdots$	$\vdots$
a	$Y_{a1\cdot}$	$Y_{a2\cdot}$	$\cdots$	$Y_{ab\cdot}$	$Y_{a\cdot\cdot}$
Sum	$Y_{\cdot 1\cdot}$	$Y_{\cdot 2\cdot}$	$\cdots$	$Y_{\cdot b\cdot}$	$Y_{\cdots}$

TABLE 6.11. Two–factorial randomized block design.

Source	SS	df	MS	F
Factor A	SS_A	$a-1$	MS_A	F_A
Factor B	SS_B	$b-1$	MS_B	F_B
$A \times B$	$SS_{A\times B}$	$(a-1)(b-1)$	$MS_{A\times B}$	$F_{A\times B}$
Block	SS_{Block}	$r-1$	MS_{Block}	F_{Block}
Error	SS_{Error}	$(r-1)(ab-1)$	MS_{Error}	
Total	SS_{Total}	$rab-1$		

TABLE 6.12. Analysis of variance table in the $A \times B$-design (6.42) with interaction and block effects.

The interpretation of the model with block effects is done in the same manner as for the model without block effects. In the case of at least one significant interaction, it is not possible to interpret the main effects—including the block effect—separately.

If $\text{H}_0 : (\alpha\beta)_{ij} = 0$ is not rejected, then an independence model with the three main effects (A, B, and block) holds, if these effects are significant. Compared to model (6.23), the parameter estimates $\hat{\alpha}$ and $\hat{\beta}$ are more precise, due to the reduction of the variance achieved by the block effect.

Example 6.3. The experiment in Example 6.2 is now designed as a randomized block design with $r = 4$ blocks. The response values are shown in Table 6.13 and the total response is given in Tables 6.14 and 6.15.

We calculate (with $C = 8740.17$)

$$\begin{aligned} SS_{\text{Block}} &= \frac{1}{2 \cdot 3}(103^2 + 115^2 + 115^2 + 125^2) - C \\ &= 8780.67 - C = 40.50 \end{aligned}$$

and

$$SS_{\text{Error}} = 98.00\,.$$

The analysis of variance table (Table 6.16) shows that with $F_{2,15;0.95} = 3.68$ the interaction effect is once again not significant. In the reduced model

$$y_{ijk} = \mu + \alpha_i + \beta_j + \rho_k + \epsilon_{ijk} \tag{6.46}$$

we test the main effects (Table 6.17).

I	II	III	IV
A_2B_2 17	A_1B_1 17	A_1B_3 31	A_2B_1 12
A_1B_3 22	A_2B_3 22	A_2B_1 8	A_1B_2 21
A_1B_1 15	A_1B_2 19	A_1B_2 20	A_2B_3 23
A_2B_1 13	A_2B_2 19	A_2B_2 18	A_1B_3 35
A_1B_2 18	A_2B_1 9	A_1B_1 14	A_2B_2 18
A_2B_3 18	A_1B_3 29	A_2B_3 24	A_1B_1 16

TABLE 6.13. Randomized block design and response in the (2×3)–factor experiment.

					Sum
Block	I	II	III	IV	
Response total	103	115	115	125	458

TABLE 6.14. Total response $Y_{\cdot\cdot k}$ per block.

	B_1	B_2	B_3	
A_1	62	78	117	257
A_2	42	72	87	201
	104	150	204	458

TABLE 6.15. Total response $Y_{ij\cdot}$ for each factor combination (Example 6.3).

Because of $F_{3,17;0.95} = 3.20$, the block effect is not significant. Hence we return to model (6.23) with the two main effects A and B which are significant according to Table 6.10.

6.5 Two–Factorial Model with Fixed Effects—Confidence Intervals and Elementary Tests

In a two–factorial experiment with fixed effects there are three different types of means: A–levels, B–levels, and $(A \times B)$–levels. In the case of a

Source	SS	df	MS	F	
Factor A	130.66	1	130.66	20.01	*
Factor B	626.33	2	313.17	47.96	*
$A \times B$	36.34	2	18.17	2.78	
Block	40.50	3	13.50	2.07	
Error	98.00	15	6.53		
Total	931.83	23			

TABLE 6.16. Analysis of variance table in model (6.42.)

Source	SS	df	MS	F	
Factor A	130.66	1	130.66	16.54	*
Factor B	626.33	2	313.17	39.64	*
Block	40.50	3	13.50	1.71	
Error	134.34	17	7.90		
Total	931.83	23			

TABLE 6.17. Analysis of variance table in model (6.46).

nonrandom block effect, the fourth type of means is that of the blocks. In the following, we assume fixed block effects.

(i) Factor A

The means of the A–levels are

$$y_{i\cdot\cdot} = \frac{1}{br}\sum_{j=1}^{b}\sum_{k=1}^{r} y_{ijk} \sim N\left(\mu + \alpha_i, \frac{\sigma^2}{br}\right). \tag{6.47}$$

The variance σ^2 is estimated by $s^2 = MS_{\text{Error}}$ with df degrees of freedom. Here MS_{Error} is computed from the model which holds after testing for interaction and block effects.

The confidence intervals for $\mu + \alpha_i$ are now of the following form ($t_{df,1-\alpha/2}$: two–sided quantile)

$$y_{i\cdot\cdot} \pm t_{df,1-\alpha/2}\sqrt{\frac{s^2}{br}}. \tag{6.48}$$

The standard error of the difference between two A–levels is $\sqrt{2s^2/br}$, so that the test statistic for $\text{H}_0 : \alpha_{i_1} = \alpha_{i_2}$ is of the following form:

$$t_{df} = \frac{y_{i_1\cdot\cdot} - y_{i_2\cdot\cdot}}{\sqrt{2s^2/br}}. \tag{6.49}$$

(ii) Factor B

Similarly, we have

$$y_{\cdot j \cdot} = \frac{1}{ar} \sum_{i=1}^{a} \sum_{k=1}^{r} y_{ijk} \sim N\left(\mu + \beta_j, \frac{\sigma^2}{ar}\right) . \qquad (6.50)$$

The $(1-\alpha)$–confidence interval for $\mu + \beta_j$ is

$$y_{\cdot j \cdot} \pm t_{df, 1-\alpha/2} \sqrt{\frac{s^2}{ar}} \qquad (6.51)$$

and the test statistic for the comparison of means ($\mathrm{H}_0 : \beta_{j_1} = \beta_{j_2}$) is

$$t_{df} = \frac{y_{\cdot j_1 \cdot} - y_{\cdot j_2 \cdot}}{\sqrt{2s^2/ar}} . \qquad (6.52)$$

(iii) Factor $A \times B$

Here we have

$$y_{ij\cdot} = \frac{1}{r} \sum_{k=1}^{r} y_{ijk} \sim N\left(\mu + \alpha_i + \beta_j + (\alpha\beta)_{ij}, \frac{\sigma^2}{r}\right) . \qquad (6.53)$$

The $(1-\alpha)$–confidence interval for $\mu + \alpha_i + \beta_j + (\alpha\beta)_{ij}$ is

$$y_{ij\cdot} \pm t_{df, 1-\alpha/2} \sqrt{s^2/r} \qquad (6.54)$$

and the test statistic for the comparison of two $(A \times B)$–effects is

$$t_{df} = \frac{y_{i_1 j_1 \cdot} - y_{i_2 j_2 \cdot}}{\sqrt{2s^2/r}} . \qquad (6.55)$$

The significance of single effects is tested by:

(i) $\mathrm{H}_0 : \mu + \alpha_i = \mu_0$:

$$t_{df} = \frac{y_{i\cdot\cdot} - \mu_0}{\sqrt{s^2/br}} ; \qquad (6.56)$$

(ii) $\mathrm{H}_0 : \mu + \beta_j = \mu_0$:

$$t_{df} = \frac{y_{\cdot j \cdot} - \mu_0}{\sqrt{s^2/ar}} ; \qquad (6.57)$$

(iii) $\mathrm{H}_0 : \mu + \alpha_i + \beta_j + (\alpha\beta)_{ij} = \mu_0$:

$$t_{df} = \frac{y_{ij\cdot} - \mu_0}{\sqrt{s^2/r}} . \qquad (6.58)$$

Here the statements in Section 4.4 about elementary and multiple tests hold.

Example 6.4. (Examples 6.2 and 6.3 continued) The test procedure leads to nonsignificant interaction and block effects. Hence, the independence model holds. From the appropriate analysis of variance table (Table 6.10) we take

$$s^2 = 8.74 \quad \text{for} \quad df = 20.$$

From Table 6.8 we obtain the means of the two levels A_1 and A_2 and of the three levels B_1, B_2, and B_3:

$$\begin{aligned}
A_1: \quad y_{1\cdot\cdot} &= \frac{257}{3\cdot 4} = 21.42, \\
A_2: \quad y_{2\cdot\cdot} &= \frac{201}{3\cdot 4} = 16.75, \\
B_1: \quad y_{\cdot 1\cdot} &= \frac{104}{2\cdot 4} = 13.00, \\
B_2: \quad y_{\cdot 2\cdot} &= \frac{150}{2\cdot 4} = 18.75, \\
B_3: \quad y_{\cdot 3\cdot} &= \frac{204}{2\cdot 4} = 25.50, .
\end{aligned}$$

(i) Confidence intervals fo A–levels:

$$\begin{aligned}
A_1: \quad 21.42 \pm t_{20;0.975}\sqrt{8.74/3\cdot 4} &= 21.42 \pm 2.09 \cdot 0.85 \\
&= 21.42 \pm 1.78 \\
\Rightarrow [19.64; 23.20], &
\end{aligned}$$

$$\begin{aligned}
A_2: \quad & 16.75 \pm 1.78 \\
& \Rightarrow [14.97; 18.53].
\end{aligned}$$

Test for $\mathrm{H}_0 : \alpha_1 = \alpha_2$ against $\mathrm{H}_1 : \alpha_1 > \alpha_2$:

$$\begin{aligned}
t_{20} &= \frac{21.42 - 16.75}{\sqrt{2\cdot 8.74/3\cdot 4}} = \frac{4.67}{1.21} = 3.86 \\
&> 1.73 = t_{20;0.95} \text{ (one–sided)}
\end{aligned}$$

$\Rightarrow \mathrm{H}_0$ is rejected.

(ii) Confidence intervals for B–levels:

With $t_{20;0.975}\sqrt{8.74/2\cdot 4} = 2.09 \cdot 1.05 = 2.19$, we obtain

$$\begin{aligned}
B_1: &\quad 13.00 \pm 2.19 \quad \Rightarrow \quad [10.81; 15.19], \\
B_2: &\quad 18.75 \pm 2.19 \quad \Rightarrow \quad [16.56; 20.94], \\
B_3: &\quad 25.50 \pm 2.19 \quad \Rightarrow \quad [23.31; 27.69].
\end{aligned}$$

The pairwise comparisons of means reject the hypothesis of identity.

6.6 Two–Factorial Model with Random or Mixed Effects

The first part of Chapter 6 has assumed the effects of Factors A and B to be fixed. This means that the factor levels of A and B are specified before the experiment and, hence, the conclusions of the analysis of variance are only valid for these factor levels. Alternative designs allow Factors A and B to act randomly (model with random effects) or keep one factor fixed and choose the other factor at random (model with mixed effects).

6.6.1 Model with Random Effects

We assume that the levels of both Factors A and B are chosen at random from populations A and B. The inferences will then be valid about all levels in the (two-dimensional) population. The response values in the model with random effects (or components of variance model) are

$$y_{ijk} = \mu + \alpha_i + \beta_j + (\alpha\beta)_{ij} + \epsilon_{ijk}\,, \tag{6.59}$$

with $i = 1, \ldots, a$, $j = 1, \ldots, b$, $k = 1, \ldots, r$ and where α_i, β_j, $(\alpha\beta)_{ij}$ are random variables independent of each other and of ϵ_{ijk}. We assume

$$\left.\begin{array}{l} \alpha = (\alpha_1, \ldots, \alpha_a)' \sim N(\mathbf{0}, \sigma_\alpha^2\, I), \\ \beta = (\beta_1, \ldots, \beta_b)' \sim N(\mathbf{0}, \sigma_\beta^2\, I), \\ (\alpha\beta) = ((\alpha\beta)_{11}, \ldots, (\alpha\beta)_{ab})' \sim N(\mathbf{0}, \sigma_{\alpha\beta}^2\, I), \\ \epsilon = (\epsilon_1, \ldots, \epsilon_{abr})' \sim N(\mathbf{0}, \sigma^2\, I)\,. \end{array}\right\} \tag{6.60}$$

In matrix notation, the covariance structure is as follows:

$$\mathrm{E}\begin{pmatrix} \alpha \\ \beta \\ (\alpha\beta) \\ \epsilon \end{pmatrix} (\alpha, \beta, (\alpha\beta), \epsilon)' = \begin{pmatrix} \sigma_\alpha^2\, I & \mathbf{0} & \mathbf{0} & \mathbf{0} \\ \mathbf{0} & \sigma_\beta^2\, I & \mathbf{0} & \mathbf{0} \\ \mathbf{0} & \mathbf{0} & \sigma_{\alpha\beta}^2\, I & \mathbf{0} \\ \mathbf{0} & \mathbf{0} & \mathbf{0} & \sigma^2\, I \end{pmatrix}.$$

Hence the variance of the response values is

$$\mathrm{Var}(y_{ijk}) = \sigma_\alpha^2 + \sigma_\beta^2 + \sigma_{\alpha\beta}^2 + \sigma^2\,. \tag{6.61}$$

σ_α^2, σ_β^2, $\sigma_{\alpha\beta}^2$, σ^2 are called *variance components*. The hypotheses that we are interested in testing are: $\mathrm{H}_0 : \sigma_\alpha^2 = 0$, $\mathrm{H}_0 : \sigma_\beta^2 = 0$, and $\mathrm{H}_0 : \sigma_{\alpha\beta}^2 = 0$.

The formulas for the decomposition of the variance SS_{Total} into SS_A, SS_B, $SS_{A\times B}$, and SS_{Error} and for the calculation of the variance remain unchanged, that is, all sums of squares are calculated as in the fixed effects case. However, to form the test statistics we must examine the expectation of the appropriate mean squares. We have

$$SS_A \;=\; \frac{1}{br}\sum_{i=1}^{a}(Y_{i\cdot\cdot} - Y_{\cdot\cdot\cdot})^2$$

$$= \sum_{i=1}^{a}\sum_{j=1}^{b}\sum_{k=1}^{r}(y_{i\cdot\cdot} - y_{\cdot\cdot\cdot})^2 . \tag{6.62}$$

With $\alpha = 1/a\sum_{i=1}^{a}\alpha_i$, $\beta = 1/b\sum_{j=1}^{b}\beta_j$, $(\alpha\beta)_{i\cdot} = 1/b\sum_{j=1}^{b}(\alpha\beta)_{ij}$, and $(\alpha\beta)_{\cdot\cdot} = 1/(ab)\sum\sum(\alpha\beta)_{ij}$, we compute, from model (6.59),

$$\begin{aligned} y_{i\cdot\cdot} &= \mu + \alpha_i + \beta + (\alpha\beta)_{i\cdot} + \epsilon_{i\cdot\cdot} , \\ y_{\cdot\cdot\cdot} &= \mu + \alpha + \beta + (\alpha\beta)_{\cdot\cdot} + \epsilon_{\cdot\cdot\cdot} , \end{aligned}$$

so that

$$y_{i\cdot\cdot} - y_{\cdot\cdot\cdot} = (\alpha_i - \alpha) + [(\alpha\beta)_{i\cdot} - (\alpha\beta)_{\cdot\cdot}] + (\epsilon_{i\cdot\cdot} - \epsilon_{\cdot\cdot\cdot}) . \tag{6.63}$$

Because of the mutual independence of the random effects and of the error, we have

$$\mathrm{E}(y_{i\cdot\cdot} - y_{\cdot\cdot\cdot})^2 = \mathrm{E}(\alpha_i - \alpha)^2 + \mathrm{E}[(\alpha\beta)_{i\cdot} - (\alpha\beta)_{\cdot\cdot}]^2 + \mathrm{E}(\epsilon_{i\cdot\cdot} - \epsilon_{\cdot\cdot\cdot})^2 . \tag{6.64}$$

For the three components, we observe that

$$\begin{aligned} \mathrm{E}(\alpha_i - \alpha)^2 &= \mathrm{E}(\alpha_i^2) + \mathrm{E}(\alpha^2) - 2\mathrm{E}(\alpha_i\alpha) \\ &= \sigma_\alpha^2\left[1 + \frac{1}{a} - \frac{2}{a}\right] \\ &= \sigma_\alpha^2\left[1 - \frac{1}{a}\right] = \sigma_\alpha^2\left(\frac{a-1}{a}\right) , \quad (6.65) \\ \mathrm{E}[(\alpha\beta)_{i\cdot} - (\alpha\beta)_{\cdot\cdot}]^2 &= \mathrm{E}[(\alpha\beta)_{i\cdot}^2] + \mathrm{E}[(\alpha\beta)_{\cdot\cdot}^2] - 2\mathrm{E}[(\alpha\beta)_{i\cdot}(\alpha\beta)_{\cdot\cdot}] \\ &= \sigma_{\alpha\beta}^2\left[\frac{1}{b} + \frac{1}{ab} - \frac{2}{ab}\right] \\ &= \sigma_{\alpha\beta}^2\left(\frac{a-1}{ab}\right) , \quad (6.66) \\ \mathrm{E}(\epsilon_{i\cdot\cdot} - \epsilon_{\cdot\cdot\cdot})^2 &= \mathrm{E}(\epsilon_{i\cdot\cdot}^2) + \mathrm{E}(\epsilon_{\cdot\cdot\cdot}^2) - 2\mathrm{E}(\epsilon_{i\cdot\cdot}\epsilon_{\cdot\cdot\cdot}) \\ &= \sigma^2\left[\frac{1}{br} + \frac{1}{abr} - \frac{2}{abr}\right] \\ &= \sigma^2\left(\frac{a-1}{abr}\right) , \quad (6.67) \end{aligned}$$

whence we find (cf. (6.62) and (6.64))

$$\begin{aligned} \mathrm{E}(MS_A) &= \frac{1}{a-1}\mathrm{E}(SS_A) \\ &= \sigma^2 + r\sigma_{\alpha\beta}^2 + br\sigma_\alpha^2 . \quad (6.68) \end{aligned}$$

Similarly, we find

$$\begin{aligned} \mathrm{E}(MS_B) &= \sigma^2 + r\sigma_{\alpha\beta}^2 + ar\sigma_\beta^2 , \quad (6.69) \\ \mathrm{E}(MS_{A\times B}) &= \sigma^2 + r\sigma_{\alpha\beta}^2 , \quad (6.70) \end{aligned}$$

$$\mathrm{E}(MS_{\text{Error}}) = \sigma^2 . \tag{6.71}$$

Estimation of the Variance Components

The estimates $\hat{\sigma}^2$, $\hat{\sigma}^2_\alpha$, $\hat{\sigma}^2_\beta$, and $\hat{\sigma}^2_{\alpha\beta}$ of the variance components σ^2, σ^2_α, σ^2_β, and $\sigma^2_{\alpha\beta}$ are computed from the equating system (6.68)–(6.71) in its sample version, that is, from the system

$$\left.\begin{array}{lclcl} MS_A & = & br\hat{\sigma}^2_\alpha & & + \; r\hat{\sigma}^2_{\alpha\beta} + \hat{\sigma}^2, \\ MS_B & = & & ar\hat{\sigma}^2_\beta & + \; r\hat{\sigma}^2_{\alpha\beta} + \hat{\sigma}^2, \\ MS_{A\times B} & = & & & r\hat{\sigma}^2_{\alpha\beta} + \hat{\sigma}^2, \\ MS_{\text{Error}} & = & & & \hat{\sigma}^2, \end{array}\right\} \tag{6.72}$$

i.e.,

$$\begin{pmatrix} MS_A \\ MS_B \\ MS_{A\times B} \\ MS_{\text{Error}} \end{pmatrix} = \begin{pmatrix} br & 0 & r & 1 \\ 0 & ar & r & 1 \\ 0 & 0 & r & 1 \\ 0 & 0 & 0 & 1 \end{pmatrix} \begin{pmatrix} \hat{\sigma}^2_\alpha \\ \hat{\sigma}^2_\beta \\ \hat{\sigma}^2_{\alpha\beta} \\ \hat{\sigma}^2 \end{pmatrix} .$$

The coefficient matrix of this linear inhomogeneous system is of triangular shape with its determinant as

$$abr^3 \neq 0 .$$

This yields the unique solution

$$\hat{\sigma}^2 = MS_{\text{Error}}, \tag{6.73}$$

$$\hat{\sigma}^2_{\alpha\beta} = \frac{1}{r}(MS_{A\times B} - MS_{\text{Error}}), \tag{6.74}$$

$$\hat{\sigma}^2_\beta = \frac{1}{ar}(MS_B - MS_{A\times B}), \tag{6.75}$$

$$\hat{\sigma}^2_\alpha = \frac{1}{br}(MS_A - MS_{A\times B}) . \tag{6.76}$$

Testing of Hypotheses about the Variance Components

(i) $\mathrm{H}_0 : \sigma^2_{\alpha\beta} = 0$
From the system (6.68)–(6.71) of the expectations of the MS's it can be seen that for $\mathrm{H}_0 : \sigma^2_{\alpha\beta} = 0$ (no interaction) we have $\mathrm{E}(MS_{A\times B}) = \sigma^2$. Hence the test statistic is of the form

$$F_{A\times B} = \frac{MS_{A\times B}}{MS_{\text{Error}}} . \tag{6.77}$$

If $\mathrm{H}_0 : \sigma^2_{\alpha\beta} = 0$ does not hold (i.e., H_0 is rejected in favor of $\mathrm{H}_1 : \sigma^2_{\alpha\beta} \neq 0$), then we have $\mathrm{E}(MS_{A\times B}) > \mathrm{E}(MS_{\text{Error}})$. Hence H_0 is rejected if

$$F_{A\times B} > F_{(a-1)(b-1),ab(r-1);1-\alpha} \tag{6.78}$$

holds.

(ii) $\mathrm{H}_0 : \sigma_\alpha^2 = 0$
The comparison of $\mathrm{E}(MS_A)$ [(6.68)] and $\mathrm{E}(MS_{A\times B})$ [(6.70)] shows that both expectations are identical under $\mathrm{H}_0 : \sigma_\alpha^2 = 0$, but $\mathrm{E}(MS_A) > \mathrm{E}(MS_{A\times B})$ holds in the case of $\mathrm{H}_1 : \sigma_\alpha^2 \neq 0$. The test statistic is then

$$F_A = \frac{MS_A}{MS_{A\times B}} \tag{6.79}$$

and H_0 is rejected if

$$F_A > F_{a-1,(a-1)(b-1);1-\alpha} \tag{6.80}$$

holds.

(iii) $\mathrm{H}_0 : \sigma_\beta^2 = 0$
Similarly, the test statistic for $\mathrm{H}_0 : \sigma_\beta^2 = 0$ against $H_1 : \sigma_\beta^2 \neq 0$ is

$$F_B = \frac{MS_B}{MS_{A\times B}}, \tag{6.81}$$

and H_0 is rejected if

$$F_B > F_{b-1,(a-1)(b-1);1-\alpha} \tag{6.82}$$

holds.

Source	SS	df	MS	F
Factor A	SS_A	$df_A = a - 1$	$MS_A = \frac{SS_A}{df_A}$	$F_A = \frac{MS_A}{MS_{A\times B}}$
Factor B	SS_B	$df_B = b - 1$	$MS_B = \frac{SS_B}{df_B}$	$F_B = \frac{MS_B}{MS_{A\times B}}$
Interaction $A \times B$	$SS_{A\times B}$	$df_{A\times B} = (a-1)(b-1)$	$MS_{A\times B} = \frac{SS_{A\times B}}{df_{A\times B}}$	$F_{A\times B} = \frac{MS_{A\times B}}{MS_{\text{Error}}}$
Error	SS_{Error}	$df_{\text{Error}} = ab(r-1)$	$MS_{\text{Error}} = \frac{SS_{\text{Error}}}{df_{\text{Error}}}$	
Total	SS_{Total}	$df_{\text{Total}} = abr - 1$		

TABLE 6.18. Analysis of variance table (two–factorial with interaction and random effects.)

	SS	df	MS	F
Factor A	130.66	1	130.66	$F_A = 130.66/18.17 = 7.19$
Factor B	626.33	2	313.17	$F_B = 313.17/18.17 = 17.24$
$A \times B$	36.34	2	18.17	$F_{A\times B} = 18.17/7.69 = 2.36$
Error	138.50	18	7.69	
Total	931.83	23		

TABLE 6.19. Analysis of variance table for Table 6.8 in the case of random effects.

Remark. In the random effects model the test statistics F_A and F_B are formed with $MS_{A\times B}$ in the denominator. In the model with fixed effects, we have MS_{Error} in the denominator.

Example 6.5. We now consider the experiment in Example 6.2 as a two–factorial experiment with random effects. For this, we assume that the two types of beans (Factor A) are chosen at random from a population, instead of being fixed effects. Similarly, we assume that the three phosphate fertilizers are chosen at random from a population. We assume the same response values as in Table 6.8 and adopt the first three columns from Table 6.9 for our analysis (Table 6.19). The estimated variance components are

$$\begin{aligned}\hat{\sigma}^2 &= 7.69,\\ \hat{\sigma}^2_{\alpha\beta} &= 1/4(18.17 - 7.69) = 2.62,\\ \hat{\sigma}^2_{\beta} &= \frac{1}{2\cdot 4}(313.17 - 18.17) = 36.88,\\ \hat{\sigma}^2_{\alpha} &= \frac{1}{3\cdot 4}(130.66 - 18.17) = 9.37\,.\end{aligned}$$

The three variance components $\sigma^2_{\alpha\beta}$, σ^2_{α}, and σ^2_{β} are not significant at the 5% level (critical values: $F_{1,2;0.95} = 18.51$; $F_{2,2;0.95} = 19.00$; $F_{2,18;0.95} = 3.55$).

Owing to the nonsignificance of $\sigma^2_{\alpha\beta}$, we return to the independence model. The analysis of variance table of this model is identical with Table 6.10 so that the two variance components σ^2_{α} and σ^2_{β} are significant.

6.6.2 *Mixed Model*

We now consider the situation where one factor (e.g., Factor A) is fixed and the other Factor B is random. The appropriate linear model in the *standard version by Scheffé* (1956; 1959) is

$$y_{ijk} = \mu + \alpha_i + \beta_j + (\alpha\beta)_{ij} + \epsilon_{ijk} \tag{6.83}$$

with $i = 1, \ldots, a$, $j = 1, \ldots, b$, $k = 1, \ldots, r$, and the following assumptions:

$$\alpha_i : \quad \text{fixed effect}, \quad \sum_{i=1}^{a} \alpha_i = 0, \tag{6.84}$$

$$\beta_j : \quad \text{random effect}, \quad \beta_j \overset{\text{i.i.d.}}{\sim} N(0, \sigma^2_{\beta}), \tag{6.85}$$

$$(\alpha\beta)_{ij} : \quad \text{random effect}, \quad (\alpha\beta)_{ij} \overset{\text{i.d.}}{\sim} N\left(0, \frac{a-1}{a}\sigma^2_{\alpha\beta}\right), \tag{6.86}$$

$$\sum_{i=1}^{a} (\alpha\beta)_{ij} = (\alpha\beta)_{\cdot j} = 0 \quad (j = 1, \ldots, b)\,. \tag{6.87}$$

We assume that the random variable groups β_j, $(\alpha\beta)_{ij}$, and ϵ_{ijk} are mutually independent, that is, we have $\mathrm{E}(\beta_j(\alpha\beta)_{ij}) = 0$, etc. As in the above models, we have $\mathrm{E}(\epsilon) = \sigma^2\mathbf{I}$.

The last assumption (6.87) means that the interaction effects between two different A–levels are correlated. For all $j = 1, \ldots, b$, we have

$$\mathrm{Cov}[(\alpha\beta)_{i_1 j}\,,\,(\alpha\beta)_{i_2 j}] = -\frac{1}{a}\sigma^2_{\alpha\beta} \quad (i_1 \neq i_2)\,, \tag{6.88}$$

but

$$\mathrm{Cov}[(\alpha\beta)_{i_1 j_1}\,,\,(\alpha\beta)_{i_2 j_2}] = 0 \quad (j_1 \neq j_2,\ \text{any } i_1, i_2)\,. \tag{6.89}$$

For $a = 3$, we provide a short outline of the proof. Using (6.87), we obtain

$$\begin{aligned}
\mathrm{Cov}[(\alpha\beta)_{1j}\,,\,(\alpha\beta)_{2j}] &= \mathrm{Cov}[(\alpha\beta)_{1j}\,,\,[-(\alpha\beta)_{1j} - (\alpha\beta)_{3j}]] \\
&= -\mathrm{Var}(\alpha\beta)_{1j} - \mathrm{Cov}[(\alpha\beta)_{1j}\,,\,(\alpha\beta)_{3j}]\,,
\end{aligned}$$

whence

$$\begin{aligned}
\mathrm{Cov}[(\alpha\beta)_{1j}\,,\,(\alpha\beta)_{2j}] + \mathrm{Cov}[(\alpha\beta)_{1j}\,,\,(\alpha\beta)_{3j}] &= -\mathrm{Var}(\alpha\beta)_{1j} \\
&= -\frac{3-1}{3}\sigma^2_{\alpha\beta}\,.
\end{aligned}$$

Since $\mathrm{Cov}[(\alpha\beta)_{i_1 j}\,,\,(\alpha\beta)_{i_2 j}]$ is identical for all pairs, (6.88) holds. If $a = b = 2$ and $r = 1$, then the model (6.83) with all assumptions has a four–dimensional normal distribution

$$\begin{pmatrix} y_{11} \\ y_{21} \\ y_{12} \\ y_{22} \end{pmatrix} \sim N\left(\begin{pmatrix} \mu+\alpha_1 \\ \mu+\alpha_2 \\ \mu+\alpha_1 \\ \mu+\alpha_2 \end{pmatrix}, \begin{pmatrix} \tilde{\sigma}^2 & \sigma_*^2 & 0 & 0 \\ \sigma_*^2 & \tilde{\sigma}^2 & 0 & 0 \\ 0 & 0 & \tilde{\sigma}^2 & \sigma_*^2 \\ 0 & 0 & \sigma_*^2 & \tilde{\sigma}^2 \end{pmatrix} \right) \tag{6.90}$$

with

$$\begin{aligned}
\mathrm{Var}(y_{ij}) &= \tilde{\sigma}^2 = \sigma^2_\beta + \sigma^2_{\alpha\beta}\frac{a-1}{a} + \sigma^2 \qquad (6.91)\\
&= (\sigma^2_{\alpha\beta} + \sigma^2) + \sigma^2_*\,,
\end{aligned}$$

using the identity $\sigma^2_* = \sigma^2_\beta - (1/a)\sigma^2_{\alpha\beta}$. The covariance matrix (6.90) can now be written as

$$\Sigma = I \otimes ((\sigma^2_{\alpha\beta} + \sigma^2)I_2 + \sigma^2_* J_2)\,,$$

where $\otimes$ is the Kronecker product. However, the second matrix has a compound symmetrical structure (3.178) so that the parameter estimates of the fixed effects are computed according to the OLS method (cf. Theorem 3.22):

$$\begin{array}{llll}
r = 1: & \hat{\mu} = y_{\cdot\cdot} & \text{and} & \hat{\alpha}_i = y_{i\cdot} - y_{\cdot\cdot}\ , \\
r > 1: & \hat{\mu} = y_{\cdot\cdot\cdot} & \text{and} & \hat{\alpha}_i = y_{i\cdot\cdot} - y_{\cdot\cdot\cdot}\ .
\end{array}$$

Expectations of the MS's

The specification of the A–effects and the reparametrization of the variance of $(\alpha\beta)_{ij}$ in $\sigma^2_{\alpha\beta}[(a-1)/a]$, as well as the constraints (6.87), have an effect on the expected mean squares. The expectations of the MS's are now

$$\begin{aligned} \mathrm{E}(MS_A) &= \sigma^2 + r\sigma^2_{\alpha\beta} + \frac{br\sum_{i=1}^a \alpha_i^2}{a-1}\,, && (6.92)\\ \mathrm{E}(MS_B) &= \sigma^2 + ar\sigma^2_\beta\,, && (6.93)\\ \mathrm{E}(MS_{A\times B}) &= \sigma^2 + r\sigma^2_{\alpha\beta}\,, && (6.94)\\ \mathrm{E}(MS_{\text{Error}}) &= \sigma^2\,. && (6.95) \end{aligned}$$

The test statistic for testing H_0 : *no A–effect*, i.e., $\mathrm{H}_0 : \alpha_i = 0$ (for all i), is

$$F_A = F_{a-1,(a-1)(b-1)} = \frac{MS_A}{MS_{A\times B}}\,. \qquad (6.96)$$

The test statistic for $\mathrm{H}_0 : \sigma^2_\beta = 0$ is

$$F_B = F_{b-1,ab(r-1)} = \frac{MS_B}{MS_{\text{Error}}}\,. \qquad (6.97)$$

The test statistic for $\mathrm{H}_0 : \sigma^2_{\alpha\beta} = 0$ is

$$F_{A\times B} = F_{(a-1)(b-1),ab(r-1)} = \frac{MS_{A\times B}}{MS_{\text{Error}}}\,. \qquad (6.98)$$

Estimation of the Variance Components

The variance components may be estimated by solving the following system (6.92)–(6.95) in its sample version:

$$\begin{array}{lclcccc} MS_A &=& [br/(a-1)]\sum\alpha_i^2 & & + & r\hat\sigma^2_{\alpha\beta} & + & \hat\sigma^2,\\ MS_B &=& & ar\hat\sigma^2_\beta & & & + & \hat\sigma^2,\\ MS_{A\times B} &=& & & & r\hat\sigma^2_{\alpha\beta} & + & \hat\sigma^2,\\ MS_{\text{Error}} &=& & & & & & \hat\sigma^2, \end{array}$$

$$\begin{aligned} \Longrightarrow\quad \hat\sigma^2 &= MS_{\text{Error}}, && (6.99)\\ \hat\sigma^2_{\alpha\beta} &= \frac{MS_{A\times B} - MS_{\text{Error}}}{r}\,, && (6.100)\\ \hat\sigma^2_\beta &= \frac{MS_B - MS_{\text{Error}}}{ar}\,. && (6.101) \end{aligned}$$

In addition to the standard model with intraclass correlation structure, several other versions of the mixed model exist (cf. Hocking, 1973). An important version is the model with independent interaction effects that assumes

$$(\alpha\beta)_{ij} \overset{\text{i.i.d.}}{\sim} N(0, \sigma^2_{\alpha\beta}) \quad (\text{for all } i, j)\,. \qquad (6.102)$$

Source	SS	df	$\mathrm{E}(MS)$	F
Factor A	SS_A	$a-1$	$\sigma^2 + r\sigma^2_{\alpha\beta} +$	$F_A = MS_A/MS_{A\times B}$
			$+[br/(a-1)]\sum \alpha_i^2$	
Factor B	SS_B	$b-1$	$\sigma^2 + ar\sigma^2_\beta$	$F_B = MS_B/MS_{\text{Error}}$
$A\times B$	$SS_{A\times B}$	$(a-1)(b-1)$	$\sigma^2 + r\sigma^2_{\alpha\beta}$	$F_{A\times B} = MS_{A\times B}/MS_{\text{Error}}$
Error	SS_{Error}	$ab(r-1)$	σ^2	
Total	SS_{Total}	$abr-1$		

TABLE 6.20. Analysis of variance table in the mixed model (standard model, dependent interaction effects).

Furthermore, independence of the $(\alpha\beta)_{ij}$ from the β_j and the ϵ_{ij} is assumed as in the standard model.

$\mathrm{E}(MS_B)$ now changes to

$$\mathrm{E}(MS_B) = \sigma^2 + r\sigma^2_{\alpha\beta} + ar\sigma^2_\beta \tag{6.103}$$

and the test statistic for $\mathrm{H}_0 : \sigma^2_\beta = 0$ changes to

$$F_B = F_{b-1,(a-1)(b-1)} = \frac{MS_B}{MS_{A\times B}} . \tag{6.104}$$

Source	SS	df	$\mathrm{E}(MS)$	F
A	SS_A	$a-1$	$\sigma^2 + r\sigma^2_{\alpha\beta} +$	$F_A = MS_A/MS_{A\times B}$
			$+[br/(a-1)]\sum \alpha_i^2$	
B	SS_B	$b-1$	$\sigma^2 + r\sigma^2_{\alpha\beta} + ar\sigma^2_\beta$	$F_B = MS_B/MS_{A\times B}$
$A\times B$	$SS_{A\times B}$	$(a-1)(b-1)$	$\sigma^2 + r\sigma^2_{\alpha\beta}$	$F_{A\times B} = MS_{A\times B}/MS_{\text{Error}}$
Error	SS_{Error}	$ab(r-1)$	σ^2	
Total	SS_{Total}	$abr-1$		

TABLE 6.21. Analysis of variance table in the mixed model with independent interaction effects.

The choice of mixed models should always be dictated by the data. In model (6.83) we have, for the covariance within the response values,

$$\mathrm{Cov}(y_{i_1 j_1 k_1}, y_{i_2 j_2 k_2}) = \delta_{j_1 j_2}\sigma^2_\beta + \mathrm{Cov}[(\alpha\beta)_{i_1 j_1}, (\alpha\beta)_{i_2 j_2}] + \sigma^2 . \tag{6.105}$$

If Factor B represents, for example, b time intervals (24–hour measure of blood pressure) and if Factor A represents the fixed effect placebo/medicament (p/m), then the assumption $\mathrm{Cov}[(\alpha\beta)_{Pj}, (\alpha\beta)_{Mj}] = 0$ would be reasonable, which is the opposite of (6.88). Similarly, (6.89) would have to be changed to

$$\mathrm{Cov}[(\alpha\beta)_{Pj_1}, (\alpha\beta)_{Pj_2}] \neq 0$$

or

$$\text{Cov}[(\alpha\beta)_{Mj_1}\,, (\alpha\beta)_{Mj_2}] \quad \neq \quad 0 \quad (j_1 \neq j_2),$$

respectively. These models are described in Chapter 8.

6.7 Three–Factorial Designs

The inclusion of a third factor in the experiment increases the number of parameters to be estimated. At the same time, the interpretation also becomes more difficult.

We denote the three factors (treatments) by A, B, and C and their factor levels by $i = 1, \ldots, a$, $j = 1, \ldots, b$, and $k = 1, \ldots, c$. Furthermore, we assume r replicates each, e.g., the randomized block design with r blocks and abc observations each. The appropriate model is the following additive model

$$\begin{aligned} y_{ijkl} &= \mu + \alpha_i + \beta_j + \gamma_k + (\alpha\beta)_{ij} + (\alpha\gamma)_{ik} + (\beta\gamma)_{jk} + (\alpha\beta\gamma)_{ijk} \\ &\quad + \tau_l + \epsilon_{ijkl} \; (l = 1, \ldots, r) \,. \end{aligned} \tag{6.106}$$

In addition to the two–way interactions $(\alpha\beta)_{ij}$, $(\beta\gamma)_{jk}$, and $(\alpha\gamma)_{ik}$, we now have the three–way interaction $(\alpha\beta\gamma)_{ijk}$. We assume the usual constraints for the main effects and the two–way interactions. Additionally, we assume

$$\sum_i (\alpha\beta\gamma)_{ijk} = \sum_j (\alpha\beta\gamma)_{ijk} = \sum_k (\alpha\beta\gamma)_{ijk} = 0\,. \tag{6.107}$$

The test strategy is similar to the two–factorial model, that is, the three–way interaction is tested first. If $\text{H}_0 : (\alpha\beta\gamma)_{ijk} = 0$ is rejected, then all of the two–way interactions and the main effects cannot be interpreted separately. The test strategy and, especially, the interpretation of submodels will be discussed in detail in Chapter 7 for models with categorical response. The results of Chapter 7 are valid for models with continuous response analogously.

The total response values are given in Table 6.22. The sums of squares are as follows:

$$\begin{aligned} C &= \frac{Y_{\cdots\cdot}^2}{abcr} \text{ (correction term)}, \\ SS_{\text{Total}} &= \sum\sum\sum\sum y_{ijkl}^2 - C, \\ SS_{\text{Block}} &= \frac{1}{abc}\sum_{l=1}^{r} Y_{\cdots l}^2 - C, \\ SS_A &= \frac{1}{bcr}\sum_i Y_{i\cdots}^2 - C, \end{aligned}$$

$$
\begin{aligned}
SS_B &= \frac{1}{acr}\sum_j Y_{\cdot j\cdot\cdot}^2 - C, \\
SS_{A\times B} &= \frac{1}{cr}\sum_i\sum_j Y_{ij\cdot\cdot}^2 - C - SS_A - SS_B, \\
SS_C &= \frac{1}{abr}\sum_k Y_{\cdot\cdot k\cdot}^2 - C, \\
SS_{A\times C} &= \frac{1}{br}\sum_i\sum_k Y_{i\cdot k\cdot}^2 - C - SS_A - SS_C, \\
SS_{B\times C} &= \frac{1}{ar}\sum_j\sum_k Y_{\cdot jk\cdot}^2 - C - SS_B - SS_C, \\
SS_{A\times B\times C} &= \frac{1}{r}\sum_i\sum_j\sum_k Y_{ijk\cdot}^2 - C, \\
&\quad - SS_A - SS_B - SS_C \\
&\quad - SS_{A\times B} - SS_{A\times C} - SS_{B\times C}, \\
SS_{\text{Error}} &= SS_{\text{Total}} - SS_{\text{Block}} \\
&\quad - SS_A - SS_B - SS_C \\
&\quad - SS_{A\times B} - SS_{A\times C} - SS_{B\times C} \\
&\quad - SS_{A\times B\times C}\,.
\end{aligned}
$$

As in the above models with fixed effects, $MS = SS/df$ holds (cf. Table 6.23). The test statistics, in general, are

$$
F_{\text{Effect}} = \frac{MS_{\text{Effect}}}{MS_{\text{Error}}}\,. \tag{6.108}
$$

Example 6.6. The firmness Y of a ceramic material is dependent on the pressure (A), on the temperature (B), and on an additive (C). A three–factorial experiment, that includes all three factors at two levels, low/high, is to analyze the influence on the response Y. A randomized block design is chosen with $r = 2$ blocks of workpieces that are homogeneous within the blocks and heterogeneous between the blocks. The results are shown in Table 6.24.

We compute ($N = abcr = 2^4 = 16$)

$$
\begin{aligned}
C &= \frac{Y_{\cdot\cdot\cdot\cdot}^2}{N} = \frac{249^2}{16} = 3875.06, \\
SS_{\text{Total}} &= 5175 - C = 1299.94, \\
SS_{\text{Block}} &= 1/8(108^2 + 141^2) - C = 3943.13 - C = 68.07, \\
SS_A &= 1/8(116^2 + 133^2) - C = 3893.13 - C = 18.07, \\
SS_B &= 1/8((42+54)^2 + (74+79)^2) - C = 4078.13 - C = 203.07, \\
SS_{A\times B} &= 1/4(42^2 + 74^2 + 54^2 + 79^2) - C - SS_A - SS_B
\end{aligned}
$$

Factor A	Factor B	Factor C 1	2	$\cdots$	c	Sum
1	1	$Y_{111\cdot}$	$Y_{112\cdot}$	$\cdots$	$Y_{11c\cdot}$	$Y_{11\cdot\cdot}$
	2	$Y_{121\cdot}$	$Y_{122\cdot}$	$\cdots$	$Y_{12c\cdot}$	$Y_{12\cdot\cdot}$
	$\vdots$	$\vdots$	$\vdots$		$\vdots$	$\vdots$
	b	$Y_{1b1\cdot}$	$Y_{1b2\cdot}$	$\cdots$	$Y_{1bc\cdot}$	$Y_{1b\cdot\cdot}$
	Sum	$Y_{1\cdot1\cdot}$	$Y_{1\cdot2\cdot}$	$\cdots$	$Y_{1\cdot c\cdot}$	$Y_{1\cdots}$
$\vdots$	$\vdots$		$\vdots$			$\vdots$
a	1	$Y_{a11\cdot}$	$Y_{a12\cdot}$	$\cdots$	$Y_{a1c\cdot}$	$Y_{a1\cdot\cdot}$
	2	$Y_{a21\cdot}$	$Y_{a22\cdot}$	$\cdots$	$Y_{a2c\cdot}$	$Y_{a2\cdot\cdot}$
	$\vdots$	$\vdots$	$\vdots$		$\vdots$	$\vdots$
	b	$Y_{ab1\cdot}$	$Y_{ab2\cdot}$	$\cdots$	$Y_{abc\cdot}$	$Y_{ab\cdot\cdot}$
	Sum	$Y_{a\cdot1\cdot}$	$Y_{a\cdot2\cdot}$	$\cdots$	$Y_{a\cdot c\cdot}$	$Y_{a\cdots}$
Sum		$Y_{\cdot\cdot1\cdot}$	$Y_{\cdot\cdot2\cdot}$	$\cdots$	$Y_{\cdot\cdot c\cdot}$	$Y_{\cdots\cdot}$

TABLE 6.22. Total response per block of the (A, B, C)–factor combinations.

Source	SS	df	MS	F
Block	SS_{Block}	$r-1$	MS_{Block}	F_{Block}
Factor A	SS_A	$a-1$	MS_A	F_A
Factor B	SS_B	$b-1$	MS_B	F_B
Factor C	SS_C	$c-1$	MS_C	F_C
$A \times B$	$SS_{A\times B}$	$(a-1)(b-1)$	$MS_{A\times B}$	$F_{A\times B}$
$A \times C$	$SS_{A\times C}$	$(a-1)(c-1)$	$MS_{A\times C}$	$F_{A\times C}$
$B \times C$	$SS_{B\times C}$	$(b-1)(c-1)$	$MS_{B\times C}$	$F_{B\times C}$
$A \times B \times C$	$SS_{A\times B\times C}$	$(a-1)(b-1)(c-1)$	$MS_{A\times B\times C}$	$F_{A\times B\times C}$
Error	SS_{Error}	$(r-1)(abc-1)$	MS_{Error}	
Total	SS_{Total}	$abcr-1$		

TABLE 6.23. Three–factorial analysis of variance table.

$$
\begin{aligned}
&= 4099.25 - C - SS_A - SS_B = 3.05, \\
SS_C &= 1/8(105^2 + 144^2) - C = 3970.13 - C = 95.07, \\
SS_{A\times C} &= 1/4(48^2 + 68^2 + 57^2 + 76^2) - C - SS_A - SS_C = 0.05, \\
SS_{B\times C} &= 1/4((14+16+18+20)^2 + (4+8+6+10)^2 \\
&\quad + (7+11+9+10)^2 + (24+32+26+34)^2) \\
&\quad - C - SS_B - SS_C = 885.05, \\
SS_{A\times B\times C} &= 1/2((14+16)^2 + \cdots + (26+34)^2) - C \\
&\quad - SS_A - SS_B - SS_{A\times B} - SS_C - SS_{A\times C} - SS_{B\times C} \\
&= 3.08, \\
SS_{\text{Error}} &= 24.43\,.
\end{aligned}
$$

Result: The F–tests with $F_{1,7;0.95} = 5.99$ show significance for the following effects: *block*, B, C, and $B \times C$. The influence of A is significant for none of the effects, hence the analysis can be done in a two–factorial $/B \times C)$–design (Table 6.26, $F_{1,11;0.95} = 4.84$). The response Y is maximized for the combination $B_2 \times C_2$.

		Block 1	Block 2	Block 1	Block 2	Sum
		C_1		C_2		
A_1	B_1	14	16	4	8	42
	B_2	7	11	24	32	74
		48		68		116
A_2	B_1	18	20	6	10	54
	B_2	9	10	26	34	79
		57		76		133
	Sum	105		144		249

$$Y_{\cdot\cdot\cdot 1} = 108, \quad Y_{\cdot\cdot\cdot 2} = 141$$

TABLE 6.24. Response values for Example 6.6.

	SS	df	MS	F
Block	68.07	1	68.07	19.50 *
Factor A	18.07	1	18.07	5.18
Factor B	203.07	1	203.07	58.19 *
Factor C	95.07	1	95.07	27.24 *
$A \times B$	3.05	1	3.05	0.87
$A \times C$	0.05	1	0.05	0.01
$B \times C$	885.05	1	885.05	253.60 *
$A \times B \times C$	3.08	1	3.08	0.88
Error	24.43	7	3.49	
Total	1299.94	15		

TABLE 6.25. Analysis of variance in the $(A \times B \times C)$–design for Example 6.6.

Remark: Three-factorial design models with random effects are discussed in Burdick (1994). Confidence intervals are used for testing the significance of variance components.

	SS	df	MS	F
Block	68.07	1	68.07	15.37 *
Factor B	203.07	1	203.07	45.84 *
Factor C	95.07	1	95.07	21.46 *
$B \times C$	885.05	1	885.05	199.79 *
Error	48.68	11	4.43	
Total	1299.94	15		

TABLE 6.26. Analysis of variance in the $(B \times C)$–design for Example 6.6.

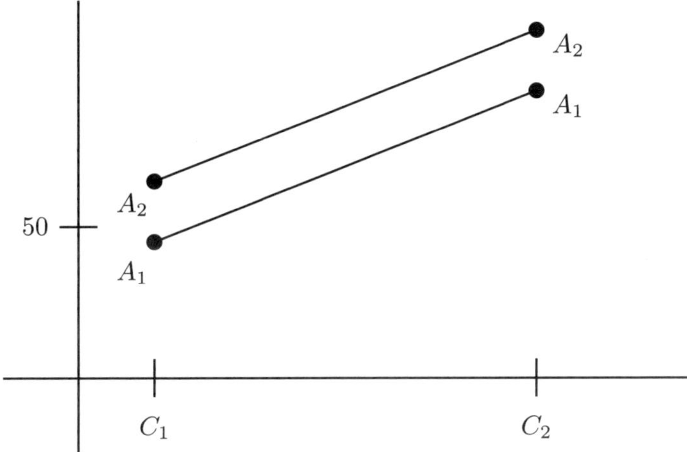

FIGURE 6.5. $(A \times C)$–response

6.8 Split–Plot Design

In many practical applications of the randomized block design it is not possible to arrange all factor combinations at random within one block. This is the case if the factors require different sizes of experimental units, e.g., because of technical reasons. Consider some examples (cf. Montgomery, 1976, pp. 292–300; Petersen, 1985, pp. 134–145):

- Employment of various drill machines (Factor B, only possible on larger fields) and of various fertilizers (Factor C, may be employed on smaller fields as well). In this case Factor B is set and only Factor C is randomized in the blocks.

- Combination of three different paper pulp preparation methods and of four different temperatures in paper manufacturing. Each replicate of the experiment requires 12 observations. In a completely randomized design, a factor combination (pulp i, temperature j) would have to be chosen at random within the block. In this example, however, this procedure may not be economical. Hence, the three types of pulp

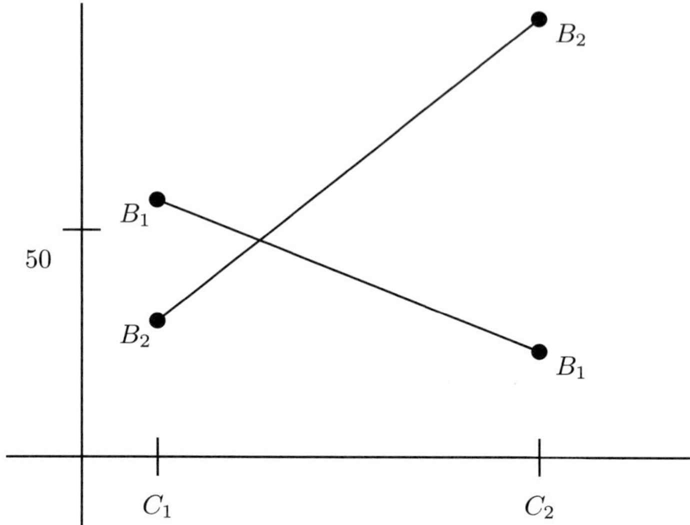

FIGURE 6.6. $(B \times C)$–response

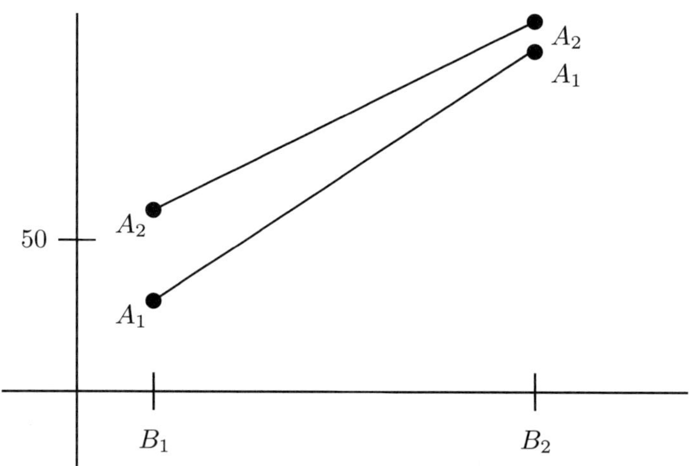

FIGURE 6.7. $(A \times B)$–response

are divided in four sample units and the temperature is randomized within these units.

Split–plot designs are used if the possibilities for randomization are restricted. The large units are called *whole–plots* while the smaller units are called *subplots* (or split–plots).

In this design of experiment, the whole–plot factor effects are estimated from the large units while the subplot effects and the interaction whole–plot – subplot is estimated from the small units. This design, however, leads to two experimental errors. The error associated with the subplot is the smaller one. The reason for this is the larger number of degrees of freedom of the subplot error, as well as the fact that the units in the subplots tend to be positively correlated in the response.

In our examples:

- the drill machine is the whole–plot and the fertilizer the subplot; and
- the type of pulp is the whole–plot and the temperature is the subplot.

The linear model for the two–factorial split–plot design is (Montgomery, 1976, p. 293)

$$y_{ijk} = \mu + \tau_i + \beta_j + (\tau\beta)_{ij} + \gamma_k + (\tau\gamma)_{ik} + (\beta\gamma)_{jk} + (\tau\beta\gamma)_{ijk} + \epsilon_{ijk}$$
$$(i = 1, \ldots, a,\ j = 1, \ldots, b,\ k = 1, \ldots, c)\,, \tag{6.109}$$

where the parameters

τ_i: random block effect (Factor A);
β_j: whole–plot effect (Factor B);
$(\tau\beta)_{ij}$: whole–plot error (= $(A \times B)$–interaction);

are the whole–plot parameters and the subplot parameters are

γ_k: treatment effect factor C;
$(\tau\gamma)_{ik}$: $(A \times C)$–interaction;
$(\beta\gamma)_{jk}$: $(B \times C)$–interaction;
$(\tau\beta\gamma)_{ijk}$: subplot error (= $A \times B \times C$)–interaction).

The sums of squares are computed as in the three–factorial model without replication (i.e., $r = 1$ in the SS's of the previous section).

The test statistics are given in Table 6.27. The effects to be tested are the main effects of Factor B and Factor C as well as the interaction $B \times C$. The test strategy starts out as in the two–factorial model, that is, with the $(B \times C)$–interaction.

Source	SS	df	MS	F
Block(A)	SS_A	$a-1$	MS_A	
Factor B	SS_B	$b-1$	MS_B	$F_B = MS_B/MS_{A\times B}$
Error($A \times B$)	$SS_{A\times B}$	$(a-1)(b-1)$	$MS_{A\times B}$	
Factor C	SS_C	$c-1$	MS_C	$F_C = MS_C/MS_{A\times B\times C}$
$A \times C$	$SS_{A\times C}$	$(a-1)(c-1)$	$MS_{A\times C}$	
$B \times C$	$SS_{B\times C}$	$(b-1)(c-1)$	$MS_{B\times C}$	$F_{B\times C} = MS_{B\times C}/MS_{A\times B\times C}$
Error($A\times B\times C$)	$SS_{A\times B\times C}$	$a(b-1)(c-1)$	$MS_{A\times B\times C}$	
Totel	SS_{Total}	$abc-1$		

TABLE 6.27. Analysis of variance in the split–plot design.

Example 6.7. A laboratory has two furnaces of which one can only be heated up to 500 °C. The hardness of a ceramic, having dependence upon two additives, and the temperature is to be tested in a split–plot design.

Factor A (block): replication on $r = 3$ days.
Factor B (whole–plot): temperature:
B_1: 500 °C (furnace I),
B_2: 750 °C (furnace II).
Factor C (subplot): additive:
C_1: 10%,
C_2: 20%.

Because of $F_{1,2;0.95} = 18.51$, only Factor C is significant (Table 6.29). Hence the experiment can be conducted with a single–factor additive (Table 6.30).

I

B_1	B_2
C_1 4	C_2 6
C_2 7	C_1 5

II

B_1	B_2
C_2 7	C_2 7
C_1 5	C_1 6

III

B_1	B_2
C_2 9	C_1 9
C_1 4	C_2 10

Block	B_1	B_2	Sum
I	11	11	22
II	12	13	25
III	13	19	32
Sum	36	43	79

	B_1	B_2	
C_1	13	20	33
C_2	23	23	46
	36	43	79

TABLE 6.28. Response tables.

	SS	df	MS	F
Block (A)	13.17	2	6.58	
Factor B	4.08	1	4.08	$F_B = 1.58$
Error ($A \times B$)	5.17	2	2.58	
Factor C	14.08	1	14.08	$F_C = 24.14$ *
$A \times C$	1.17	2	0.58	
$B \times C$	4.08	1	4.08	$F_{B \times C} = 7.00$
Error ($A \times B \times C$)	1.17	2	0.58	
Total	42.92	11		

TABLE 6.29. Analysis of variance table for Example 6.7.

Source	SS	df	MS	F
Factor C	14.08	1	14.08	$F_C = 4.88$
Error	28.83	10	2.88	
Total	42.92	11		

TABLE 6.30. One–factor analysis of variance table (Example 6.7).

Remark: Generalizations in model (6.109) are discussed in Algina (1995), Algina, (1997), especially with respect to unequal group dispersion matrices. The analysis of covariance in various types of split–plot design is presented by Brzeskwiniewicz and Wagner (1991).

6.9 2^k-Factorial Design

Especially in the industrial area, factorial designs at the first stage of an analysis are usually conducted with only two factor levels for each of the included factors. The idea of this procedure is to make the important effects identifiable so that the analysis in the following stages can test factor combinations more specifically and more cost–effectively. A complete analysis with k factors, each of two levels, requires 2^k replications for one trial. This fact leads to the nomenclature of the design: the 2^k experiment. The restriction to two levels for all factors makes a minimum of observations possible for a complete factorial experiment with all two–way and higher–order interactions. We assume fixed effects and complete randomization. The same linear models and constraints, as for the previous two– and three–factorial designs, are valid in the 2^k design, too. The advantage of this design is the immediate computation of the sums of squares from special constraints which are linked to the effects.

6.9.1 The 2^2 Design

The 2^2 design has already been introduced in Section 6.1. Two factors A and B are run at two levels each (e.g., low and high). The chosen parametrization is usually

$$\text{low: } 0, \qquad \text{high: } 1 \quad .$$

The high levels of the factors are represented by a or b, respectively, and the low level is denoted by the absence of the corresponding letter. If both factors are at the low level, (1) is used as representation:

$$\begin{aligned} (0,0) &\longrightarrow (1), \\ (1,0) &\longrightarrow a, \\ (0,1) &\longrightarrow b, \\ (1,1) &\longrightarrow ab\,. \end{aligned}$$

Here (1), a, b, ab denote the response for all r replicates. The average effect of a factor is defined as the reaction of the response to a change of level of this factor, averaged over the levels of the other factor. The effect of A at the low level of B is $[a-(1)]/r$ and the effect of A at the high level of B is $[ab-b]r$. The average effect of A is then

$$A = \frac{1}{2r}\left[ab + a - b - (1)\right] . \tag{6.110}$$

The average effect of B is

$$B = \frac{1}{2r}\left[ab + b - a - (1)\right] . \tag{6.111}$$

The interaction effect AB is defined as the average difference between the effect of A at the high level of B and the effect of A at the low level of B. Thus

$$\begin{aligned} AB &= \frac{1}{2r}\left[(ab-b) - (a-(1))\right] \\ &= \frac{1}{2r}\left[ab + (1) - a - b\right] . \end{aligned} \tag{6.112}$$

Similarly, the effect BA may be defined as the average difference between the effect of B at the high level of A (i.e., $(ab-a)/r$) and the effect of B at the low level of A (i.e., $(b-(1))/r$). We obviously have $AB = BA$. Hence, the average effects A, B, and AB are linear orthogonal contrasts in the total response values (1), a, b, ab, except for the factor $1/2r$.

Let $Y_* = ((1), a, b, ab)'$ be the vector of the total response values. Then

$$\left.\begin{aligned} &A = 1/2rc_A'Y_*, \quad B = 1/2rc_B'Y_* , \\ &AB = 1/2rc_{AB}'Y_* , \end{aligned}\right\} \tag{6.113}$$

holds where the contrasts c_A, c_B, c_{AB} are taken from Table 6.31.

	(1)	a	b	ab	Contrast
Factor A	-1	+1	-1	+1	c_A'
Factor B	-1	-1	+1	+1	c_B'
AB	+1	-1	-1	+1	c_{AB}'

TABLE 6.31. Contrasts in the 2^2 design.

We have $c_A'c_A = c_B'c_B = c_{AB}'c_{AB} = 4$.

From Section 4.3.2, we find the following sums of squares:

$$SS_A = \frac{(c_A'Y_*)^2}{(rc_A'c_A)} = \frac{(ab + a - b - (1))^2}{4r} , \tag{6.114}$$

$$SS_B = \frac{(c_B'Y_*)^2}{(rc_B'c_B)} = \frac{(ab + b - a - (1))^2}{4r} , \tag{6.115}$$

$$SS_{AB} = \frac{(c'_{AB}Y_*)^2}{(rc'_{AB}c_{AB})} = \frac{(ab + (1) - a - b)^2}{4r}. \tag{6.116}$$

The sum of squares SS_{Total} is computed as usual

$$SS_{\text{Total}} = \sum_{i=1}^{2}\sum_{j=1}^{2}\sum_{k=1}^{r} y_{ijk}^2 - \frac{Y_{\cdots}^2}{4r} \tag{6.117}$$

and has $(2 \cdot 2 \cdot r) - 1$ degrees of freedom. As usual, we have

$$SS_{\text{Error}} = SS_{\text{Total}} - SS_A - SS_B - SS_{AB}. \tag{6.118}$$

We now illustrate this procedure with an example.

Example 6.8. We wish to investigate the influence of Factors A (temperature, 0 : low, 1 : high) and B (catalytic converter, 0 : not used, 1 : used) on the response Y (hardness of a ceramic material). The response is shown in Table 6.32.

Combination	Replication 1	Replication 2	Total response	Coding
(0, 0)	86	92	178	(1)
(1, 0)	47	39	86	a
(0, 1)	104	114	218	b
(1, 1)	141	153	294	ab
			$Y_{\cdots} = 776$	

TABLE 6.32. Response in Example 6.8.

From Table 6.32, we obtain the average effects

$$\begin{aligned} A &= 1/4\,[294 + 86 - 218 - 178] = -4, \\ B &= 1/4\,[294 + 218 - 86 - 178] = 62, \\ AB &= 1/4\,[294 + 178 - 86 - 218] = 42, \end{aligned}$$

and from these the sums of squares

$$\begin{aligned} SS_A &= \frac{(4A)^2}{4 \cdot 2} = 32, \\ SS_B &= \frac{(4B)^2}{4 \cdot 2} = 7688, \\ SS_{AB} &= \frac{(4AB)^2}{4 \cdot 2} = 3528. \end{aligned}$$

Furthermore, we have

$$\begin{aligned} SS_{\text{Total}} &= (86^2 + \ldots + 153^2) - \frac{776^2}{8} = 86692 - 75272 = 11420, \\ SS_{\text{Error}} &= 172. \end{aligned}$$

The analysis of variance table is shown in Table 6.33.

	SS	df	MS	F
Factor A	32	1	32	$F_A = 0.74$
Factor B	7688	1	7688	$F_B = 178.79$ *
AB	3528	1	3528	$F_{AB} = 82.05$ *
Error	172	4	43	
Total	11420	7		

TABLE 6.33. Analysis of variance for Example 6.8.

6.9.2 The 2^3 Design

Suppose that in a complete factorial experiment three binary factors A, B, C are to be studied. The number of combinations is eight and with r replicates we have $N = 8r$ observations that are to be analyzed for their influence on a response.

Assume the total response values are (in the so–called *standard order*)

$$Y_* = [(1), a, b, ab, c, ac, bc, abc]' \ . \qquad (6.119)$$

In the coding 0: low and 1: high, this corresponds to the triples $(0,0,0), (1,0,0), (0,1,0), (1,1,0), \ldots, (1,1,1)$. The response values can be arranged as a three–dimensional contingency table (cf. Table 6.35). The effects are determined by linear contrasts

$$c'_{\text{Effect}} \cdot ((1), a, b, ab, c, ac, bc, abc) = c'_{\text{Effect}} \cdot Y_* \qquad (6.120)$$

(cf. Table 6.34).

Factorial effect	Factor combination							
	(1)	a	b	ab	c	ac	bc	abc
I	+	+	+	+	+	+	+	+
A	-	+	-	+	-	+	-	+
B	-	-	+	+	-	-	+	+
AB	+	-	-	+	+	-	-	+
C	-	-	-	-	+	+	+	+
AC	+	-	+	-	-	+	-	+
BC	+	+	-	-	-	-	+	+
ABC	-	+	+	-	+	-	-	+

TABLE 6.34. Algebraic structure for the computation of the effects from the total response values.

The first row in Table 6.34 is a basic element. With this element, the total response $Y_{....} = \mathbf{1}'Y_*$ can be computed. If the other rows are multiplied with the first row, they stay unchanged (therefore I for identity). Every other row has the same numbers of + and - signs. If + is replaced by 1 and - is replaced by -1, we obtain vectors of orthogonal contrasts with the norm 8.

If each row is multiplied by itself, we obtain I (row 1). The product of any two rows leads to a different row of Table 6.34. For example, we have

$$\begin{aligned} A \cdot B &= AB, \\ (AB) \cdot (B) &= A \cdot B^2 = A, \\ (AC) \cdot (BC) &= A \cdot C^2 B = AB\,. \end{aligned}$$

The sums of squares in the 2^3 design are

$$SS_{\text{Effect}} = \frac{(\text{Contrast})^2}{8r}\,. \tag{6.121}$$

Estimation of the Effects

The algebraic structure of Table 6.34 immediately leads to the estimates of the average effects. For instance, the average effect A is

$$A = \frac{1}{4r}\left[a - (1) + ab - b + ac - c + abc - bc\right]. \tag{6.122}$$

Explanation. The average effect of A at the low level of B and C is

$$(1\,0\,0) - (0\,0\,0): \quad [a - (1)]/r\,.$$

The average effect of A at the high level of B and the low level of C is

$$(1\,1\,0) - (0\,1\,0): \quad [ab - b]/r\,.$$

The average effect of A at the low level of B and the high level of C is

$$(1\,0\,1) - (0\,0\,1): \quad [ac - c]/r\,.$$

The average effect of A at the high level of B and C is

$$(1\,1\,1) - (0\,1\,1): \quad [abc - bc]/r\,.$$

Hence for all combinations of B and C the average effect of A is the average of these four values, which equals (6.122). Similarly, we obtain the other average effects

$$B = \frac{1}{4r}\left[b + ab + bc + abc - (1) - a - c - ac\right], \tag{6.123}$$

$$C = \frac{1}{4r}\left[c + ac + bc + abc - (1) - a - b - ab\right], \tag{6.124}$$

$$AB = \frac{1}{4r}\left[(1) + ab + c + abc - a - b - ac - bc\right], \tag{6.125}$$

$$AC = \frac{1}{4r}\left[(1) + b + ac + abc - a - ab - c - bc\right], \tag{6.126}$$

$$BC = \frac{1}{4r}\left[(1) + a + bc + abc - b - ab - c - ac\right], \tag{6.127}$$

$$\begin{aligned} ABC &= \frac{1}{4r}\left[(abc - bc) - (ac - c) - (ab - b) + (a - (1))\right] \\ &= \frac{1}{4r}\left[abc + a + b + c - ab - ac - bc - (1)\right] . \end{aligned} \tag{6.128}$$

Example 6.9. We demonstrate the analysis by means of Table 6.35. We have $r = 2$.

	Factor B			
	0		1	
	Factor C		Factor C	
Factor A	0	1	0	1
0	4	7	20	10
	5	9	14	6
	$9 = (1)$	$16 = c$	$34 = b$	$16 = bc$
1	4	2	4	14
	11	7	6	16
	$15 = a$	$9 = ac$	$10 = ab$	$30 = abc$

TABLE 6.35. Example for a 2^3 design with $r = 2$ replicates.

Average Effects

$$\begin{aligned} A &= 1/8\,[15 - 9 + 10 - 34 + 9 - 16 + 30 - 16] = 1/8[64 - 75] \\ &= -11/8 = -1.375, \\ B &= 1/8\,[34 + 10 + 16 + 30 - (9 + 15 + 16 + 9)] = 1/8[90 - 49] \\ &= 41/8 = 5.125, \\ C &= 1/8\,[16 + 9 + 16 + 30 - (9 + 15 + 34 + 10)] = 1/8[71 - 68] \\ &= 3/8 = 0.375, \\ AB &= 1/8\,[9 + 10 + 16 + 30 - (15 + 34 + 9 + 16)] = 1/8[65 - 74] \\ &= -9/8 = -1.125, \\ AC &= 1/8\,[9 + 34 + 9 + 30 - (15 + 10 + 16 + 16)] = 1/8[82 - 57] \\ &= 25/8 = 3.125, \\ BC &= 1/8\,[9 + 15 + 16 + 30 - (34 + 10 + 16 + 9)] = 1/8[70 - 69] \\ &= 1/8 = 0.125, \\ ABC &= 1/8\,[30 + 15 + 34 + 16 - (10 + 9 + 16 + 9)] = 1/8[95 - 44] \\ &= 51/8 = 6.375\,. \end{aligned}$$

	SS	df	MS	F	
Factor A	7.56	1	7.56	0.87	
Factor B	105.06	1	105.06	12.09	*
AB	5.06	1	5.06	0.58	
Factor C	0.56	1	0.56	0.06	
AC	39.06	1	39.06	4.49	
BC	0.06	1	0.06	0.01	
ABC	162.56	1	162.56	18.71	*
Error	69.52	8	8.69		
Total	389.44	15			

TABLE 6.36. Analysis of variance for Table 6.35.

The sums of squares are (cf. (6.121))

$$\begin{aligned} SS_A &= 11^2/16 = 7.56, & SS_{AB} &= 9^2/16 = 5.06, \\ SS_B &= 41^2/16 = 105.06, & SS_{AC} &= 25^2/16 = 39.06, \\ SS_C &= 3^2/16 = 0.56, & SS_{BC} &= 1^2/16 = 0.06\,. \\ SS_{ABC} &= 51^2/16 = 162.56, \\ SS_{\text{Total}} &= (4^2 + 5^2 + \ldots + 14^2 + 16^2) \\ &\quad -139^2/16 \\ &= 1597 - 1207.56 = 389.44, \\ SS_{\text{Error}} &= 69.52, \end{aligned}$$

The critical value for the F–statistics is $F_{1,8;0.95} = 5.32$ (cf. Table 6.36). Since the ABC effect is significant, no reduction to a two–factorial model is possible.

Remark: For further examples and other multifactor designs we refer to the overview given by Draper and Pukelsheim (1996).

6.10 Exercises and Questions

6.10.1 What advantages does a two–factorial experiment (A, B) have, compared to two one–factor experiments (A) and (B)?

6.10.2 Name the score function for parameter estimation in a two–factorial model with interaction. Name the parameter estimates of the overall mean and of the two main effects.

6.10.3 Fill in the degrees of freedom and the F–statistics (A in a levels, B in b levels, r replicates) in the two–factorial design with fixed effects:

		df	MS	F
Factor A	SS_A			
Factor B	SS_B			
$A \times B$	$SS_{A\times B}$			
Error	SS_{Error}			
Total	SS_{Total}			

6.10.4 At least how many replicates r are needed in order to be able to show interaction?

6.10.5 What is meant by a saturated model and what is meant by the independence model?

6.10.6 How are the following test results to be interpreted (i.e., which model corresponds to the two–factorial design with fixed effects)?

(a)

F_A	*
F_B	*
$F_{A\times B}$	*

(b)

F_A	*
F_B	*
$F_{A\times B}$	

(c)

F_A	
F_B	
$F_{A\times B}$	*

(d)

F_A	*
F_B	
$F_{A\times B}$	*

(e)

F_A	
F_B	*
$F_{A\times B}$	

6.10.7 Of what rank is the design matrix X in the two–factorial model ($A : a$, $B : b$ levels, r replicates)?

6.10.8 Let $a = b = 2$ and $r = 1$. Describe the two–factorial model with interaction in effect coding.

6.10.9 Of what form is the covariance matrix of the OLS estimate in the two–factorial model with fixed effects in effect coding?

$$\mathrm{V}(\hat{\mu}, \hat{\alpha}, \hat{\beta}, (\widehat{\alpha\beta})) = \sigma^2\,?$$

In what way do the parameter estimates $\hat{\mu}$, $\hat{\alpha}$, and $\hat{\beta}$ change if $F_{A\times B}$ is not significant? How does the estimate $\hat{\sigma}^2$ change? In what way do the confidence intervals for $\hat{\alpha}$, and $\hat{\beta}$, and the test statistics F_A and F_B, change? Is the test more conservative than in the model with significant interaction?

6.10.10 Carry out the following test in the two–factorial model with fixed effects and define the final model:

		df	MS	F
SS_A	130	1		
SS_B	630	2		
$SS_{A\times B}$	40	2		
SS_{Error}	150	18		
SS_{Total}		23		

6.10.11 Assume the two–factorial experiment with fixed effects to be designed as a randomized block design. Specify the model. In what way do the parameter estimates and the SS's for the other parameters or effects change, compared to the model without block effects? Name the SS_{Error}. What meaning does a significant block effect have?

6.10.12 Analyze the following two–factorial experiment with $a = b = 2$ and $r = 2$ replicates (randomized design, no block design):

	B_1	B_2	
	17	4	
A_1	18	6	
	35	10	45
	6	15	
A_2	4	10	
	10	25	35
	45	35	80

$$\begin{aligned}
C &= \frac{?^2}{N}, \\
SS_{\text{Total}} &= \sum\sum\sum y_{ijk}^2 - C, \\
SS_A &= \frac{1}{br}\sum_i Y_{i..}^2 - C, \\
SS_B &= ?, \\
SS_{\text{Subtotal}} &= 1/2(35^2 + 10^2 + 10^2 + 25^2) - C, \\
SS_{A\times B} &= SS_{\text{Subtotal}} - SS_A - SS_B, \\
SS_{\text{Error}} &= ?.
\end{aligned}$$

6.10.13 Name the assumptions for μ, α_i, β_j, and $(\alpha\beta)_{ij}$ in the two–factorial model with random effects. Complete the following:

- $\text{Var}(y_{ijk}) = ?$.
- $\text{E}\begin{pmatrix} \alpha \\ \beta \\ \alpha\beta \\ \epsilon \end{pmatrix} (\alpha, \beta, \alpha\beta, \epsilon)' = ?$.

– Solve the following system:

$$\begin{array}{lcccccc} MS_A & = & br\hat{\sigma}_\alpha^2 & & + & r\hat{\sigma}_{\alpha\beta}^2 & + & \hat{\sigma}^2, \\ MS_B & = & & ar\hat{\sigma}_\beta^2 & + & r\hat{\sigma}_{\alpha\beta}^2 & + & \hat{\sigma}^2, \\ MS_{A\times B} & = & & & & r\hat{\sigma}_{\alpha\beta}^2 & + & \hat{\sigma}^2, \\ MS_{\text{Error}} & = & & & & & + & \hat{\sigma}^2. \end{array}$$

– Compute the test statistics

$$\begin{aligned} F_{A\times B} &= ?, \\ F_A &= ?, \\ F_B &= ?. \end{aligned}$$

– Name the test statistics if $F_{A\times B}$ is not significant.

6.10.14 The covariance matrix in the mixed two–factorial model (A fixed, B random) has a compound symmetric structure, i.e., $\Sigma = ?$ Therefore, we have a generalized linear regression model. According to which method are the estimates of the fixed effects obtained? The test statistics in the model with the interactions correlated over the A–levels are

$$\begin{aligned} F_{A\times B} &= \frac{MS_{A\times B}}{MS_{\text{Error}}}, \\ F_B &= \frac{MS_B}{?}, \\ F_A &= \frac{MS_A}{?}, \end{aligned}$$

and in the model with independent interactions

$$F_B = \frac{MS_B}{?}.$$

6.10.15 Name the test statistics for the three–factorial ($A \times B \times C$)–design with fixed effects

$$F_{\text{Effect}} = \frac{\quad}{\quad}.$$

(Effect, e.g., A, B, C, $A \times B$, $A \times B \times C$) ?

6.10.16 The following table is used in the 2^2 design with fixed effects and r replications:

	(1)	a	b	ab
A	-1	+1	-1	+1
B	-1	-1	+1	+1
AB	+1	-1	-1	+1

Here (1) is the total response for (0, 0) (A low, B high), (a) for (1, 0), (b) for (0, 1) and (ab) for (1, 1). Hence, the vector of the total response values is $Y_* = ((1), a, b, ab)'$. Compute the average effects A, B, and AB in the following 2^2 design.

	Replications		Total
	1	2	response
(0, 0)	85	93	
(1, 0)	46	40	
(0, 1)	103	115	
(1, 1)	140	154	

How are SS_A, SS_B, and $SS_{A \times B}$ computed? (Hint: Use the contrasts).

7

Models for Categorical Response Variables

7.1 Generalized Linear Models

7.1.1 Extension of the Regression Model

Generalized linear models (GLMs) are a generalization of the classical linear models of regression analysis and analysis of variance, which model the relationship between the expectation of a response variable and unknown predictor variables according to

$$\begin{aligned} \mathrm{E}(y_i) &= x_{i1}\beta_1 + \ldots + x_{ip}\beta_p \\ &= x_i'\beta \,. \end{aligned} \tag{7.1}$$

The parameters are estimated according to the principle of least squares and are optimal according to the minimum dispersion theory or, in the case of a normal distribution, are optimal according to the ML theory (cf. Chapter 3).

Assuming an additive random error ϵ_i, the density function can be written as

$$f(y_i) = f_{\epsilon_i}(\, y_i - x_i'\beta) \,, \tag{7.2}$$

where $\eta_i = x_i'\beta$ is the linear predictor. Hence, for continuous normally distributed data, we have the following distribution and mean structure:

$$y_i \sim N(\mu_i, \sigma^2), \quad \mathrm{E}(y_i) = \mu_i \,, \quad \mu_i = \eta_i = x_i'\beta \,. \tag{7.3}$$

In analyzing categorical response variables, three major distributions may arise: the binomial, multinomial, and Poisson distributions, which belong to the natural exponential family (along with the normal distribution).

In analogy to the normal distribution, the effect of covariates on the expectation of the response variables may be modeled by linear predictors for these distributions as well.

Binomial Distribution

Assume that I predictors $\eta_i = x_i'\beta$ $(i = 1, \ldots, I)$ and N_i realizations y_{ij} $(j = 1, \ldots, N_i)$, respectively, are given and, furthermore, assume that the response has a binomial distribution

$$y_i \sim B(N_i, \pi_i) \quad \text{with} \quad \mathrm{E}(y_i) = N_i \pi_i = \mu_i \, .$$

Let $g(\pi_i) = \mathrm{logit}(\pi_i)$ be the chosen link function between μ_i and η_i:

$$\begin{aligned} \mathrm{logit}(\pi_i) &= \ln\left(\frac{\pi_i}{1 - \pi_i}\right) \\ &= \ln\left(\frac{N_i \pi_i}{N_i - N_i \pi_i}\right) = x_i'\beta \, . \end{aligned} \tag{7.4}$$

With the inverse function $g^{-1}(x_i'\beta)$ we then have

$$N_i \pi_i = \mu_i = N_i \frac{\exp(x_i'\beta)}{1 + \exp(x_i'\beta)} = g^{-1}(\eta_i) \, . \tag{7.5}$$

Poisson Distribution

Let y_i $(i = 1, \ldots, I)$ have a Poisson distribution with $\mathrm{E}(y_i) = \mu_i$:

$$P(y_i) = \frac{e^{-\mu_i} \mu_i^{y_i}}{y_i!} \quad \text{for } y_i = 0, 1, 2, \ldots \, . \tag{7.6}$$

The link function can then be chosen as $\ln(\mu_i) = x_i'\beta$.

Contingency Tables

The cell frequencies y_{ij} of an $(I \times J)$–contingency table of two categorical variables can have a Poisson, multinomial, or binomial distribution (depending on the sampling design). By choosing appropriate design vectors x_{ij}, the expected cell frequencies can be described by a loglinear model

$$\begin{aligned} \ln(m_{ij}) &= \mu + \alpha_i^A + \beta_j^B + (\alpha\beta)_{ij}^{AB} \\ &= x_{ij}'\beta \end{aligned} \tag{7.7}$$

and, hence, we have

$$\mu_{ij} = m_{ij} = \exp(x_{ij}'\beta) = \exp(\eta_{ij}) \, . \tag{7.8}$$

In contrast to the classical model of regression analysis, where $E(y)$ is linear in the parameter vector β, so that $\mu = \eta = x'\beta$ holds, the generalized models are of the following form:

$$\mu = g^{-1}(x'\beta), \tag{7.9}$$

where g^{-1} is the inverse function of the link function. Furthermore, the additivity of the random error is no longer a necessary assumption, so that, in general,

$$f(y) = f(y, x'\beta) \tag{7.10}$$

is assumed, instead of (7.2).

7.1.2 Structure of the Generalized Linear Model

The *generalized linear model (GLM)* (cf. Nelder and Wedderburn, 1972) is defined as follows. A GLM consists of three components:

- the *random component*, which specifies the probability distribution of the response variable;
- the *systematic component*, which specifies a linear function of the explanatory variables; and
- the *link function*, which describes a functional relationship between the systematic component and the expectation of the random component.

The three components are specified as follows:

1. The random component Y consists of N independent observations $y' = (y_1, y_2, \ldots, y_N)$ of a distribution belonging to the natural exponential family (cf. Agresti, 1990, p. 80). Hence, each observation y_i has—in the simplest case of a one–parametric exponential family—the following probability density function:

$$f(y_i, \theta_i) = a(\theta_i)\, b(y_i) \exp\left(y_i\, Q(\theta_i)\right) . \tag{7.11}$$

Remark. The parameter θ_i can vary over $i = 1, 2, \ldots, N$, depending on the value of the explanatory variable, which influences y_i through the systematic component.

Special distributions of particular importance in this family are the Poisson and the binomial distribution. $Q(\theta_i)$ is called the *natural parameter* of the distribution. Likewise, if the y_i are independent, the joint distribution is a member of the exponential family.

A more general parametrization allows inclusion of scaling or nuisance variables. For example, an alternative parametrization with an additional scaling parameter ϕ (the so–called dispersion parameter) is given by

$$f(y_i \mid \theta_i, \phi) = \exp\left\{\frac{y_i\theta_i - b(\theta_i)}{a(\phi)} + c(y_i, \phi)\right\}, \tag{7.12}$$

where θ_i is called the natural parameter. If ϕ is known, (7.12) represents a linear exponential family. If, on the other hand, ϕ is unknown, then (7.12) is called an *exponential dispersion model*. With ϕ and θ_i, (7.12) is a two–parametric distribution for $i = 1, \ldots, N$, which, for instance, is used for normal or gamma distributions. Introducing y_i and θ_i as vector–valued parameters rather than scalars leads to multivariate generalized models, which include multinomial response models as a special case (cf. Fahrmeir and Tutz, 2001, Chapter 3).

2. The systematic component relates a vector $\eta = (\eta_1, \eta_2, \ldots, \eta_N)$ to a set of explanatory variables through a linear model

$$\eta = X\beta\,. \tag{7.13}$$

Here η is called the linear predictor, $X : N \times p$ is the matrix of observations on the explanatory variables, and β is the $(p \times 1)$–vector of parameters.

3. The link function connects the systematic component with the expectation of the random component. Let $\mu_i = \mathrm{E}(y_i)$; then μ_i is linked to η_i by $\eta_i = g(\mu_i)$. Here g is a monotonic and differentiable function

$$g(\mu_i) = \sum_{j=1}^{p} \beta_j x_{ij}, \quad i = 1, 2, \ldots, N\,. \tag{7.14}$$

Special cases:

(i) $g(\mu) = \mu$ is called the *identity link*. We get $\eta_i = \mu_i$.

(ii) $g(\mu) = Q(\theta_i)$ is called the *canonical natural link*. We have $Q(\theta_i) = \sum_{j=1}^{p} \beta_j x_{ij}$.

Properties of the Density Function (7.12)

Let

$$l_i = l(\theta_i, \phi; y_i) = \ln f(y_i; \theta_i, \phi) \tag{7.15}$$

be the contribution of the ith observation y_i to the loglikelihood. Then

$$l_i = [y_i\theta_i - b(\theta_i)]/a(\phi) + c(y_i; \phi) \tag{7.16}$$

holds and we get the following derivatives with respect to θ_i:

$$\frac{\partial l_i}{\partial \theta_i} = \frac{[y_i - b'(\theta_i)]}{a(\phi)}, \tag{7.17}$$

$$\frac{\partial^2 l_i}{\partial \theta_i^2} = \frac{-b''(\theta_i)}{a(\phi)}, \tag{7.18}$$

where $b'(\theta_i) = \partial b(\theta_i)/\partial\theta_i$ and $b''(\theta_i) = \partial^2 b(\theta_i)/\partial\theta_i^2$ are the first and second derivatives of the function $b(\theta_i)$, assumed to be known. By equating (7.17) to zero, it becomes obvious that the solution of the likelihood equations is independent of $a(\phi)$. Since our interest lies with the estimation of θ and β in $\eta = x'\beta$, we could assume $a(\phi) = 1$ without any loss of generality (this corresponds to assuming $\sigma^2 = 1$ in the case of a normal distribution). For the present, however, we retain $a(\phi)$.

Under certain assumptions of regularity, the order of integration and differentiation may be interchangeable, so that

$$\mathrm{E}\left(\frac{\partial l_i}{\partial\theta_i}\right) = 0, \tag{7.19}$$

$$-\mathrm{E}\left(\frac{\partial^2 l_i}{\partial\theta_i^2}\right) = \mathrm{E}\left(\frac{\partial l_i}{\partial\theta_i}\right)^2. \tag{7.20}$$

Hence, we have, from (7.17) and (7.19),

$$\mathrm{E}(y_i) = \mu_i = b'(\theta_i). \tag{7.21}$$

Similarly, from (7.18) and (7.20), we find

$$\begin{aligned} \frac{b''(\theta_i)}{a(\phi)} &= \mathrm{E}\left\{\frac{[y_i - b'(\theta_i)]^2}{a^2(\phi)}\right\} \\ &= \frac{\mathrm{var}(y_i)}{a^2(\phi)}, \end{aligned} \tag{7.22}$$

since $\mathrm{E}[y_i - b'(\theta_i)] = 0$ and, hence,

$$V(\mu_i) = \mathrm{var}(y_i) = b''(\theta_i)a(\phi). \tag{7.23}$$

Under the assumption that the y_i ($i = 1, \ldots, N$) are independent, the loglikelihood of $y' = (y_1, \ldots, y_N)$ equals the sum of $l_i(\theta_i, \phi; y_i)$. Let

$$\theta' = (\theta_1, \ldots, \theta_N), \quad \mu' = (\mu_1, \ldots, \mu_N), \quad X = \begin{pmatrix} x_1' \\ \vdots \\ x_N' \end{pmatrix},$$

and

$$\eta = (\eta_1, \ldots, \eta_N)' = X\beta.$$

We then have, from (7.21),

$$\mu = \frac{\partial b(\theta)}{\partial\theta} = \left(\frac{\partial b(\theta_1)}{\partial\theta_1}, \ldots, \frac{\partial b(\theta_1)}{\partial\theta_N}\right)', \tag{7.24}$$

and, in analogy to (7.23) for the covariance matrix of $y' = (y_1, \ldots, y_N)$,

$$\mathrm{cov}(y) = V(\mu) = \frac{\partial^2 b(\theta)}{\partial\theta\,\partial\theta'} = a(\phi)\ \mathrm{diag}(b''(\theta_1), \ldots, b''(\theta_N)). \tag{7.25}$$

These relations hold in general, as we show in the following discussion.

7.1.3 Score Function and Information Matrix

The likelihood of the random sample is the product of the density functions

$$L(\theta, \phi; y) = \prod_{i=1}^{N} f(y_i; \theta_i, \phi) . \tag{7.26}$$

The loglikelihood $\ln L(\theta, \phi; y)$ for the sample y of independent y_i (for $i = 1, \ldots, N$) is of the form

$$l = l(\theta, \phi; y) = \sum_{i=1}^{N} l_i = \sum_{i=1}^{N} \left\{ \frac{(y_i\theta_i - b(\theta_i))}{a(\phi)} + c(y_i; \phi) \right\} . \tag{7.27}$$

The vector of first derivatives of l with respect to θ_i is needed for determining the ML estimates. This vector is called the *score function.* For now, we neglect the parametrization with ϕ in the representation of l and L and thus get the score function as

$$s(\theta; y) = \frac{\partial}{\partial \theta} l(\theta; y) = \frac{1}{L(\theta; y)} \frac{\partial}{\partial \theta} L(\theta; y) . \tag{7.28}$$

Let

$$\frac{\partial^2 l}{\partial \theta \, \partial \theta'} = \left(\frac{\partial^2 l}{\partial \theta_i \, \partial \theta_j} \right)_{\substack{i=1,\ldots,N \\ j=1,\ldots,N}}$$

be the matrix of the second derivatives of the loglikelihood. Then

$$F_{(N)}(\theta) = \mathrm{E} \left(\frac{-\partial^2 l(\theta; y)}{\partial \theta \, \partial \theta'} \right) \tag{7.29}$$

is called the expected *Fisher–information matrix* of the sample with $y' = (y_1, \ldots, y_N)$, where the expectation is to be taken with respect to the following density function

$$f(y_1, \ldots, y_N | \theta_i) = \prod f(y_i | \theta_i) = L(\theta; y) .$$

In the case of regular likelihood functions (where regular means that the exchange of integration and differentiation is possible), which the exponential families belong to, we have

$$\mathrm{E}(s(\theta; y)) = 0 \tag{7.30}$$

and

$$F_{(N)}(\theta) = \mathrm{E}(s(\theta; y) s'(\theta; y)) = \mathrm{cov}(s(\theta; y)) , \tag{7.31}$$

Relation (7.30) follows from

$$\int f(y_1, \ldots, y_N | \theta) \, \mathrm{d}y_1 \cdots \mathrm{d}y_N = \int L(\theta; y) \, \mathrm{d}y = 1 , \tag{7.32}$$

by differentiating with respect to θ, using (7.28),

$$\begin{aligned}\int \frac{\partial L(\theta; y)}{\partial \theta}\, \mathrm{d}y &= \int \frac{\partial l(\theta; y)}{\partial \theta} L(\theta; y)\mathrm{d}y \\ &= \mathrm{E}(s(\theta; y)) = 0\,. \end{aligned} \tag{7.33}$$

Differentiating (7.33) with respect to θ', we get

$$\begin{aligned} 0 &= \int \frac{\partial^2 l(\theta; y)}{\partial \theta\, \partial \theta'} L(\theta; y)\, \mathrm{d}y \\ &\quad + \int \frac{\partial l(\theta; y)}{\partial \theta} \frac{\partial l(\theta; y)}{\partial \theta'} L(\theta; y)\, \mathrm{d}y \\ &= -F_{(N)}(\theta) + \mathrm{E}(s(\theta; y) s'(\theta; y))\,, \end{aligned}$$

and hence (7.31), because $\mathrm{E}(s(\theta; y)) = 0$.

7.1.4 Maximum Likelihood Estimation

Let $\eta_i = x_i'\beta = \sum_{j=1}^{p} x_{ij}\beta_j$ be the predictor of the ith observation of the response variable ($i = 1, \ldots, N$) or, in matrix representation,

$$\eta = \begin{pmatrix} \eta_1 \\ \vdots \\ \eta_N \end{pmatrix} = \begin{pmatrix} x_1'\beta \\ \vdots \\ x_N'\beta \end{pmatrix} = X\beta\,. \tag{7.34}$$

Assume that the predictors are linked to $\mathrm{E}(y) = \mu$ by a monotonic differentiable function $g(\cdot)$:

$$g(\mu_i) = \eta_i \quad (i = 1, \ldots, N)\,, \tag{7.35}$$

or, in matrix representation,

$$g(\mu) = \begin{pmatrix} g(\mu_1) \\ \vdots \\ g(\mu_N) \end{pmatrix} = \eta\,. \tag{7.36}$$

The parameters θ_i and β are then linked by relation (7.21), that is, $\mu_i = b'(\theta_i)$, with $g(\mu_i) = x_i'\beta$. Hence we have $\theta_i = \theta_i(\beta)$. Since we are interested only in estimating β, we write the loglikelihood (7.27) as a function of β:

$$l(\beta) = \sum_{i=1}^{N} l_i(\beta)\,. \tag{7.37}$$

We can find the derivatives $\partial l_i(\beta)/\partial \beta_j$ according to the chain rule

$$\frac{\partial l_i(\beta)}{\partial \beta_j} = \frac{\partial l_i}{\partial \theta_i} \frac{\partial \theta_i}{\partial \mu_i} \frac{\partial \mu_i}{\partial \eta_i} \frac{\partial \eta_i}{\partial \beta_j}\,. \tag{7.38}$$

The partial results are as follows:

$$\begin{aligned} \frac{\partial l_i}{\partial \theta_i} &= \frac{[y_i - b'(\theta_i)]}{a(\phi)} && \text{[cf. (7.17)]} \\ &= \frac{[y_i - \mu_i]}{a(\phi)} && \text{[cf. (7.21)]}, \end{aligned} \tag{7.39}$$

$$\begin{aligned} \mu_i &= b'(\theta_i)\,, \\ \frac{\partial \mu_i}{\partial \theta_i} &= b''(\theta_i) = \frac{\text{var}(y_i)}{a(\phi)} && \text{[cf. (7.23)]}, \end{aligned} \tag{7.40}$$

$$\frac{\partial \eta_i}{\partial \beta_j} = \frac{\partial \sum_{k=1}^{p} x_{ik}\beta_k}{\partial \beta_j} = x_{ij}\,. \tag{7.41}$$

Because $\eta_i = g(\mu_i)$, the derivative $\partial \mu_i / \partial \eta_i$ is dependent on the link function $g(\cdot)$, or rather its inverse $g^{-1}(\cdot)$. Hence, it cannot be specified until the link is defined.

Summarizing, we now have

$$\frac{\partial l_i}{\partial \beta_j} = \frac{(y_i - \mu_i)x_{ij}}{\text{var}(y_i)} \frac{\partial \mu_i}{\partial \eta_i}\,, \quad j = 1, \ldots, p, \tag{7.42}$$

using the rule

$$\frac{\partial \theta_i}{\partial \mu_i} = \left(\frac{\partial \mu_i}{\partial \theta_i} \right)^{-1}$$

for inverse functions ($\mu_i = b'(\theta_i), \theta_i = (b')^{-1}(\mu_i)$). The likelihood equations for finding the components β_j are now

$$\sum_{i=1}^{N} \frac{(y_i - \mu_i)x_{ij}}{\text{var}(y_i)} \frac{\partial \mu_i}{\partial \eta_i} = 0, \quad j = 1, \ldots, p. \tag{7.43}$$

The loglikelihood is nonlinear in β. Hence, the solution of (7.43) requires iterative methods. For the second derivative with respect to components of β, we have, in analogy to (7.20), with (7.42),

$$\begin{aligned} \mathrm{E}\left(\frac{\partial^2 l_i}{\partial \beta_j \partial \beta_h} \right) &= -\mathrm{E}\left(\frac{\partial l_i}{\partial \beta_j} \right) \left(\frac{\partial l_i}{\partial \beta_h} \right) \\ &= -\mathrm{E}\left[\frac{(y_i - \mu_i)(y_i - \mu_i)x_{ij}x_{ih}}{(\text{var}(y_i))^2} \left(\frac{\partial \mu_i}{\partial \eta_i} \right)^2 \right] \\ &= -\frac{x_{ij}x_{ih}}{\text{var}(y_i)} \left(\frac{\partial \mu_i}{\partial \eta_i} \right)^2, \end{aligned} \tag{7.44}$$

and, hence,

$$\mathrm{E}\left(-\frac{\partial^2 l(\beta)}{\partial \beta_j \partial \beta_h} \right) = \sum_{i=1}^{N} \frac{x_{ij}x_{ih}}{\text{var}(y_i)} \left(\frac{\partial \mu_i}{\partial \eta_i} \right)^2 \tag{7.45}$$

and, in matrix representation for all (j, h) combinations,

$$F_{(N)}(\beta) = \mathrm{E}\left(-\frac{\partial^2 l(\beta)}{\partial\beta\partial\beta'}\right) = X'WX \tag{7.46}$$

with

$$W = \mathrm{diag}(w_1, \ldots, w_N) \tag{7.47}$$

and the weights

$$w_i = \frac{(\partial\mu_i/\partial\eta_i)^2}{\mathrm{var}(y_i)}\,. \tag{7.48}$$

Fisher–Scoring Algorithm

For the iterative determination of the ML estimate of β, the method of iterative reweighted least squares is used. Let $\beta^{(k)}$ be the kth approximation of the ML estimate $\hat\beta$. Furthermore, let $q^{(k)}(\beta) = \partial l(\beta)/\partial\beta$ be the vector of the first derivatives at $\beta^{(k)}$ (cf. (7.42)). Analogously, we define $W^{(k)}$. The formula of the Fisher–scoring algorithm is then

$$(X'W^{(k)}X)\beta^{(k+1)} = (X'W^{(k)}X)\beta^{(k)} + q^{(k)}\,. \tag{7.49}$$

The vector on the right side of (7.49) has the components (cf. (7.45) and (7.42))

$$\sum_h \left[\sum_i \frac{x_{ij}x_{ih}}{\mathrm{var}(y_i)}\left(\frac{\partial\mu_i}{\partial\eta_i}\right)^2 \beta_h^{(k)}\right] + \sum_i \frac{(y_i - \mu_i^{(k)})x_{ij}}{\mathrm{var}(y_i)}\left(\frac{\partial\mu_i}{\partial\eta_i}\right) \tag{7.50}$$
$$(j = 1, \ldots, p).$$

The entire vector (7.50) can now be written as

$$X'W^{(k)}z^{(k)}\,, \tag{7.51}$$

where the $(N \times 1)$–vector $z^{(k)}$ has the jth element as follows:

$$\begin{aligned} z_i^{(k)} &= \sum_{j=1}^{p} x_{ij}\beta_j^{(k)} + (y_i - \mu_i^{(k)})\left(\frac{\partial\eta_i^{(k)}}{\partial\mu_i^{(k)}}\right) \\ &= \eta_i^{(k)} + (y_i - \mu_i^{(k)})\left(\frac{\partial\eta_i^{(k)}}{\partial\mu_i^{(k)}}\right)\,. \end{aligned} \tag{7.52}$$

Hence, the equation of the Fisher–scoring algorithm (7.49) can now be written as

$$(X'W^{(k)}X)\beta^{(k+1)} = X'W^{(k)}z^{(k)}\,. \tag{7.53}$$

This is the likelihood equation of a generalized linear model with the response vector $z^{(k)}$ and the random error covariance matrix $(W^{(k)})^{-1}$. If $\mathrm{rank}(X) = p$ holds, we obtain the ML estimate $\hat\beta$ as the limit of

$$\hat\beta^{(k+1)} = (X'W^{(k)}X)^{-1}X'W^{(k)}z^{(k)} \tag{7.54}$$

for $k \to \infty$, with the asymptotic covariance matrix

$$\mathrm{V}(\hat{\beta}) = (X'\hat{W}X)^{-1} = F_{(N)}^{-1}(\hat{\beta}), \tag{7.55}$$

where $\hat{W}$ is determined at $\hat{\beta}$. Once a solution is found, then $\hat{\beta}$ is consistent for β, asymptotically normal, and asymptotically efficient (see Fahrmeir and Kaufmann (1985) and Wedderburn (1976) for existence and uniqueness of the solutions). Hence we have $\hat{\beta} \overset{\text{a.s.}}{\sim} N(\beta, \mathrm{V}(\hat{\beta}))$.

Remark. In the case of a canonical link function, that is for $g(\mu_i) = \theta_i$, the ML equations simplify and the Fisher–scoring algorithm is identical to the Newton–Raphson algorithm (cf. Agresti, 1990, p. 451). If the values $a(\phi)$ are identical for all observations, then the ML equations are

$$\sum_i x_{ij} y_i = \sum_i x_{ij} \mu_i \,. \tag{7.56}$$

If, on the other hand, $a(\phi) = a_i(\phi) = a_i \phi$ $(i = 1, \ldots, N)$ holds, then the ML equations are

$$\sum_i \frac{x_{ij} y_i}{a_i} = \sum_i \frac{x_{ij} \mu_i}{a_i} \,. \tag{7.57}$$

As starting values for the Fisher–scoring algorithm the estimates $\hat{\beta}^{(0)} = (X'X)^{-1}X'y$ or $\hat{\beta}^{(0)} = (X'X)^{-1}X'g(y)$ may be used.

7.1.5 Testing of Hypotheses and Goodness of Fit

A generalized linear model $g(\mu_i) = x_i'\beta$ is—besides the distributional assumptions—determined by the link function $g(\cdot)$ and the explanatory variables $X_1, \ldots, X_p$, as well as their number p, which determines the length of the parameter vector β to be estimated. If $g(\cdot)$ is chosen, then the model is defined by the design matrix X.

Testing of Hypotheses

Let X_1 and X_2 be two design matrices (models), and assume that the hierarchical order $X_1 \subset X_2$ holds; that is, we have $X_2 = (X_1, X_3)$ with some matrix X_3 and hence $\mathcal{R}(X_1) \subset \mathcal{R}(X_2)$. Let β_1, β_2, and β_3 be the corresponding parameter vectors to be estimated. Further, let $g(\hat{\mu}_1) = \hat{\eta}_1 = X_1\hat{\beta}_1$ and $g(\hat{\mu}_2) = \hat{\eta}_2 = X_2\tilde{\beta}_2 = X_1\tilde{\beta}_1 + X_3\tilde{\beta}_3$, where $\hat{\beta}_1$ and $\tilde{\beta}_2 = (\tilde{\beta}_1', \tilde{\beta}_3')'$ are the maximum–likelihood estimates under the two models, and $\operatorname{rank}(X_1) = r_1$, $\operatorname{rank}(X_2) = r_2$, and $(r_2 - r_1) = r = df$. The likelihood ratio statistic, which compares a larger model X_2 with a (smaller) submodel X_1, is then defined as follows (where L is the likelihood function)

$$\Lambda = \frac{\max_{\beta_1} L(\beta_1)}{\max_{\beta_2} L(\beta_2)} \,. \tag{7.58}$$

Wilks (1938) showed that $-2\ln\Lambda$ has a limiting χ^2_{df}–distribution where the degrees of freedom *df* equal the difference in the dimensions of the two models. Transforming (7.58) according to $-2\ln\Lambda$, with l denoting the loglikelihood, and inserting the maximum likelihood estimates gives

$$-2\ln\Lambda = -2[l(\hat{\beta}_1) - l(\tilde{\beta}_2)] \,. \tag{7.59}$$

In fact, one tests the hypotheses $H_0 : \beta_3 = 0$ against $H_1 : \beta_3 \neq 0$. If H_0 holds, then $-2\ln\Lambda \sim \chi^2_r$. Therefore, H_0 is rejected if the loglikelihood is significantly higher under the greater model using X_2. According to Wilks, we write

$$G^2 = -2\ln\Lambda \,.$$

Goodness of Fit

Let X be the design matrix of the saturated model that contains the same number of parameters as observations. Denote by $\tilde{\theta}$ the estimate of θ that belongs to the estimates $\tilde{\mu}_i = y_i$ $(i = 1, \ldots, N)$ in the saturated model. For every submodel X_j that is not saturated, we then have (assuming again that $a(\phi) = a_i(\phi) = a_i\phi$)

$$\begin{aligned} G^2(X_j|X) &= 2\sum \frac{1}{a_i} \frac{y_i(\tilde{\theta}_i - \hat{\theta}_i) - b(\tilde{\theta}_i) + b(\hat{\theta}_i)}{\phi} \\ &= \frac{D(y;\hat{\mu}_j)}{\phi} \end{aligned} \tag{7.60}$$

as a measure for the loss in goodness of fit of the model X_j compared to the perfect fit achieved by the saturated model. The statistic $D(y;\hat{\mu}_j)$ is called the *deviance* of the model X_j. We then have

$$G^2(X_1 \mid X_2) = G^2(X_1 \mid X) - G^2(X_2 \mid X) = \frac{D(y;\hat{\mu}_1) - D(y;\hat{\mu}_2)}{\phi} \,. \tag{7.61}$$

That is, the test statistic for comparing the model X_1 with the larger model X_2 equals the difference of the goodness–of–fit statistics of the two models, weighted with $1/\phi$.

7.1.6 *Overdispersion*

In samples of a Poisson or multinomial distribution, it may occur that the elements show a larger variance than that given by the distribution. This may be due to a violation of the assumption of independence, as, for example, a positive correlation in the sample elements. A frequent cause for this is the cluster structure of the sample. Examples are:

- the behavior of families of insects in the case of the influence of insecticides (Agresti, 1990, p. 42), where the family (cluster, batch) shows a *collective* (correlated) survivorship (many survive or most of them die) rather than an independent survivorship, due to dependence on cluster-specific covariates such as the temperature;

- the survivorship of dental implants when two or more implants are incorporated for each patient;
- the developement of diseases, or the social behavior of the members of a family; and
- heterogeneity is not taken into account, which is, for example, caused by not having measured important covariates for the linear predictor.

The existence of a larger variation (inhomogeneity) in the sample than in the sample model is called *overdispersion*. Overdispersion is, in the simplest way, modeled by multiplying the variance with a constant $\phi > 1$, where ϕ is either known (e.g., $\phi = \sigma^2$ for a normal distribution), or has to be estimated from the sample (Fahrmeir and Tutz, 2001).

Example (McCullagh and Nelder, 1989, p. 125): Let N individuals be divided into N/k clusters of equal cluster size k. Assume that the individual response is binary with $P(Y_i = 1) = \pi_i$, so that the total response

$$Y = Z_1 + Z_2 + \cdots + Z_{N/k}$$

equals the sum of independent $B(k; \pi_i)$–distributed binomial variables Z_i $(i = 1, \ldots, N/k)$. The π_i's vary across the clusters and we assume that $\mathrm{E}(\pi_i) = \pi$ and $\mathrm{var}(\pi_i) = \tau^2\pi(1-\pi)$ with $0 \le \tau^2 \le 1$. We then have

$$\begin{aligned} \mathrm{E}(Y) &= N\pi, \\ \mathrm{var}(Y) &= N\pi(1-\pi)\{1 + (k-1)\tau^2\} \\ &= \phi N\pi(1-\pi). \end{aligned} \tag{7.62}$$

The dispersion parameter $\phi = 1 + (k-1)\tau^2$ is dependent on the cluster size k and on the variability of the π_i, but not on the sample size N. This fact is essential for interpreting the variable Y as the sum of the binomial variables Z_i and for estimating the dispersion parameter ϕ from the residuals. Because of $0 \le \tau^2 \le 1$, we have

$$1 \le \phi \le k \le N. \tag{7.63}$$

Relationship (7.62) means that

$$\frac{\mathrm{var}(Y)}{N\pi(1-\pi)} = 1 + (k-1)\tau^2 = \phi \tag{7.64}$$

is constant. An alternative model—the beta–binomial distribution—has the property that the quotient in (7.64), i.e., ϕ, is a linear function of the sample size N. By plotting the residuals against N, it is easy to recognize which of the two models is more likely. Rosner (1984) used the beta–binomial distribution for estimation in clusters of size $k = 2$.

7.1.7 Quasi Loglikelihood

The generalized models assume a distribution of the natural exponential family for the data as the random component (cf. (7.11)). If this assumption does not hold, an alternative approach can be used to specify the functional relationship between the mean and the variance. For exponential families, the relationship (7.23) between variance and expectation holds. Assume the general approach

$$\operatorname{var}(Y) = \phi V(\mu)\,, \tag{7.65}$$

where $V(\cdot)$ is an appropriately chosen function.

In the quasi–likelihood approach (Wedderburn, 1974), only assumptions about the first and second moments of the random variables are made. It is not necessary for the distribution itself to be specified. The starting point in estimating the influence of covariates is the score function (7.28), or rather the system of ML equations (7.43). If the general specification (7.65) is inserted into (7.43), we get the system of *estimating equations* for β:

$$\sum_{i=1}^{N} \frac{(y_i - \mu_i)}{V(\mu_i)} x_{ij} \frac{\partial \mu_i}{\partial \eta_i} = 0 \quad (j = 1, \ldots, p)\,, \tag{7.66}$$

which is of the same form as the likelihood equations (7.43) for GLMs. However, system (7.66) is an ML equation system only if the y_i's have a distribution of the natural exponential family.

In the case of independent response, the modeling of the influence of the covariates X on the mean response $\mathrm{E}(y) = \mu$ is done according to McCullagh and Nelder (1989, p. 324) as follows. Assume that for the response vector we have

$$y \sim (\mu, \phi V(\mu))\,, \tag{7.67}$$

where $\phi > 0$ is an unknown dispersion parameter and $V(\mu)$ is a matrix of known functions. Expression $\phi V(\mu)$ is called the *working variance.*

If the components of y are assumed to be independent, the covariance matrix $\phi V(\mu)$ has to be diagonal, that is,

$$V(\mu) = \operatorname{diag}(V_1(\mu), \ldots, V_N(\mu))\,. \tag{7.68}$$

Here it is realistic to assume that the variance of each random variable y_i is dependent only on the ith component μ_i of μ, meaning thereby

$$V(\mu) = \operatorname{diag}(V_1(\mu_1), \ldots, V_N(\mu_N)). \tag{7.69}$$

A dependency on all components of μ according to (7.68) is difficult to interpret in practice, if independence of the y_i is demanded as well. (Nevertheless, situations as in (7.68) are possible.) In many applications it is reasonable to assume, in addition to the functional independency (7.69), that the V_i functions are identical, so that

$$V(\mu) = \text{diag}(v(\mu_1), \ldots, v(\mu_N)) \tag{7.70}$$

holds, with $V_i = v(\cdot)$.

Under the above assumptions, the following function for a component y_i of y:

$$U = u(\mu_i, y_i) = \frac{y_i - \mu_i}{\phi v(\mu_i)} \tag{7.71}$$

has the properties

$$\begin{aligned} \text{E}(U) &= 0, & (7.72)\\ \text{var}(U) &= \frac{1}{\phi v(\mu_i)}, & (7.73)\\ \frac{\partial U}{\partial \mu_i} &= \frac{-\phi v(\mu_i) - (y_i - \mu_i)\phi \partial v(\mu_i)/\partial \mu_i}{\phi^2 v^2(\mu_i)}, & \\ -\text{E}\left(\frac{\partial U}{\partial \mu_i}\right) &= \frac{1}{\phi v(\mu_i)}. & (7.74) \end{aligned}$$

Hence U has the same properties as the derivative of a loglikelihood, which, of course, is the score function (7.28). Property (7.47) corresponds to (7.31), whereas property (7.74), in combination with (7.73), corresponds to (7.31). Therefore,

$$Q(\mu; y) = \sum_{i=1}^{N} Q_i(\mu_i; y_i) \tag{7.75}$$

with

$$Q_i(\mu_i; y_i) = \int_{y_i}^{\mu_i} \frac{\mu_i - t}{\phi v(t)} \text{d}t \tag{7.76}$$

(cf. McCullagh and Nelder, 1989, p. 325) is the analog of the loglikelihood function. $Q(\mu; y)$ is called *quasi loglikelihood.* Hence, the *quasi–score function*, which is obtained by differentiating $Q(\mu; y)$, equals

$$U(\beta) = \phi^{-1} D' V^{-1} (y - \mu), \tag{7.77}$$

with $D = (\partial \mu_i / \partial \beta_j)$ $(i = 1, \ldots, N, j = 1, \ldots, p)$ and $V = \text{diag}(v_1, \ldots, v_N)$. The quasi–likelihood estimate $\hat{\beta}$ is the solution of $U(\hat{\beta}) = 0$. It has the asymptotic covariance matrix

$$\text{cov}(\hat{\beta}) = \phi (D' V^{-1} D)^{-1}. \tag{7.78}$$

The dispersion parameter ϕ is estimated by

$$\hat{\phi} = \frac{1}{N-p} \frac{\sum (y_i - \hat{\mu}_i)^2}{v(\hat{\mu}_i)} = \frac{X^2}{N-p}, \tag{7.79}$$

where X^2 is the so–called Pearson statistic. In the case of overdispersion (or assumed overdispersion), the influence of covariates (i.e., of the vector β) is to be estimated by a quasi–likelihood approach (7.66) rather than by a likelihood approach.

7.2 Contingency Tables

7.2.1 Overview

This section deals with contingency tables and the appropriate models. We first consider so–called two–way contingency tables. In general, a bivariate relationship is described by the joint distribution of the two associated random variables. The two marginal distributions are obtained by integrating (summing) the joint distribution over the respective variables. Likewise, the conditional distributions can be derived from the joint distribution.

Definition 7.1 (Contingency Table). Let X and Y denote two categorical variables, with X at I levels and Y at J levels. When we observe subjects with the variables X and Y, there are $I \times J$ possible combinations of classifications. The outcomes $(X;Y)$ of a sample with sample size n are displayed in an $I \times J$ (contingency) table. (X,Y) are realizations of the joint two–dimensional distribution

$$P(X=i, Y=j) = \pi_{ij}\,. \tag{7.80}$$

The set $\{\pi_{ij}\}$ forms the joint distribution of X and Y. The marginal distributions are obtained by summing over rows or columns

		Y				Marginal
		1	2	...	J	distribution of X
	1	π_{11}	π_{12}	...	π_{1J}	π_{1+}
	2	π_{21}	π_{22}	...	π_{2J}	π_{2+}
X	$\vdots$	$\vdots$	$\vdots$		$\vdots$	$\vdots$
	I	π_{I1}	π_{I2}	...	π_{IJ}	π_{I+}
Marginal distribution of Y		π_{+1}	π_{+2}	...	π_{+J}	

$$\begin{aligned}
\pi_{+j} &= \sum_{i=1}^{I} \pi_{ij}\,, \quad j = 1, \ldots, J\,, \\
\pi_{i+} &= \sum_{j=1}^{J} \pi_{ij}\,, \quad i = 1, \ldots, I\,, \\
\sum_{i=1}^{I} \pi_{i+} &= \sum_{j=1}^{J} \pi_{+j} = 1\,.
\end{aligned}$$

In many contingency tables the explanatory variable X is fixed, and only the response Y is a random variable. In such cases, the main interest is not the joint distribution, but rather the conditional distribution. $\pi_{j|i} = P(Y = j \mid X = i)$ is the conditional probability, and $\{\pi_{1|i}, \pi_{2|i}, \ldots, \pi_{J|i}\}$, with $\sum_{j=1}^{J} \pi_{j|i} = 1$, is the conditional distribution of Y, given $X = i$.

A general aim of many studies is the comparison of the conditional distributions of Y at various levels i of X.

Suppose that X as well as Y are random response variables, so that the joint distribution describes the association of the two variables. Then, for the conditional distribution $Y|X$, we have

$$\pi_{j|i} = \frac{\pi_{ij}}{\pi_{i+}} \quad \forall i, j\,. \tag{7.81}$$

Definition 7.2. Two variables are called independent if

$$\pi_{ij} = \pi_{i+}\pi_{+j} \quad \forall i, j. \tag{7.82}$$

If X and Y are independent, we obtain

$$\pi_{j|i} = \frac{\pi_{ij}}{\pi_{i+}} = \frac{\pi_{i+}\pi_{+j}}{\pi_{i+}} = \pi_{+j}\,. \tag{7.83}$$

The conditional distribution is equal to the marginal distribution and thus is independent of i.

Let $\{p_{ij}\}$ denote the sample joint distribution. They have the following properties, with n_{ij} being the cell frequencies and $n = \sum_{i=1}^{I}\sum_{j=1}^{J} n_{ij}$:

$$\left.\begin{array}{lllllll}
p_{ij} & = & \dfrac{n_{ij}}{n}\,, & & & \\[2ex]
p_{j|i} & = & \dfrac{p_{ij}}{p_{i+}} = \dfrac{n_{ij}}{n_{i+}}\,, & & p_{i|j} & = & \dfrac{p_{ij}}{p_{+j}} = \dfrac{n_{ij}}{n_{+j}}\,, \\[2ex]
p_{i+} & = & \dfrac{\sum_{j=1}^{J} n_{ij}}{n}\,, & & p_{+j} & = & \dfrac{\sum_{i=1}^{I} n_{ij}}{n}\,, \\[2ex]
n_{i+} & = & \sum_{j=1}^{J} n_{ij} = np_{i+}\,, & & n_{+j} & = & \sum_{i=1}^{I} n_{ij} = np_{+j}\,.
\end{array}\right\} \tag{7.84}$$

7.2.2 Ways of Comparing Proportions

Suppose that Y is a binary response variable (Y can take only the values 0 or 1), and let the outcomes of X be grouped. When row i is fixed, $\pi_{1|i}$ is the probability for response ($Y = 1$), and $\pi_{2|i}$ is the probability for nonresponse ($Y = 0$). The conditional distribution of the binary response variable Y, given $X = i$, then is

$$(\pi_{1|i}; \pi_{2|i}) = (\pi_{1|i}, (1 - \pi_{1|i})). \tag{7.85}$$

We can now compare two rows, say i and h, by calculating the difference in proportions for response, or nonresponse, respectively,

$$\text{response:} \quad \pi_{1|h} - \pi_{1|i}$$

and

$$\begin{aligned} \text{nonresponse:} \quad \pi_{2|h} - \pi_{2|i} &= (1 - \pi_{1|h}) - (1 - \pi_{1|i}) \\ &= -\left(\pi_{1|h} - \pi_{1|i}\right). \end{aligned}$$

The differences have different signs, but their absolute values are identical. Additionally, we have

$$-1.0 \leq \pi_{1|h} - \pi_{1|i} \leq 1.0\,. \tag{7.86}$$

The difference equals zero if the conditional distributions of the two rows i and h coincide. From this, one may conjecture that the response variable Y is independent of the row classification when

$$\pi_{1|h} - \pi_{1|i} = 0 \quad \forall (h,i), \quad i,h = 1,2,\ldots,I\,, \quad i \neq h\,. \tag{7.87}$$

In a more general setting, with the response variable Y having J categories, the variables X and Y are independent if

$$\pi_{j|h} - \pi_{j|i} = 0 \quad \forall j\,, \forall (h,i), \quad i,h = 1,2,\ldots,I\,, \quad i \neq h\,. \tag{7.88}$$

Definition 7.3 (Relative Risk). Let Y denote a binary response variable. The ratio $\pi_{1|h}/\pi_{1|i}$ is called the relative risk for response of category h in relation to category i.

For 2×2 tables the relative risk (for response) is

$$0 \leq \frac{\pi_{1|1}}{\pi_{1|2}} < \infty\,. \tag{7.89}$$

The relative risk is a nonnegative real number. A relative risk of 1 corresponds to independence. For nonresponse, the relative risk is

$$\frac{\pi_{2|1}}{\pi_{2|2}} = \frac{1 - \pi_{1|1}}{1 - \pi_{1|2}}\,. \tag{7.90}$$

Definition 7.4 (Odds). The odds are defined as the ratio of the probability of response in relation to the probability of nonresponse, within one category of X.

For 2×2 tables, the odds in row 1 equal

$$\Omega_1 = \frac{\pi_{1|1}}{\pi_{2|1}}\,. \tag{7.91}$$

Within row 2, the corresponding odds equal

$$\Omega_2 = \frac{\pi_{1|2}}{\pi_{2|2}}\,. \tag{7.92}$$

Hint. For the joint distribution of two binary variables, the definition is

$$\Omega_i = \frac{\pi_{i1}}{\pi_{i2}}, \quad i = 1, 2. \tag{7.93}$$

In general, Ω_i is nonnegative. When $\Omega_i > 1$, response is more likely than nonresponse. If, for instance, $\Omega_1 = 4$, then response in the first row is four times as likely as nonresponse. The *within–row conditional distributions* are independent when $\Omega_1 = \Omega_2$. This implies that the two variables are independent:

$$X, Y \text{ independent} \quad \Leftrightarrow \quad \Omega_1 = \Omega_2. \tag{7.94}$$

Definition 7.5 (Odds Ratio). The *odds ratio* is defined as

$$\theta = \frac{\Omega_1}{\Omega_2}. \tag{7.95}$$

From the definition of the odds using joint probabilities, we have

$$\theta = \frac{\pi_{11}\pi_{22}}{\pi_{12}\pi_{21}}. \tag{7.96}$$

Another terminology for θ is the cross–product ratio. X and Y are independent when the odds ratio equals 1:

$$X, Y \text{ independent} \quad \Leftrightarrow \quad \theta = 1. \tag{7.97}$$

When all the cell probabilities are greater than 0 and $1 < \theta < \infty$, response for the subjects in the first row is more likely than for the subjects in the second row, that is, $\pi_{1|1} > \pi_{1|2}$. For $0 < \theta < 1$, we have $\pi_{1|1} < \pi_{1|2}$ (with a reverse interpretation).

The sample version of the odds ratio for the 2×2 table

		Y 1	Y 2	
X	1	n_{11}	n_{12}	n_{1+}
	2	n_{21}	n_{22}	n_{2+}
		n_{+1}	n_{+2}	n

is

$$\hat{\theta} = \frac{n_{11}n_{22}}{n_{12}n_{21}}. \tag{7.98}$$

Odds Ratios for $I \times J$ Tables

From any given $I \times J$ table, 2×2 tables can be constructed by picking two different rows and two different columns. There are $I(I-1)/2$ pairs of rows and $J(J-1)/2$ pairs of columns; hence an $I \times J$ table contains $IJ(I-1)(J-1)/4$ tables. The set of all 2×2 tables contains much redundant information; therefore, we consider only neighboring 2×2 tables with the local odds ratios

$$\theta_{ij} = \frac{\pi_{i,j}\pi_{i+1,j+1}}{\pi_{i,j+1}\pi_{i+1,j}}, \quad i = 1, 2, \ldots, I-1, \quad j = 1, 2, \ldots, J-1. \tag{7.99}$$

These $(I-1)(J-1)$ odds ratios determine all possible odds ratios formed from all pairs of rows and all pairs of columns.

7.2.3 Sampling in Two–Way Contingency Tables

Variables having nominal or ordinal scale are denoted as categorical variables. In most cases, statistical methods assume a multinomial or a Poisson distribution for categorical variables. We now elaborate these two sample models. Suppose that we observe counts n_i $(i = 1, 2, \ldots, N)$ in the N cells of a contingency table with a single categorical variable or in $N = I \times J$ cells of a two–way contingency table.

We assume that the n_i are random variables with a distribution in $\mathbb{R}^+$ and the expected values $\mathrm{E}(n_i) = m_i$, which are called expected frequencies.

Poisson Sample

The Poisson distribution is used for counts of events (such as response to a medical treatment) that occur randomly over time when outcomes in disjoint periods are independent. The Poisson distribution may be interpreted as the limit distribution of the binomial distribution $B(n; p)$ if $\lambda = n \cdot p$ is fixed for increasing n. For each of the N cells of a contingency table $\{n_i\}$, we have

$$P(n_i) = \frac{e^{-m_i} m_i^{n_i}}{n_i!}, \quad n_i = 0, 1, 2, \ldots, \quad i = 1, \ldots, N. \tag{7.100}$$

This is the probability mass function of the Poisson distribution with the parameter m_i. This satisfies the identities $\mathrm{var}(n_i) = E(n_i) = m_i$.

The Poisson model for $\{n_i\}$ assumes that the n_i are independent. The joint distribution for $\{n_i\}$ then is the product of the distributions for n_i in the N cells. The total sample size $n = \sum_{i=1}^N n_i$ also has a Poisson distribution with $E(n) = \sum_{i=1}^N m_i$ (the rule for summing up independent random variables with Poisson distribution).

The Poisson model is used if rare events are independently distributed over disjoint classes.

Let $n = \sum_{i=1}^N n_i$ be fixed. The conditional probability of a contingency table $\{n_i\}$ that satisfies this condition is

$$\begin{aligned}
&P\left(n_i \text{ observations in cell } i, i = 1, 2, \ldots, N \mid \textstyle\sum_{i=1}^N n_i = n\right) \\
&\quad = \frac{P(n_i \text{ observations in cell } i, i = 1, 2, \ldots, N)}{P(\sum_{i=1}^N n_i = n)} \\
&\quad = \frac{\prod_{i=1}^N e^{-m_i}[(m_i^{n_i})/n_i!]}{\exp(-\sum_{j=1}^N m_j)[(\sum_{j=1}^N m_j)^n/n!]}
\end{aligned}$$

$$= \left(\frac{n!}{\prod_{i=1}^{N} n_i!}\right) \cdot \prod_{i=1}^{N} \pi_i^{n_i}, \quad \text{with} \quad \pi_i = \frac{m_i}{\sum_{i=1}^{N} m_i}. \tag{7.101}$$

For $N = 2$, this is the binomial distribution. For the multinomial distribution for $(n_1, n_2, \ldots, n_N)$, the marginal distribution for n_i is a binomial distribution with $E(n_i) = n\pi_i$ and $\text{var}(n_i) = n\pi_i(1 - \pi_i)$.

Independent Multinomial Sample

Suppose we observe on a categorical variable Y at various levels of an explanatory variable X. In the cell $(X = i, Y = j)$ we have n_{ij} observations. Suppose that $n_{i+} = \sum_{j=1}^{J} n_{ij}$, the number of observations of Y for fixed level i of X, is fixed in advance (and thus not random) and that the n_{i+} observations are independent and have the distribution $(\pi_{1|i}, \pi_{2|i}, \ldots, \pi_{J|i})$. Then the cell counts in row i have the multinomial distribution

$$\left(\frac{n_{i+}!}{\prod_{j=1}^{J} n_{ij}!}\right) \cdot \prod_{j=1}^{J} \pi_{j|i}^{n_{ij}}. \tag{7.102}$$

Furthermore, if the samples are independent for different i, then the joint distribution for the n_{ij} in the $I \times J$ table is the product of the multinomial distributions (7.102). This is called *product multinomial sampling* or *independent multinomial sampling*.

7.2.4 Likelihood Function and Maximum Likelihood Estimates

For the observed cell counts $\{n_i, i = 1, 2, \ldots, N\}$, the likelihood function is defined as the probability of $\{n_i, i = 1, 2, \ldots, N\}$ for a given sampling model. This function, in general, is dependent on an unknown parameter θ—here, for instance, $\theta = \{\pi_{j|i}\}$. The maximum–likelihood estimate for this vector of parameters is the value for which the likelihood function of the observed data takes its maximum.

To illustrate, we now look at the estimates of the category probabilities $\{\pi_i\}$ for multinomial sampling. The joint distribution $\{n_i\}$ is (cf. (7.102) and the notation $\{\pi_i\}$, $i = 1, \ldots, N$, $N = I \cdot J$, instead of $\pi_{j|i}$)

$$\frac{n!}{\prod_{i=1}^{N} n_i!} \underbrace{\prod_{i=1}^{N} \pi_i^{n_i}}_{\text{kernel}}. \tag{7.103}$$

It is proportional to the so–called kernel of the likelihood function. The kernel contains all unknown parameters of the model. Hence, maximizing the likelihood is equivalent to maximizing the kernel of the loglikelihood function

$$\ln(\text{kernel}) = \sum_{i=1}^{N} n_i \ln(\pi_i) \to \max_{\pi_i} . \tag{7.104}$$

Under the condition $\pi_i > 0$ $(i = 1, 2, \ldots, N)$, $\sum_{i=1}^{N} \pi_i = 1$, we have $\pi_N = 1 - \sum_{i=1}^{N-1} \pi_i$ and, hence,

$$\frac{\partial \pi_N}{\partial \pi_i} = -1 , \quad i = 1, 2, \ldots, N-1 , \tag{7.105}$$

$$\frac{\partial \ln \pi_N}{\partial \pi_i} = \frac{1}{\pi_N} \cdot \frac{\partial \pi_N}{\partial \pi_i} = \frac{-1}{\pi_N} , \quad i = 1, 2, \ldots, N-1 , \tag{7.106}$$

$$\frac{\partial L}{\partial \pi_i} = \frac{n_i}{\pi_i} - \frac{n_N}{\pi_N} = 0 , \quad i = 1, 2, \ldots, N-1 . \tag{7.107}$$

From (7.107) we get

$$\frac{\hat{\pi}_i}{\hat{\pi}_N} = \frac{n_i}{n_N} , \quad i = 1, 2, \ldots, N-1 , \tag{7.108}$$

and thus

$$\hat{\pi}_i = \hat{\pi}_N \frac{n_i}{n_N} . \tag{7.109}$$

Using

$$\sum_{i=1}^{N} \hat{\pi}_i = 1 = \frac{\hat{\pi}_N \sum_{i=1}^{N} n_i}{n_N} , \tag{7.110}$$

we obtain the solutions

$$\hat{\pi}_N = \frac{n_N}{n} = p_N , \tag{7.111}$$

$$\hat{\pi}_i = \frac{n_i}{n} = p_i , \quad i = 1, 2, \ldots, N-1 . \tag{7.112}$$

The ML estimates are the proportions (relative frequencies) p_i.

For contingency tables we have, for independent X and Y,

$$\pi_{ij} = \pi_{i+} \pi_{+j} . \tag{7.113}$$

The ML estimates under this condition are

$$\hat{\pi}_{ij} = p_{i+} p_{+j} = \frac{n_{i+} n_{+j}}{n^2} \tag{7.114}$$

with the expected cell frequencies

$$\hat{m}_{ij} = n \hat{\pi}_{ij} = \frac{n_{i+} n_{+j}}{n} . \tag{7.115}$$

Because of the similarity of the likelihood functions, the ML estimates for Poisson, multinomial, and product multinomial sampling are identical (as long as no further assumptions are made).

7.2.5 Testing the Goodness of Fit

A principal aim of the analysis of contingency tables is to test whether the observed and the expected cell frequencies (specified by a model) coincide. For instance, Pearson's χ^2–statistic compares the observed and the expected cell frequencies from (7.115) for independent X and Y.

Testing a Specified Multinomial Distribution (Theoretical Distribution)

We first want to compare a multinomial distribution, specified by $\{\pi_{i0}\}$, with the observed distribution $\{n_i\}$ for N classes.

The hypothesis for this problem is

$$H_0 : \pi_i = \pi_{i0}\,, \quad i = 1, 2, \ldots, N\,, \tag{7.116}$$

whereas for the π_i we have the restriction

$$\sum_{i=1}^{N} \pi_i = 1\,. \tag{7.117}$$

When H_0 is true, the expected cell frequencies are

$$m_i = n\pi_{i0}\,, \quad i = 1, 2, \ldots, N\,. \tag{7.118}$$

The appropriate test statistic is Pearson's χ^2, where

$$\chi^2 = \sum_{i=1}^{N} \frac{(n_i - m_i)^2}{m_i} \overset{\text{approx.}}{\sim} \chi^2_{N-1}\,. \tag{7.119}$$

This can be justified as follows: Let $p = (n_1/n, \ldots, n_{N-1}/n)$ and $\pi_0 = (\pi_{1_0}, \ldots, \pi_{N-1_0})$. By the central limit theorem we then have, for $n \to \infty$,

$$\sqrt{n}\,(p - \pi_0) \to N\,(0, \Sigma_0)\;, \tag{7.120}$$

and so

$$n\,(p - \pi_0)'\,\Sigma_0^{-1}\,(p - \pi_0) \to \chi^2_{N-1}\,. \tag{7.121}$$

The asymptotic covariance matrix has the form

$$\Sigma_0 = \Sigma_0(\pi_0) = \text{diag}(\pi_0) - \pi_0\pi_0'\,. \tag{7.122}$$

Its inverse can be written as

$$\Sigma_0^{-1} = \frac{1}{\pi_{N0}} 11' + \text{diag}\left(\frac{1}{\pi_{10}}, \ldots, \frac{1}{\pi_{N-1\,,0}}\right). \tag{7.123}$$

The equivalence of (7.119) and (7.121) is proved by direct calculation. To illustrate, we choose $N = 3$. Using the relationship $\pi_1 + \pi_2 + \pi_3 = 1$, we have

$$\Sigma_0 = \begin{pmatrix} \pi_1 & 0 \\ 0 & \pi_2 \end{pmatrix} - \begin{pmatrix} \pi_1^2 & \pi_1\pi_2 \\ \pi_1\pi_2 & \pi_2^2 \end{pmatrix},$$

$$
\begin{aligned}
\Sigma_0^{-1} &= \begin{pmatrix} \pi_1(1-\pi_1) & -\pi_1\pi_2 \\ -\pi_1\pi_2 & \pi_2(1-\pi_2) \end{pmatrix}^{-1} \\
&= \frac{1}{\pi_1\pi_2\pi_3} \begin{pmatrix} \pi_2(1-\pi_2) & \pi_1\pi_2 \\ \pi_1\pi_2 & \pi_1(1-\pi_1) \end{pmatrix} \\
&= \begin{pmatrix} 1/\pi_1 + 1/\pi_3 & 1/\pi_3 \\ 1/\pi_3 & 1/\pi_2 + 1/\pi_3 \end{pmatrix}.
\end{aligned}
$$

The left side of (7.121) now is

$$
\begin{aligned}
& n\left(\frac{n_1}{n} - \frac{m_1}{n}, \frac{n_2}{n} - \frac{m_2}{n}\right) \begin{pmatrix} \frac{n}{m_1} + \frac{n}{m_3} & \frac{n}{m_3} \\ \frac{n}{m_3} & \frac{n}{m_2} + \frac{n}{m_3} \end{pmatrix} \begin{pmatrix} \frac{n_1}{n} - \frac{m_1}{n} \\ \frac{n_2}{n} - \frac{m_2}{n} \end{pmatrix} \\
&= \frac{(n_1 - m_1)^2}{m_1} + \frac{(n_2 - m_2)^2}{m_2} + \frac{1}{m_3}\left[(n_1 - m_1) + (n_2 - m_2)\right]^2 \\
&= \sum_{i=1}^{3} \frac{(n_i - m_i)^2}{m_i}.
\end{aligned}
$$

Goodness of Fit for Estimated Expected Frequencies

When the unknown parameters are replaced by the ML estimates for a specified model, the test statistic is again approximately distributed as χ^2 with the number of degrees of freedom reduced by the number of estimated parameters.

The degrees of freedom are $(N-1) - t$, if t parameters are estimated.

Testing for Independence

In two–way contingency tables with multinomial sampling, the hypothesis H_0 : *X and Y are statistically independent* is equivalent to $H_0 : \pi_{ij} = \pi_{i+}\pi_{+j} \; \forall i, j$. The test statistic is Pearson's χ^2 in the following form:

$$
\chi^2 = \sum_{\substack{i=1,2,\ldots,I \\ j=1,2,\ldots,J}} \frac{(n_{ij} - m_{ij})^2}{m_{ij}}, \tag{7.124}
$$

where $m_{ij} = n\pi_{ij} = n\pi_{i+}\pi_{+j}$ (expected cell frequencies under H_0) are unknown.

Given the estimates $\hat{m}_{ij} = np_{i+}p_{+j}$, the χ^2–statistic then equals

$$
\chi^2 = \sum_{\substack{i=1,2,\ldots,I \\ j=1,2,\ldots,J}} \frac{(n_{ij} - \hat{m}_{ij})^2}{\hat{m}_{ij}} \tag{7.125}
$$

with $(I-1)(J-1) = (IJ-1) - (I-1) - (J-1)$ degrees of freedom. The numbers $(I-1)$ and $(J-1)$ correspond to the $(I-1)$ independent row proportions $(\pi_{i+})'$ and $(J-1)$ independent column proportions (π_{+j}) estimated from the sample.

Likelihood–Ratio Test

The likelihood–ratio test (LRT) is a general–purpose method for testing H_0 against H_1. The main idea is to compare $\max_{H_0} L$ and $\max_{H_1 \vee H_0} L$ with the corresponding parameter spaces $\omega \subseteq \Omega$. As a test statistic, we have

$$\Lambda = \frac{\max_\omega L}{\max_\Omega L} \leq 1 . \tag{7.126}$$

It follows that, for $n \to \infty$ (Wilks, 1932),

$$G^2 = -2 \ln \Lambda \to \chi^2_d \tag{7.127}$$

with $d = \dim(\Omega) - \dim(\omega)$ as the degrees of freedom.

For multinomial sampling in a contingency table, the kernel of the likelihood function is

$$K = \prod_{i=1}^{I} \prod_{j=1}^{J} \pi_{ij}^{n_{ij}} , \tag{7.128}$$

with the constraints for the parameters

$$\pi_{ij} \geq 0 \quad \text{and} \quad \sum_{i=1}^{I} \sum_{j=1}^{J} \pi_{ij} = 1 . \tag{7.129}$$

Under the null hypothesis $H_0 : \pi_{ij} = \pi_{i+}\pi_{+j}$, K is maximum for $\hat{\pi}_{i+} = n_{i+}/n$, $\hat{\pi}_{+j} = n_{+j}/n$, and $\hat{\pi}_{ij} = n_{i+}n_{+j}/n^2$. Under $H_0 \vee H_1$, K is maximum for $\hat{\pi}_{ij} = n_{ij}/n$. We then have

$$\Lambda = \frac{\prod_{i=1}^{I} \prod_{j=1}^{J} (n_{i+}n_{+j})^{n_{ij}}}{n^n \prod_{i=1}^{I} \prod_{j=1}^{J} n_{ij}^{n_{ij}}} . \tag{7.130}$$

It follows that Wilks's G^2 is given by

$$G^2 = -2 \ln \Lambda = 2 \sum_{i=1}^{I} \sum_{j=1}^{J} n_{ij} \ln \left(\frac{n_{ij}}{\hat{m}_{ij}} \right) \sim \chi^2_{(I-1)(J-1)} \tag{7.131}$$

with $\hat{m}_{ij} = n_{i+}n_{+j}/n$ (estimate under H_0).

If H_0 holds, Λ will be large, i.e., near 1, and G^2 will be small. This means that H_0 is to be rejected for large G^2.

7.3 Generalized Linear Model for Binary Response

7.3.1 *Logit Models and Logistic Regression*

Let Y be a binary random variable, that is, Y has only two categories (for instance, success/failure or case/control). Hence the response variable

Y can always be coded as $(Y = 0, Y = 1)$. Y_i has a Bernoulli distribution, with $P(Y_i = 1) = \pi_i = \pi_i(x_i)$ and $P(Y_i = 0) = 1 - \pi_i$, where $x_i = (x_{i1}, x_{i2}, \ldots, x_{ip})'$ denotes a vector of *prognostic factors*, which we believe influence the success probability $\pi(x_i)$, and $i = 1, \ldots, N$ denotes individuals as usual. With these assumptions it immediately follows that

$$\begin{aligned} E(Y_i) &= 1 \cdot \pi_i + 0 \cdot (1 - \pi_i) = \pi_i \,, \\ E(Y_i^2) &= 1^2 \cdot \pi_i + 0^2 \cdot (1 - \pi_i) = \pi_i \,, \\ \operatorname{var}(Y_i) &= E(Y_i^2) - (E(Y_i))^2 = \pi_i - \pi_i^2 = \pi_i(1 - \pi_i) \,. \end{aligned}$$

The likelihood contribution of an individual i is further given by

$$\begin{aligned} f(y_i; \pi_i) &= \pi_i^{y_i} (1 - \pi_i)^{1 - y_i} \\ &= (1 - \pi_i) \left(\frac{\pi_i}{1 - \pi_i} \right)^{y_i} \\ &= (1 - \pi_i) \exp\left(y_i \ln\left(\frac{\pi_i}{1 - \pi_i} \right) \right) . \end{aligned}$$

The natural parameter $Q(\pi_i) = \ln[\pi_i/(1 - \pi_i)]$ is the log odds of response 1 and is called the logit of π_i.

A GLM with the *logit link* is called a logit model or *logistic regression model*. The model is, on an individual basis, given by

$$\ln\left(\frac{\pi_i}{1 - \pi_i} \right) = x_i' \beta \,. \tag{7.132}$$

This parametrization guarantees a monotonic course (S–curve) of the probability π_i, under inclusion of the linear approach $x_i'\beta$ over the range of definition [0, 1]:

$$\pi_i = \frac{\exp(x_i'\beta)}{1 + \exp(x_i'\beta)} \,. \tag{7.133}$$

Grouped Data

If possible (e.g., if prognostic factors are themselves categorical), patients can be grouped along the strata defined by the number of possible factor combinations. Let n_j, $j = 1, \ldots, G$, $G \leq N$, be the number of patients falling in strata j. Then we observe y_j patients having response $Y = 1$ and $n_j - y_j$ patients with response $Y = 0$. Then a natural estimate for π_j is $\hat{\pi}_j = y_j/n_j$. This corresponds to a saturated model, that is, a model in which main effects and all interactions between the factors are included. But one should note that this is reasonable only if the number of strata is low compared to N so that n_j is not too low. Whenever $n_j = 1$ these estimates degenerate, and more smoothing of the probabilities and thus a more parsimonious model is necessary.

j	Age Group	Loss yes	Loss No	n_j
1	< 40	4	70	74
2	40–50	28	147	175
3	50–60	38	207	245
4	60–70	51	202	253
5	> 70	32	92	124
		153	718	871

TABLE 7.1. (5×2)–Table of loss of abutment teeth by age groups (Example 7.1).

The Simplest Case and an Example

For simplicity, we assume now that $p = 1$, that is, we consider only one explanatory variable. The model in this simplest case is given by

$$\ln\left(\frac{\pi_i}{1-\pi_i}\right) = \alpha + \beta x_i \,. \tag{7.134}$$

For this special situation, we get for the odds,

$$\frac{\pi_i}{1-\pi_i} = \exp(\alpha + \beta x_i) = e^{\alpha}\left(e^{\beta}\right)^{x_i} , \tag{7.135}$$

that is, if x_i increases by one unit, the odds increase by e^{β}.

An advantage of this link is that the effects of X can be estimated, whether the study of interest is retrospective or prospective (cf. Toutenburg, 1992b, Chapter 5). The effects in the logistic model refer to the odds. For two different x–values, $\exp(\alpha + \beta x_1)/\exp(\alpha + \beta x_2)$ is an odds ratio.

To find the appropriate form for the systematic component of the logistic regression, the sample logits are plotted against x.

Remark. Let x_j be chosen (j being a group index). For n_j observations of the response variable Y, let 1 be observed y_j times at this setting. Hence $\hat{\pi}(x_j) = y_j/n_j$ and $\ln[\hat{\pi}_j/(1-\hat{\pi}_j)] = \ln[y_j/(n_j - y_j)]$ is the sample logit.

This term, however, is not defined for $y_j = 0$ or $n_j = 0$. Therefore, a correction is introduced, and we utilize the smoothed logit

$$\ln\Big[(y_j + 1/2)/(n_j - y_j + 1/2)\Big] .$$

Example 7.1. We examine the risk (Y) for the loss of abutment teeth by extraction in dependence on age (X) (Walther and Toutenburg, 1991). From Table 7.1, we calculate $\chi^2_4 = 15.56$, which is significant at the 5% level ($\chi^2_{4;0.95} = 9.49$). Using the unsmoothed sample logits results in the following table:

i	Sample logits	$\hat{\pi}_{1\|j} = y_j/n_j$
1	−2.86	0.054
2	−1.66	0.160
3	−1.70	0.155
4	−1.38	0.202
5	−1.06	0.258

x_1 x_2 x_3 x_4 x_5

0 −0.5 −1 −1.5 −2 −2.5 −3

$\hat{\pi}_{1|j}$ is the estimated risk for loss of abutment teeth. It increases linearly with age group. For instance, age group 5 has five times the risk of age group 1.

Modeling with the *logistic regression*

$$\ln\left(\frac{\hat{\pi}_1(x_j)}{1-\hat{\pi}_1(x_j)}\right) = \alpha + \beta x_j$$

results in

x_j	Sample logits	Fitted logits	$\hat{\pi}_1(x_j)$	Expected $n_j\hat{\pi}_1(x_j)$	Observed y_j
35	−2.86	−2.22	0.098	7.25	4
45	−1.66	−1.93	0.127	22.17	28
55	−1.70	−1.64	0.162	39.75	38
65	−1.38	−1.35	0.206	51.99	51
75	−1.06	−1.06	0.257	31.84	32

with the ML estimates

$$\begin{aligned} \hat{\alpha} &= -3.233\,, \\ \hat{\beta} &= 0.029\,. \end{aligned}$$

7.3.2 Testing the Model

Under general conditions the maximum–likelihood estimates are asymptotically normal. Hence tests of significance and the setting up of confidence limits can be based on the normal theory.

The significance of the effect of the variable X on π is equivalent to the significance of the parameter β. The hypothesis *β is significant* or $\beta \neq 0$ is tested by the statistical hypothesis $H_0 : \beta = 0$ against $H_1 : \beta \neq 0$. For this test, we compute the Wald statistic $Z^2 = \hat{\beta}'(\mathrm{cov}_{\hat{\beta}})^{-1}\hat{\beta} \sim \chi^2_{df}$, where *df* is the number of components of the vector β.

In the above Example 7.1, we have $Z^2 = 13.06 > \chi^2_{1;0.95} = 3.84$ (the upper 5% value), which leads to a rejection of $H_0 : \beta = 0$ so that the trend is seen to be significant.

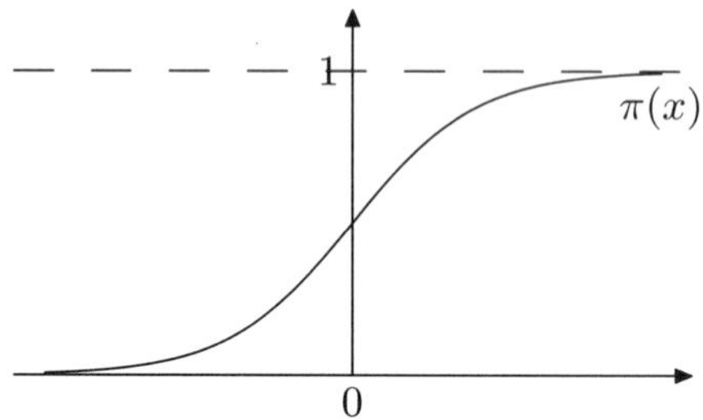

FIGURE 7.1. Logistic function $\pi(x) = \exp(x)/(1+\exp(x))$.

7.3.3 Distribution Function as a Link Function

The logistic function has the shape of the cumulative distribution function of a continuous random variable.

This suggests a class of models for binary responses having the form

$$\pi(x) = F(\alpha + \beta x) \,, \tag{7.136}$$

where F is a standard, continuous, cumulative distribution function. If F is strictly monotonically increasing over the entire real line, we have

$$F^{-1}(\pi(x)) = \alpha + \beta x \,. \tag{7.137}$$

This is a GLM with F^{-1} as the link function. F^{-1} maps the $[0,1]$ range of probabilities onto $(-\infty, \infty)$.

The cumulative distribution function of the logistic distribution is

$$F(x) = \frac{\exp\left(\frac{x-\mu}{\tau}\right)}{1+\exp\left(\frac{x-\mu}{\tau}\right)}\,, \quad -\infty < x < \infty\,, \tag{7.138}$$

with μ as the location parameter and $\tau > 0$ as the scale parameter.

The distribution is symmetric with mean μ and standard deviation $\tau\pi/\sqrt{3}$ (bell–shaped curve, similar to the standard normal distribution). The logistic regression $\pi(x) = F(\alpha + \beta x)$ belongs to the standardized logistic distribution F with $\mu = 0$ and $\tau = 1$. Thus, the logistic regression has mean $-\alpha/\beta$ and standard deviation $\pi/|\beta|\sqrt{3}$.

If F is the standard normal cumulative distribution function, $\pi(x) = F(\alpha + \beta x) = \Phi(\alpha + \beta x)$, $\pi(x)$ is called the *probit model.*

7.4 Logit Models for Categorical Data

The explanatory variable X can be continuous or categorical. Assume X to be categorical and choose the logit link; then the logit models are equivalent to *loglinear models (categorical regression)*, which are discussed in detail in

Section 7.6. For the explanation of this equivalence we first consider the logit model.

Logit Models for $I \times 2$ Tables

Let X be an explanatory variable with I categories. If response/nonresponse is the Y factor, we then have an $I \times 2$ table. In row i the probability for response is $\pi_{1|i}$ and for nonresponse $\pi_{2|i}$, with $\pi_{1|i} + \pi_{2|i} = 1$.

This leads to the following logit model:

$$\ln\left(\frac{\pi_{1|i}}{\pi_{2|i}}\right) = \alpha + \beta_i \,. \tag{7.139}$$

Here the x–values are not included explicitly but only through the category i. β_i describes the effect of category i on the response. When $\beta_i = 0$, there is no effect. This model resembles the one–way analysis of variance and, likewise, we have the constraints for identifiability $\sum \beta_i = 0$ or $\beta_I = 0$. Then $I - 1$ of the parameters $\{\beta_i\}$ suffice for characterization of the model. For the constraint $\sum \beta_i = 0$, α is the overall mean of the logits and β_i is the deviation from this mean for row i. The higher β_i is, the higher is the logit in row i, and the higher is the value of $\pi_{1|i}$ (= chance for response in category i).

When the factor X (in I categories) has no effect on the response variable, the model simplifies to the model of statistical independence of the factor and response

$$\ln\left(\frac{\pi_{1|i}}{\pi_{2|i}}\right) = \alpha \quad \forall i \,,$$

We now have $\beta_1 = \beta_2 = \cdots = \beta_I = 0$, and thus $\pi_{1|1} = \pi_{1|2} = \cdots = \pi_{1|I}$.

Logit Models for Higher Dimensions

As a generalization to two or more categorical factors that have an effect on the binary response, we now consider the two factors A and B with I and J levels. Let $\pi_{1|ij}$ and $\pi_{2|ij}$ denote the probabilities for response and nonresponse for the combination ij of factors so that $\pi_{1|ij} + \pi_{2|ij} = 1$. For the $I \times J \times 2$ table, the logit model

$$\ln\left(\frac{\pi_{1|ij}}{\pi_{2|ij}}\right) = \alpha + \beta_i^A + \beta_j^B \tag{7.140}$$

represents the effects of A and B without interaction. This model is equivalent to the two–way analysis of variance without interaction.

7.5 Goodness of Fit—Likelihood Ratio Test

For a given model M, we can use the estimates of the parameters $(\widehat{\alpha + \beta_i})$ and $(\hat{\alpha}, \hat{\beta})$ to predict the logits, to estimate the probabilities of response $\hat{\pi}_{1|i}$, and hence to calculate the expected cell frequencies $\hat{m}_{ij} = n_{i+}\hat{\pi}_{j|i}$.

We can now test the goodness of fit of a model M with Wilks' G^2–statistic

$$G^2(M) = 2\sum_{i=1}^{I}\sum_{j=1}^{J} n_{ij} \ln\left(\frac{n_{ij}}{\hat{m}_{ij}}\right) . \tag{7.141}$$

The $\hat{m}_{ij}$ are calculated by using the estimated model parameters. The degrees of freedom equal the number of logits minus the number of independent parameters in the model M.

We now consider three models for binary response (cf. Agresti, 1990, p. 95).

(1) Independence model:

$$M = I: \quad \ln\left(\frac{\pi_{1|i}}{\pi_{2|i}}\right) = \alpha . \tag{7.142}$$

Here we have I logits and one parameter, that is, $I - 1$ degrees of freedom.

(2) Logistic model:

$$M = L: \quad \ln\left(\frac{\pi_{1|i}}{\pi_{2|i}}\right) = \alpha + \beta x_i . \tag{7.143}$$

The number of degrees of freedom equals $I - 2$.

(3) Logit model:

$$M = S: \quad \ln\left(\frac{\pi_{1|i}}{\pi_{2|i}}\right) = \alpha + \beta_i . \tag{7.144}$$

The model has I logits and I independent parameters. The number of degrees of freedom is 0, so it has perfect fit. This model, with equal numbers of parameters and observations, is called a *saturated model.*

As mentioned earlier, the likelihood–ratio test compares a model M_1 with a simpler model M_2 (in which a few parameters equal zero). The test statistic here is then

$$\Lambda = \frac{L(M_2)}{L(M_1)} , \tag{7.145}$$

or

$$G^2\left(M_2|M_1\right) = -2\left(\ln L(M_2) - \ln L(M_1)\right) . \tag{7.146}$$

The statistic $G^2(M)$ is a special case of this statistic, in which $M_2 = M$ and M_1 is the saturated model. If we want to test the goodness of fit with $G^2(M)$, this is equivalent to testing whether all the parameters that are in the saturated model, but not in the model M, are equal to zero.

Let l_S denote the maximized loglikelihood function for the saturated model. Then we have

$$\begin{aligned} G^2(M_2|M_1) &= -2\,(\ln L(M_2) - \ln L(M_1)) \\ &= -2\,(\ln L(M_2) - l_S) - [-2(\ln L(M_1) - l_S)] \\ &= G^2(M_2) - G^2(M_1)\,. \end{aligned} \quad (7.147)$$

That is, the statistic $G^2(M_2|M_1)$ for comparing two models is identical to the difference of the goodness–of–fit statistics for the two models.

Example 7.2. In Example 7.1 "Loss of abutment teeth/age" for the logistic model we have:

Age	Loss		No loss	
group	Observed	Expected	Observed	Expected
1	4	7.25	70	66.75
2	28	22.17	147	152.83
3	38	39.75	207	205.25
4	51	51.99	202	201.01
5	32	31.84	92	92.16

and get $G^2(L) = 3.66$, $df = 5 - 2 = 3$.

For the independence model, we get $G^2(I) = 17.25$ with $df = 4 = (I-1)(J-1) = (5-1)(2-1)$. The test statistic for testing $H_0 : \beta = 0$ in the logistic model is then

$$G^2(I|L) \;=\; G^2(I) - G^2(L) = 17.25 - 3.66 = 13.59, \quad df = 4 - 3 = 1\,.$$

This value is significant, which means that the logistic model, compared to the independence model, holds.

7.6 Loglinear Models for Categorical Variables

7.6.1 Two–Way Contingency Tables

The previous models focused on bivariate response, that is, on $I \times 2$ tables. We now generalize this set–up to $I \times J$ and later to $I \times J \times K$ tables.

Suppose that we have a realization (sample) of two categorical variables with I and J categories and sample size n. This yields observations in $N = I \times J$ cells of the contingency table. The number in the (i, j)th cell is denoted by n_{ij}.

The probabilities π_{ij} of the multinomial distribution form the joint distribution. Independence of the variables is equivalent to

$$\pi_{ij} = \pi_{i+}\pi_{+j} \quad \text{(for all } i, j). \tag{7.148}$$

If this is applied to the expected cell frequencies $m_{ij} = n\pi_{ij}$, the condition of independence is equivalent to

$$m_{ij} = n\pi_{i+}\pi_{+j} \, . \tag{7.149}$$

The modeling of the $I \times J$ table is based on this relation as an independence model on the logarithmic scale

$$\ln(m_{ij}) = \ln n + \ln \pi_{i+} + \ln \pi_{+j} \, . \tag{7.150}$$

Hence, the effects of the rows and columns on $\ln(m_{ij})$ are additive. An alternative expression, following the models of analysis of variance of the form,

$$y_{ij} = \mu + \alpha_i + \beta_j + \varepsilon_{ij} \, , \quad \left(\sum \alpha_i = \sum \beta_j = 0\right) , \tag{7.151}$$

is given by

$$\ln m_{ij} = \mu + \lambda_i^X + \lambda_j^Y \tag{7.152}$$

with

$$\lambda_i^X = \ln \pi_{i+} - \frac{1}{I}\left(\sum_{k=1}^{I} \ln \pi_{k+}\right) , \tag{7.153}$$

$$\lambda_j^Y = \ln \pi_{+j} - \frac{1}{J}\left(\sum_{k=1}^{J} \ln \pi_{+k}\right) , \tag{7.154}$$

$$\mu = \ln n + \frac{1}{I}\left(\sum_{k=1}^{I} \ln \pi_{k+}\right) + \frac{1}{J}\left(\sum_{k=1}^{J} \ln \pi_{+k}\right) . \tag{7.155}$$

The parameters satisfy the constraints

$$\sum_{i=1}^{I} \lambda_i^X = \sum_{j=1}^{J} \lambda_j^Y = 0 \, , \tag{7.156}$$

which make the parameters identifiable.

Model (7.152) is called a *loglinear model of independence* in a two–way contingency table.

The related saturated model contains the additional interaction parameters λ_{ij}^{XY}:

$$\ln m_{ij} = \mu + \lambda_i^X + \lambda_j^Y + \lambda_{ij}^{XY} \, . \tag{7.157}$$

This model describes the perfect fit. The interaction parameters satisfy

$$\sum_{i=1}^{I} \lambda_{ij}^{XY} = \sum_{j=1}^{J} \lambda_{ij}^{XY} = 0 . \tag{7.158}$$

Given the λ_{ij} in the first $(I-1)(J-1)$ cells, these constraints determine the λ_{ij} in the last row or the last column. Thus, the saturated model contains

$$\underbrace{1}_{\mu} + \underbrace{(I-1)}_{\lambda_i^X} + \underbrace{(J-1)}_{\lambda_j^Y} + \underbrace{(I-1)(J-1)}_{\lambda_{ij}^{XY}} = IJ \tag{7.159}$$

independent parameters.

For the independence model, the number of independent parameters equals

$$1 + (I-1) + (J-1) = I + J - 1 . \tag{7.160}$$

Interpretation of the Parameters

Loglinear models estimate the effects of rows and columns on $\ln m_{ij}$. For this, no distinction is made between explanatory and response variables. The information of the rows or columns influence m_{ij} symmetrically.

Consider the simplest case—the $I \times 2$ table (independence model). According to (7.160), the logit of the binary variable equals

$$\begin{aligned} \ln\left(\frac{\pi_{1|i}}{\pi_{2|i}}\right) &= \ln\left(\frac{m_{i1}}{m_{i2}}\right) \\ &= \ln(m_{i1}) - \ln(m_{i2}) \\ &= (\mu + \lambda_i^X + \lambda_1^Y) - (\mu + \lambda_i^X + \lambda_2^Y) \\ &= \lambda_1^Y - \lambda_2^Y . \end{aligned} \tag{7.161}$$

The logit is the same in every row and hence independent of X or the categories $i = 1, \ldots, I$, respectively.

For the constraints

$$\begin{aligned} \lambda_1^Y + \lambda_2^Y = 0 \quad &\Rightarrow \quad \lambda_1^Y = -\lambda_2^Y , \\ &\Rightarrow \quad \ln\left(\frac{\pi_{1|i}}{\pi_{2|i}}\right) = 2\lambda_1^Y \;\; (i = 1, \ldots, I) . \end{aligned}$$

Hence we obtain

$$\frac{\pi_{1|i}}{\pi_{2|i}} = \exp(2\lambda_1^Y) \quad (i = 1, \ldots, I) . \tag{7.162}$$

In each category of X, the odds that Y is in category 1 rather than in category 2 are equal to $\exp(2\lambda_1^Y)$, when the independence model holds.

The following relationship exists between the odds ratio in a 2×2 table and the saturated loglinear model

$$\ln \theta = \ln\left(\frac{m_{11}\, m_{22}}{m_{12}\, m_{21}}\right)$$

Age group	Form of construction	Endodontic treatment Yes	No
< 60	H	62	1041
	B	23	463
≥ 60	H	70	755
	B	30	215
Σ		185	2474

TABLE 7.2. $2 \times 2 \times 2$ Table for endodontic risk.

$$\begin{aligned}
&= \ln(m_{11}) + \ln(m_{22}) - \ln(m_{12}) - \ln(m_{21}) \\
&= (\mu + \lambda_1^X + \lambda_1^Y + \lambda_{11}^{XY}) + (\mu + \lambda_2^X + \lambda_2^Y + \lambda_{22}^{XY}) \\
&\quad - (\mu + \lambda_1^X + \lambda_2^Y + \lambda_{12}^{XY}) - (\mu + \lambda_2^X + \lambda_1^Y + \lambda_{21}^{XY}) \\
&= \lambda_{11}^{XY} + \lambda_{22}^{XY} - \lambda_{12}^{XY} - \lambda_{21}^{XY} .
\end{aligned}$$

Since $\sum_{i=1}^2 \lambda_{ij}^{XY} = \sum_{j=1}^2 \lambda_{ij}^{XY} = 0$, we have $\lambda_{11}^{XY} = \lambda_{22}^{XY} = -\lambda_{12}^{XY} = -\lambda_{21}^{XY}$ and thus $\ln \theta = 4\lambda_{11}^{XY}$. Hence the odds ratio in a 2×2 table equals

$$\theta = \exp(4\lambda_{11}^{XY}) , \tag{7.163}$$

and is dependent on the association parameter in the saturated model. When there is no association, i.e., $\lambda_{ij} = 0$, we have $\theta = 1$.

7.6.2 Three-Way Contingency Tables

We now consider three categorical variables X, Y, and Z. The frequencies of the combinations of categories are displayed in the $I \times J \times K$ contingency table. We are especially interested in $I \times J \times 2$ contingency tables, where the last variable is a bivariate risk or response variable. Table 7.2 shows the risk for an endodontic treatment depending on the age of patients and the type of construction of the denture (Walther and Toutenburg, 1991).

In addition to the bivariate associations, we want to model an overall association. The three variables are mutually independent if the following independence model for the cell frequencies m_{ijk} (on a logarithmic scale) holds:

$$\ln(m_{ijk}) = \mu + \lambda_i^X + \lambda_j^Y + \lambda_k^Z . \tag{7.164}$$

(In the above example, we have X : age group, Y : type of construction, and Z : endodontic treatment.) The variable Z is independent of the joint distribution of X and Y (jointly independent) if

$$\ln(m_{ijk}) = \mu + \lambda_i^X + \lambda_j^Y + \lambda_k^Z + \lambda_{ij}^{XY} . \tag{7.165}$$

A third type of independence (conditional independence of two variables given a fixed category of the third variable) is expressed by the following model (j fixed!):

$$\ln(m_{ijk}) = \mu + \lambda_i^X + \lambda_j^Y + \lambda_k^Z + \lambda_{ij}^{XY} + \lambda_{jk}^{YZ} . \tag{7.166}$$

This is the approach for the conditional independence of X and Z at level j of Y. If they are conditionally independent for all $j = 1, \ldots, J$, then X and Z are called conditionally independent, given Y. Similarly, if X and Y are conditionally independent at level k of Z, the parameters λ_{ij}^{XY} and λ_{jk}^{YZ} in (7.166) are replaced by the parameters λ_{ik}^{XZ} and λ_{jk}^{YZ}. The parameters with two subscripts describe two–way interactions. The appropriate conditions for the cell probabilities are:

(a) mutual independence of X, Y, Z:

$$\pi_{ijk} = \pi_{i++}\pi_{+j+}\pi_{++k} \quad \text{(for all } i, j, k). \tag{7.167}$$

(b) joint independence:
Y is jointly independent of X and Z when

$$\pi_{ijk} = \pi_{i+k}\pi_{+j+} \quad \text{(for all } i, j, k). \tag{7.168}$$

(c) conditional independence:
X and Y are conditionally independent of Z when

$$\pi_{ijk} = \frac{\pi_{i+k}\pi_{+jk}}{\pi_{++k}} \quad \text{(for all } i, j, k). \tag{7.169}$$

The most general loglinear model (saturated model) for three–way tables is the following:

$$\ln(m_{ijk}) = \mu + \lambda_i^X + \lambda_j^Y + \lambda_k^Z + \lambda_{ij}^{XY} + \lambda_{ik}^{XZ} + \lambda_{jk}^{YZ} + \lambda_{ijk}^{XYZ} . \tag{7.170}$$

The last parameter describes the three–factor interaction.

All association parameters,, describing the deviation from the general mean μ, satisfy the constraints

$$\sum_{i=1}^{I} \lambda_{ij}^{XY} = \sum_{j=1}^{J} \lambda_{ij}^{XY} = \ldots = \sum_{k=1}^{K} \lambda_{ijk}^{XYZ} = 0 . \tag{7.171}$$

Similarly, for the main factor effects we have

$$\sum_{i=1}^{I} \lambda_i^X = \sum_{j=1}^{J} \lambda_j^Y = \sum_{k=1}^{K} \lambda_k^Z = 0 . \tag{7.172}$$

From the general model (7.170), submodels can be constructed. For this, the hierarchical principle of construction is preferred. A model is called hierarchical when, in addition to significant higher–order effects, it contains all lower–order effects of the variables included in the higher–order effects, even if these parameter estimates are not statistically significant. For instance, if the model contains the association parameter λ_{ik}^{XZ}, it must also contain λ_i^X and λ_k^Z:

$$\ln(m_{ijk}) = \mu + \lambda_i^X + \lambda_k^Z + \lambda_{ik}^{XZ} . \tag{7.173}$$

		Loglinear model	Symbol
$\ln(m_{ij+})$	=	$\mu + \lambda_i^X + \lambda_j^Y$	(X, Y)
$\ln(m_{i+k})$	=	$\mu + \lambda_i^X + \lambda_k^Z$	(X, Z)
$\ln(m_{+jk})$	=	$\mu + \lambda_j^Y + \lambda_k^Z$	(Y, Z)
$\ln(m_{ijk})$	=	$\mu + \lambda_i^X + \lambda_j^Y + \lambda_k^Z$	(X, Y, Z)
$\ln(m_{ijk})$	=	$\mu + \lambda_i^X + \lambda_j^Y + \lambda_k^Z + \lambda_{ij}^{XY}$	(XY, Z)
	$\vdots$		$\vdots$
$\ln(m_{ijk})$	=	$\mu + \lambda_i^X + \lambda_j^Y + \lambda_{ij}^{XY}$	(XY)
	$\vdots$		$\vdots$
$\ln(m_{ijk})$	=	$\mu + \lambda_i^X + \lambda_j^Y + \lambda_k^Z + \lambda_{ij}^{XY} + \lambda_{ik}^{XZ}$	(XY, XZ)
	$\vdots$		$\vdots$
$\ln(m_{ijk})$	=	$\mu + \lambda_i^X + \lambda_j^Y + \lambda_k^Z + \lambda_{ij}^{XY} + \lambda_{ik}^{XZ} + \lambda_{jk}^{YZ}$	(XY, XZ, YZ)
	$\vdots$		$\vdots$
$\ln(m_{ijk})$	=	$\mu + \lambda_i^X + \lambda_j^Y + \lambda_k^Z + \lambda_{ij}^{XY} + \lambda_{ik}^{XZ} + \lambda_{jk}^{YZ} + \lambda_{ijk}^{XYZ}$	(XYZ)

TABLE 7.3. Symbols of the hierarchical models for three–way contingency tables (Agresti, 1990, p. 144).

A symbol is assigned to the various hierarchical models (Table 7.3).

Similar to 2×2 tables, a close relationship exists between the parameters of the model and the odds ratios. Given a $2 \times 2 \times 2$ table, we have, under the constraints (7.171) and (7.172), for instance,

$$\begin{aligned} \frac{\theta_{11(1)}}{\theta_{11(2)}} &= [(\pi_{111}\pi_{221})/(\pi_{211}\pi_{121})]/[(\pi_{112}\pi_{222})/(\pi_{212}\pi_{122})] \\ &= exp(8\lambda_{111}^{XYZ}) . \end{aligned} \tag{7.174}$$

This is the conditional odds ratio of X and Y given the levels $k = 1$ (numerator) and $k = 2$ (denominator) of Z. The same holds for X and Z under Y and for Y and Z under X. In the population, we thus have- for the three–way interaction λ_{111}^{XYZ},

$$\frac{\theta_{11(1)}}{\theta_{11(2)}} = \frac{\theta_{1(1)1}}{\theta_{1(2)1}} = \frac{\theta_{(1)11}}{\theta_{(2)11}} = \exp(8\lambda_{111}^{XYZ}) . \tag{7.175}$$

In the case of independence in the equivalent subtables, the odds ratios (of the population) equal 1. The sample odds ratio gives a first hint at a deviation from independence.

Consider the conditional odds ratio (7.175) for Table 7.2 assuming that X is the variable "age group," Y is the variable "form of construction," and Z is the variable "endodontic treatment."

We then have a value of 1.80. This indicates a positive tendency for an increased risk of endodontic treatment in comparing the following subtables for endodontic treatment (left) versus no endodontic treatment (right):

	H	B
< 60	62	23
≥ 60	70	30

	H	B
< 60	1041	463
≥ 60	755	215

The relationship (7.102) is also valid for the sample version. Thus a comparison of the following subtables for < 60 (left) versus ≥ 60 (right):

	Treatment Yes	Treatment No
H	62	1041
B	23	463

	Treatment Yes	Treatment No
H	70	755
B	30	215

or for H (left) versus B (right):

	Treatment Yes	Treatment No
< 60	62	1041
≥ 60	70	755

	Treatment Yes	Treatment No
< 60	23	463
≥ 60	30	215

leads to the same sample value 1.80 and hence $\hat{\lambda}_{111}^{XYZ} = 0.073$.

Calculations for Table 7.2:

$$\frac{\hat{\theta}_{11(1)}}{\hat{\theta}_{11(2)}} = \frac{\frac{n_{111}n_{221}}{n_{211}n_{121}}}{\frac{n_{112}n_{222}}{n_{212}n_{122}}} = \frac{\frac{62\cdot 30}{70\cdot 23}}{\frac{1041\cdot 215}{755\cdot 463}} = \frac{1.1553}{0.6403} = 1.80\,,$$

$$\frac{\hat{\theta}_{(1)11}}{\hat{\theta}_{(2)11}} = \frac{\frac{n_{111}n_{122}}{n_{121}n_{112}}}{\frac{n_{211}n_{222}}{n_{221}n_{212}}} = \frac{\frac{62\cdot 463}{23\cdot 1041}}{\frac{70\cdot 215}{30\cdot 755}} = \frac{1.1989}{0.6645} = 1.80\,,$$

$$\frac{\hat{\theta}_{1(1)1}}{\hat{\theta}_{1(2)1}} = \frac{\frac{n_{111}n_{212}}{n_{211}n_{112}}}{\frac{n_{121}n_{222}}{n_{221}n_{122}}} = \frac{\frac{62\cdot 755}{70\cdot 1041}}{\frac{23\cdot 215}{30\cdot 463}} = \frac{0.6424}{0.3560} = 1.80\,.$$

7.7 The Special Case of Binary Response

If one of the variables is a binary response variable (in our example, Z : endodontic treatment) and the others are explanatory categorical variables (in our example X : age group and Y : type of construction), these models lead to the already known logit model.

Given the independence model

$$\ln(m_{ijk}) = \mu + \lambda_i^X + \lambda_j^Y + \lambda_k^Z\,, \tag{7.176}$$

we then have, for the logit of the response variable Z,

$$\ln\left(\frac{m_{ij1}}{m_{ij2}}\right) = \lambda_1^Z - \lambda_2^Z\,. \tag{7.177}$$

With the constraint $\sum_{k=1}^{2} \lambda_k^Z = 0$ we thus have

$$\ln\left(\frac{m_{ij1}}{m_{ij2}}\right) = 2\lambda_1^Z \quad (\text{for all } i, j). \tag{7.178}$$

The higher the value of λ_1^Z is, the higher is the risk for category $Z = 1$ (endodontic treatment), independent of the values of X and Y.

In case the other two variables are also binary, implying a $2 \times 2 \times 2$ table, and if the constraints

$$\lambda_2^X = -\lambda_1^X\,, \quad \lambda_2^Y = -\lambda_1^Y\,, \quad \lambda_2^Z = -\lambda_1^Z\,,$$

hold, then the model (7.176) can be expressed as follows:

$$\begin{pmatrix} \ln(m_{111}) \\ \ln(m_{112}) \\ \ln(m_{121}) \\ \ln(m_{122}) \\ \ln(m_{211}) \\ \ln(m_{212}) \\ \ln(m_{221}) \\ \ln(m_{222}) \end{pmatrix} = \begin{pmatrix} 1 & 1 & 1 & 1 \\ 1 & 1 & 1 & -1 \\ 1 & 1 & -1 & 1 \\ 1 & 1 & -1 & -1 \\ 1 & -1 & 1 & 1 \\ 1 & -1 & 1 & -1 \\ 1 & -1 & -1 & 1 \\ 1 & -1 & -1 & -1 \end{pmatrix} \begin{pmatrix} \mu \\ \lambda_1^X \\ \lambda_1^Y \\ \lambda_1^Z \end{pmatrix}, \tag{7.179}$$

which is equivalent to $\ln(m) = X\beta$.

This corresponds to the effect coding of categorical variables (Section 7.8). The ML equation is

$$X'n = X'\hat{m}\,. \tag{7.180}$$

The estimated asymptotic covariance matrix for Poisson sampling reads as

$$\widehat{\text{cov}}(\hat{\beta}) = \left[X'(\text{diag}(\hat{m}))X\right]^{-1}, \tag{7.181}$$

where $\text{diag}(\hat{m})$ has the elements $\hat{m}$ on the main diagonal. The solution of the ML equation (7.180) is obtained by the Newton–Raphson or any other iterative algorithm, for instance, the iterative proportional fitting (IPF).

The IPF method (Deming and Stephan, 1940; cf. Agresti, 1990, p. 185) adjusts initial estimates $\{\hat{m}_{ijk}^{(0)}\}$ successively to the respective expected marginal table of the model until a prespecified accuracy is achieved. For the independence model the steps of iteration are

$$\begin{aligned} \hat{m}_{ijk}^{(1)} &= \hat{m}_{ijk}^{(0)} \left(\frac{n_{i++}}{\hat{m}_{i++}^{(0)}}\right), \\ \hat{m}_{ijk}^{(2)} &= \hat{m}_{ijk}^{(1)} \left(\frac{n_{+j+}}{\hat{m}_{+j+}^{(1)}}\right), \\ \hat{m}_{ijk}^{(3)} &= \hat{m}_{ijk}^{(2)} \left(\frac{n_{++k}}{\hat{m}_{++k}^{(2)}}\right). \end{aligned}$$

Example 7.3 (Tartar Smoking Analysis). A study cited in Toutenburg (1992b, p. 42) investigates to what extent smoking influences the development of tartar. The 3×3 contingency table (Table 7.5) is modeled by the loglinear model

$$\ln(m_{ij}) = \mu + \lambda_i^{\text{Smoking}} + \lambda_j^{\text{Tartar}} + \lambda_{ij}^{\text{Smoking/Tartar}},$$

with $i, j = 1, 2$. Here we have

$$\begin{aligned} \lambda_1^{\text{Smoking}} &= \text{effect nonsmoker}, \\ \lambda_2^{\text{Smoking}} &= \text{effect light smoker}, \\ \lambda_3^{\text{Smoking}} = -(\lambda_1^{\text{Smoking}} + \lambda_2^{\text{Smoking}}) &= \text{effect heavy smoker}. \end{aligned}$$

For the development of tartar, analogous expressions are valid:

(i) Model of independence. For the null hypothesis

$$H_0 : \ln(m_{ij}) = \mu + \lambda_i^{\text{Smoking}} + \lambda_j^{\text{Tartar}},$$

we receive $G^2 = 76.23 > 9.49 = \chi^2_{4;0.95}$. This leads to a clear rejection of this model.

(ii) Saturated model. Here we have $G^2 = 0$. The estimates of the parameters are (values in parantheses are standardized values)

$$\begin{aligned} \lambda_1^{\text{Smoking}} &= -1.02 \quad (-25.93), \\ \lambda_2^{\text{Smoking}} &= 0.20 \quad (7.10), \\ \lambda_3^{\text{Smoking}} &= 0.82 \quad (\text{—}), \\ \lambda_1^{\text{Tartar}} &= 0.31 \quad (11.71), \\ \lambda_2^{\text{Tartar}} &= 0.61 \quad (23.07), \\ \lambda_3^{\text{Tartar}} &= -0.92 \quad (\text{—}). \end{aligned}$$

All single effects are highly significant. The interaction effects are shown in Table 7.4.

		Tartar			
		1	2	3	$\sum$
	1	0.34	-0.14	-0.20	0
Smoking	2	-0.12	0.06	0.06	0
	3	-0.22	0.08	0.14	0
$\sum$		0	0	0	

TABLE 7.4. Interaction effects

The main diagonal is very well marked, which is an indication for a trend. The standardized interaction effects are significant as well:

	1	2	3
1	7.30	-3.05	—
2	-3.51	1.93	—
3	—	—	—

		Tartar		
		None	Middle	Heavy
	None	284	236	48
Smoking	Middle	606	983	209
	Heavy	1028	1871	425

TABLE 7.5. Smoking and development of tartar.

7.8 Coding of Categorical Explanatory Variables

7.8.1 *Dummy and Effect Coding*

If a bivariate response variable Y is connected to a linear model $x'\beta$, with x being categorical, by an appropriate link, the parameters β are always to be interpreted in terms of their dependence on the x scores. To eliminate this arbitrariness, an appropriate coding of x is chosen. Here two ways of coding are suggested (partly in analogy to the analysis of variance).

Dummy Coding

Let A be a variable in I categories. Then the $I-1$ dummy variables are defined as follows:

$$x_i^A = \begin{cases} 1 & \text{for category } i \text{ of variable } A, \\ 0 & \text{for others,} \end{cases} \tag{7.182}$$

with $i = 1, \ldots, I-1$.

The category I is implicitly taken into account by $x_1^A = \ldots = x_{I-1}^A = 0$. Thus, the vector of explanatory variables belonging to variable A is of the following form:

$$x^A = (x_1^A, x_2^A, \ldots, x_{I-1}^A)' \,. \tag{7.183}$$

The parameters β_i, which go into the final regression model proportional to $x'^A\beta$, are called the main effects of A.

Example:

(i) Sex male/female, with male : category 1, female : category 2,

$$\begin{aligned} x_1^{\text{Sex}} &= (1) \quad \Rightarrow \quad \text{person is male,} \\ x_2^{\text{Sex}} &= (0) \quad \Rightarrow \quad \text{person is female.} \end{aligned}$$

(ii) Age groups $i = 1, \ldots, 5$,

$$\begin{aligned} x^{\text{Age}} &= (1, 0, 0, 0)' \quad \Rightarrow \quad \text{age group is 1,} \\ x^{\text{Age}} &= (0, 0, 0, 0)' \quad \Rightarrow \quad \text{age group is 5.} \end{aligned}$$

Let y be a bivariate response variable. The probability of response ($y = 1$) dependent on a categorical variable A in I categories can be modeled as follows:

$$P(y = 1 \mid x^A) = \beta_0 + \beta_1 x_1^A + \cdots + \beta_{I-1} x_{I-1}^A . \tag{7.184}$$

Given category i (age group i), we have

$$P(y = 1 \mid x^A \textit{ represents the ith age group}) = \beta_0 + \beta_i ,$$

as long as $i = 1, 2, \ldots, I - 1$ and, for the implicitly coded category I, we get

$$P(y = 1 \mid x^A \textit{ represents the Ith age group}) = \beta_0 . \tag{7.185}$$

Hence, for each category i, another probability of response $P(y = 1 \mid x^A)$ is possible.

Effect Coding

For an explanatory variable A in I categories, effect coding is defined as follows:

$$x_i^A = \begin{cases} 1 & \text{for category } i, \quad i = 1, \ldots, I-1, \\ -1 & \text{for category } I, \\ 0 & \text{for others.} \end{cases} \tag{7.186}$$

Consequently, we have

$$\beta_I = -\sum_{i=1}^{I-1} \beta_i , \tag{7.187}$$

which is equivalent to

$$\sum_{i=1}^{I} \beta_i = 0 . \tag{7.188}$$

In analogy to the analysis of variance, the model for the probability of response has the following form:

$$P(y = 1 \mid x^A \textit{ represents the ith age group}) = \beta_0 + \beta_i \tag{7.189}$$

for $i = 1, \ldots, I$ and with the constraint (7.188).

Example: $I = 3$ age groups A1, A2, A3. A person in A1 is coded $(1, 0)$, a person in A2 is coded $(0, 1)$ for both dummy and effect coding. A person in A3 is coded $(0, 0)$ using dummy coding or $(-1, -1)$ using effect coding. The two ways of coding categorical variables generally differ only for category I.

Inclusion of More than One Variable

If more than one explanatory variable is included in the model, the categories of A, B, and C (with I, J, and K categories, respectively), for example, are combined in a common vector

$$x' = (x_1^A, \ldots, x_{I-1}^A, x_1^B, \ldots, x_{J-1}^B, x_1^C, \ldots, x_{K-1}^C) . \tag{7.190}$$

In addition to these main effects, the interaction effects $x_{ij}^{AB}, \ldots, x_{ijk}^{ABC}$ can be included. The codings of the $x_{ij}^{AB}, \ldots, x_{ijk}^{ABC}$ are chosen in consideration of constraints (7.171).

Example: In the case of effect coding, we obtain, for the saturated model (7.157) with binary variables A and B,

$$\begin{pmatrix} \ln(m_{11}) \\ \ln(m_{12}) \\ \ln(m_{21}) \\ \ln(m_{22}) \end{pmatrix} = \begin{pmatrix} 1 & 1 & 1 & 1 \\ 1 & 1 & -1 & -1 \\ 1 & -1 & 1 & -1 \\ 1 & -1 & -1 & 1 \end{pmatrix} \begin{pmatrix} \mu \\ \lambda_1^A \\ \lambda_1^B \\ \lambda_{11}^{AB} \end{pmatrix},$$

from which we receive the following values for x_{ij}^{AB}, recoded for parameter λ_{11}^{AB}:

(i, j)		Parameter	Constraints	Recoding for λ_{11}^{AB}
(1, 1)	$x_{11}^{AB} = 1$	λ_{11}^{AB}		
(1, 2)	$x_{12}^{AB} = 1$	λ_{12}^{AB}	$\lambda_{12}^{AB} = -\lambda_{11}^{AB}$	$x_{12}^{AB} = -1$
(2, 1)	$x_{21}^{AB} = 1$	λ_{21}^{AB}	$\lambda_{21}^{AB} = \lambda_{12}^{AB} = -\lambda_{11}^{AB}$	$x_{21}^{AB} = -1$
(2, 2)	$x_{22}^{AB} = 1$	λ_{22}^{AB}	$\lambda_{22}^{AB} = -\lambda_{21}^{AB} = \lambda_{11}^{AB}$	

Thus the interaction effects develop from multiplying the main effects.

Let L be the number of possible (different) combinations of variables. If, for example, we have three variables A, B, C in I, J, K categories, L equals IJK.

Consider a complete factorial experimental design (as in an $I \times J \times K$ contingency table). Now L is known, and the design matrix X (in effect or dummy coding) for the main effects can be specified (independence model).

Example (Fahrmeir and Hamerle, 1984, p. 507): The reading habits of women (preference for a specific magazine: yes/no) are to be analyzed in terms of dependence on employment (A: yes/no), age group (B: three categories), and education (C: four categories). The complete design matrix X (Figure 7.2) is of dimension $IJK \times \{1 + (I - 1) + (J - 1) + (K - 1)\}$,

$$X = \begin{array}{c} \begin{array}{ccccccc} \beta_0 & x_1^A & x_1^B & x_2^B & x_1^C & x_2^C & x_3^C \end{array} \\ \left(\begin{array}{ccccccc}
1 & 1 & 1 & 0 & 1 & 0 & 0 \\
1 & 1 & 1 & 0 & 0 & 1 & 0 \\
1 & 1 & 1 & 0 & 0 & 0 & 1 \\
1 & 1 & 1 & 0 & -1 & -1 & -1 \\
1 & 1 & 0 & 1 & 1 & 0 & 0 \\
1 & 1 & 0 & 1 & 0 & 1 & 0 \\
1 & 1 & 0 & 1 & 0 & 0 & 1 \\
1 & 1 & 0 & 1 & -1 & -1 & -1 \\
1 & 1 & -1 & -1 & 1 & 0 & 0 \\
1 & 1 & -1 & -1 & 0 & 1 & 0 \\
1 & 1 & -1 & -1 & 0 & 0 & 1 \\
1 & 1 & -1 & -1 & -1 & -1 & -1 \\
1 & -1 & 1 & 0 & 1 & 0 & 0 \\
1 & -1 & 1 & 0 & 0 & 1 & 0 \\
1 & -1 & 1 & 0 & 0 & 0 & 1 \\
1 & -1 & 1 & 0 & -1 & -1 & -1 \\
1 & -1 & 0 & 1 & 1 & 0 & 0 \\
1 & -1 & 0 & 1 & 0 & 1 & 0 \\
1 & -1 & 0 & 1 & 0 & 0 & 1 \\
1 & -1 & 0 & 1 & -1 & -1 & -1 \\
1 & -1 & -1 & -1 & 1 & 0 & 0 \\
1 & -1 & -1 & -1 & 0 & 1 & 0 \\
1 & -1 & -1 & -1 & 0 & 0 & 1 \\
1 & -1 & -1 & -1 & -1 & -1 & -1
\end{array}\right) \end{array}$$

FIGURE 7.2. Design matrix for the main effects of a $2 \times 3 \times 4$ contingency table.

therefore $(2 \cdot 3 \cdot 4) \times (1 + 1 + 2 + 3) = 24 \times 7$. In this case, the number of columns m is equal to the number of parameters in the independence model (cf. Figure 7.2).

7.8.2 Coding of Response Models

Let

$$\pi_i = P(y = 1 \mid x_i)\,, \quad i = 1, \ldots, L\,,$$

be the probability of response dependent on the level x_i of the vector of covariates x. Summarized in matrix representation we then have

$$\underset{L,1}{\pi} = \underset{L,m}{X} \; \underset{m,1}{\beta} \; . \tag{7.191}$$

N_i observations are made for the realization of covariates coded by x_i. Thus, the vector $\{y_i^{(j)}\}(j = 1, \ldots, N_i)$ is observed, and we get the ML estimate

$$\hat{\pi}_i = \hat{P}(y = 1 \mid x_i) = \frac{1}{N_i} \sum_{j=1}^{N_i} y_i^{(j)} \tag{7.192}$$

for π_i $(i = 1, \dots, L)$. For contingency tables the cell counts with binary response $N_i^{(1)}$ and $N_i^{(0)}$ are given from which $\hat{\pi}_i = N_i^{(1)}/(N_i^{(1)} + N_i^{(0)})$ is calculated.

The problem of finding an appropriate link function $h(\hat{\pi})$ for estimating

$$h(\hat{\pi}) = X\beta + \varepsilon \tag{7.193}$$

has already been discussed in several previous sections. If model (7.191) is chosen, i.e., the identity link, the parameters β_i are to be interpreted as the percentages with which the categories contribute to the conditional probabilities.

The logit link

$$h(\hat{\pi}_i) = \ln\left(\frac{\hat{\pi}_i}{1 - \hat{\pi}_i}\right) = x_i'\beta \tag{7.194}$$

is again equivalent to the logistic model for $\hat{\pi}_i$:

$$\hat{\pi}_i = \frac{\exp(x_i'\beta)}{1 + \exp(x_i'\beta)} \,. \tag{7.195}$$

The design matrices under inclusion of various interactions (up to the saturated model) are obtained as an extension of the designs for effect–coded main effects.

7.8.3 Coding of Models for the Hazard Rate

The analysis of lifetime data, given the variables $Y = 1$ (event) and $Y = 0$ (censored), is an important special case of the application of binary response in long–term studies.

The Cox model is often used as a semiparametric model for the modeling of failure time. Under inclusion of the vector of covariates x, this model can be written as follows:

$$\lambda(t \mid x) = \lambda_0(t) \exp(x'\beta) \,. \tag{7.196}$$

If the hazard rates of two vectors of covariates x_1, x_2 are to be compared with each other (e.g., stratification according to therapy x_1, x_2), the following relation is valid:

$$\frac{\lambda(t \mid x_1)}{\lambda(t \mid x_2)} = \exp((x_1 - x_2)'\beta) \,. \tag{7.197}$$

In order to be able to realize tests for quantitative or qualitative interactions between types of therapy and groups of patients, J subgroups of patients are defined (e.g., stratification according to prognostic factors). Let therapy Z be bivariate, i.e., $Z = 1$ (therapy A) and $Z = 0$ (therapy B). For a fixed group of patients the hazard rate $\lambda_j(t \mid Z)$ $(j = 1, \ldots, J)$, for instance, is determined according to the Cox approach

$$\lambda_j(t \mid Z) = \lambda_{0j}(t) \exp(\beta_j Z) . \tag{7.198}$$

In the case of $\hat{\beta}_j > 0$, the risk is higher for $Z = 1$ than for $Z = 0$ (jth stratum).

Test for Quantitative Interaction

We test H_0 : effects of therapy is identical across the J strata, i.e., $\mathrm{H}_0 : \beta_1 = \ldots = \beta_J = \beta$, against the alternative $\mathrm{H}_1 : \beta_i \lessgtr \beta_j$ for at least one pair (i, j). Under H_0, the test statistic

$$\chi^2_{J-1} = \sum_{j=1}^{J} \frac{\left(\hat{\beta}_j - \bar{\hat{\beta}}\right)^2}{\operatorname{var}(\hat{\beta}_j)} \tag{7.199}$$

with

$$\bar{\hat{\beta}} = \frac{\sum_{j=1}^{J} [\hat{\beta}_j / \operatorname{var}(\hat{\beta}_j)]}{\sum\limits_{j=1}^{J} [1 / \operatorname{var}(\hat{\beta}_j)]} \tag{7.200}$$

is distributed according to χ^2_{J-1}.

Test for Qualitative Differences

The null hypothesis H_0 : therapy B $(Z = 0)$ is better than therapy A $(Z = 1)$ means $\mathrm{H}_0 : \beta_j \le 0 \; \forall j$. We define the sum of squares of the standardized estimates

$$Q^- = \sum_{j:\beta_j < 0} \frac{(\hat{\beta}_j)^2}{\operatorname{var}(\hat{\beta}_j)} \tag{7.201}$$

and

$$Q^+ = \sum_{j:\beta_j > 0} \left[\frac{\hat{\beta}_j}{\operatorname{var}(\hat{\beta}_j)} \right]^2 , \tag{7.202}$$

as well as the test statistic

$$Q = \min(Q^-, Q^+) . \tag{7.203}$$

H_0 is rejected if $Q > c$ (Table 7.6).

Starting with the logistic model for the probability of response

$$P(Y = 1 \mid x) = \frac{\exp(\theta + x'\beta)}{1 + \exp(\theta + x'\beta)} , \tag{7.204}$$

J	2	3	4	5
c	2.71	4.23	5.43	6.50

TABLE 7.6. Critical values for the Q–test for $\alpha = 0.05$ (Gail and Simon, 1985).

and

$$P(Y = 0 \mid x) = 1 - P(Y = 1 \mid x) = \frac{1}{1 + \exp(\theta + x'\beta)} \tag{7.205}$$

with the binary variable

$$\begin{array}{llll} Y = 1: & \{T = t \mid T \geq t, x\} & \Rightarrow & \text{failure at time } t, \\ Y = 0: & \{T > t \mid T \geq t, x\} & \Rightarrow & \text{no failure,} \end{array}$$

we obtain the model for the hazard function

$$\lambda(t \mid x) = \frac{\exp(\theta + x'\beta)}{1 + \exp(\theta + x'\beta)} \quad \text{for } t = t_1, \ldots, t_T \tag{7.206}$$

(Cox, 1972b; cf. Doksum and Gasko, 1990; Lawless, 1982; Hamerle and Tutz, 1989). Thus the contribution of a patient to the likelihood (x fixed) with failure time t is

$$P(T = t \mid x) = \frac{\exp(\theta_t + x'\beta)}{\prod_{i=1}^{t}(1 + \exp(\theta_i + x'\beta))} . \tag{7.207}$$

Example 7.4. Assume that a patient has an event in the four failure times (e.g., loss of abutment teeth by extraction). Let the patient have the following categories of the covariates: sex = 1 and age group = 5 (60–70 years). The model is then $l = \theta + x'\beta$:

$$\begin{pmatrix} 0 \\ 0 \\ 0 \\ 1 \end{pmatrix} = \begin{pmatrix} 1 & 0 & 0 & 0 & 1 & 5 \\ 0 & 1 & 0 & 0 & 1 & 5 \\ 0 & 0 & 1 & 0 & 1 & 5 \\ 0 & 0 & 0 & 1 & \underbrace{1 & 5}_{x} \end{pmatrix} \begin{pmatrix} \theta_1 \\ \theta_2 \\ \theta_3 \\ \theta_4 \\ \beta_{11} \\ \beta_{12} \end{pmatrix} \begin{matrix} \left.\vphantom{\begin{matrix}1\\1\\1\\1\end{matrix}}\right\} \theta_t \\ \left.\vphantom{\begin{matrix}1\\1\end{matrix}}\right\} \beta. \end{matrix} \tag{7.208}$$

(Column labels above the last two columns: Sex, Age.)

For N patients we have the model

$$\begin{pmatrix} l_1 \\ l_2 \\ \vdots \\ l_N \end{pmatrix} = \begin{pmatrix} I_1 & x_1 \\ I_2 & x_2 \\ \vdots & \\ I_N & x_N \end{pmatrix} \begin{pmatrix} \theta \\ \beta \end{pmatrix},$$

The dimension of the identity matrices I_j (patient j) is the number of survived failure times plus 1 (failure time of the jth patient). The vectors l_j for the jth patient contain as many zeros as the number of survived failure times of the other patients and the value 1 at the failure time of the jth patient.

The numerical solutions (for instance, according to Newton–Raphson) for the ML estimates $\hat{\theta}$ and $\hat{\beta}$ are obtained from the product of the likelihood functions (7.207) of all patients.

7.9 Extensions to Dependent Binary Variables

Although loglinear models are sufficiently rich to model any dependence structure between categorical variables, if one is interested in a regression of multivariate binary responses on a set of possibly continuous covariates, alternative models exist which are better suited and have easier parameter interpretation. Two often–used models in applications are marginal models and random effects models. In the following, we emphasize the idea of marginal models, because these seem to be a natural extension of the logistic regression model to more than one response variable. The first approach we describe in detail is called the quasi–likelihood approach (cf. Section 7.1.7), because the distribution of the binary response variables is not fully specified. We start by describing these models in detail in Section 7.9.3. Then the generalized estimating equations (GEEs) approach (Liang and Zeger, 1986) is introduced and two examples are given. The third approach is a full likelihood approach (Section 7.9.12). Section 7.9.12 mainly gives an overview of the recent literature.

7.9.1 Overview

We now extend the problems of categorical response to the situations of correlation within the response values. These correlations are due to classification of the individuals into clusters of "related" elements. As already mentioned in Section 7.1.6, a positive correlation among related elements in a cluster leads to overdispersion, if independence among these elements is falsely assumed.

Examples:

- Two or more implants or abutment teeth in dental reconstructions (Walther and Toutenburg, 1991).
- Response of a patient in cross–over in the case of a significant carry–over effect.
- Repeated categorical measurement of a response such as function of the lungs, blood pressure, or performance in training (repeated measures design or panel data).

- Measurement of paired organs (eyes, kidneys, etc.)
- Response of members of a family.

Let y_{ij} be the categorical response of the jth individual in the ith cluster

$$y_{ij}, \quad i = 1, \ldots, N, \quad j = 1, \ldots, n_i \,. \tag{7.209}$$

We assume that the expectation of the response y_{ij} is dependent on prognostic variables (covariates) x_{ij} by a regression, that is,

$$\mathrm{E}(y_{ij}) = \beta_0 + \beta_1 x_{ij} \,. \tag{7.210}$$

Assume $\mathrm{var}(y_{ij}) = \sigma^2$ and

$$\mathrm{cov}(y_{ij}, y_{ij'}) = \sigma^2 \rho \quad (j \neq j'). \tag{7.211}$$

The response of individuals from different clusters is assumed to be uncorrelated. Let us assume that the covariance matrix for the response of every cluster equals

$$\mathrm{V}\begin{pmatrix} y_{i1} \\ \vdots \\ y_{in_i} \end{pmatrix} = \mathrm{V}(y_i) = \sigma^2(1-\rho)I_{n_i} + \sigma^2 \rho J_{n_i} \tag{7.212}$$

and thus has a compound symmetric structure. Hence, the covariance matrix of the entire sample vector is block–diagonal

$$W = \mathrm{V}\begin{pmatrix} y_1 \\ \vdots \\ y_N \end{pmatrix} = \mathrm{diag}(\mathrm{V}(y_1), \ldots, \mathrm{V}(y_N)) \,. \tag{7.213}$$

Notice that the matrix W itself does not have a compound symmetric structure. Hence, we have a generalized regression model. The best linear unbiased estimate of $\beta = (\beta_0, \beta_1)'$ is given by the Gauss–Markov–Aitken estimator [(3.168)]

$$b = (X'W^{-1}X)^{-1}X'W^{-1}y \tag{7.214}$$

and does not coincide with the OLS estimator. The choice of an incorrect covariance structure leads, according to our remarks in Section 3.9.2, to a bias in the estimate of the variance. On the other hand, the unbiasedness or consistency of the estimator of β stays untouched even in the case of an incorrect choice of the covariance matrix. Liang and Zeger (1993) examined the bias of $\mathrm{var}(\hat{\beta}_1)$ for the wrong choice of $\rho = 0$. In the case of positive correlation within the cluster, the variance is underestimated. This corresponds to the results of Goldberger (1964) for positive autocorrelation in econometric models.

The following problems arise in practice:

(i) identification of the covariance structure;

(ii) estimation of the correlation; and

(iii) application of an Aitken-type estimate.

However, it is no longer possible to assume the usual GLM approach, because this does not take the correlation structure into consideration. Various approaches were developed as extensions of the GLM approach, in order to be able to include the correlation structure in the response:

- the marginal model;
- the random–effects model;
- the observation–driven model; and
- the conditional model.

For binary response, simplifications arise (Section 7.9.8). Liang and Zeger (1989) proved that the joint distribution of the y_{ij} can be descibed by n_i logistic models for y_{ij} given y_{ik} $(k \neq j)$. Rosner (1984) used this approach and developed beta–binomial models.

7.9.2 Modeling Approaches for Correlated Response

The modeling approaches can be ordered according to diverse criteria.

Population–Averaged versus Subject–Specific Models

The essential difference between population–averaged (PA) and subject–specific (SS) models lies in the answer to the question of whether the regression coefficients vary for the individuals. In PA models, the β's are independent of the specific individual i. Examples are the marginal and conditional models. In SS models, the β's are dependent on the specific i and are therefore written as β_i. An example for an SS model is the random–effects model.

Marginal, Conditional, and Random–Effects Models

In the marginal model, the regression is modeled separately from the dependence within the measurement in contrast to the two other approaches. The marginal expectation $\mathrm{E}(y_{ij})$ is modeled as a function of the explanatory variables and is interpreted as the mean response over the population of individuals with the same x. Hence, marginal models are mainly suitable for the analysis of covariate effects in a population.

The random–effects model, often also titled the mixed model, assumes that there are fixed effects, as in the marginal model, as well as individual specific effects. The dependent observations on each individual are assumed to be *conditionally independent given the subject–specific effects.*

Hence random–effects models are useful if one is interested in subject–specific behavior. But, concerning interpretation, only the *linear mixed model* allows an easy interpretation of fixed effect parameters as population–averaged effects and the others as subject–specific effects. *Generalized linear mixed models* are more complex, and even if a parameter is estimated as a fixed effect it may not be easily interpreted as a population–averaged effect.

For the conditional model (observation–driven model), a time–dependent response y_{it} is modeled as a function of the covariates and of the past response values $y_{it-1}, \ldots, y_{i1}$. This is done by assuming a specific correlation structure among the response values. Conditional models are useful if the main point of interest is the conditional probability of a state or the transition of states.

7.9.3 Quasi–Likelihood Approach for Correlated Binary Response

The following sections are dedicated to binary response variables and especially the bivariate case (i.e., cluster size $n_i = 2$ for all $i = 1, \ldots, N$).

In the case of a violation of independence or in the case of a missing distribution assumption of the natural exponential family, the core of the ML method, namely, the score function, may be used, nevertheless, for parameter estimation. We now want to specify the so–called quasi–score function (7.77) for the binary response (cf. Section 7.1.7).

Let $y_i' = (y_{i1}, \ldots, y_{in_i})$ be the response vector of the ith cluster $(i = 1, \ldots, N)$ with the true covariance matrix $\operatorname{cov}(y_i)$ and let x_{ij} be the $(p \times 1)$–vector of the covariate corresponding to y_{ij}. Assume the variables y_{ij} are binary with values 1 and 0, and assume $P(y_{ij} = 1) = \pi_{ij}$. We then have $\mu_{ij} = \pi_{ij}$. Let $\pi_i' = (\pi_{i1}, \ldots, \pi_{in_i})$. Suppose that the link function is $g(\cdot)$, that is,

$$g(\pi_{ij}) = \eta_{ij} = x_{ij}'\beta \,.$$

Let $h(\cdot)$ be the inverse function, that is,

$$\mu_{ij} = \pi_{ij} = h(\eta_{ij}) = h(x_{ij}'\beta) \,.$$

For the canonical link

$$\operatorname{logit}(\pi_{ij}) = \ln\left(\frac{\pi_{ij}}{1-\pi_{ij}}\right) = g(\pi_{ij}) = x_{ij}'\beta$$

we have

$$\pi_{ij} = h(\eta_{ij}) = \frac{\exp(\eta_{ij})}{1+\exp(\eta_{ij})} = \frac{\exp(x_{ij}'\beta)}{1+\exp(x_{ij}'\beta)} \,.$$

Hence

$$D = \left(\frac{\partial \mu_{ij}}{\partial \beta}\right) = \left(\frac{\partial \pi_{ij}}{\partial \beta}\right) .$$

We have

$$\frac{\partial \pi_{ij}}{\partial \beta} = \frac{\partial \pi_{ij}}{\partial \eta_{ij}} \frac{\partial \eta_{ij}}{\partial \beta} = \frac{\partial h(\eta_{ij})}{\partial \eta_{ij}} x_{ij} \,,$$

and, hence, for $i = 1, \ldots, N$ and the $(p \times n_i)$–matrix $X_i' = (x_{i1}, \ldots, x_{in_i})$:

$$D_i = \tilde{D}_i \, X_i \quad \text{with} \quad \tilde{D}_i = \left(\frac{\partial h(\eta_{ij})}{\partial \eta_{ij}} \right) .$$

For the quasi–score function for all N clusters, we now get

$$U(\beta) = \sum_{i=1}^{N} X_i' \tilde{D}_i' \mathrm{V}_i^{-1} (y_i - \pi_i) \,, \tag{7.215}$$

where V_i is the matrix of the working variances and covariances of the y_{ij} of the ith cluster. The solution of $U(\hat{\beta}) = 0$ is found iteratively under further specifications, which we describe in the next section.

7.9.4 The Generalized Estimating Equation Method by Liang and Zeger

The variances are modeled as a function of the mean, that is,

$$v_{ij} = \mathrm{var}(y_{ij}) = v(\pi_{ij}) \phi \,. \tag{7.216}$$

(In the binary case, the form of the variance of the binomial distribution is often chosen: $v(\pi_{ij}) = \pi_{ij}(1 - \pi_{ij})$.) With these, the following matrix is formed

$$A_i = \mathrm{diag}(v_{i1}, \ldots, v_{in_i}) \,. \tag{7.217}$$

Since the structure of dependence is not known, an $(n_i \times n_i)$–*quasi–correlation matrix* $R_i(\alpha)$ is chosen for the vector of the ith cluster $y_i' = (y_{i1}, \ldots, y_{in_i})$ according to

$$R_i(\alpha) = \begin{pmatrix} 1 & \rho_{i12}(\alpha) & \cdots & \rho_{i1n_i}(\alpha) \\ \rho_{i21}(\alpha) & 1 & \cdots & \rho_{i2n_i}(\alpha) \\ \vdots & & & \vdots \\ \rho_{in_i1}(\alpha) & \rho_{in_i2}(\alpha) & \cdots & 1 \end{pmatrix} , \tag{7.218}$$

where the $\rho_{ikl}(\alpha)$ are the correlations as function of α (α may be a scalar or a vector). $R_i(\alpha)$ may vary for the clusters.

By multiplying the quasi–correlation matrix $R_i(\alpha)$ with the root diagonal matrix of the variances A_i, we obtain a working covariance matrix

$$\mathrm{V}_i(\beta, \alpha, \phi) = A_i^{1/2} R_i(\alpha) A_i^{1/2} \,, \tag{7.219}$$

which is no longer completely specified by the expectations, as in the case of independent response. We have $\mathrm{V}_i(\beta, \alpha, \phi) = \mathrm{cov}(y_i)$ if and only if $R_i(\alpha)$ is the true correlation matrix of y_i.

If the matrices V_i in (7.215) are replaced by the matrices $V_i(\beta, \alpha, \phi)$ from (7.219), we get the *generalized estimating equations* by Liang and Zeger (1986), that is,

$$U(\beta, \alpha, \phi) = \sum_{i=1}^{N} \left(\frac{\partial \pi_i}{\partial \beta} \right)' V_i^{-1}(\beta, \alpha, \phi)(y_i - \pi_i) = 0 . \qquad (7.220)$$

The solutions are denoted by $\hat{\beta}_G$. For the quasi–Fisher matrix, we have

$$F_G(\beta, \alpha) = \sum_{i=1}^{N} \left(\frac{\partial \pi_i}{\partial \beta} \right)' V_i^{-1}(\beta, \alpha, \phi) \left(\frac{\partial \pi_i}{\partial \beta} \right) . \qquad (7.221)$$

To avoid the dependence of α in determining $\hat{\beta}_G$, Liang and Zeger (1986) proposed replacing α by a $N^{1/2}$–consistent estimate $\hat{\alpha}(y_1, \ldots, y_N, \beta, \phi)$ and ϕ by $\hat{\phi}$ (7.79) and determining $\hat{\beta}_G$ from $U(\beta, \hat{\alpha}, \hat{\phi}) = 0$.

Remark. The iterative estimating procedure for GEE is described in detail in Liang and Zeger (1986). For the computational translation, an SAS macro by Karim and Zeger (1988) and a program by Kastner, Fieger and Heumann (1997) exist.

If $R_i(\alpha) = I_{n_i}$ $(i = 1, \ldots, N)$, is chosen, then the GEEs are reduced to the *independence estimating equations* (IEEs) . The IEEs are

$$U(\beta, \phi) = \sum_{i=1}^{N} \left(\frac{\partial \pi_i}{\partial \beta} \right)' A_i^{-1}(y_i - \pi_i) = 0 \qquad (7.222)$$

with $A_i = \text{diag}(v(\pi_{ij})\phi)$. The solution is denoted by $\hat{\beta}_I$. Under some weak conditions, we have (Theorem 1 in Liang and Zeger, 1986) that $\hat{\beta}_I$ *is asymptotically consistent if the expectation* $\pi_{ij} = h(x'_{ij}\beta)$ *is correctly specified and the dispersion parameter* ϕ *is consistently estimated.*

$\hat{\beta}_I$ is asymptotically normal

$$\hat{\beta}_I \overset{\text{a.s.}}{\sim} N(\beta; F_Q^{-1}(\beta, \phi) F_2(\beta, \phi) F_Q^{-1}(\beta, \phi)), \qquad (7.223)$$

where

$$F_Q^{-1}(\beta, \phi) = \left[\sum_{i=1}^{N} \left(\frac{\partial \pi_i}{\partial \beta} \right)' A_i^{-1} \left(\frac{\partial \pi_i}{\partial \beta} \right) \right]^{-1} ,$$

$$F_2(\beta, \phi) = \sum_{i=1}^{N} \left(\frac{\partial \pi_i}{\partial \beta} \right)' A_i^{-1} \operatorname{cov}(y_i) A_i^{-1} \left(\frac{\partial \pi_i}{\partial \beta} \right) ,$$

and $\operatorname{cov}(y_i)$ is the true covariance matrix of y_i.

A consistent estimate for the variance of $\hat{\beta}_I$ is found by replacing β_I by $\hat{\beta}_I$, $\operatorname{cov}(y_i)$ by its estimate $(y_i - \hat{\pi}_i)(y_i - \hat{\pi}_i)'$, and ϕ by $\hat{\phi}$ from (7.79), if ϕ is an unknown nuisance parameter. The consistency is independent of the correct specification of the covariance.

The advantages of $\hat{\beta}_I$ are that $\hat{\beta}_I$ is easy to calculate using software for generalized linear models and that in the case of correct specification of the regression model, $\hat{\beta}_I$ and $\text{cov}(\hat{\beta}_I)$ are consistent estimates. However, $\hat{\beta}_I$ loses in efficiency if the correlation between the clusters is large.

7.9.5 *Properties of the Generalized Estimating Equation Estimate $\hat{\beta}_G$*

Liang and Zeger (1986, Theorem 2) state that under some weak assumptions, and under the conditions:

(i) $\hat{\alpha}$ is $N^{1/2}$–consistent for α, given β and ϕ;

(ii) $\hat{\phi}$ is a $N^{1/2}$–consistent estimate for ϕ; and given β

(iii) the derivation $\partial\hat{\alpha}(\beta,\phi)/\partial\phi$ is independent of ϕ and α and is of stochastic order $O_p(1)$;

the estimate $\hat{\beta}_G$ is consistent and asymptotically normal

$$\hat{\beta}_G \overset{\text{a.s.}}{\sim} N(\beta,\ \text{V}_G) \tag{7.224}$$

with the asymptotic covariance matrix

$$\text{V}_G = F_Q^{-1}(\beta,\alpha)F_2(\beta,\alpha)F_Q^{-1}(\beta,\alpha), \tag{7.225}$$

where

$$\begin{aligned} F_Q^{-1}(\beta,\alpha) &= \left(\sum_{i=1}^{N}\left(\frac{\partial\pi_i}{\partial\beta}\right)' \text{V}_i^{-1}\left(\frac{\partial\pi_i}{\partial\beta}\right)\right)^{-1}, \\ F_2(\beta,\alpha) &= \sum_{i=1}^{N}\left(\frac{\partial\pi_i}{\partial\beta}\right)' \text{V}_i^{-1}\,\text{cov}(y_i)\text{V}_i^{-1}\left(\frac{\partial\pi_i}{\partial\beta}\right) \end{aligned}$$

and $\text{cov}(y_i) = \text{E}[(y_i-\pi_i)(y_i-\pi_i)']$ is the true covariance matrix of y_i. A short outline of the proof may be found in the Appendix of Liang and Zeger (1986).

The asymptotic properties hold only for $N \to \infty$. Hence, it should be remembered that the estimation procedure should be used only for a large number of clusters.

An estimate $\hat{\text{V}}_G$ for the covariance matrix V_G may be found by replacing β, ϕ, and α by their consistent estimates in (7.225), or by replacing $\text{cov}(y_i)$ by $(y_i-\hat{\pi}_i)(y_i-\hat{\pi}_i)'$.

If the covariance structure is specified correctly, so that $\text{V}_i = \text{cov}(y_i)$, then the covariance of $\hat{\beta}_G$ is the inverse of the expected Fisher–information matrix

$$\text{V}_G = \left(\sum_{i=1}^{N}\left(\frac{\partial\pi_i}{\partial\beta}\right)' \text{V}_i^{-1}\left(\frac{\partial\pi_i}{\partial\beta}\right)\right)^{-1} = F^{-1}(\beta,\alpha).$$

The estimate of this matrix is more stable than that of (7.225), but it has a loss in efficiency if the correlation structure is specified incorrectly (cf. Prentice, 1988, p. 1040).

The method of Liang and Zeger leads to an asymptotic variance of $\hat{\beta}_G$ that is independent of the choice of the estimates $\hat{\alpha}$ and $\hat{\phi}$ within the class of the $N^{1/2}$–consistent estimates. This is true for the asymptotic distribution of $\hat{\beta}_G$ as well.

In the case of correct specification of the regression model, the estimates $\hat{\beta}_G$ and $\hat{\mathrm{V}}_G$ are consistent, independent of the choice of the quasi–correlation matrix $R_i(\alpha)$. This means that even if $R_i(\alpha)$ is specified incorrectly, $\hat{\beta}_G$ and $\hat{\mathrm{V}}_G$ stay consistent as long as $\hat{\alpha}$ and $\hat{\phi}$ are consistent. This robustness of the estimates is important, because the admissibility of the working covariance matrix V_i is difficult to check for small n_i. An incorrect specification of $R_i(\alpha)$ can reduce the efficiency of $\hat{\beta}_G$.

If the identity matrix is assumed for $R_i(\alpha)$, i.e., $R_i(\alpha) = I$ $(i = 1, \ldots, N)$, then the estimating equations for β are reduced to the IEE. If the variances of the binomial distribution are chosen, as is usually done in the binary case, then the IEE and the ML score function (with binomially distributed variables) lead to the same estimates for β. However, the IEE method should be preferred in general, because the ML estimation procedure leads to incorrect variances for $\hat{\beta}_G$ and hence, for example, incorrect test statistics and *p–values*. This leads to incorrect conclusions, for instance, related to the significance or nonsignificance of the covariates (cf. Liang and Zeger, 1993).

Diggle, Liang and Zeger (1994, Chapter 7.5) have proposed checking the consistency of $\hat{\beta}_G$ by fitting an appropriate model with various covariance structures. The estimates $\hat{\beta}_G$ and their consistent variances are then compared. If these differ too much, the modeling of the covariance structure calls for more attention.

7.9.6 Efficiency of the Generalized Estimating Equation and Independence Estimating Equation Methods

Liang and Zeger (1986) stated the following about the comparison of $\hat{\beta}_I$ and $\hat{\beta}_G$. $\hat{\beta}_I$ is almost as efficient as $\hat{\beta}_G$ if the true correlation α is small. $\hat{\beta}_I$ is very efficient if α is small and the data are binary.

If α is large, then $\hat{\beta}_G$ is more efficient than $\hat{\beta}_I$, and the efficiency of $\hat{\beta}_G$ can be increased if the correlation matrix is specified correctly.

In the case of a high correlation within the blocks, the loss of efficiency of $\hat{\beta}_I$ compared to $\hat{\beta}_G$ is larger if the number of subunits n_i $(i = 1, \ldots, N)$, varies between the clusters than if the clusters are all of the same size.

7.9.7 Choice of the Quasi–Correlation Matrix $R_i(\alpha)$

The working correlation matrix $R_i(\alpha)$ is chosen according to considerations such as simplicity, efficiency, and amount of existing data. Furthermore, assumptions about the structure of the dependence among the data should be considered by the choice. As mentioned before, the importance of the correlation matrix is due to the fact that it influences the variance of the estimated parameters.

The simplest specification is the assumption that the repeated observations of a cluster are uncorrelated, that is,

$$R_i(\alpha) = I, \quad i = 1, \dots, N.$$

This assumption leads to the IEE equations for uncorrelated response variables.

Another special case, which is the most efficient according to Liang and Zeger (1986, Section 4) but may be used only if the number of observations per cluster is small and is the same for all clusters (e.g., equals n), is given by the choice

$$R_i(\alpha) = R(\alpha),$$

where $R(\alpha)$ is left totally unspecified and may be estimated by the empirical correlation matrix. The $n(n-1)/2$ parameters have to be estimated.

If it is assumed that the same pairwise dependencies exist among all the response variables of one cluster, then the *exchangeable correlation structure* may be chosen:

$$\operatorname{Corr}(y_{ik}, y_{il}) = \alpha, \quad k \neq l, \quad i = 1, \dots, N\,.$$

This corresponds to the correlation assumption in random–effects models.

If $\operatorname{Corr}(y_{ik}, y_{il}) = \alpha(|k-l|)$ is chosen, then the correlations are stationary. The specific form $\alpha(|k-l|) = \alpha^{|l-k|}$ corresponds to the autocorrelation function of an AR(1)–process.

Further methods for parameter estimation in quasi–likelihood approaches are: the GEE1 method by Prentice (1988) that estimates the α and β simultaneously from the GEE for α and β; the modified GEE1 method by Fitzmaurice, Laird and Rotnitzky (1993) based on conditional odds ratios; those by Lipsitz, Laird and Harrington (1991) and Liang, Zeger and Qaqish (1992) based on marginal odds ratios for modeling the cluster correlation; the GEE2 method by Liang et al. (1992) that estimates $\delta' = (\beta', \alpha)$ simultaneously as a joint parameter; and the pseudo–ML method by Zhao and Prentice (1990) and Prentice and Zhao (1991).

7.9.8 Bivariate Binary Correlated Response Variables

The previous sections introduced various methods developed for regression analysis of correlated binary data. They were described in a general form for N blocks (clusters) of size n_i. These methods may, of course, be used for bivariate binary data as well. This has the advantage that it simplifies the matter.

In this section, the GEE and IEE methods are developed for the bivariate binary case. Afterward, an example demonstrates, for the case of bivariate binary data, the difference between a naive ML estimate and the GEE method of Liang and Zeger (1986).

We have $y_i = (y_{i1}, y_{i2})'$ $(i = 1, \ldots, N)$. Each response variable y_{ij} $(j = 1, 2)$, has its own vector of covariates $x_{ij}' = (x_{ij1}, \ldots, x_{ijp})$. The chosen link function for modeling the relationship between $\pi_{ij} = P(y_{ij} = 1)$ and x_{ij} is the logit link

$$\text{logit}(\pi_{ij}) = \ln\left(\frac{\pi_{ij}}{1 - \pi_{ij}}\right) = x_{ij}'\beta\,. \tag{7.226}$$

Let

$$\pi_i' = (\pi_{i1}, \pi_{i2})\,, \quad \eta_{ij} = x_{ij}'\beta\,, \quad \eta' = (\eta_{i1}, \eta_{i2})\,. \tag{7.227}$$

The logistic regression model has become the standard method for regression analysis of binary data.

7.9.9 The Generalized Estimating Equation Method

From Section 7.9.4 it can be seen that the form of the estimating equations for β is as follows:

$$U(\beta, \alpha, \phi) = S(\beta, \alpha) = \sum_{i=1}^{N} \left(\frac{\partial \pi_i}{\partial \beta}\right)' \mathrm{V}_i^{-1}(y_i - \pi_i) = 0\,, \tag{7.228}$$

where $\mathrm{V}_i = A_i^{1/2} R_i(\alpha) A_i^{1/2}$, $A_i = \text{diag}(v(\pi_{ij})\phi)$ $(j = 1, 2)$, and $R_i(\alpha)$ is the working correlation matrix. Since only one correlation coefficient $\rho_i = \text{Corr}(y_{i1}, y_{i2})$ $(i = 1, \ldots, N)$, has to be specified for bivariate binary data, and this is assumed to be constant, we have, for the correlation matrix,

$$R_i(\alpha) = \begin{pmatrix} 1 & \rho \\ \rho & 1 \end{pmatrix}, \quad i = 1, \ldots, N\,. \tag{7.229}$$

For the matrix of derivatives we have

$$\begin{aligned} \left(\frac{\partial \pi_i}{\partial \beta}\right)' &= \left(\frac{\partial h(\eta_i)}{\partial \beta}\right)' = \left(\frac{\partial \eta_i}{\partial \beta}\right)' \left(\frac{\partial h(\eta_i)}{\partial \eta_i}\right)' \\ &= \begin{pmatrix} x_{i1}' \\ x_{i2}' \end{pmatrix}' \begin{pmatrix} \partial h(\eta_{i1})/\partial \eta_{i1} & 0 \\ 0 & \partial h(\eta_{i2})/\partial \eta_{i2} \end{pmatrix}. \end{aligned}$$

Since

$$h(\eta_{i1}) = \pi_{i1} = (\exp(x_{i1}'\beta))/(1 + \exp(x_{i1}'\beta))$$

and

$$\exp(x'_{i1}\beta) = \pi_{i1}/(1-\pi_{i1}),$$

we have

$$1 + \exp(x'_{i1}\beta) = 1 + \pi_{i1}/(1-\pi_{i1}) = 1/(1-\pi_{i1}),$$

and

$$\frac{\partial h(\eta_{i1})}{\partial \eta_{i1}} = \frac{\pi_{i1}}{1+\exp(x'_{i1}\beta)} = \pi_{i1}(1-\pi_{i1}) \tag{7.230}$$

holds. Analogously, we have

$$\frac{\partial h(\eta_{i2})}{\partial \eta_{i2}} = \pi_{i2}(1-\pi_{i2}). \tag{7.231}$$

If the variance is specified as $\text{var}(y_{ij}) = \pi_{ij}(1-\pi_{ij})$, $\phi = 1$, then we get

$$\left(\frac{\partial \pi_i}{\partial \beta}\right)' = x'_i \begin{pmatrix} \text{var}(y_{i1}) & 0 \\ 0 & \text{var}(y_{i2}) \end{pmatrix} = x'_i \Delta_i$$

with $x'_i = (x_{i1}, x_{i2})$ and $\Delta_i = \begin{pmatrix} \text{var}(y_{i1}) & 0 \\ 0 & \text{var}(y_{i2}) \end{pmatrix}$. For the covariance matrix V_i we have:

$$\begin{aligned} \text{V}_i &= \begin{pmatrix} \text{var}(y_{i1}) & 0 \\ 0 & \text{var}(y_{i2}) \end{pmatrix}^{1/2} \begin{pmatrix} 1 & \rho \\ \rho & 1 \end{pmatrix} \begin{pmatrix} \text{var}(y_{i1}) & 0 \\ 0 & \text{var}(y_{i2}) \end{pmatrix}^{1/2} \\ &= \begin{pmatrix} \text{var}(y_{i1}) & \rho(\text{var}(y_{i1})\,\text{var}(y_{i2}))^{1/2} \\ \rho(\text{var}(y_{i1})\,\text{var}(y_{i2}))^{1/2} & \text{var}(y_{i2}) \end{pmatrix} \end{aligned} \tag{7.232}$$

and for the inverse of V_i:

$$\begin{aligned} \text{V}_i^{-1} &= \frac{1}{(1-\rho^2)\,\text{var}(y_{i1})\,\text{var}(y_{i2})} \\ &\quad \begin{pmatrix} \text{var}(y_{i2}) & -\rho(\text{var}(y_{i1})\,\text{var}(y_{i2}))^{1/2} \\ -\rho(\text{var}(y_{i1})\,\text{var}(y_{i2}))^{1/2} & \text{var}(y_{i1}) \end{pmatrix} \\ &= \frac{1}{1-\rho^2} \begin{pmatrix} [\text{var}(y_{i1})]^{-1} & -\rho(\text{var}(y_{i1})\,\text{var}(y_{i2}))^{-1/2} \\ -\rho(\text{var}(y_{i1})\,\text{var}(y_{i2}))^{-1/2} & [\text{var}(y_{i2})]^{-1} \end{pmatrix}. \end{aligned} \tag{7.233}$$

If Δ_i is multiplied by ${\text{V}_i}^{-1}$, we obtain

$$W_i = \Delta_i {\text{V}_i}^{-1} = \frac{1}{1-\rho^2} \begin{pmatrix} 1 & -\rho\left(\frac{\text{var}(y_{i1})}{\text{var}(y_{i2})}\right)^{1/2} \\ -\rho\left(\frac{\text{var}(y_{i2})}{\text{var}(y_{i1})}\right)^{\frac{1}{2}} & 1 \end{pmatrix} \tag{7.234}$$

and for the GEE method for β in the bivariate binary case

$$S(\beta, \alpha) = \sum_{i=1}^{N} x_i' W_i (y_i - \pi_i) = 0. \tag{7.235}$$

According to Liang and Zeger (1986, Theorem 2), under some weak conditions, and under the assumption that the correlation parameter was consistently estimated, the solution $\hat{\beta}_G$ is consistent and asymptotically normal with expectation β and covariance matrix (7.225).

7.9.10 The Independence Estimating Equation Method

If it is assumed that the response variables of each of the blocks are independent, i.e., $R_i(\alpha) = I$ and $\mathrm{V}_i = A_i$, then the GEE method is reduced to the IEE method,

$$U(\beta, \phi) = S(\beta) = \sum_{i=1}^{N} \left(\frac{\partial \pi_i}{\partial \beta}\right)' A_i{}^{-1} (y_i - \pi_i) = 0. \tag{7.236}$$

As we have just shown, we have, for the bivariate binary case,

$$\left(\frac{\partial \pi_i}{\partial \beta}\right)' = x_i' \Delta_i = x_i' \begin{pmatrix} \mathrm{var}(y_{i1}) & 0 \\ 0 & \mathrm{var}(y_{i2}) \end{pmatrix} \tag{7.237}$$

with $\mathrm{var}(y_{ij}) = \pi_{ij}(1 - \pi_{ij})$, $\phi = 1$, and

$$A_i{}^{-1} = \begin{pmatrix} [\mathrm{var}(y_{i1})]^{-1} & 0 \\ 0 & [\mathrm{var}(y_{i2})]^{-1} \end{pmatrix}.$$

The IEE method then simplifies to

$$S(\beta) = \sum_{i=1}^{N} x_i' (y_i - \pi_i) = 0. \tag{7.238}$$

The solution $\hat{\beta}_I$ is consistent and asymptotically normal, according to Liang and Zeger (1986, Theorem 1).

7.9.11 An Example from the Field of Dentistry

In this section, we demonstrate the procedure of the GEE method by means of a "twin" data set that was documented by the Dental Clinic in Karlsruhe, Germany (Walther, 1992). The focal point is to show the difference between a robust estimate (GEE method), that takes the correlation of the response variables into account, and the naive ML estimate. For the parameter estimation with the GEE method, an SAS macro is available (Karim and Zeger, 1988), as well as a procedure by Kastner et al. (1997).

Description of the "Twin" Data Set

During the examined interval, 331 patients were provided with two conical crowns each in the Dental Clinic in Karlsruhe. Since 50 conical crowns showed missing values, and since the SAS macro for the GEE method needs complete data sets, these patients were excluded. Hence, for the estimation of the regression parameters, the remaining 612 completely observed twin data sets were used. In this example, the twin pairs make up the clusters, and the twins themselves (1.twin, 2.twin) are the subunits of the clusters.

The Response Variable

For all twin pairs in this study, the lifetime of the conical crowns was recorded in days. This lifetime is chosen as the response and is transformed into a binary response variable y_{ij} of the jth twin ($j = 1, 2$) in the ith cluster with

$$y_{ij} = \begin{cases} 1\,, & \text{if the conical crown is in function longer than } x \text{ days} \\ 0\,, & \text{if the conical crown is in function no longer than } x \text{ days.} \end{cases}$$

Different values may be defined for x. In the example, the values, in days, of 360 (1 year), 1100 (3 years), and 2000 (5 years) were chosen. Because the response variable is binary, the response probability of y_{ij} is modeled by the logit link (logistic regression). The model for the log–odds (i.e., the logarithm of the odds $\pi_{ij}/(1-\pi_{ij})$ of the response $y_{ij} = 1$) is linear in the covariates, and in the model for the odds itself, the covariates have a multiplicative effect on the odds. The aim of the analysis is to find whether the prognostic factors have a significant influence on the response probability.

Prognostic Factors

The covariates that were included in the analysis with the SAS macro, are:

- age (in years);
- sex (1 : male, 2 : female);
- jaw (1 : upper jaw, 2 : lower jaw); and
- type (1 : dentoalveolar design, 2 : transversal design).

All covariates, except for the covariate age, are dichotomous. The two types of conical crown constructions, dentoalveolar and transversal design, are distinguished as follows (cf. Walther, 1992):

- The dentoalveolar design connects all abutments exclusively by a rigid connection that runs on the alveolar ridge.
- The transversal design is used if parts of the reconstruction have to be connected by a transversal bar. This is the case if teeth in the front area are not included in the construction.

A total of 292 conical crowns were included in a dentoalveolar design and 320 in a transversal design. Of these, 258 conical crowns were placed in the upper jaw, and 354 in the lower jaw.

The GEE Method

A problem that arises for the twin data is that the twins of a block are correlated. If this correlation is not taken into account, then the estimates $\hat{\beta}$ stay unchanged but the variance of the $\hat{\beta}$ is underestimated. In the case of positive correlation in a cluster, we have

$$\text{var}(\hat{\beta})_{\text{naive}} < \text{var}(\hat{\beta})_{\text{robust}}.$$

Therefore,

$$\frac{\hat{\beta}}{\sqrt{\text{var}(\hat{\beta})_{\text{naive}}}} > \frac{\hat{\beta}}{\sqrt{\text{var}(\hat{\beta})_{\text{robust}}}},$$

which leads to incorrect tests and possibly to significant effects that might not be significant in a correct analysis (e.g., GEE). For this reason, appropriate methods that estimate the variance correctly should be chosen if the response variables are correlated.

The following regression model without interaction is assumed:

$$\begin{aligned}\ln \frac{P(\text{lifetime} \geq x)}{P(\text{lifetime} < x)} &= \beta_0 + \beta_1 \cdot \text{age} + \beta_2 \cdot \text{sex} \\ &\quad + \beta_3 \cdot \text{jaw} + \beta_4 \cdot \text{type}\,.\end{aligned}$$

Additionally, we assume that the dependencies between the twins are identical and hence the exchangeable correlation structure is suitable for describing the dependencies.

To demonstrate the effects of various correlation assumptions on the estimation of the parameters, the following logistic regression models, which differ only in the assumed association parameter, are compared:

Model 1: Naive (incorrect) ML estimation.

Model 2: Robust (correct) estimation, where independence is assumed, i.e., $R_i(\alpha) = I$.

Model 3: Robust estimation with exchangeable correlation structure ($\rho_{ikl} = \text{Corr}(y_{ik}, y_{il}) = \alpha,\ k \neq l$).

Model 4: Robust estimation with unspecified correlation structure ($R_i(\alpha) = R(\alpha)$).

As a test statistic (z–naive and z–robust) the ratio of estimate and standard error is calculated.

	Model 2 (Independence assump.)		Model 3 (Exchangeable)		Model 4 (Unspecified)	
Age	0.017[1]	(0.012)[2]	0.017	(0.012)	0.017	(0.012)
	1.330[3]	(0.185)[4]	1.330	(0.185)	1.330	(0.185)
Sex	−0.117	(0.265)	−0.117	(0.265)	−0.117	(0.265)
	−0.440	(0.659)	−0.440	(0.659)	−0.440	(0.659)
Jaw	0.029	(0.269)	0.029	(0.269)	0.029	(0.269)
	0.110	(0.916)	0.110	(0.916)	0.110	(0.916)
Type	−0.027	(0.272)	−0.027	(0.272)	−0.027	(0.272)
	−0.100	(0.920)	−0.100	(0.920)	−0.100	(0.920)

[1] Estimated regression values $\hat{\beta}$. [2] Standard errors of $\hat{\beta}$.
[3] z–Statistic. [4] p–Value.

TABLE 7.7. Results of the robust estimates for Models 2, 3, and 4 for $x = 360$.

Results

Table 7.7 summarizes the estimated regression parameters, the standard errors, the z–statistics, and the p–values of Models 2, 3, and 4 of the response variables

$$y_{ij} = \begin{cases} 1, & \text{if the conical crown is in function longer than 360 days,} \\ 0, & \text{if the conical crown is in function no longer than 360 days.} \end{cases}$$

It turns out that the $\hat{\beta}$–values and the z–statistics are identical, independent of the choice of R_i, even though a high correlation between the twins exists. The exchangeable correlation model yields the value 0.9498 for the estimated correlation parameter $\hat{\alpha}$. In the model with the unspecified correlation structure, ρ_{i12} and ρ_{i21} were estimated as 0.9498 as well. The fact that the estimates of Models 2, 3, and 4 coincide was observed in the analyses of the response variables with $x = 1100$ and $x = 2000$ as well. This means that the choice of R_i has no influence on the estimation procedure in the case of bivariate binary response. The GEE method is robust with respect to various correlation assumptions.

Table 7.8 compares the results of Models 1 and 2. A striking difference between the two methods is that the covariate age in the case of a naive ML estimation (Model 1) is significant at the 10% level, even though this significance does not turn up if the robust method with the assumption of independence (Model 2) is used. In the case of coinciding estimated regression parameters, the robust variances of $\hat{\beta}$ are larger and, accordingly, the robust z–statistics are smaller than the naive z–statistics. This result shows clearly that the ML method, which is incorrect in this case, underestimates the variances of $\hat{\beta}$ and hence leads to an incorrect age effect.

Tables 7.9 and 7.10 summarize the results with x–values 1100 and 2000. Table 7.9 shows that if the response variable is modeled with $x = 1100$, then none of the observed covariates is significant. As before, the estimated

	Model 1 (naive)			Model 2 (robust)		
	σ	z	p–value	σ	z	p–value
Age	0.008	1.95	0.051*	0.012	1.33	0.185
Sex	0.190	−0.62	0.538	0.265	−0.44	0.659
Jaw	0.192	0.15	0.882	0.269	0.11	0.916
Type	0.193	−0.14	0.887	0.272	−0.10	0.920

* Indicates significance at the 10% level.

TABLE 7.8. Comparison of the standard errors, the z–statistics, and the p–values of Models 1 and 2 for $x = 360$.

	$\hat{\beta}$	Model 1 (naive)			Model 2 (robust)		
		σ	z	p–value	σ	z	p–value
Age	0.0006	0.008	0.08	0.939	0.010	0.06	0.955
Sex	−0.0004	0.170	−0.00	0.998	0.240	−0.00	0.999
Jaw	0.1591	0.171	0.93	0.352	0.240	0.66	0.507
Type	0.0369	0.172	0.21	0.830	0.242	0.15	0.878

TABLE 7.9. Comparison of the standard errors, the z–statistics, and the p–values of models 1 and 2 for $x = 1100$.

correlation parameter $\hat{\alpha} = 0.9578$ indicates a strong dependency between the twins. In Table 7.10, the covariate "type" has significant influence in the case of naive estimation. In the case of the GEE method ($R = I$), it might be significant with a p–value $= 0.104$ (10% level). The result $\hat{\beta}_{\text{type}} = 0.6531$ indicates that a dentoalveolar design significantly increases the log–odds of the response variable

$$y_{ij} = \begin{cases} 1, & \text{if the conical crown is in function longer than 2000 days,} \\ 0, & \text{if the conical crown is in function no longer than 2000 days.} \end{cases}$$

Assuming the model

$$\frac{P(\text{lifetime} \geq 2000)}{P(\text{lifetime} < 2000)} = \exp(\beta_0 + \beta_1 \cdot \text{age} + \beta_2 \cdot \text{sex} + \beta_3 \cdot \text{jaw} + \beta_4 \cdot \text{type})$$

	$\hat{\beta}$	Model 1 (naive)			Model 2 (robust)		
		σ	z	p–value	σ	z	p–value
Age	−0.0051	0.013	−0.40	0.691	0.015	−0.34	0.735
Sex	−0.2177	0.289	−0.75	0.452	0.399	−0.55	0.586
Jaw	0.0709	0.287	0.25	0.805	0.412	0.17	0.863
Type	0.6531	0.298	2.19	0.028*	0.402	1.62	0.104

* Indicates significance at the 10% level.

TABLE 7.10. Comparison of the standard errors, the z–statistics, and the p–values of Models 1 and 2 for $x = 2000$.

the odds $P(\text{lifetime} \geq 2000)/P(\text{lifetime} < 2000)$ for a dentoalveolar design are higher than the odds for a transversal design by the factor $\exp(\beta_4) = \exp(0.6531) = 1.92$ or, alternatively, the odds ratio equals 1.92. The correlation parameter yields the value 0.9035.

In summary, it can be said that age and type are significant but not time–dependent covariates. The robust estimation yields no significant interaction, and a high correlation α exists between the twins of a pair.

Problems

The GEE estimations, which were carried out stepwise, have to be compared with caution, because they are not independent due to the time effect in the response variables. In this context, time–adjusted GEE methods that could be applied in this example are still missing. Therefore, further efforts are necessary in the field of survivorship analysis, in order to be able to complement the standard procedures, such as the Kaplan–Meier estimate and log–rank test, which are based on the independence of the response variables.

7.9.12 Full Likelihood Approach for Marginal Models

A useful full likelihood approach for marginal models in the case of multivariate binary data was proposed by Fitzmaurice et al. (1993). Their starting point is the joint density

$$\begin{aligned} f(y; \Psi, \Omega) &= P(Y_1 = y_1, \ldots, Y_T = y_T; \Psi, \Omega) \\ &= \exp\{y'\Psi + w'\Omega - A(\Psi, \Omega)\} \end{aligned} \quad (7.239)$$

with $y = (y_1, \ldots, y_T)'$, $w = (y_1y_2, y_1y_3, \ldots, y_{T-1}y_T, \ldots, y_1y_2\cdots y_T)'$, $\Psi = (\Psi_1, \ldots, \Psi_T)'$, and $\Omega = (\omega_{12}, \omega_{13}, \ldots, \omega_{T-1T}, \ldots, \omega_{12\cdots T})'$. Further

$$\exp\{A(\Psi, \Omega)\} = \sum_{y=(0,0,\ldots,0)}^{y=(1,1,\ldots,1)} \exp\{y'\Psi + w'\Omega\}$$

is a normalizing constant. Note that this is essentially the saturated parametrization in a loglinear model for T binary responses, since interactions of order 2 to T are included. A model that considers only all pairwise interactions, i.e., $w = (y_1y_2), \ldots, (y_{T-1}y_T)$ and $\Omega = (\omega_{12}, \omega_{13}, \ldots, \omega_{T-1,T})$, was already proposed by Cox (1972b) and by Zhao and Prentice (1990). The models are special cases of the so–called partial exponential families that were introduced by Zhao, Prentice and Self (1992). The idea of Fitzmaurice et al. (1993) was then to make a one–to–one transformation of the canonical parameter vector Ψ to the mean vector μ, which then can be linked to covariates via link functions such as in logistic regression. This idea of transforming canonical parameters one–to–one into (eventually centralized) moment parameters can be generalized to higher moments and to dependent categorical variables with more than two categories. Because the details, theoretically and computationally, are somewhat complex, we refer the reader to Lang and Agresti (1994), Molenberghs and Lesaffre (1994), Glonek (1996), Heagerty and Zeger (1996), and Heumann (1998). Each of these sources gives different possibilities on how to model the pairwise and higher interactions.

7.10 Exercises and Questions

7.10.1 Let two models be defined by their design matrices X_1 and $X_2 = (X_1, X_3)$. Name the test statistic for testing H_0 : "Model X_1 holds" and its distribution.

7.10.2 What is meant by overdispersion? How is it parametrized in the case of a binomial distribution?

7.10.3 Why would a quasi–loglikelihood approach be chosen? How is the correlation in cluster data parametrized?

7.10.4 Compare the models of two–way classification for continuous, normal data (ANOVA) and for categorical data. What are the reparametrization conditions in each case?

7.10.5 Given the following G^2 analysis of a two–way model with all submodels:

Model	G^2	p–value
A	200	0.00
B	100	0.00
$A+B$	20	0.10
$A*B$	0	1.00

which model is valid?

7.10.6 Given the following $I \times 2$ table for X : age group and Y : binary response:

	1	0
< 40	10	8
40–50	15	12
50–60	20	12
60–70	30	20
> 70	30	25

analyze the trend of the sample logits.

8
Repeated Measures Model

8.1 The Fundamental Model for One Population

In contrast to the previous chapters, we now assume that instead of having only one observation per object/subject (e.g., patient) we now have repeated observations. These repeated measurements are collected at previously exact defined times. The principle idea is that these observations give information about the development of a response Y. This response might, for instance, be the blood pressure (measured every hour) for a fixed therapy (treatment A), the blood sugar level (measured every day of the week), or the monthly training performance of sprinters for training method A, etc., i.e., variables which change with time (or a different scale of measurement). The aim of a design like this is not so much the description of the average behavior of a group (with a fixed treatment), rather the comparison of two or more treatments and their effect across the scale of measurement (e.g., time), i.e., the treatment or therapy comparison.

First of all, before we deal with this interesting question, let us introduce the model for one treatment, i.e., for one sample from one population.

The Model

We index the I elements (e.g., patients) with $i = 1, \ldots, I$ and the measurements with $j = 1, \ldots, p$, so that the response of the jth measurement on the ith element (individual) is denoted by y_{ij}. The general basis for many analyses is the specific modeling approach of a mixed model

$$y_{ij} = \mu_{ij} + \alpha_{ij} + \epsilon_{ij} \tag{8.1}$$

with the three components:

(i) μ_{ij} is the average response of y_{ij} over hypothetical repetitions with randomly chosen individuals from the population. Thus, μ_{ij} would stay unchanged if the ith element is substituted by any other element of the sample.

(ii) α_{ij} represents the deviation between y_{ij} and μ_{ij} for the particular individual of the sample that was selected as the ith element. Thus, under hypothetical repetitions, this indiviual would have mean $\mu_{ij} + \alpha_{ij}$.

(iii) ϵ_{ij} describes the random deviation of the ith individual from the hypothetical mean $\mu_{ij} + \alpha_{ij}$.

μ_{ij} is a fixed effect. α_{ij}, on the other hand, is a random effect that varies over the index i (i.e., over the individuals, e.g., patients), hence α_{ij} is a specific characteristic of the individual. *"To be poetic, μ_{ij} is an immutable constant of the universe, α_{ij} is a lasting characteristic of the individual"* (Crowder and Hand, 1990, p. 15). Since μ_{ij} does not vary over the individuals, the index i could be dropped. However, we retain this index in order to be able to identify the individuals.

The vector $\mu_i = (\mu_{i1}, \ldots, \mu_{ip})'$ is called the **μ–profile of the individual**. The following assumptions are made:

(A1) The α_{ij} are random effects that vary over the population for given j according to

$$\mathrm{E}(\alpha_{ij}) = 0 \quad (\text{for all } i, j), \tag{8.2}$$
$$\mathrm{var}(\alpha_{ij}) = \sigma^2_{\alpha_{ij}} . \tag{8.3}$$

(A2) The errors ϵ_{ij} vary over the individuals for given j according to

$$\mathrm{E}(\epsilon_{ij}) = 0 \quad (\text{for all } i, j), \tag{8.4}$$
$$\mathrm{var}(\epsilon_{ij}) = \sigma^2_j . \tag{8.5}$$

(A3) For different individuals $i \neq i'$ the α–profiles are uncorrelated, i.e.,

$$\mathrm{cov}(\alpha_{ij}, \alpha_{i'j'}) = 0 \quad (i \neq i') . \tag{8.6}$$

However, for different measurements $j \neq j'$, the α–profiles of an individual i are correlated

$$\mathrm{cov}(\alpha_{ij}, \alpha_{ij'}) = \sigma^2_{\alpha_{jj'}} \quad (j \neq j') . \tag{8.7}$$

This assumption is essential for the repeated measures model, since it models the natural assumption that the response of an element over the j is an individual interdependent characteristic of the individual.

(A4) The random errors are uncorrelated according to

$$\mathrm{E}(\epsilon_{ij}\epsilon_{i'j'}) = 0 \quad (\text{for all } i, i', j, j')\,. \tag{8.8}$$

(A5) The random components α_{ij} and ϵ_{ij} are uncorrelated according to

$$\mathrm{E}(\alpha_{ij}\epsilon_{i'j'}) = 0 \quad (\text{for all } i, i', j, j')\,. \tag{8.9}$$

(A6) The α_{ij} and ϵ_{ij} are normally distributed.

From these assumptions it follows that

$$\mathrm{E}(y_{ij}) = \mu_{ij} \tag{8.10}$$

and (with δ_{ij} the Kronecker symbol)

$$\begin{aligned} \mathrm{cov}(y_{ij},\, y_{i'j'}) &= \mathrm{E}\left((\alpha_{ij} + \epsilon_{ij})(\alpha_{i'j'} + \epsilon_{i'j'})\right) \\ &= \mathrm{E}(\alpha_{ij}\alpha_{i'j'} + \alpha_{ij}\epsilon_{i'j'} + \epsilon_{ij}\alpha_{i'j'} + \epsilon_{ij}\epsilon_{i'j'}) \\ &= \delta_{ii'}(\sigma^2_{\alpha_{jj'}} + \delta_{jj'}\sigma_j^2)\,. \end{aligned} \tag{8.11}$$

If homogeneity of the variance over j is called for, i.e.,

$$\sigma^2_{\alpha_{jj'}} = \sigma^2_\alpha \tag{8.12}$$

and

$$\sigma_j^2 = \sigma^2\,, \tag{8.13}$$

then the covariance (8.11) simplifies to

$$\mathrm{cov}(y_{ij},\, y_{i'j'}) = \delta_{ii'}(\sigma^2_\alpha + \delta_{jj'}\sigma^2)\,. \tag{8.14}$$

Thus, the variance is

$$\mathrm{var}(y_{ij}) = \sigma^2_\alpha + \sigma^2\,. \tag{8.15}$$

The relation (8.14) expresses that two different individuals $i \neq i'$ are uncorrelated, although the observations of an individual i are correlated over the measurements

$$\mathrm{cov}(y_{ij},\, y_{i'j'}) = 0 \quad (i \neq i'), \tag{8.16}$$

$$\mathrm{cov}(y_{ij},\, y_{ij'}) = \sigma^2_\alpha \quad (j \neq j')\,. \tag{8.17}$$

If the intraclass correlation coefficient for one individual over different measurements is taken, then

$$\rho(j, j') = \rho = \frac{\mathrm{cov}(y_{ij}, y_{ij'})}{\sqrt{\mathrm{var}(y_{ij})\mathrm{var}(y_{ij'})}} = \frac{\sigma^2_\alpha}{\sigma^2_\alpha + \sigma^2}\,. \tag{8.18}$$

The covariance matrix of every individual i ($i = 1, \ldots, I$) is then of the following form

$$\mathrm{var}\begin{pmatrix} y_{i1} \\ \vdots \\ y_{ip} \end{pmatrix} = \mathrm{var}(y_i) = \Sigma = \sigma^2 I_p + \sigma^2_\alpha J_p \tag{8.19}$$

with $J_p = \mathbf{1}_p \mathbf{1}'_p$ (cf. DefinitionA.7). This matrix, that we already became acquainted with in Section 3.9, is called compound symmetric.

Remark. The designs of Chapters 4 to 6 always had a covariance structure $\sigma^2 I$, with the exception of the mixed model from Section 6.6.2 (cf. (6.91)). Hence, the assumptions of the classical linear regression model (3.23) were valid.

Because of the compound symmetry, we now have a generalized linear regression model and the parameter vector β has to be estimated according to the Gauss–Markov–Aitken theorem by the generalized least–squares estimate

$$b = (X'\Sigma^{-1}X)^{-1}X'\Sigma^{-1}y.$$

However (according to Theorem 3.22 by McElroy (1967)), the ordinary and the generalized least–squares estimates are identical if and only if Σ has the structure (8.19), under the assumption that the model contains the constant $\mathbf{1}$. The error structure Σ from (8.19) is ignored if the OLS estimate is applied, i.e., it does not have to be estimated. Hence, more degrees of freedom are available for the residual variance. This explains the preference given to the univariate ANOVA compared to the MANOVA for the comparison of therapies in two groups, if they are treated according to the repeated measures design, and if the assumption of compound symmetry holds for both groups separately or, rather, if an assumption derived from this holds for the difference in response. This will be discussed in detail further on.

8.2 The Repeated Measures Model for Two Populations

We assume that two treatments, I and II, are to be compared with the repeated measures design. Additionally, we assume:

- n_1 individuals receive treatment I;
- n_2 individuals receive treatment II;
- both groups are homogeneous relating to all essential prognostic factors for a response variable Y of interest; and
- realization of repeated measurements for both at $j = 1, \ldots, p$.

This results in two matrices of sample vectors

$$
\begin{array}{ccc}
 & & \text{occasions} & \\
 & & 1 \quad \dots \quad p & \\
Y(I) & = & \begin{pmatrix} y_{111} & \dots & y_{11p} \\ & \dots & \\ y_{1n_11} & \dots & y_{1n_1p} \end{pmatrix} & \begin{array}{c} \text{individual } I_1 \\ \dots \\ \text{individual } I_{n_1} \end{array}
\end{array}
$$

$$
\begin{array}{ccc}
 & & \text{occasions} & \\
 & & 1 \quad \dots \quad p & \\
Y(II) & = & \begin{pmatrix} y_{211} & \dots & y_{21p} \\ & \dots & \\ y_{2n_21} & \dots & y_{2n_2p} \end{pmatrix} & \begin{array}{c} \text{individual } II_1 \\ \dots \\ \text{individual } II_{n_2} \end{array}
\end{array}
$$

The subscripts of y_{ijk} stand for

$k = 1$ or 2:	treatment I or II ,
$i = 1, \dots, n_i$:	individual,
$j = 1, \dots, p$:	occasion (time of measurement) .

The response matrices $Y(I)$ and $Y(II)$ are assumed to be independent. We introduce the fixed factor "treatment" into the model (8.1) and choose the following parametrization

$$y_{kij} = \mu + \alpha_k + \beta_j + (\alpha\beta)_{kj} + a_{ki} + \epsilon_{kij} \,. \tag{8.20}$$

These components have the following meaning:

μ	is the overall mean;
α_k	is the treatment effect;
β_j	is the occasion effect (= time effect);
$(\alpha\beta)_{kj}$	is the treatment $\times$ time interaction;
a_{ki}	is the random effect of the ith individual in the kth treatment; and
ϵ_{kij}	is the random error.

The effects α_k, β_j, $(\alpha\beta)_{kj}$ are assumed to be fixed with the usual constraints for fixed effects, i.e., $\sum \alpha_k = 0$, $\sum \beta_j = 0$, and $\sum_i (\alpha\beta)_{ij} = \sum_j (\alpha\beta)_{ij} = 0$. The effects α_{ki} and the errors ϵ_{kij}, however, are random. Hence, (8.20) is a mixed model.

For the random variables the following assumptions hold:

(i) The vector $\epsilon_k = (\epsilon_{k11}, \dots, \epsilon_{kn_kp})'$, $k = 1, 2$, is normally distributed according to

$$\epsilon_k \;\sim\; N(\mathbf{0}, \sigma^2 I) \,. \tag{8.21}$$

(ii) The vector $a_k = (\alpha_{k1}, \ldots, \alpha_{kn_k})'$, $k = 1, 2$, is normally distributed according to

$$a_k \quad \sim \quad N(\mathbf{0}, \sigma_\alpha^2 I) \,. \tag{8.22}$$

(iii) Both random variables are independent

$$\mathrm{E}(\epsilon_k a'_{k'}) = \mathbf{0} \quad (k, k' = 1, 2) \,. \tag{8.23}$$

With these assumptions, we obtain the expectation of y_{kij}:

$$\mathrm{E}(y_{kij}) = \mu_{kj} = \mu + \alpha_k + \beta_j + (\alpha\beta)_{kj} \,, \tag{8.24}$$

and for the expectation vector of the ith individual in the kth treatment, i.e., for $y_{ki} = (y_{ki1}, \ldots, y_{kip})'$, we obtain

$$\mathrm{E}(y_{ki}) = \mu_k = (\mu_{k1}, \ldots, \mu_{kp})', \quad k = 1, 2 \,. \tag{8.25}$$

The vector μ_k, that represents the mean vector over the p observations of an individual and that is identical for all n_k individuals of a group, is called the **μ_k–profile** of the individuals (Crowder and Hand, 1990, p. 26; Morrison, 1983, p. 153). The observation vector y_{ki}, on the other hand, is called the **curve of progress** of the ith individual in the kth treatment group.

With (8.24) and the assumptions (8.21)–(8.23), we have

$$\mathrm{cov}(y_{kij}, y_{k'i'j'}) = \begin{cases} \sigma_\alpha^2 + \sigma^2, & \text{if } k = k',\, i = i',\, j = j', \\ \sigma_\alpha^2, & \text{if } k = k',\, i = i',\, j \neq j', \\ 0, & \text{otherwise.} \end{cases} \tag{8.26}$$

Hence, the $(p \times p)$–covariance matrix Σ_k $(k = 1, 2)$ of the ith observation vector y_{ki}, $k = 1, 2$ $(i = 1, \ldots, n_k)$ is of the form

$$\Sigma_k = \sigma^2 I_p + \sigma_\alpha^2 J_p \tag{8.27}$$

(cf. (8.19)), which is the structure of compound symmetry.

Remark. The reparametrization of (8.1) into (8.20) maintained all the assumptions of Section 8.1. Model (8.20) has the advantage that it can adopt the structure of the mixed models, as well as the estimation and interpretation of the parameters. For the correlation between the observations

$$\rho(y_{kij}, y_{k'i'j'}) = \begin{cases} \dfrac{\sigma_\alpha^2}{\sigma_\alpha^2 + \sigma^2}, & \text{if } k = k',\, i = i',\, j \neq j', \\ 1, & \text{if } k = k',\, i = i',\, j = j', \\ 0, & \text{otherwise,} \end{cases} \tag{8.28}$$

we find:

(1) The observations, and hence the observation vectors, of individuals from different groups are uncorrelated. Due to the normal distribution they are independent as well.

(2) Observations, or rather observation vectors, of different individuals of the same group are uncorrelated (independent).

(3) Observations of an individual at different times of measurement are correlated (dependent) with the so–called intraclass correlation

$$\rho = \frac{\sigma_\alpha^2}{\sigma_\alpha^2 + \sigma^2} \,. \tag{8.29}$$

8.3 Univariate and Multivariate Analysis

Parametric procedures for analyzing continuous data require the assumption of a distribution. Here the normal distribution as an extensive and, after the elimination of outliers or smoothing, an adequate class of distributions is available. Often, however, the variables have to be transformed first. The comparison of therapies is part of the complex of general mean comparisons of normally distributed populations. However, therapy comparison requires only the far more weak assumption that the distances (differences) of the populations are normal.

Multivariate procedures for the mean comparison of two independent normal distributions are constructed in analogy to univariate procedures. The major principles will be explained in the following section.

8.3.1 The Univariate One–Sample Case

Given a sample $(y_1, \ldots, y_n)$ from $N(\mu, \sigma^2)$ with y_i independent identically distributed. Then $\overline{y} \sim N(\mu, \sigma^2/n)$ and $s^2(n-1)/\sigma^2 \sim \chi^2_{n-1}$. The t–test for $\mathrm{H}_0 : \mu = \mu_0$ is given by $t_{n-1} = [(\overline{y} - \mu_0)/s]\sqrt{n}$.

8.3.2 The Multivariate One–Sample Case

We assume that not only *one* random variable is observed, but a p–dimensional vector of random variables. The sample size is n. The sample is then of the form

$$\underset{n,p}{Y} = \begin{pmatrix} y_1' \\ \vdots \\ y_n' \end{pmatrix} = \begin{pmatrix} y_{11}, \ldots, y_{1p} \\ \vdots \\ y_{n1}, \ldots, y_{np} \end{pmatrix}$$

and we assume for every vector $y_i \overset{\text{i.i.d}}{\sim} N_p(\mu, \Sigma)$, with $\mu' = (\mu_1, \dots, \mu_p)$ and Σ positive definite. Hence

$$Y \sim N_p\left(\begin{pmatrix} \mu \\ \vdots \\ \mu \end{pmatrix}, \begin{pmatrix} \Sigma & & \mathbf{0} \\ & \ddots & \\ \mathbf{0} & & \Sigma \end{pmatrix}\right). \tag{8.30}$$

The sample mean vector is

$$y_{..} = (y_{.1}, \dots, y_{.p})' \tag{8.31}$$

with

$$y_{.j} = \frac{1}{n}\sum_{i=1}^{n} y_{ij} \quad (j = 1, \dots, p) \tag{8.32}$$

and the sample covariance matrix is

$$S = (S_{jh}) = \frac{1}{n-1}\sum_{i=1}^{n}(y_i - y_{..})(y_i - y_{..})' \tag{8.33}$$

with the elements

$$S_{jh} = (n-1)^{-1}\sum_{i=1}^{n}(y_{ij} - y_{j.})(y_{ih} - y_{h.})\,. \tag{8.34}$$

Hence

$$y_{..} \sim N_p(\mu, \Sigma/n) \tag{8.35}$$

with $\mu' = (\mu_1, \dots, \mu_p)$ and

$$(n-1)S \sim W_p(\Sigma, n-1) \tag{8.36}$$

distributed independently, where W_p denotes the p–dimensional Wishart distribution with $(n-1)$ degrees of freedom.

Definition 8.1. Let $X = (x_1, \dots, x_n)'$ be an $(n \times p)$–data matrix from an $N_p(\mathbf{0}, \Sigma)$, where $x_1, \dots, x_n$ are independent and identically $N_p(\mathbf{0}, \Sigma)$–distributed. The $(p \times p)$–matrix

$$W = X'X = \sum_{i=1}^{n} x_i x_i' \sim W_p(\Sigma, n)$$

then has a Wishart distribution with n degrees of freedom.

For $p = 1$, we have $X'X = \sum_{i=1}^{n} x_i^2 = x'x \sim W_1(\sigma^2, n)$ so that $W_1(\sigma^2, n) = \sigma^2\chi_n^2$ holds. Hence, the Wishart distribution is the multivariate analog of the χ^2–distribution.

Definition 8.2. A random variable u has a Hotelling T^2–distribution with the parameters p and n if it can be expressed as

$$u = nx'W^{-1}x \tag{8.37}$$

with

$$x \sim N_p(\mathbf{0}, I) \quad \text{and} \quad W \sim W_p(I, n)$$

being independent. We write

$$u \sim T^2(p, n). \tag{8.38}$$

Remark. If $x \sim N_p(\mu, \Sigma)$ and $W \sim W_p(\Sigma, n)$ and x and W are independent, then

$$n(x - \mu)' W^{-1} (x - \mu) \sim T^2(p, n). \tag{8.39}$$

The T^2–distribution is equivalent to the F–distribution (Mardia, Kent and Bibby, 1979, p. 74):

$$T^2(p, n) \sim \frac{np}{n - p + 1} F_{p, n-p+1}. \tag{8.40}$$

The multivariate two–sided hypothesis

$$\mathrm{H}_0 : \mu = \mu_0 \quad \text{against} \quad \mathrm{H}_1 : \mu \neq \mu_0 \tag{8.41}$$

is tested in analogy to the t–test with the test statistic by Hotelling

$$T^2 = n(y_{..} - \mu_0)' S^{-1} (y_{..} - \mu_0), \tag{8.42}$$

where $(y_{..} - \mu_0)' S^{-1} (y_{..} - \mu_0)$ is the Mahalanobis–D^2 statistic. If H_0 holds, then the test statistic

$$F = \frac{n - p}{p(n - 1)} T^2 \tag{8.43}$$

has an $F_{p,n-p}$–distribution, according to (8.36) and (8.40) (replace n by $n - 1$). The decision rule is as follows:

do not reject $\mathrm{H}_0 : \mu = \mu_0$ if

$$T^2 \leq \frac{p(n - 1)}{n - p} F_{p, n-p; 1-\alpha}. \tag{8.44}$$

Idea of Proof. This test procedure is dealt with in detail in the standard literature for multivariate analysis (cf., e.g., Timm, 1975, pp. 158–166; Morrison, 1983, pp. 128–134). Hence, we only want to give a short outline of the proof.

The decision rule (8.44) is derived by the union–intersection principle that dates back to Roy (1953; 1957) . Assume $y \sim N_p(\mu, \Sigma)$ and let $a \neq \mathbf{0}$ be any $(p \times 1)$–vector. Hence (cf. A.55)

$$a'y \sim N_1(a'\mu, a'\Sigma a) = N_1(\mu_a, \sigma_a^2). \tag{8.45}$$

If $\mathrm{H}_0 : \mu = \mu_0$ [(8.41)] is true, then $\mathrm{H}_{0a} : \mu_a = a'\mu_0 = \mu_{0a}$ is true for all vectors a as well. If, on the other hand, H_{0a} is true for every $a \neq \mathbf{0}$, H_0 is true as well.

Hence, the multivariate hypothesis $H_0 : \mu = \mu_0$ is the intersection of the univariate hypotheses

$$H_0 = \bigcap_{\mathbf{a} \neq 0} H_{0a} \,. \tag{8.46}$$

Let Y $(n \times p)$ be a sample from $N(\mu, \Sigma)$ with $y'_{..} = (y_{1.}, \ldots, y_{p.})$ and S from (8.33). Every univariate hypothesis $H_{0a} : a'\mu = a'\mu_0$ is tested against its two–sided alternative $H_{1a} : a'\mu \neq a'\mu_0$ by the t–statistic

$$t(a) = \frac{a'(y_{..} - \mu_0)}{\sqrt{a'Sa}} \sqrt{n} \,, \tag{8.47}$$

and the acceptance region for H_0 is given by

$$t^2(a) \leq t^2_{n-1,1-\alpha/2} \,. \tag{8.48}$$

Hence, the multivariate acceptance region is the intersection of all univariate acceptance regions

$$\bigcap_{a \neq \mathbf{0}} (t^2(a) \leq t^2_{n-1,1-\alpha/2}) \,. \tag{8.49}$$

Therefore, this area has to contain the largest $t^2(a)$, so that (8.49) is equivalent to

$$\max_a t^2(a) \leq t^2_{n-1,1-\alpha/2} \,. \tag{8.50}$$

Hence, the multivariate test for $H_0 : \mu = \mu_0$ can be based on $t^2(a)$. Since $t^2(a)$ is dimensionless and unaffected by a change of scale of the elements of a, this indeterminacy can be eliminated by a constraint as, for instance,

$$a'Sa = 1 \,. \tag{8.51}$$

The optimization problem $\max_a\{t^2(a) \mid a'Sa = 1\}$ is now equivalent to

$$\max_a \{a'(y_{..} - \mu_0)(y_{..} - \mu_0)'an - \lambda(a'Sa - 1)\} \,. \tag{8.52}$$

Differentiation with respect to a, and to the Lagrangian multiplier λ (Theorems A.63–A.67), yields the system of normal equations

$$[(y_{..} - \mu_0)(y_{..} - \mu_0)'n - \lambda S]\, a = \mathbf{0} \tag{8.53}$$

and

$$a'Sa = 1 \,. \tag{8.54}$$

Premultiplication of (8.53) by a', and taking (8.54) and (8.47) into account, gives

$$\begin{aligned} \hat{\lambda} &= a'(y_{..} - \mu_0)(y_{..} - \mu_0)'an \\ &= t^2(a \mid a'Sa = 1) \,. \end{aligned} \tag{8.55}$$

On the other hand, (8.53), as a homogeneous system in a, has a nontrivial solution $a \neq \mathbf{0}$, as long as the determinant of the matrix equals zero. The matrix $(y_{..} - \mu_0)(y_{..} - \mu_0)'$ is of rank 1. With the determinantal constraint (S is assumed to be regular), (8.53) yields according to

$$\begin{aligned} 0 &= |(y_{..} - \mu_0)(y_{..} - \mu_0)'n - \lambda S| \\ &= |S^{-1/2}(y_{..} - \mu_0)(y_{..} - \mu_0)'S^{-1/2}n - \lambda I_p| \cdot |S| \end{aligned}$$

the characteristic equation for the first matrix, which is symmetric and of rank 1 as well.

The only nontrivial eigenvalue of a matrix of rank 1 is the trace of this matrix (corollary to Theorem A.10):

$$\begin{aligned} \hat{\lambda} &= \mathrm{tr}\{S^{-1/2}(y_{..} - \mu_0)(y_{..} - \mu_0)'S^{-1/2}n\} \\ &= (y_{..} - \mu_0)'S^{-1}(y_{..} - \mu_0)n \,. \end{aligned} \tag{8.56}$$

Hence $t^2(a \mid a'Sa = 1)$ equals Hotelling's T^2 from (8.42).

The test statistic derived according to the union–intersection principle is equivalent to the likelihood–ratio statistic. However, this equivalence is not true in general. The advantage of the union–intersection test is that in the case of a rejection of H_0, it is possible to test which one of the rejection regions caused this. By choosing $a = e_i$, it can be tested for which components of μ are responsible for the rejection of $H_0 : \mu = \mu_0$. This is not possible for the likelihood–ratio test. Furthermore, the importance of the union–intersection principle also lies in the fact that simultaneous confidence intervals for μ can be computed (Fahrmeir and Hamerle, 1984, p. 81). With

$$\begin{aligned} \max_{a \neq 0} t^2(a) &= n(y_{..} - \mu_0)'S^{-1}(y_{..} - \mu_0) \\ &= T^2 \end{aligned} \tag{8.57}$$

and (cf. (8.43))

$$T^2 = \frac{p(n-1)}{n-p} F_{p,n-p} \tag{8.58}$$

we have for $\mu = \mu_0$

$$P\left\{ \frac{n-p}{p(n-1)} T^2 \leq F_{p,n-p,1-\alpha} \right\} = 1 - \alpha \tag{8.59}$$

or, equivalently,

$$P\left(\bigcap_{a \neq 0} \frac{(n-p)n}{p(n-1)} \frac{a'(y_{..} - \mu)^2}{a'Sa} \leq F_{p,n-p,1-\alpha} \right) = 1 - \alpha \,. \tag{8.60}$$

These confidence regions are simultaneously true for all $a'\mu$ with $a \in \mathcal{R}^p$. If only a few comparisons are of interest, i.e., only a few a_i, then we have

$$P\left(a_i'y_{..} - c \leq a_i'\mu \leq a_i'y_{..} + c\right) \geq 1 - \alpha \tag{8.61}$$

with

$$c^2 = F_{p,n-p,1-\alpha} \frac{p(n-1)}{(n-p)n} a'Sa \,. \tag{8.62}$$

In order to assure the confidence coefficient $1 - \alpha$ for the chosen comparisons, i.e., for $a'_1\mu, \ldots, a'_k\mu$ with $k \le p$, and to simultaneously shorten the length of the interval, the **Bonferroni method** is applied. Assume E_i $(i = 1, \ldots, k)$ is the event that the ith confidence interval covers the parameter $a'_i\mu$, and also assume that $\alpha_i = 1 - P(E_i) = P(\overline{E}_i)$ is the corresponding significance level. Let $\overline{E_i}$ be the appropriate complementary event, then

$$P\left(\bigcap_{i=1}^{k} E_i\right) = 1 - P\left(\bigcup_{i=1}^{k} \overline{E_i}\right) \ge 1 - \sum_{i=1}^{k} P(\overline{E}_i) = 1 - \sum_{i=1}^{k} \alpha_i \,. \tag{8.63}$$

Hence, $(1 - \sum \alpha_i)$ is a lower limit for the real simultaneous confidence coefficient

$$1 - \delta = P\left(\bigcap_{i=1}^{k} E_i\right) .$$

If $\alpha_i = \alpha/k$ is chosen, then

$$P\left(\bigcap_{i=1}^{k} E_i\right) \ge 1 - \alpha \,.$$

The corresponding simultaneous confidence intervals are

$$a'_i y_{..} \pm \sqrt{F_{1,n-1,1-\alpha/k} \frac{a'Sa}{n}} \,. \tag{8.64}$$

8.4 The Univariate Two–Sample Case

Suppose that we are given two independent samples

$$(x_1, \ldots, x_{n_1}) \quad \text{from} \quad N(\mu_1, \sigma^2) \tag{8.65}$$

and

$$(y_1, \ldots, y_{n_2}) \quad \text{from} \quad N(\mu_2, \sigma^2) \,. \tag{8.66}$$

In the case of equal variances, the test statistic for $\mathrm{H}_0 : \mu_1 = \mu_2$ is

$$t_{n_1+n_2-2} = \frac{(\overline{x} - \overline{y})}{s\sqrt{1/n_1 + 1/n_2}} \tag{8.67}$$

with the pooled sample variance

$$s^2 = \frac{(n_1 - 1)s_x^2 + (n_2 - 1)s_y^2}{n_1 + n_2 - 2} \,. \tag{8.68}$$

The assumption of equal variances has to be tested with the F–test. In the case of a rejection of $\mathrm{H}_0 : \sigma_x^2 = \sigma_y^2$, no exact solution exists. This is called the Behrens–Fisher problem. The comparison of means in the case of $\sigma_x \neq \sigma_y$ is done approximately by a t_v–statistic, where the sample variances influence the degrees of freedom v.

8.5 The Multivariate Two–Sample Case

The multivariate analog of the t–test for testing $\mathrm{H}_0 : \mu_x = \mu_y$ ($(p \times 1)$–vectors each) is defined as Hotelling's two–sample T^2:

$$T^2 = (n_1^{-1} + n_2^{-1})^{-1}(x_{..} - y_{..})' S^{-1} (x_{..} - y_{..}) \tag{8.69}$$

with the pooled sample covariance matrix (within–groups)

$$(n_1 + n_2 - 2)S = (n_1 - 1)S_x + (n_2 - 1)S_y \,. \tag{8.70}$$

The statistic T^2 is, in fact, an estimate of the Mahalanobis distance $D^2 = (\mu_x - \mu_y)'\Sigma^{-1}(\mu_x - \mu_y)$ of both populations. Under $\mathrm{H}_0 : \mu_x = \mu_y$, T^2 has the following relationship to the central F–distribution

$$F_{p,v} = \frac{n_1 + n_2 - p - 1}{(n_1 + n_2 - 2)p} T^2 \tag{8.71}$$

with the degrees of freedom of the denominator

$$v = n_1 + n_2 - p - 1 \,. \tag{8.72}$$

The decision rule based on the union–intersection principle (Roy, 1953, 1957)—or, equivalently, on the likelihood–ratio principle—yields the rejection region for $\mathrm{H}_0 : \mu_x = \mu_y$ as

$$T^2 > \frac{(n_1 + n_2 - 2)p}{v} F_{p,v,1-\alpha} \,. \tag{8.73}$$

Hotelling's T^2–statistic for the model with fixed effects assumes the equality of the covariance matrices Σ_x and Σ_y, in analogy to the univariate comparison of means. This equality can be tested by various measures.

Remark.

(i) If $\mathrm{H}_0 : \mu_x = \mu_y$ is replaced by $\mathrm{H}_0 : C(\mu_x - \mu_y) = \mathbf{0}$ where C is a contrast matrix for differences, then the statistic F [(8.71)] has one degree of freedom less in the numerator as well as in the denominator, i.e., p is to be replaced by $p - 1$.

(ii) The performance of Hotelling's T^2 and four nonparametric tests were investigated by Harwell and Serlin (1994) with respect to type I error distributions with varying skewness and sample size.

8.6 Testing of $H_0 : \Sigma_x = \Sigma_y$

Box (1949) has given the following generalization of Bartlett's test for the equality of two univariate variances to $H_0 : \Sigma_x = \Sigma_y$ in the multivariate (p–dimensional) case.

Assume that S [(8.70)] is the pooled sample covariance matrix of the two p–variate normal distributions. The **Box–M statistic** is αM with

$$M = (n_1 - 1) \ln \left(\frac{|S|}{|S_x|} \right) + (n_2 - 1) \ln \left(\frac{|S|}{|S_y|} \right) \tag{8.74}$$

and α according to

$$1 - 1/6(2p^2 + 3p - 1)(p+1)^{-1} \left\{ \frac{1}{n_1 - 1} + \frac{1}{n_2 - 1} - \frac{1}{n_1 + n_2 - 2} \right\} . \tag{8.75}$$

Under $H_0 : \Sigma_x = \Sigma_y$, we have the following approximate distribution

$$\alpha M \sim \chi^2_{p(p+1)/2} . \tag{8.76}$$

Remark. Box (1949) developed this statistic for the general comparison of $g \geq 2$ normal distributions and gave equivalent representations as an F–statistic. For the comparison of g independent normal distributions $N_p(\mu_1, \Sigma_1), \ldots, N_p(\mu_g, \Sigma_g)$, the test problem is

$$H_0 : \Sigma_1 = \ldots = \Sigma_g \tag{8.77}$$

against

$$H_1 : H_0 \text{ not true} .$$

Let S_i be the unbiased estimates (i.e., the appropriate sample covariance matrices) of Σ_i ($i = 1, \ldots, g$) and let n_i be the corresponding sample size. We assume

$$N = \sum_{i=1}^{g} n_i, \quad v_i = n_i - 1, \tag{8.78}$$

and denote the pooled sample covariance matrix by S;

$$S = \frac{1}{N - g} \sum_{i=1}^{g} v_i S_i . \tag{8.79}$$

The test statistic is then of the form αM (cf. Timm, 1975, p. 252) with

$$M = (N - g) \ln |S| - \sum_{i=1}^{g} v_i \ln |S_i| \tag{8.80}$$

and

$$\alpha = 1 - C, \tag{8.81}$$

$$C = \frac{2p^2 + 3p - 1}{6(p+1)(g-1)} \left(\sum_{i=1}^{g} \frac{1}{v_i} - \frac{1}{N - g} \right) . \tag{8.82}$$

The approximate distribution is

$$\alpha M \sim \chi^2_v \quad \text{with} \quad v = \frac{p(p+1)(g-1)}{2} \,. \tag{8.83}$$

For $g = 2$, we have α specified by (8.75).

8.7 Univariate Analysis of Variance in the Repeated Measures Model

8.7.1 Testing of Hypotheses in the Case of Compound Symmetry

Consider the model (8.20) formulated in Section 8.2

$$y_{kij} = \mu + \alpha_k + \beta_j + (\alpha\beta)_{kj} + a_{ki} + \epsilon_{kij} \,, \tag{8.84}$$

which can be interpreted as a mixed model, i.e., as a two–factorial model (fixed factors: treatments $k = 1, 2$ and occasions $j = 1, \ldots, p$), with interaction and one random effect α_{ki} (individual).

The univariate analysis of variance assumes equal covariance matrices of the two subpopulations ($k = 1$ and 2). Furthermore, the structure of compound symmetry [(8.19)] is required for both covariance matrices. This assumption is sufficient for the validity of the univariate F–tests. Compound symmetry is a special case of a more general covariance structure which ensures the exact F–distribution. This situation, which occurs often in practice, will be discussed in detail in Section 8.7.2

In the mixed model, the following hypotheses, tailored to the situation of the repeated measures model, are tested:

(i) The null hypothesis of homogeneous levels of both treatments

$$\mathrm{H}_0 : \alpha_1 = \alpha_2 \,. \tag{8.85}$$

(ii) The null hypothesis of homogeneous occasions (cf. Figure 8.1)

$$\mathrm{H}_0 : \beta_1 = \ldots = \beta_p \,. \tag{8.86}$$

(iii) The null hypothesis of no interaction between the treatment and time effects (cf. Figure 8.2)

$$\mathrm{H}_0 : (\alpha\beta)_{ij} = 0 \quad (k = 1, 2,\ j = 1, \ldots, p) \,. \tag{8.87}$$

We define the correction term once again as

$$C = \frac{Y^2_{\cdots}}{N}$$

with $N = (n_1 + n_2)p = np$. Taking the possibly unbalanced sample sizes ($n_1 \neq n_2$) into consideration, we obtain the following sums of squares

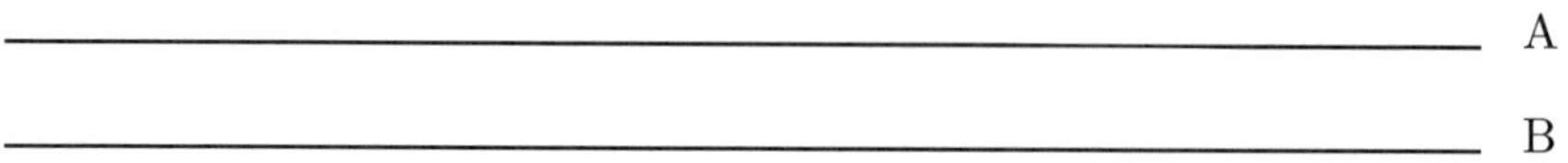

A

B

FIGURE 8.1. No interaction and no time effect.

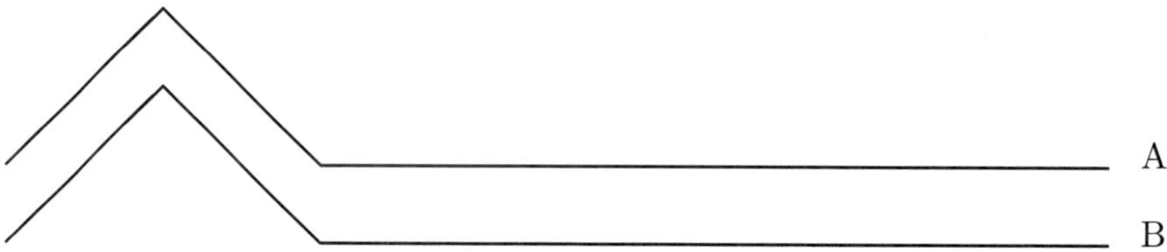

FIGURE 8.2. No interaction ($\mathrm{H}_0 : (\alpha\beta)_{ij} = 0$ not rejected) and a time effect.

(cf. (6.17)–(6.22) and Morrison, 1983, p. 213):

$$\begin{aligned}
SS_{\text{Total}} &= \sum\sum\sum (y_{kij} - y_{...})^2 \\
&= \sum\sum\sum y_{kij}^2 - C, && (8.88)\\
SS_A = SS_{\text{Treat}} &= \sum\sum\sum (y_{k..} - y_{...})^2 \\
&= \frac{1}{p}\sum_{k=1}^{2}\frac{1}{n_k}Y_{k..}^2 - C, && (8.89)\\
SS_B = SS_{\text{Time}} &= \sum\sum\sum (y_{..j} - y_{...})^2 \\
&= \frac{1}{n_1+n_2}\sum_{j=1}^{p} Y_{..j}^2 - C, && (8.90)\\
SS_{\text{Subtotal}} &= \sum\sum\sum (y_{k.j} - y_{...})^2 \\
&= \sum_k \frac{1}{n_k}\sum_j Y_{k.j}^2 - C, && (8.91)\\
SS_{A\times B} &= SS_{\text{Treat} \times \text{Time}} \\
&= SS_{\text{Subtotal}} - SS_{\text{Treat}} - SS_{\text{Time}}, && (8.92)\\
SS_{\text{Ind}} &= \sum\sum\sum (y_{.i.} - y_{k..})^2
\end{aligned}$$

$$= \frac{1}{p}\sum_{k=1}^{2}\sum_{i=1}^{n_k} Y_{.i.}^2 - \frac{1}{p}\sum_{k=1}^{2}\frac{1}{n_k}Y_{k..}^2 \tag{8.93}$$

$$SS_{\text{Error}} = SS_{\text{Total}} - SS_{\text{Subtotal}} - SS_{\text{Ind}}, . \tag{8.94}$$

The test statistics are (cf. Greenhouse and Geisser, 1959)

$$F_{\text{Treat}} = \frac{MS_{\text{Treat}}}{MS_{\text{Ind}}}, \tag{8.95}$$

$$F_{\text{Time}} = \frac{MS_{\text{Time}}}{MS_{\text{Error}}}, \tag{8.96}$$

$$F_{\text{Treat} \times \text{Time}} = \frac{MS_{\text{Treat} \times \text{Time}}}{MS_{\text{Error}}}. \tag{8.97}$$

Source	*SS*	*df*	*MS*	*F*–Values
Treatment	SS_{Treat}	1	SS_{Treat}	$F_{\text{Treat}} = \frac{MS_{\text{Treat}}}{MS_{\text{Ind}}}$
Occasion	SS_{Time}	$p-1$	$\frac{SS_{\text{Time}}}{p-1}$	$F_{\text{Time}} = \frac{MS_{\text{Time}}}{MS_{\text{Error}}}$
Treatment × Occasion	$SS_{\text{Treat} \times \text{Time}}$	$p-1$	$\frac{SS_{\text{Treat} \times \text{Time}}}{p-1}$	$F_{\text{Treat} \times \text{Time}} = \frac{MS_{\text{Treat} \times \text{Time}}}{MS_{\text{Error}}}$
Individual	SS_{Ind}	$n-2$	$\frac{SS_{\text{Ind}}}{n-2}$	
Error	SS_{Error}	$(p-1)(n-2)$	$\frac{SS_{\text{Error}}}{(p-1)(n-2)}$	
Total	SS_{Total}	$np-1$		

TABLE 8.1. Table of the univariate analysis of variance in the repeated measures model.

These F–tests are called **unadjusted univariate F–tests**—as opposed to the **adjusted F–tests** named according to the Greenhouse–Geisser strategy.

Remark. The assumption of a compound symmetric structure is not very realistic in the repeated measures model, since this requirement means that the correlation of the response between two occasions is identical. This assumption, however, cannot be expected for all situations. Hence, the question of interest is whether and when univariate tests may be applied in the case of a more general covariance structure (sphericity of the contrast covariance matrix) (cf. Girden, 1992).

8.7.2 Testing of Hypotheses in the Case of Sphericity

We assume that the two populations have an identical covariance matrix Σ. The comparison of therapies, i.e., the testing of the linear hypotheses (8.85)–(8.87), is done by means of linear contrasts. The comparison of the

p means of the p occasions requires a system of $p-1$ orthogonal contrasts. The test statistic follows an F–distribution, if and only if the covariance matrix of the orthogonal contrasts is a scalar multiple of the identity matrix. This condition is called the **circularity** or **sphericity condition**. It can be expressed in a number of alternative ways.

For example, it can be demanded that all the variances of pairwise differences of the response values of an individual are equal. For any random variables x_i and x_j, the following is valid

$$\text{var}(x_i - x_j) = \text{var}(x_i) + \text{var}(x_j) - 2\text{cov}(x_i,\, x_j)\,.$$

If $\text{var}(x_i) = \text{var}(x_j)$ and $\text{cov}(x_i, x_j)$ is constant (for all i, j), then compound symmetry holds. However, more general dependent structures exist, under which the condition

$$\text{var}(x_i - x_j) = \text{const}$$

is valid, from which sphericity of every contrast covariance matrix follows, as long as sphericity is proven for one specific covariance matrix.

The necessary and sufficient condition is known as the *Huynh–Feldt condition* (Huynh and Feldt, 1970). It can be expressed in three equivalent (alternative) forms.

Huynh–Feldt Condition (H Pattern)

(i) The common covariance matrix Σ of both populations is $\Sigma = (\sigma_{jj'})$ with

$$\sigma_{jj'} = \begin{cases} \alpha_j + \alpha_{j'} + \lambda, & j = j'\,, \\ \alpha_j + \alpha_{j'}, & j \neq j'\,. \end{cases} \quad . \tag{8.98}$$

(ii) All possible differences $y_{kij} - y_{kij'}$ of the response variables have equal variance, i.e., $\text{var}(y_{kij} - y_{kij'}) = 2\lambda$ is valid for every individual i from each of the two groups.

(iii) For the Huynh–Feldt epsilon $\varepsilon_{HF} = 1$ holds, where

$$\varepsilon_{HF} = \frac{p^2(\overline{\sigma}_d - \overline{\sigma}_{\cdot\cdot})^2}{(p-1)(\sum\sum \sigma_{rs}^2 - 2p\sum \overline{\sigma}_{r\cdot}^2 + p^2\overline{\sigma}_{\cdot\cdot}^2)}\,. \tag{8.99}$$

Here $\Sigma = (\sigma_{rs})$ is the population covariance matrix where
$\overline{\sigma}_d$ is the average of the diagonal elements;
$\overline{\sigma}_{\cdot\cdot}$ is the overall mean of the σ_{rs}; and
$\overline{\sigma}_{r\cdot}$ is the average of the rth row.

Testing the Huynh–Feldt Condition

Huynh and Feldt (1970) proved that the necessary and sufficient conditions (i), (ii), or (iii) are valid, if

$$\tilde{C}_H \Sigma \tilde{C}_H' = \lambda I \tag{8.100}$$

holds where $\tilde{C}_H$ is the normalized form of C_H. C_H is the suborthogonal $((p-1) \times p)$–submatrix of the orthogonal Helmert matrix

$$\begin{pmatrix} \mathbf{1}'_p/\sqrt{p} \\ C_H \end{pmatrix}, \tag{8.101}$$

that is formed from the Helmert contrasts. The Helmert matrix C_H in (8.101) contains the following elements:

$$\underset{p-1,p}{C_H} = \begin{pmatrix} c'_1 \\ c'_2 \\ \vdots \\ c'_p \end{pmatrix} = \begin{pmatrix} (p-1) & -1 & -1 & \dots & -1 & -1 \\ 0 & (p-2) & -1 & \dots & -1 & -1 \\ \vdots & & & & & \vdots \\ 0 & 0 & 0 & \dots & 1 & -1 \end{pmatrix}. \tag{8.102}$$

The vectors c'_s $(s = 1, \dots, p-1)$ are called *Helmert contrasts.* They are orthogonal

$$c'_{s_1} c_{s_2} = 0 \quad (s_1 \neq s_2)$$

and $\sum_{j=1}^{p} c_{sj} = 0$, i.e.,

$$c'_s \mathbf{1}_p = 0.$$

However, the c_s are not normed $(c'_s c_s \neq 1)$. Therefore, the vector $\mathbf{1}'_p$ or its standardized version $\mathbf{1}_p/\sqrt{p}$ is included in the contrast matrix as the first row, although strictly speaking this is not a contrast $(\mathbf{1}'_p \mathbf{1}_p = p \neq 0$, i.e., the second property of contrasts is not fulfilled).

Standard software is available that converts the contrasts C_H into orthonormal contrasts $\tilde{C}_H$.

Remark. Based on the standardized Helmert matrix $\tilde{C}_H$, we give a short outline of how to prove the equivalence of (ii) and (8.100):

Case $p = 2$
The Helmert matrix is $C_H = (1, -1)$, hence $\tilde{C}_H = (1/\sqrt{2}, -1/\sqrt{2})$. Thus, (8.100) is

$$(1/\sqrt{2} \;\; -1/\sqrt{2}) \begin{pmatrix} \sigma_1^2 & \sigma_{12} \\ \sigma_{12} & \sigma_2^2 \end{pmatrix} \begin{pmatrix} 1/\sqrt{2} \\ -1/\sqrt{2} \end{pmatrix} = \lambda$$

$$\iff \quad \sigma_1^2 + \sigma_2^2 - 2\sigma_{12} = 2\lambda\,.$$

Case $p = 3$
We obtain $\tilde{C}_H \Sigma \tilde{C}'_H = \lambda I$ as

$$\begin{pmatrix} 2/\sqrt{6} & -1/\sqrt{6} & -1/\sqrt{6} \\ 0 & 1/\sqrt{2} & -1/\sqrt{2} \end{pmatrix} \begin{pmatrix} \sigma_1^2 & \sigma_{12} & \sigma_{13} \\ \sigma_{12} & \sigma_2^2 & \sigma_{23} \\ \sigma_{13} & \sigma_{23} & \sigma_3^2 \end{pmatrix} \begin{pmatrix} 2/\sqrt{6} & 0 \\ -1/\sqrt{6} & 1/\sqrt{2} \\ -1/\sqrt{6} & -1/\sqrt{2} \end{pmatrix} = \lambda I_2$$

$$\iff$$

$$\begin{array}{ll}
\text{Element } (1,1): & \frac{1}{6}\left[4\sigma_1^2 + \sigma_2^2 + \sigma_3^2 - 4\sigma_{12} - 4\sigma_{13} + 2\sigma_{23}\right] = \lambda; \\
\text{Element } (1,2): & \sigma_3^2 - \sigma_2^2 + 2\sigma_{12} - 2\sigma_{13} = 0; \\
(= \text{Element } (2,1)): & \Longrightarrow \quad \sigma_2^2 = \left[\sigma_3^2 + 2\sigma_{12} - 2\sigma_{13}\right]; \\
\text{Element } (2,2): & \frac{1}{2}\left[\sigma_2^2 + \sigma_3^2 - 2\sigma_{23}\right] = \lambda\,.
\end{array}$$

Form

$$\begin{aligned}
\sigma_1^2 + \sigma_2^2 - 2\sigma_{12} &= \sigma_1^2 + \left[\sigma_3^2 + 2\sigma_{12} - 2\sigma_{13}\right] - 2\sigma_{12} \\
&= \sigma_1^2 + \sigma_3^2 - 2\sigma_{13}\,.
\end{aligned}$$

Equate $(1,1) = (2,2)$ (since the right–hand sides are equal) $\Longrightarrow$

$$(\sigma_1^2 + \sigma_2^2 - 2\sigma_{12}) + (\sigma_1^2 + \sigma_3^2 - 2\sigma_{13}) = 2(\sigma_2^2 + \sigma_3^2 - 2\sigma_{23}) = 4\lambda\,.$$

Both terms on the left are identical

$$\Longrightarrow \quad \sigma_j^2 + \sigma_{j'}^2 - 2\sigma_{jj'} = 2\lambda\,(j \neq j') \quad .$$

The Condition of Sphericity or Circularity

Compound symmetry is a special case of covariance structures, for which the univariate F–tests are valid. Let us first consider the case of a therapy group measured on p occasions. We can apply $(p-1)$ orthonormal contrasts for testing the differences in the p occasions.

The univariate statistics $(c_j' y_{ki})^2$ follow exact F–distributions if and only if the covariance matrix of the contrasts has equal variances and zero covariances, i.e., if it has the form $\sigma^2 I$ (circularity or sphericity). This corresponds to the assumption of the mixed model that the differences in the y_{ki} are caused only by unequal means and not by variance inhomogeneity.

The model of compound symmetry is a special case of the model of sphericity of the orthonormal contrasts. Compound symmetry is equivalent to the intraclass correlation structure, i.e., the diagonal elements being σ^2 and the off–diagonal elements being σ_α^2 [(8.19)]. Every term on the main diagonal of the covariance matrix of orthonormal contrasts estimates the denominator in the univariate F–statistic of the corresponding contrast. Thus, **when sphericity holds**, each element estimates the same thing. Hence, a better statistic is the average of these elements. This is called the **averaged F–test**. If sphericity does not hold, the denominators of the F–statistics may become too large or too small so that the test is biased.

Comparison of Two or More Therapy Groups—Test for Sphericity

Similar to the above arguments, univariate F–tests only stay valid if the covariance matrix of orthonormal contrasts within therapy groups are spherical and—additionally—are identical across the therapy groups so that global sphericity holds. This assumption may be weakened, for instance, by demanding sphericity only for the main effects (e.g., j fixed, comparison of two therapies by means of a linear contrast).

For the test of **global sphericity** [(8.100)], the equality of the covariance matrices in the therapy groups is tested first. This is done by the Box–M statistic [(8.74)]. If H_0: $\Sigma_1 = \Sigma_2$ is not rejected, then the **test for sphericity by Mauchly** (1940) may be applied. According to Morrison (1983), p. 251 the test statistic is

$$W = \frac{q^q |R|}{(\mathrm{tr}\{R\})^q} \tag{8.103}$$

with $q = p - 1$,

$$R = \tilde{C}_H S \tilde{C}'_H \tag{8.104}$$

and $\tilde{C}_H$ is the $(q \times p)$–matrix of orthonormal Helmert contrasts. In addition to the exact critical values (cf. tables in Kres (1983)), the approximate distribution

$$-\left[(N-1) - \frac{2p^2 - 3p + 3}{6(p-1)}\right] \ln W \sim \chi^2_v \tag{8.105}$$

with

$$v = 1/2(p-2)(p+1) = 1/2p(p-1) - 1 \tag{8.106}$$

may be used in the case of equal sample sizes $n_1 = n_2 = N$.

Tests relating to the covariance structure—especially the Box–M test and the Mauchly test—are sensitive to nonnormality in general. Huynh and Mandeville (1979) analyzed the robustness of the Mauchly test to a such departure by means of simulation studies. The following conclusions are drawn:

(i) the W–test tends to err on the conservative side for light–tailed distributions, the difference between the empirical type I error and the nominal significance level α increases for large samples and for small α; and

(ii) for heavy–tailed distributions the reverse is true, i.e., H_0 : *sphericity* is rejected earlier, even though H_0 is true.

8.7.3 The Problem of Nonsphericity

After the pretests (univariate F–tests, Box–M test, Mauchly test) are carried out, the following questions have to be settled (cf. Crowder and Hand, 1990, pp. 50–56):

(i) Which effect occurs if the F–test is applied in spite of a rejection of sphericity?

(ii) What is to be done if the assumptions seem unjustifiable altogether?

To (i): If sphericity does not hold, then the actual level of significance $\hat{\alpha}$ of the univariate F–tests will exceed the nominal level α, with the effect that too many true null hypotheses are rejected. For tests with complete systems of orthonormal contrasts, this effect can be analyzed by studying the ε correction factor. Rouanet and Lepine (1970), Mitzel and Games (1981), and Boik (1981) discuss the effect of nonsphericity on single contrasts. Boik concludes that the type I error is out of control. Rouanet and Lepine (1970) recommend using all relevant statistics.

To (ii): What is to be done in the case of nonsphericity? The multivariate analysis only assumes the equality of the covariance matrices, but not any specific form of the (common) covariance matrix. If however sphericity holds, then the MANOVA has a relatively low power compared to the univariate approach.

Hence, the direct application of a multivariate analysis, i.e., without previously testing the possibility of sphericity, is not the best strategy.

8.7.4 Application of Univariate Modified Approaches in the Case of Nonsphericity

Let $\{c\}$ be a set of $(p-1)$–orthonormal contrasts with the covariance matrix Σ_c. The **Greenhouse–Geisser epsilon** is then defined as

$$\varepsilon_{G-H} = \frac{(\text{tr } \Sigma_c)^2}{(p-1)\,\text{tr}(\Sigma_c^2)} = \frac{(\sum \theta_j)^2}{(p-1)\sum \theta_j^2}, \tag{8.107}$$

where θ_j are the eigenvalues of Σ_c. If $\Sigma_c = I$, then all $\theta_j = 1$ and ε is equal to 1. Otherwise, we have $\varepsilon_{G-H} < 1$. The overall F–tests for an occasion effect, and for interaction in the case of two therapy groups with $n = n_1 + n_2$ individuals and p measures, involves the $F_{p-1,(p-1)(n-2)}$–distribution (cf. test statistics (8.96) and (8.97)). In the case of non-sphericity, the $F_{\varepsilon_{G-H}(p-1),\varepsilon_{G-H}(p-1)(n-2)}$–distribution is used for testing. Hence, for $\varepsilon_{G-H} < 1$, the critical values increase, i.e., the null hypotheses are rejected less often. This counteracts the previously described effect (answer to (i)).

Since ε_{G-H} will not be known, it will have to be estimated. Hence the question arises: What influence does the estimation error of $\hat{\varepsilon}_{G-H}$ have on the power of the F–test corrected by $\hat{\varepsilon}_{G-H}$?

Greenhouse–Geisser Test Strategy

In order to avoid this problem, Greenhouse and Geisser (1959) suggest a conservative approach. This strategy consists of the following steps:

- standard F–test (unmodified). If H_0 is not rejected, then stop.

- If H_0 is rejected, then the smallest ε–value is chosen (lower bound epsilon)

$$\varepsilon_{\min} = 1/(p-1) \tag{8.108}$$

and tested with the modified F–test. If H_0 is rejected by this most conservative test, then the decision is accepted and stop.

If H_0 is not rejected, then ε_{G-H} is estimated [(8.107)] and the $\hat{\varepsilon}_{G-H}$–F–test is conducted and its decision is accepted.

As a universal answer for the entire problem, we conclude:

If strong prior reasons favor the assumption of sphericity (i.e., for the independence of the univariate distributions of the contrasts), then the univariate F–tests should be conducted. Otherwise, either a modified ε–F–test or a multivariate test or a nonparametric approach should be applied. It is obvious that this problem cannot be solved academically, but only on the basis of the data.

Test Procedure in the Two–Sample Case in the Mixed Model

1. Testing for interaction and for occasions effects (H_0 from (8.87) and (8.86)):

   (a) $\Sigma_1 = \Sigma_2 \Rightarrow$ MANOVA;
   (b) $\tilde{C}_H(\Sigma_1 - \Sigma_2)\tilde{C}'_H = \lambda I \Rightarrow$ ANOVA (averaged F–test); and
   (c) $\tilde{C}_H(\Sigma_1 - \Sigma_2)\tilde{C}'_H \neq \lambda I \Rightarrow$ ANOVA (modified) or MANOVA.

   Comment. If sphericity holds, then the ANOVA (unmodified) is more powerful than the MANOVA.

   If we have nonsphericity, the power of the ANOVA (modified) compared to the MANOVA depends on the ϵ–values (Huynh–Feldt ϵ or Greenhouse–Geisser ϵ) or, rather, on the estimation errors in $\hat{\epsilon}$.

2. Testing for the main effect $H_0 : \alpha_1 = \alpha_2$ [(8.85)] under the assumption of $H_0 : (\alpha\beta)_{ij} = 0$:

$$\begin{aligned} \Sigma_1 = \Sigma_2 \quad &\Rightarrow \quad \text{univariate } F\text{–test} \\ &\quad\quad (\text{MANOVA} = \text{unmodified ANOVA}) \\ \Sigma_1 \neq \Sigma_2 \quad &\Rightarrow \quad \text{nonparametric approach.} \end{aligned}$$

8.7.5 *Multiple Tests*

If a global treatment effect is proven, i.e., if $H_0{:}\mu_1 = \mu_2$ is rejected, then the question of interest is whether regions with a multiple treatment effect exist. Multiple treatment effect means that $\mu_{1j} \neq \mu_{2j}$ for some j.

Of special interest are connected regions with *local multiple treatment effects* as, for example,

$$\mu_{1j} \neq \mu_{2j}, \quad j = 1, \ldots, \tilde{p}, \quad \tilde{p} < p, \tag{8.109}$$

i.e., treatment effects from the first occasion until a specific occasion $\tilde{p}$. For this, a multiple testing procedure is performed that meets the multiple α–level. This is done by defining so–called Holm–adjusted quantiles (cf. Lehmacher, 1987, p. 29), starting out with Bonferroni's inequality.

Holm–Procedure for Local Multiple Treatment Effects

To begin with, the global treatment effect is tested, i.e., $H_0 : \mu_1 = \mu_2$ is tested with Hotelling's T^2 (cf. (8.69)). If H_0 is not significant the procedure stops. If, however, H_0 is rejected, then the Holm–procedure is conducted, which sorts all p univariate t–statistics of the p single occasions by their size (thus, in analogy to the size of the p–values, starting with the smallest p–value). These p–values are compared to the Holm–adjusted sequence:

$j = 1$	$j = 2$	$j = 3$	$j = 4$	$\ldots$	$j = p - 1$	$j = p$
$\alpha/p - 1$	$\alpha/p - 1$	$\alpha/p - 2$	$\alpha/p - 3$	$\ldots$	$\alpha/2$	α

As soon as one p–value of a t_j lies above its appropriate Holm limit, the procedure is terminated and $H_0 : \mu_{1j} = \mu_{2j}$ $(j = 1, \ldots, \tilde{p})$, is rejected in favor of H_1 (8.109).

Interpretation. A local multiple treatment effect exists for all occasions j with a p–value of $t_j \leq j$th Holm limit. This means that all univariate hypotheses $H_{0j} : \mu_{1j} = \mu_{2j}$, whose test statistics have p–values below the appropriate Holm limit, are rejected in favor of a local multiple treatment effect.

8.7.6 Examples

Example 8.1. Two treatments, 1 and 2, over $p = 3$ measures with $n_1 = n_2 = 4$ individuals each are compared in Table 8.2.
Call in SPSS:

```
MANOVA A B C by Treat (1,2)
/ws factors = Time(3)
/contrast(Time) = difference
/ws design
/print = homogeneity(boxm) transform error (cor)
         signig(averf) param(estim)
/design .
```

The steps of the test are:

Treatment	Occasion A	B	C	$Y_{ki.}$
1	10	19	27	56
	9	13	25	47
	4	10	20	34
	5	6	12	23
2	13	16	19	48
	11	18	28	57
	17	28	25	70
	20	23	29	72

TABLE 8.2. Repeated measures design for the treatment comparison.

(i) $\text{H}_0 : \Sigma_1 = \Sigma_2$:
The Box–M statistic is $\alpha M = 3.93638$, i.e., approximately (cf. (8.76))

$$\chi^2_{p(p+1)/2} = \chi^2_6 = 1.80417 \quad (p\text{–value } 0.937) .$$

Hence H_0 is not rejected. After the test procedure, the MANOVA may be performed. Before doing this, however, it should be tested whether sphericity holds for the contrast covariance matrix.

(ii) $\text{H}_0 : \tilde{C}_H \Sigma \tilde{C}'_H = \lambda I$: We have

$$\tilde{C}_H = \begin{pmatrix} 2/\sqrt{6} & -1/\sqrt{6} & -1/\sqrt{6} \\ 0 & 1/\sqrt{2} & -1/\sqrt{2} \end{pmatrix} .$$

```
Test involving 'Time' Within Subject Effect

Mauchly sphericity test, W =       .90352
Chi-square approx. =               .50728  with 2 D.F.
Significance =                     .776
Greenhouse-Geisser Epsilon =       .91201
Huynh-Feldt Epsilon =             1.00000
```

Hence H_0 : *Sphericity* is not rejected and we may conduct the unadjusted F–tests of the ANOVA.

According to the test strategy in the mixed model, we first test

$$\text{H}_0 : (\alpha\beta)_{ij} = 0$$

with (cf. (8.97) and Table 8.1)

$$F_{\text{Treat} \times \text{Time}} = F_{(p-1);(p-1)(p-2)} = \frac{MS_{\text{Treat}}}{MS_{\text{Error}}} .$$

From Table 8.2, we get

	A	B	C	
$Y_{1.j}$	28	48	84	$Y_{1..} = 160$
$Y_{2.j}$	61	85	101	$Y_{2..} = 247$
$Y_{..j}$	89	133	185	$Y_{...} = 407$

$$\begin{aligned}
N &= 2 \cdot 3 \cdot 4 = 24, \\
C &= \frac{Y_{...}^2}{N} = \frac{407^2}{24} = 6902.04, \\
SS_{\text{Total}} &= 8269 - C = 1366.96, \\
SS_{\text{Treat}} &= 1/12(160^2 + 247^2) - C \\
&= 7217.42 - C = 315.38, \\
SS_{\text{Time}} &= 1/8(89^2 + 133^2 + 185^2) - C \\
&= 7479.38 - C = 577.33, \\
SS_{\text{Subtotal}} &= 1/4(28^2 + 48^2 + 84^2 + 61^2 + 85^2 + 101^2) - C \\
&= 7822.75 - C = 920.71, \\
SS_{\text{Treat} \times \text{Time}} &= SS_{\text{Subtotal}} - SS_{\text{Treat}} - SS_{\text{Time}} \\
&= 920.71 - 315.38 - 577.33 \\
&= 28.00, \\
SS_{\text{Ind}} &= 1/3(56^2 + 47^2 + \ldots + 70^2 + 72^2) - 1/12(160^2 + 247^2) \\
&= 7555.67 - 7217.42 = 338.25, \\
SS_{\text{Error}} &= SS_{\text{Total}} - SS_{\text{Subtotal}} - SS_{\text{Ind}} \\
&= 108.00\,.
\end{aligned}$$

	SS	df	MS	F	p–value
Treat	315.38	1	315.38	5.59	0.056
Time	577.33	2	288.67	32.07	0.000
Treat × Time	28.00	2	14.00	1.56	0.251
Ind	338.25	6	56.38		
Error	108.00	12	9.00		
Total	1366.96	23			

TABLE 8.3. Analysis of variance table in the model with interaction.

We have

$$F_{\text{Treat} \times \text{Time}} = \frac{MS_{\text{Treat} \times \text{Time}}}{MS_{\text{Error}}} = 1.56\,.$$

Because of $1.56 < F_{2,12;0.95} = 3.88$, $\text{H}_0 : (\alpha\beta)_{ij} = 0$ is not rejected. Hence we return to the independence model for testing the main effect "Time". $SS_{\text{Treat} \times \text{Time}}$ is added to SS_{Error}. The treatment effect (p–value, 0.056) is not significant; the time effect is significant. The test statistic of the treatment effect is identical in both tables: $F_{\text{Treat}} = MS_{\text{Treat}}/MS_{\text{Ind}}$.

	SS	df	MS	F	p–value	
Treat	315.38	1	315.38	5.59	0.056	
Time	577.33	2	288.67	29.73	0.000	*
Ind	338.25	6	56.38			
Error	136.00	14	9.71			
Total	1366.96	23				

TABLE 8.4. Analysis of variance table in the independence model.

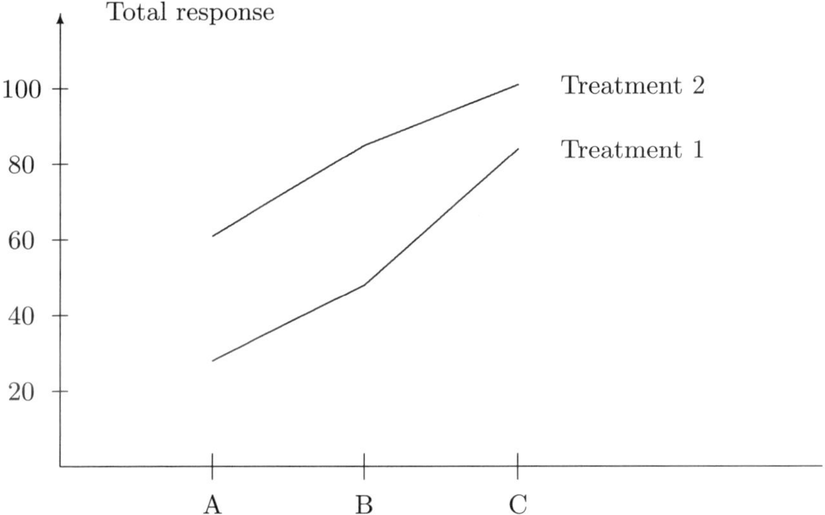

FIGURE 8.3. Total response treatment 1 and treatment 2 (Example 8.1).

Example 8.2. Two blood pressure lowering drugs, B and a combination of B and another drug, are to be compared. On 3 control days, the diastolic blood pressure is measured in intervals of 2 hours. The last day is then analyzed. This results in a repeated measures design with $p = 12$ measures. The sample sizes are $n_1 = 24$ (B) and $n_2 = 27$ (combination).

The analysis is done with SPSS.

```
MANOVA X1 TO X12 by Treat(1,2)
/wsfactors=Interval(12)
/contrast(Interval)=Difference
/Print=Homogeneity(BoxM)
/Design=Treat .
```

(i) Test of the homogeneity of variance, i.e., $H_0 : \Sigma_B = \Sigma_{\text{comb.}}$:

```
BoxsM =                       109.59084
F with (78,7357)DF =            1.03211   ,   P ≐ .401
Chi-square with 78 DF =        81.66664   ,   P ≐ .366
```

With $p = 12$, we have $p(p+1)/2 = 78$, so that the Box-M statistic αM follows a χ^2_{78} (cf. (8.76)).

Hence, the null hypothesis $H_0 : \Sigma_B = \Sigma_{\text{comb}} = \Sigma$ is not rejected. The univariate unadjusted F–tests require, in addition to the assumption of the homogeneity of variance, the special structure of compound symmetry. This assumption is included in the sphericity of the contrast covariance matrix as a special case.

(ii) Testing of $H_0 : \tilde{C}_H \Sigma \tilde{C}_H' = \lambda I$:

The test statistic by Mauchly is (cf. (8.103)) $W \sim \chi^2_v$ with $v = \frac{1}{2}(p-2)(p+1) = \frac{1}{2}(12-2)(12+1) = 65$ degrees of freedom.

```
Mauchly sphericity test, W =        .00478
Chi-square approx. =             241.17785   with 65 D.F.
Significance                        .00000
```

Hence, sphericity (and, of course, compound symmetry as well) is rejected and the unadjusted (averaged) univariate F–tests may not be applied. However, the adjusted univariate F–tests according to the Greenhouse–Geisser strategy can now be conducted.

(iii) Greenhouse–Geisser strategy:

The measures for sphericity/nonsphericity are:

Greenhouse–Geisser epsilon (8.107):	$\hat{\epsilon}_{G-H} = 0.41$,
Huynh–Feldt epsilon (8.99):	$\hat{\epsilon} = 0.46$.

They are distinctly smaller than 1 and indicate nonsphericity of the contrast covariance matrix $\tilde{C}_H \Sigma \tilde{C}_H'$. The Greenhouse–Geisser strategy now corrects the univariate test statistics according to their degrees of freedom.

The null hypothesis $H_0 : (\alpha\beta)_{ij} = 0$ is rejected by the unadjusted univariate F–test. The test value of $F_{\text{Treat} \times \text{Time}} = 2.74$ is now assessed with respect to the $F_{\epsilon(p-1),\epsilon(p-1)(n-2)}$–distribution, where we start with the lower–bound epsilon $\epsilon_{\min} = 1/(p-1) = 1/11$. We have $2.74 < F_{1,49;0.95} = 4.04$, hence the interaction is not significant, i.e., $H_0 : (\alpha\beta)_{ij} = 0$ is not rejected.

Now the next step of the Greenhouse–Geisser strategy is to be carried out. The value estimated with SPSS is $\hat{\epsilon}_{G-H} = 0.41$, hence the adjusted F–statistic has $11 \cdot 0.41 = 4.5$ degrees of freedom in the numerator, and

Source	SS	df	MS	F	p–value	
Treat	5014.49	1	5014.49	4.24	0.045	*
Time	32414.11	11	2946.74	41.64	0.000	*
Treat × Time	2135.01	11	194.09	2.74	0.002	*
Ind	57996.61	49	1183.60			
Error	38141.34	539	70.76			
Total	135701.56	611				

TABLE 8.5. Unadjusted univariate averaged F–tests.

$539 \cdot 0.41 = 221$ degrees of freedom in the denominator. Because of

$$F_{\text{Treat } \times \text{ Time}} = 2.74 > 2.32 = F_{4.5,221;0.95}$$

$H_0 : (\alpha\beta)_{ij} = 0$ is rejected. This decision is accepted.

Source				p–value
Treat×Time	$F_{11,39}$	=	1.75	0.099
Time	$F_{11,39}$	=	18.01	0.000 *
Treat	$F_{1,49}$	=	4.24	0.045 *

TABLE 8.6. Results of the MANOVA.

Results of the MANOVA and the corrected ANOVA:

At the 5% level, the model with interaction holds for the ANOVA, and for the MANOVA the independence model holds. Hence, the significant main effects "treatment" and "time" can be interpreted separately only in the case of the MANOVA. If the 10% level is chosen the independence model holds for the adjusted ANOVA as well.

Multiple Tests:

The overall treatment effect is significant. Hence the multiple test procedure from Section 8.7.5 may be applied.

From the table of the p–values of the univariate comparison of means, we find values in ascending order, which we compare with the adjusted Holm limits.

Hence the following local multiple treatment effects are significant:

j	p–values
1	0.006
2	0.003
3	0.002
4	0.061
5	0.329
6	0.374
7	0.424
8	0.893
9	0.536
10	0.117
11	0.582
12	0.024

TABLE 8.7. Ordered p–values.

	$j=3$	$j=2$	$j=1$	$j=12$	$j=4$	...
p–Values	0.002	0.003	0.006	0.024	0.061	...
Holm 5%	$\frac{0.05}{11}=0.0045$	0.0045	$\frac{0.05}{10}=0.005$	$\frac{0.05}{9}=0.0056$	$\frac{0.05}{8}=0.0063$	...
Holm 10%	0.0091	0.0091	0.010	0.011	0.0125	...

8.8 Multivariate Rank Tests in the Repeated Measures Model

In the case of continuous but not necessarily normal response values, the same hypotheses as in the previous sections may be tested by statistics that are based on ranks. The starting point is once again a multivariate two–sample problem. Assume the following observation vectors

$$y_{ki} = (y_{ki1}, \ldots, y_{kip})', \quad k = 1, 2, \quad i = 1, \ldots, n_k . \tag{8.110}$$

For the observation vectors, we assume that the y_{ki} have independent distributions with a continuous distribution function

$$F_k(y_{ki}) = G(y_{ki} - m_k), \quad k = 1, 2 , \tag{8.111}$$

where $m_k = (m_{k1}, \ldots, m_{kp})'$ is the vector of medians of the kth group for the p measures. The function G characterizes the type of distribution and m_k represents the location parameter.

The null hypothesis H_0 : *no treatment effect* means $\mathrm{H}_0 : F_1 = F_2$ and implies

$$\mathrm{H}_0 : \; m_1 = m_2 , \tag{8.112}$$

(i) 5 % level: $j = 2$ and $j = 3$;

(ii) 10 % level: $j = 1, 2$ and $j = 3$.

so that both distributions are identical. The null hypothesis H_0 : *no time effect* means (cf. Koch, 1969)

$$H_0: \; m_{k1} = \ldots = m_{kp}, \quad k = 1, 2. \tag{8.113}$$

The test procedures are to be carried out considering the fact, whether we have a significant interaction *treatment* $\times$ *time* or not. A detailed description of these tests can be found in Koch (1969) (cf. Puri and Sen, 1971). Since these nonparametric tests are quite burdensome and not implemented in standard software, we confine ourselves to a short description of the tests for one treatment effect. In the case of a continuous but not necessarily normal response, it is more practical to go over to loglinear models by applying categorical coding. These tests may then be conducted according to Chapter 7.

For the construction of a test for H_0 from (8.112), we proceed as follows. Let

$$r_{kij} := [\text{rank of } y_{kij} \text{ in } y_{11j}, \ldots, y_{1n_1j}, y_{21j}, \ldots, y_{2n_2j}] \tag{8.114}$$

($k = 1, 2,\; i = 1, \ldots, n_k,\; j = 1, \ldots, p$), i.e., for every occasion j ($j = 1, \ldots, p$) the ranks $1, \ldots, N = n_1 + n_2$ are assigned. If ties occur, then the averaged ranks are used.

Since the distribution is assumed to be continuous, we can assume

$$P(y_{kij} = y_{k'i'j}) = 0. \tag{8.115}$$

Hence, we disregard the ties in the following.

If the r_{kij} (cf. (8.114)) are combined for each individual, we get the rank observation vector of the ith individual in the kth group

$$r_{ki} = (r_{ki1}, \ldots, r_{kip})', \quad k = 1, 2, \quad i = 1, \ldots, n_k. \tag{8.116}$$

This yields N rank vectors that can be summarized by the $(p \times N)$–rank matrix

$$R = (r_{11}, \ldots, r_{1n_1}, r_{21}, \ldots, r_{2n_2}). \tag{8.117}$$

Because of the rank assignment (cf. (8.114)), each of the p rows of R is a permutation of the numbers $1, \ldots, N$.

If the columns of R are exchanged in a way that the first row of R contains the ordered ranks, we find the matrix

$$R_{\text{per}} = \begin{pmatrix} 1 & \cdots & N \\ r_{21}^{\text{per}} & \cdots & r_{2N}^{\text{per}} \\ \vdots & & \vdots \\ r_{p1}^{\text{per}} & \cdots & r_{pN}^{\text{per}} \end{pmatrix} = (r_1, \ldots, r_N), \tag{8.118}$$

which is a permutation equivalent to R (cf. (8.117)).

Since the p observations y_{kij} ($j = 1, \ldots, p$) are not independent, the common distribution of the elements of R (or of R_{per}) is dependent on the unknown distributions, even if H_0 holds.
Assume $\{R_{\text{per}}\}$ is the set of all possible realizations of R_{per}. For the size of $\{R_{\text{per}}\}$, we have

$$|\{R_{\text{per}}\}| = (N!)^{p-1}. \tag{8.119}$$

In general, the distribution of R_{per} over $\{R_{\text{per}}\}$ is dependent on the distributions F_1 and F_2.

If, however, $\mathrm{H}_0 : F_1 = F_2$ holds, then the observation vectors y_{ki} ($k = 1, 2, i = 1, \ldots, n_k$) are independent and identically distributed. Hence, their common distribution stays invariant in the case of a permutation within itself, i.e., it is of no great importance from which treatment group the vectors are derived.

This means, however, that under H_0, R is uniformly distributed over the set $\{R_{\text{per}}\}$ of the $N!$ possible realizations that we get by all possible permutations of the columns of R_{per}.

Hence, we have

$$P(R = r_S \mid \{R_{\text{per}}\}, H_0) = \frac{1}{N!} \quad \text{for all } r_S \in \{R_{\text{per}}\}. \tag{8.120}$$

Denote this (conditional) probability distribution by P_0.

Assume that the N rank observation vectors r_{ki}, $k = 1, 2, (i = 1, \ldots, n_k)$ (cf. (8.116)), are known and that these are represented by R_{per}, then the following holds (cf. Koch, 1969):

The probability that a rank observation vector r_{ki} takes the value r is

$$P(r_{ki} = r) = \frac{(N-1)!}{N!} = \frac{1}{N} \quad \text{for } r = r_1, \ldots, r_N. \tag{8.121}$$

Hence, for the expectation of r_{ki} ($k = 1, 2, i = 1, \ldots, n_k$), we have

$$\begin{aligned} \mathrm{E}(r_{ki} \mid \mathrm{H}_0) &= \sum_{j=1}^{N} \frac{1}{N} r_j \\ &= \frac{1}{N} \frac{N(N+1)}{2} \mathbf{1}_p = \frac{N+1}{2} \mathbf{1}_p. \end{aligned} \tag{8.122}$$

For the construction of an appropriate test statistic, we define the rank mean vector of the kth group

$$r_{k.} = \frac{1}{n_k} \sum_{i=1}^{n_k} r_{ki}, \quad k = 1, 2. \tag{8.123}$$

With (8.122), we obtain

$$\mathrm{E}(r_{k.}) = \frac{N+1}{2}\,\mathbf{1}_p\,. \tag{8.124}$$

The hypothesis H_0 can now be tested with the following test statistic (cf. Puri and Sen, 1971, p. 186):

$$L_I = \sum_{k=1}^{2} n_k \left(r_{k.} - \frac{N+1}{2}\mathbf{1}_p\right)' S_I^{-1} \left(r_{k.} - \frac{N+1}{2}\mathbf{1}_p\right), \tag{8.125}$$

where we assume that the empirical rank covariance matrix S_I is regular.

Remark. The matrix S_I measures the interaction *treatment* $\times$ *time*. If no interaction exists, S_I equals the identity matrix (except for a variance factor) and the multivariate test statistic L_I equals the univariate statistic by Kruskal–Wallis (cf. (4.139)).

We have

$$S_I = \frac{1}{N}\sum_{k=1}^{2}\sum_{i=1}^{n_k} \left(r_{ki} - \frac{N+1}{2}\mathbf{1}_p\right)\left(r_{ki} - \frac{N+1}{2}\mathbf{1}_p\right)'. \tag{8.126}$$

The test statistic L_I is the multivariate version of the statistic of the Kruskal–Wallis test and is equivalent to a generalized Lawley–Hotelling T^2–statistic. It can be shown that L_I has an asymptotic χ^2–distribution under H_0 with p degrees of freedom (cf. Puri and Sen, 1971, p. 193). Based on the construction of the test, large values of L_I indicate a violation of the null hypothesis H_0 from (8.112). Hence, H_0 is rejected if

$$L_I \quad \geq \quad \chi^2_{p;1-\alpha}\,. \tag{8.127}$$

Example 8.3. In the following, we demonstrate the calculation of the test statistic by means of a simple example. Suppose that we are given the following data set for $p = 3$ repeated measures:

$$\begin{array}{c} \text{Group 1} \\ \\ \text{Group 2} \end{array} \left(\begin{array}{ccc} 2 & 3 & 6 \\ 5 & 6 & 4 \\ 4 & 5 & 5 \\ \hline 8 & 14 & 10 \\ 10 & 12 & 14 \\ 12 & 13 & 12 \end{array}\right) \underset{\text{ranks}}{\Longrightarrow} \left(\begin{array}{ccc} 1 & 1 & 3 \\ 3 & 3 & 1 \\ 2 & 2 & 2 \\ \hline 4 & 6 & 4 \\ 5 & 4 & 6 \\ 6 & 5 & 5 \end{array}\right),$$

$$\begin{aligned} R &= \begin{pmatrix} 1 & 3 & 2 & 4 & 5 & 6 \\ 1 & 3 & 2 & 6 & 4 & 5 \\ 3 & 1 & 2 & 4 & 6 & 5 \end{pmatrix} \\ &= \begin{pmatrix} r_{11} & r_{12} & r_{13} & r_{21} & r_{22} & r_{23} \end{pmatrix}. \end{aligned}$$

The rank means in the two therapy groups are

$$\begin{aligned} r_{1.} &= \frac{1}{n_k}(r_{11}+r_{12}+r_{13}) \\ &= \frac{1}{3}\left[\begin{pmatrix}1\\1\\3\end{pmatrix}+\begin{pmatrix}3\\3\\1\end{pmatrix}+\begin{pmatrix}2\\2\\2\end{pmatrix}\right] \\ &= \frac{1}{3}\begin{pmatrix}6\\6\\6\end{pmatrix}=\begin{pmatrix}2\\2\\2\end{pmatrix}, \end{aligned}$$

$$\begin{aligned} r_{2.} &= \frac{1}{3}\left[\begin{pmatrix}4\\6\\4\end{pmatrix}+\begin{pmatrix}5\\4\\6\end{pmatrix}+\begin{pmatrix}6\\5\\5\end{pmatrix}\right] \\ &= \frac{1}{3}\begin{pmatrix}15\\15\\15\end{pmatrix}=\begin{pmatrix}5\\5\\5\end{pmatrix}. \end{aligned}$$

From this we calculate, according to (8.125),

$$\begin{aligned} r_{i\cdot}-\frac{N+1}{2}\mathbf{1}_p &= r_{i\cdot}-\frac{6+1}{2}\mathbf{1}_3 \quad (i=1,2), \\ (r_{1\cdot}-\frac{7}{2}\mathbf{1}_3) &= -\frac{3}{2}\mathbf{1}_3\,, \\ (r_{2\cdot}-\frac{7}{2}\mathbf{1}_3) &= \frac{3}{2}\mathbf{1}_3\,. \end{aligned}$$

This yields the covariance matrix S_I, from (8.126),

$$S_I = \frac{1}{6\cdot 4}\begin{pmatrix}70&58&50\\58&70&38\\50&38&70\end{pmatrix}$$

and

$$S_I^{-1} = \frac{24}{51840}\begin{pmatrix}3456&-2160&-1296\\-2160&2400&240\\-1296&240&1536\end{pmatrix}.$$

For L_I, from (8.125), we have

$$L_I=\sum_{k=1}^{2} n_k\left(r_{k\cdot}-\frac{N+1}{2}\mathbf{1}_3\right)' S_I^{-1}\left(r_{k\cdot}-\frac{N+1}{2}\mathbf{1}_3\right)=6.00\,.$$

Hence, the test for $\mathrm{H}_0 : m_1 = m_2$ (cf. (7.112)) with

$$L_I = 6.00 < 7.81 = \chi^2_{3;0.95}$$

does not lead to a rejection of H_0.

8.9 Categorical Regression for the Repeated Binary Response Data

8.9.1 *Logit Models for the Repeated Binary Response for the Comparison of Therapies*

Unlike the previous sections of this chapter, we now assume categorical response. In order to explain the problems, we start with binary response $y_{ijk} = 1$ or $y_{ijk} = 0$. These categories can stand for a reaction above/below an average. In an example, the blood pressure of each patient above/below the median blood pressure of a control group is measured in this way.

Let $I = 2$ (response categories) and assume two therapies (P : placebo and M : treatment) to be compared. We define the logit for the response distribution of the kth subpopulation (therapy P or M, i.e., $k = 1$ or $k = 2$) for occasion j $(j = 1, \ldots, m)$ as

$$L(j;k) = \ln\left[P_1(j;k)/P_2(j;k)\right] . \tag{8.128}$$

The independence model in effect coding

$$L(j;k) = \mu + \lambda_1^P + \lambda_j^V \quad (j = 1, \ldots, m-1) \tag{8.129}$$

contains the main effects

λ_1^P: placebo effect,

λ_j^V $(j = 1, \ldots, m-1)$: occasions effect,

where the constraints of effect coding (cf. Chapter 6) hold

$$\lambda_2^M = -\lambda_1^P \quad \text{(treatment effect)}, \tag{8.130}$$

$$\lambda_m^V = -\sum_{j=1}^{m-1} \lambda_j^V . \tag{8.131}$$

The inclusion of interaction effects λ_{1j}^{PV} is possible (saturated model).

The ML estimation of the parameters of the model (8.129) is quite complicated since marginal probabilities, that are to be estimated from the marginal frequencies, are used for the odds. These marginal frequencies, however, have no independent multinomial distributions. The ML estimation has to be achieved by maximizing the likelihood under the constraint that the marginal distributions satisfy the model [(8.129)] of the null hypothesis. For this, iterative procedures (e.g., Koch, Landis, Freeman, Freeman and Lehnen (1977); Aitchison and Silvey (1958)) have to be applied. These procedures replace the necessary nonlinear optimization under linear constraints by stepwise weighted ordinary least squares estimates, and the iterated ML estimates are again used to form the standard χ^2 or G^2 goodness–of–fit statistics.

8.9.2 First–Order Markov Chain Models

A Markov chain of the lth order $\{X_t\}$ is a stochastic process with a "memory" of length l, i.e., in the case of $l = 1$, we have, for a given occasion t,

$$P(X_{t+1} \mid X_0, \ldots, X_t) = P(X_{t+1} \mid X_t) . \tag{8.132}$$

Hence, the conditional probability for a future value X_{t+1} is only dependent on the preceding value X_t and not on the past $X_0, \ldots, X_{t-1}$. The common density of $(X_0, \ldots, X_m)$ is then of the form

$$f(x_0, \ldots, x_m) = f(x_0) \cdot f(x_1 \mid x_0) \cdot \cdots \cdot f(x_m \mid x_{m-1}) . \tag{8.133}$$

Hence the common distribution is only dependent on the starting distribution $f(x_0)$ and on the conditional transition probabilities $f(x_i \mid x_{i-1})$. This corresponds to a loglinear model with the effects

$$(X_0, (X_0, X_1), (X_1, X_2), \ldots, (X_{m-1}, X_m)) . \tag{8.134}$$

Remark. The transformation of the first–order Markov chain into categorical time–dependent response is the nonparametric counterpart of modeling the process as a time series with first–order autocorrelated errors.

Applied to our problem of binary response $\{X_j\}$ at occasions t_j $(j = 1, \ldots, m)$ in the comparison of two therapies (P and M), the probabilities

$$P_{\alpha,\beta}(j-1, j) \quad \alpha, \beta = 1, 2 \quad \text{(response)}, \tag{8.135}$$

specify the common distribution of X_{j-1} and X_j.

The conditional probability that the process is in state $\alpha = i$ at occasion j, under the condition that it was in state $\alpha = k$ $(i, k = 1, 2)$ at occasion $j - 1$, equals

$$\pi_{i/k}(j) = P(X_j = imidX_{j-1} = k) = \frac{P_{i,k}(j-1, j)}{\sum_{k=1}^{2} P_{i,k}(j-1, j)} . \tag{8.136}$$

Hence, the modeling of this process is equivalent to the loglinear model [(8.137)]. We find the estimates of the $\pi_{i/k}(j)$ by constructing a contingency table and counting the frequencies of possible events. By means of observations in the subpopulations of the prognostic factor (placebo/treatment), we get the estimates $\hat{\pi}^P_{i/k}(j)$ and $\hat{\pi}^M_{i/k}(j)$ for both subpopulations.

Example 8.4. Binary response X_j, binary prognostic factor (placebo, treatment). Assume

$$X_j^M \text{ and } X_j^P = \begin{cases} 1 & \text{blood pressure of the patient lies above the median} \\ & \text{of the placebo group at the } j\text{th occasion,} \\ 0 & \text{below.} \end{cases}$$

We choose the following fictitious numbers for a therapy group, in order to illustrate the calculation of the estimates of $\pi_{i/k}(j)$:

	$j = 1$
1	80
0	20
	100

	$j = 2$
1	60
0	40
	100

Assume the following counts of transitions for each patient:

$j = 1$	$\Rightarrow$	$j = 2$	Number of transitions
1		1	50
1		0	30
0		1	10
0		0	10
			100

This yields

$$\begin{aligned} P_{1,1}(1,2) &= \frac{50}{100} = 0.5, \\ P_{1,0}(1,2) &= \frac{30}{100} = 0.3, \\ P_{0,1}(1,2) &= \frac{10}{100} = 0.1, \\ P_{0,0}(1,2) &= \frac{10}{100} = 0.1\,. \end{aligned}$$

Hence the estimated conditional transition probabilities are

$$\left.\begin{aligned} \hat{\pi}_{1/1}(2) &= \frac{0.5}{0.8} = 0.625, \\ \hat{\pi}_{0/1}(2) &= \frac{0.3}{0.8} = 0.375, \end{aligned}\right\} \sum = 1\,,$$

$$\left.\begin{aligned} \hat{\pi}_{1/0}(2) &= \frac{0.1}{0.2} = 0.5, \\ \hat{\pi}_{0/0}(2) &= \frac{0.1}{0.2} = 0.5, \end{aligned}\right\} \sum = 1\,.$$

Remark. The separate modeling for each therapy group by a loglinear model

$$\ln(\hat{\pi}_{i/k}(j)) = \mu + \lambda_1^{X_0 X_1} + \cdots + \lambda_m^{X_{m-1} X_m} \tag{8.137}$$

gives an insight into significant transitions and filters out the best model according to the G^2 criterion.

If both therapy groups are included in one joint model, i.e., if the indicator placebo/therapy is chosen as a third dimension, then local statements within the scope of the discrete Markov chain models of the following form can be tested:

H_0 : The effects of the treatment $\lambda^M_{0,j} = -\lambda^P_{1,j}$ on the transition probabilities $\hat{\pi}_{1/0}(j)$ are significant (or significant at some occasions of the day's rhythm of blood pressure).

The actual aim—a global measure (overall superiority) or a global test for H_0 : "placebo=treatment"—cannot be achieved directly with this model, but only via an additional consideration.

8.9.3 *Multinomial Sampling and Loglinear Models for a Global Comparison of Therapies*

We assume the response of a patient to therapy A or B to be a categorical response (e.g., binary response) over m occasions. Thus, for each therapy, we have m dependent (correlated) response values. If the response is observed in I categories, then the possible response values for the m occasions can be represented in an I^m–dimensional contingency table. Table 8.8 corresponds to $I = 2$ and $m = 4$.

Example 8.5.
$I = 2$ (binary response),
$m = 4$ occasions.
Coding of the response: 1,
Coding of the nonresponse: 0.
Denote by $\underset{\sim}{i} = (i_1, \ldots, i_m)$ the cell in the table corresponding to response $i_j = 1$ or $i_j = 0$ $(j = 1, \ldots, m)$ at the occasions $t_1, \ldots, t_m$, and by $\pi_{\underset{\sim}{i}}$ the probability for this cell. We then have

$$\sum_{1}^{I^m} \pi_{\underset{\sim}{i}} = 1\,. \tag{8.138}$$

Let $m_{\underset{\sim}{i}} = n\pi_{\underset{\sim}{i}}$ be the expected cell count of the $\underset{\sim}{i}$th cell. Let the I categories be indexed by h $(h = 1, \ldots, I)$ and let $P_h(j)$ be the probability of response h at occasion j. The $\{P_h(j),\ h = 1, \ldots, I\}$ for given j are then the jth marginal distribution of the contingency table.

We now consider Table 8.8 with $m = 4$ occasions. For each therapy group (P or M), we count separately the completely crossed experimental design for the binary response (e.g., 1 : above the median blood pressure of the placebo group at occasion j, 0 : below), i.e., the 2^4 table. We now classify the response according to the independent multinomial scheme $M(n; \pi_1, \ldots, \pi_5)$:

Response	i	Occasion 1	2	3	4	Number
4 times	1	1	1	1	1	n_1
3 times	2	1	1	1	0	n_2
	3	1	1	0	1	n_3
	4	0	1	1	1	n_4
	5	1	0	1	1	n_5
2 times	6	1	1	0	0	n_6
	7	1	0	1	0	n_7
	8	1	0	0	1	n_8
	9	0	1	1	0	n_9
	10	0	1	0	1	n_{10}
	11	0	0	1	1	n_{11}
1 time	12	1	0	0	0	n_{12}
	13	0	1	0	0	n_{13}
	14	0	0	1	0	n_{14}
	15	0	0	0	1	n_{15}
0 times	16	0	0	0	0	n_{16}
						n

TABLE 8.8. 2^4 Table.

Class 1: 4–times response 1,
0–times nonresponse 0
$\Rightarrow$ row 1 of Table 8.8.

Class 2: 3–times response 1,
1–time nonresponse 0
$\Rightarrow$ rows 2–5.

Class 3: 2–times response 1,
2–times nonresponse 0
$\Rightarrow$ rows 6–11.

Class 4: 1–time response 1,
3–times nonresponse 0
$\Rightarrow$ rows 12–15.

Class 5: 0–times response 1,
4–times nonresponse 0
$\Rightarrow$ row 16.

If both therapies (P/M) are included, we receive a 5×2 table. The **disjoint categories of the rows** are often called **profiles**.

Cumulated number of response 1	P	M
0	n_{11}	n_{12}
1	n_{21}	n_{22}
2	n_{31}	n_{32}
3	n_{41}	n_{42}
4	n_{51}	n_{52}
	n_{+1}	n_{+2}

Since P and M are independent and since the columns follow the model of the independent multinomial scheme $M(n_{+1}; \pi_P)$, or $M(n_{+2}; \pi_M)$, respectively, the null hypothesis H_0 : "independent decomposition according to cumulated response and therapy" can, equivalently, be formulated by a loglinear model (m_{ij} : under H_0 expected cell frequencies)

$$\ln(m_{ij}) = \mu + \lambda_i^R + \lambda_1^P + \lambda_{i1}^{RP}, \tag{8.139}$$

where

μ	is the total mean;
λ_i^R	is the effect of the ith cumulated response category (ith profile);
λ_1^P	is the effect of the placebo; and
λ_{i1}^{RP}	is the interaction ith response category–placebo.

If effect coding is chosen, the effect of the treatment is $\lambda_1^M = -\lambda_1^P$.

Example 8.6. We illustrate the global test on a 13-hour blood pressure data set. The data set consists of measures of $n_1 = 63$ and $n_2 = 64$ patients of the therapy groups P (placebo) and M (treatment) over a stretch of $m = 13$ hours (start: $j = 0$, then 12 measures taken in 1–hour intervals). For each patient, it is recorded to which cumulated response category i ($i = 0, \ldots, 13$) he belongs, with i : number of hourly blood pressures above the median of the jth hourly measurement of the placebo group ($j = 0, \ldots, 12$).

The results are shown in Table 8.9. Table 8.10 shows these results summarized according to groups $(0, 1), (2, 3), \ldots, (12, 13)$ (in order to overcome zero–counts in the cells). The parameter estimates and the standardized parameter estimates ($*$: significance at the two–sided level of 5%, i.e., comparison with $u_{0.95\ (\text{two--sided})} = 1.96$) are shown in Table 8.11.

Remark. The calculations have been done with the newly developed software LOGGY 1.0 (cf. Heumann and Jacobsen, 1993), the standard software PCS, as well as additional programs.

i	P	M	$\sum$
0	5	30	35
1	7	7	14
2	3	6	9
3	4	6	10
4	3	5	8
5	3	3	6
6	5	2	7
7	6	0	6
8	3	2	5
9	9	0	9
10	5	0	5
11	2	2	4
12	2	1	3
13	6	0	6
$\sum$	63	64	127

TABLE 8.9. Classification of the 12–hour measures at the end point according to "i–times blood pressure values above the respective hourly median of the placebo group".

	P	M	$\sum$
0, 1	12	37	49
2, 3	8	12	20
4, 5	6	8	14
6, 7	10	2	12
8, 9	13	2	15
10, 11	7	2	9
12, 13	7	1	8
$\sum$	63	64	127

TABLE 8.10. Summary of the classes in Table 8.9.

Interpretation

(i) Saturated model

$$\ln(m_{ij}) = \mu + \lambda_i^R + \lambda_1^P + \lambda_{i1}^{RP} \,. \tag{8.140}$$

The test statistic for H$_0$: "saturated model valid" is $G^2 = 0$ (perfect fit) as usual.

The placebo effect $\hat{\lambda}_1^P = 0.35$ (2.57 standardized) is significant. Since code 1 symbolizes high blood pressure (above the respective hourly median of the placebo group), a positive λ_1^P stands for an effect toward higher blood pressure. Hence ($\lambda_1^M = -0.35$), the treatment

Parameter	Parameter estimate	Significant	Standardized
μ	1.81	*	13.42
λ_1^P	0.35	*	2.57
λ_1^R	1.24	*	6.35
λ_2^R	0.47	*	2.00
λ_3^R	0.12		0.47
λ_4^R	-0.31		-0.89
λ_5^R	-0.18		-0.53
λ_6^R	-0.49		-1.35
λ_7^R	-0.84	*	-1.98
λ_{11}^{RP}	-0.91	*	-4.67
λ_{21}^{RP}	-0.55	*	-2.34
λ_{31}^{RP}	-0.49		-1.85
λ_{41}^{RP}	0.46		1.29
λ_{51}^{RP}	0.59		1.69
λ_{61}^{RP}	0.28		0.77
λ_{71}^{RP}	0.63		1.33

TABLE 8.11. Parameter estimates and standardized values for the saturated model $\ln(m_{ij}) = \mu + \lambda_i^R + \lambda_1^P + \lambda_{i1}^{RP}$.

significantly lowers the blood pressure.

The significant response effects λ_1^R (categories 0- and 1-time above the median) and λ_2^R (2- and 3-times above the median) are positive, and λ_7^R (10- and 11-times above the median) is negative. These two results once again speak (in a qualitative way) for the blood pressure lowering effect of the treatment.
The interactions are hard to interpret separately.

The analysis of the submodels of the hierarchy lead to the following results:

(ii) Independence model

$$\mathrm{H}_0 : \ln(m_{ij}) = \mu + \lambda_i^R + \lambda_1^P \,. \tag{8.141}$$

The test value $G^2 = 37$ (p–value 0.000002) is significant, hence H_0 [(8.141)] is rejected.

(iii) Model for isolated profile effects

$$\mathrm{H}_0 : \ln(m_{ij}) = \mu + \lambda_i^R \,. \tag{8.142}$$

The test value is $G^2 = 37$ (7 *df*) is significant as well (H_0 : (8.142) is rejected).

(iv) Model for isolated treatment effect

$$\mathrm{H}_0 : \ln(m_{ij}) = \mu + \lambda_1^P \tag{8.143}$$

The test value is $G^2 = 90$ (12 df) and hence significant.

As a result, it can be stated that the saturated model is the only possible statistical model for the observed profiles of the two subpopulations placebo and treatment. This model indicates:

- a blood pressure lowering effect of the treatment;
- profile effects;

and gives evidence for:

- significant interactions.

As an interesting result, it can be stated that the therapy effect is not isolated (i.e., it is not an orthogonal component), but has a mutual effect with the time after taking the treatment.

This analysis is confirmed by the following crude–rate analysis for which the profiles 0–6 and 7–13 were combined:

	P	M	$\sum$
0–6	32	59	91
7–12	31	5	36
$\sum$	63	64	127

The saturated model

$$\ln(m_{ij}) = \mu + \lambda_1^R + \lambda_1^P + \lambda_{11}^{RP} \tag{8.144}$$

yields the significant parameter estimates

	$\hat{\mu}$		$\hat{\lambda}_1^R$		$\hat{\lambda}_1^P$		$\hat{\lambda}_{11}^{RP}$	
	3.15		0.63		0.30		-0.61	
Standardized	23.77	*	4.72	*	2.69	*	-4.60	*

In the saturated model we have, for the odds ratio,

$$\theta = \exp(4\lambda_{11}^{RP}),$$

i.e.,

$$\begin{aligned} \hat{\theta} &= 0.0036, \\ \ln \hat{\theta} &= -2.44 \quad \text{(negative interaction).} \end{aligned}$$

The crude model of the 2×2 table is regarded as a robust indicator of interactions, in general, that can be broken down by finer structures. The advantage of the 2×2 table is the estimation of a crude interaction over all levels of the categories of the rows.

Remark. The model calculations assume a Poisson sampling scheme for the contingency table, i.e., unrestricted random sampling, especially a random total sample size.

The sampling scheme is restricted to independent multinomial sampling in the case of the model of therapy comparison. Birch (1963) has proved that the ML estimates are identical for independent multinomial sampling and Poisson sampling, as long as the model contains a term (parameter) for the marginal distribution given by the experimental design. For our case of therapy comparison, this means that the marginal sums n_{+1} and n_{+2} (i.e., the number of patients in the placebo group and the treated group), have to appear as sufficient statistics in the parameter estimates. This is the case in:

(i) the saturated model (8.140);

(ii) the independence model (8.141);

(iii) the model for isolated profile effects (8.142);

but not in:

(iv) the model for the isolated treatment effect (8.143).

As our model calculations show, model (8.143) is of no interest, since a treatment effect cannot be detected isolated, but only in interaction with the profiles.

Remark. Tables 8.9 and 8.10 differ slightly due to patients whose blood pressure coincide with the hourly median.

Trend of the Profiles of the Medicated Group

As another nonparametric indicator for the blood pressure lowering effect of the treatment, we now model the crude binary risk

7–12 times over the respective placebo hourly median/
0–6 times over the median

over three observation days (i.e., $i = 1, 2, 3$) by a logistic regression. The results are shown in Table 7.11.

i	7–12	0–6	Logit
1	34	32	0.06
2	12	51	-1.45
3	5	59	-2.47

TABLE 8.12. Crude profile of the medicated group for the three observation days.

From this we calculate the model

$$\ln\left(\frac{n_{i1}}{n_{i2}}\right) = \hat{\alpha} + \hat{\beta} i \quad (i = 1, 2, 3)$$
$$= 1.243 - 1.265 \cdot i\,, \qquad (8.145)$$

with the correlation coefficient $r = 0.9938$ (p–value 0.0354, one–sided) and the residual variance $\hat{\sigma}^2 = 0.2^2$.

Hence, the negative trend to fall into the unfavorable profile group "7–12" is significant for this model (three observations, two parameters!). However, this result can only be regarded as a crude indicator. Results that are more reliable are achieved with Table 8.13, which is subdivided into seven groups instead of only two profiles.

i	0–1	2–3	4–5	6–7	8–9	10–11	12–13
1	4	10	10	13	8	13	8
2	29	14	7	4	4	2	1
3	37	12	8	2	2	2	1

TABLE 8.13. Fine profiles of the medicated group for the three observation days.

The G^2 analysis in Table 8.13 for testing H_0 : "cell counts over the profiles and days are independent" yields a significant value of $G^2_{14} = 70.50$ ($> 23.7 = \chi^2_{14;0.95}$) so that H_0 is rejected.

8.10 Exercises and Questions

8.10.1 How is the correlation of an individual over the occasions defined? In which way are two individuals correlated? Name the intraclass correlation coefficient of an individual over two different occasions.

8.10.2 What structure does the compound symmetric covariance matrix have? Name the best linear unbiased estimate of β in the model $y = X\beta + \epsilon$, $\epsilon \sim (\mathbf{0}, \sigma^2\Sigma)$, with Σ of compound symmetric structure.

8.10.3 Why is the ordinary least–squares estimate chosen instead of the Aitken estimate in the case of compound symmetry?

8.10.4 Name the repeated measures model for two independent populations. Why can it be interpreted as a mixed model and as a split–plot design?

8.10.5 What is meant by the μ_k–profile of an individual?

8.10.6 How is the Wishart distribution defined?

8.10.7 How is $H_0 : \mu = \mu_0$ (one–sample problem) tested univariate for $x_1, \ldots, x_n$ independent and identically distributed $\sim N_p(\mu, \Sigma)$?

8.10.8 How is $H_0 : \mu_x = \mu_y$ (two–sample problem) tested multivariate for $x_1, \ldots, x_{n_1} \sim N_p(\mu_x, \Sigma_x)$ and $y_1, \ldots, y_{n_2} \sim N_p(\mu_y, \Sigma_y)$? Which conditions have to hold true?

8.10.9 Describe the test strategy (univariate/multivariate) dependent on the fulfillment of the sphericity condition.

9
Cross–Over Design

9.1 Introduction

Clinical trials form an important part of the examination of new drugs or medical treatments. The drugs are usually assessed by comparing their effects on human subjects. From an ethical point of view, the risks which patients might be exposed to must be reduced to a minimum and also the number of individuals should be as small as statistically required. Cross–over trials follow the latter, treating each patient successively with two or more treatments. For that purpose, the individuals are divided into randomized groups in which the treatments are given in certain orders. In a 2×2 design, each subject receives two treatments, conventionally labeled as A and B. Half of the subjects receive A first and then *cross over* to B while the remaining subjects receive B first and then *cross over* to A. Between two treatments a suitable period of time is chosen, where no treatment is applied. This *washout period* is used to avoid the persistence of a treatment applied in one period to a subsequent period of treatment.

The aim of cross–over designs is to estimate most of the main effects using within–subject differences (or contrasts). Since it is often the case that there is considerably more variation between subjects than within subjects, this strategy leads to more powerful tests than simply comparing two independent groups using between–subject information. As each subject acts as his own control, between–subject variation is eliminated as a source of error.

If the washout periods are not chosen long enough, then a treatment may persist in a subsequent period of treatment. This *carry–over effect* will make it more difficult, or nearly impossible, to estimate *direct treatment effects*.

To avoid psychological effects, subjects are treated in a double blinded manner so that neither patients nor doctors know which of the treatments is actually applied.

9.2 Linear Model and Notations

We assume that there are s groups of subjects. Each group receives the M treatments in a different order. It is favorable to use all of the $M!$ orderings of treatments, i.e., to use the orderings AB and BA for comparison of $M = 2$ treatments and $ABC, BCA, CAB, ACB, CBA, BAC$ for $M = 3$ treatments so that $s = M!$

We generally assume that the trial lasts p periods (i.e., $p = M$ periods if all possible orderings are used). Let y_{ijk} denote the response observed on the kth subject ($k = 1, \ldots, n_i$) of group i ($i = 1, \ldots, s$) in period j ($j = 1, \ldots, p$). We first consider the following linear model (cf. Jones and Kenward, 1989, p. 9) which Ratkovsky, Evans and Alldredge (1993, pp. 81–84) label as parametrization 1:

$$y_{ijk} = \mu + s_{ik} + \pi_j + \tau_{[i,j]} + \lambda_{[i,j-1]} + \epsilon_{ijk} , \tag{9.1}$$

where

y_{ijk}:	is the response of the kth subject of group i in period j;
μ:	is the overall mean;
s_{ik}:	is the effect of subject k in group i ($i = 1, \ldots\ s,\ k = 1, \ldots, n_i$);
π_j:	is the effect of period j ($j = 1, \ldots, p$);
$\tau_{[i,j]}$:	is the direct effect of the treatment administered in period j of group i (treatment effect);
$\lambda_{[i,j-1]}$:	is the carry–over effect (effect of the treatment administered in period $j - 1$ of group i) that still persists in period j; and where $\lambda_{[i,0]} = 0$; and
ϵ_{ijk}:	is random error.

The subject effects s_{ik} are taken to be random. Sample totals will be denoted by capital letters, sample means by small letters. A dot $(\cdot)$ will replace a subscript to indicate that the data has been summed over that subscript. For example,

$$\begin{array}{llll} \text{total response:} & Y_{ij\cdot} = \sum_{k=1}^{n_i} y_{ijk}, & Y_{i\cdot\cdot} = \sum_{j=1}^{p} Y_{ij\cdot}, & Y_{\cdot\cdot\cdot} = \sum_{i=1}^{s} Y_{i\cdot\cdot}\,, \\ \text{means:} & y_{ij\cdot} = Y_{ij\cdot}/n_i, & y_{i\cdot\cdot} = Y_{i\cdot\cdot}/pn_i, & y_{\cdot\cdot\cdot} = Y_{\cdot\cdot\cdot}/(p\sum_{i=1}^{s} n_i)\,. \end{array} \tag{9.2}$$

To begin with, we assume that the response has been recorded on a continuous scale.

Remark. Model (9.1) may be called the *classical approach* and has been explored intensively since the 1960s (Grizzle, 1965). This parametrization, however, shows some inconsistencies concerning the effect caused by the order in which the treatments are given. This so–called sequence effect becomes important, especially regarding higher–order designs. For example, using the following plan in a cross–over design trial

	Period			
	1	2	3	4
Sequence	A	B	C	D
	B	D	A	C
	C	A	D	B
	D	C	B	A

,

the actual sequence (group) might have a fixed effect on the response. Then the between–subject effect s_{ik} would also be stratified by sequences (groups). This effect would have to be considered as an additional parameter γ_i ($i = 1, \ldots, s$) in model (9.1). Applying the classical approach (9.1) without this sequence effect leads to the sequence effect being confounded with other effects. We will discuss this fact later in this chapter.

9.3 2 × 2 Cross–Over (Classical Approach)

We now consider the common comparison of $M = 2$ treatments A and B (cf. Figure 9.1) using a 2×2 cross–over trial with $p = 2$ periods.

	Period 1	Period 2
Group 1	A	B
Group 2	B	A

FIGURE 9.1. 2 × 2 Cross–over design with two treatments.

As there are only four sample means $y_{11\cdot}, y_{12\cdot}, y_{21\cdot}$, and $y_{22\cdot}$ available from the 2×2 cross–over design, we can only use three degrees of freedom to estimate the period, treatment, and carry–over effects. Thus, we have to omit the direct *treatment* × *period* interaction which now has to be estimated as an aliased effect confounded with the carry–over effect. Therefore, the 2×2 cross–over design has the special parametrization

$$\tau_1 = \tau_A \quad \text{and} \quad \tau_2 = \tau_B \,. \tag{9.3}$$

The carry–over effects are simplified as

$$\left.\begin{array}{l} \lambda_1 = \lambda_{[1,1]} = \lambda_{[A,1]} \,, \\ \lambda_2 = \lambda_{[2,1]} = \lambda_{[B,1]} \,. \end{array}\right\} \tag{9.4}$$

Group	Period 1	Period 2
1 (AB)	$\mu + \pi_1 + \tau_1 + s_{1k} + \epsilon_{11k}$	$\mu + \pi_2 + \tau_2 + \lambda_1 + s_{1k} + \epsilon_{12k}$
2 (BA)	$\mu + \pi_1 + \tau_2 + s_{2k} + e_{21k}$	$\mu + \pi_2 + \tau_1 + \lambda_2 + s_{2k} + \epsilon_{22k}$

TABLE 9.1. The effects in the 2 × 2 cross–over model.

Then λ_1 and λ_2 denote the carry–over effect of treatment A (resp., B) applied in the first period so that the effects in the full model are as shown in Table 9.1. The subject effects s_{ik} are regarded as random.

The random effects are assumed to be distributed as follows:

$$\left.\begin{array}{rcl} s_{ik} & \overset{\text{i.i.d.}}{\sim} & N(0, \sigma_s^2), \\ \epsilon_{ijk} & \overset{\text{i.i.d.}}{\sim} & N(0, \sigma^2), \\ \mathrm{E}(\epsilon_{ijk} s_{ik}) & = & 0 \quad (\forall i, j, k). \end{array}\right\} \tag{9.5}$$

9.3.1 *Analysis Using t–Tests*

The analysis of data from a 2 × 2 cross–over trial using t–tests was first suggested by Hills and Armitage (1979). Jones and Kenward (1989) note that these are valid, whatever the covariance structure of the two measurements y_A and y_B taken on each subject during the active treatment periods.

Testing Carry–Over Effects, i.e., $H_0 : \lambda_1 = \lambda_2$

The first test we consider is the test on equality of the carry–over effects λ_1 and λ_2. Only if equality is not rejected, the following tests on main effects are valid, since the difference of the carry–over effects $\lambda_d = \lambda_1 - \lambda_2$ is the aliased effect of the *treatment* × *period* interaction.

We note that the subject total $Y_{1 \cdot k}$ of the kth subject in Group 1

$$Y_{1 \cdot k} = y_{11k} + y_{12k} \tag{9.6}$$

has the expectation (cf. Table 9.1)

$$\begin{aligned} \mathrm{E}(Y_{1 \cdot k}) &= \mathrm{E}(y_{11k}) + \mathrm{E}(y_{12k}) \\ &= (\mu + \pi_1 + \tau_1) + (\mu + \pi_2 + \tau_2 + \lambda_1) \\ &= 2\mu + \pi_1 + \pi_2 + \tau_1 + \tau_2 + \lambda_1 \,. \end{aligned} \tag{9.7}$$

In Group 2 (BA) we get

$$Y_{2 \cdot k} = y_{21k} + y_{22k} \tag{9.8}$$

and

$$\mathrm{E}(Y_{2\cdot k}) = 2\mu + \pi_1 + \pi_2 + \tau_1 + \tau_2 + \lambda_2 \,. \tag{9.9}$$

Under the null hypothesis,

$$\mathrm{H}_0 : \lambda_1 = \lambda_2 \,, \tag{9.10}$$

these two expectations are equal

$$\mathrm{E}(Y_{1\cdot k}) = \mathrm{E}(Y_{2\cdot k}) \quad \text{for all } k. \tag{9.11}$$

Now we can apply the two–sample t–test to the subject totals and define

$$\lambda_d = \lambda_1 - \lambda_2 \,. \tag{9.12}$$

Then

$$\hat{\lambda}_d = \frac{Y_{1\cdot\cdot}}{n_1} - \frac{Y_{2\cdot\cdot}}{n_2} = 2(y_{1\cdot\cdot} - y_{2\cdot\cdot}) \tag{9.13}$$

is an unbiased estimator for λ_d, i.e.,

$$\mathrm{E}(\hat{\lambda}_d) = \lambda_d \,. \tag{9.14}$$

Using

$$Y_{i\cdot k} - \mathrm{E}(Y_{i\cdot k}) = 2s_{ik} + \epsilon_{i1k} + \epsilon_{i2k}$$

and

$$\mathrm{Var}(Y_{i\cdot k}) = 4\sigma_s^2 + 2\sigma^2$$

we get

$$\mathrm{Var}\left(\frac{Y_{i\cdot\cdot}}{n_i}\right) = \frac{1}{n_i^2}\sum_{k=1}^{n_i} \mathrm{Var}(Y_{i\cdot k}) = \frac{4\sigma_s^2 + 2\sigma^2}{n_i} \quad (i = 1, 2) \,.$$

Therefore we have

$$\begin{aligned} \mathrm{Var}(\hat{\lambda}_d) &= 2(2\sigma_s^2 + \sigma^2)\left(\frac{1}{n_1} + \frac{1}{n_2}\right) \\ &= \sigma_d^2 \left(\frac{n_1 + n_2}{n_1 n_2}\right) \end{aligned} \tag{9.15}$$

where

$$\sigma_d^2 = 2(2\sigma_s^2 + \sigma^2) \,. \tag{9.16}$$

To estimate σ_d^2 we use the pooled sample variance

$$s^2 = \frac{(n_1 - 1)s_1^2 + (n_2 - 1)s_2^2}{n_1 + n_2 - 2} \tag{9.17}$$

which has $(n_1 + n_2 - 2)$ degrees of freedom, with s_1^2 and s_2^2 denoting the sample variances of the response totals within groups, where

$$s_i^2 = \frac{1}{n_i - 1} \sum_{k=1}^{n_i} \left(Y_{i \cdot k} - \frac{Y_{i \cdot \cdot}}{n_i} \right)^2 = \frac{1}{n_i - 1} \left(\sum_{k=1}^{n_i} Y_{i \cdot k}^2 - \frac{Y_{i \cdot \cdot}^2}{n_i} \right) \quad (i = 1, 2) \,. \tag{9.18}$$

We construct the test statistic

$$T_\lambda = \frac{\hat{\lambda}_d}{s} \sqrt{\frac{n_1 n_2}{n_1 + n_2}} \tag{9.19}$$

that follows a Student's t–distribution with $(n_1 + n_2 - 2)$ degrees of freedom under H_0 [(9.10)].

According to Jones and Kenward (1989), it is usual practice to follow Grizzle (1965) to run this test at the $\alpha = 0.1$ level. If this test does not reject H_1, we can proceed to test the main effects.

Testing Treatment Effects (Given $\lambda_1 = \lambda_2 = \lambda$)

If we can assume that $\lambda_1 = \lambda_2 = \lambda$, then the period differences

$$\begin{array}{rcll} d_{1k} & = & y_{11k} - y_{12k} & \text{(Group 1, i.e., A–B)}\,, \\ d_{2k} & = & y_{21k} - y_{22k} & \text{(Group 2, i.e., B–A)}\,, \end{array} \tag{9.20}$$

have expectations

$$\begin{array}{rcl} \mathrm{E}(d_{1k}) & = & \pi_1 - \pi_2 + \tau_1 - \tau_2 - \lambda\,, \\ \mathrm{E}(d_{2k}) & = & \pi_1 - \pi_2 + \tau_2 - \tau_1 - \lambda\,. \end{array} \tag{9.21}$$

Under the null hypothesis H_0 : no treatment effect, i.e.,

$$\mathrm{H}_0 : \tau_1 = \tau_2 \,, \tag{9.22}$$

these two expectations coincide. The difference of the treatment effects

$$\tau_d = \tau_1 - \tau_2 \tag{9.23}$$

is estimated by

$$\hat{\tau}_d = 1/2(d_{1\cdot} - d_{2\cdot}) \tag{9.24}$$

which is unbiased

$$\mathrm{E}(\hat{\tau}_d) = \tau_d \,, \tag{9.25}$$

and has variance

$$\begin{aligned} \mathrm{Var}(\hat{\tau}_d) &= \frac{2\sigma^2}{4} \left(\frac{1}{n_1} + \frac{1}{n_2} \right) \\ &= \frac{\sigma_D^2}{4} \left(\frac{1}{n_1} + \frac{1}{n_2} \right) , \end{aligned} \tag{9.26}$$

where

$$\sigma_D^2 = 2\sigma^2 \,. \tag{9.27}$$

The pooled estimate of σ_D^2, according to (9.17), replacing s_i^2 by

$$s_{iD}^2 = \frac{1}{n_i - 1} \sum_{k=1}^{n_i} (d_{ik} - d_{i\cdot})^2$$

becomes

$$s_D^2 = \frac{(n_1 - 1)s_{1D}^2 + (n_2 - 1)s_{2D}^2}{n_1 + n_2 - 2} . \qquad (9.28)$$

Under the null hypothesis $\mathrm{H}_0 : \tau_d = 0$, the statistic

$$T_\tau = \frac{\hat{\tau}_d}{\frac{1}{2} s_D} \sqrt{\frac{n_1 n_2}{n_1 + n_2}} , \qquad (9.29)$$

follows a t–distribution with $(n_1 + n_2 - 2)$ degrees of freedom.

Testing Period Effects (Given $\lambda_1 + \lambda_2 = 0$)

Finally we test for period effects using the null hypothesis

$$\mathrm{H}_0 : \pi_1 = \pi_2 . \qquad (9.30)$$

The "cross–over" differences

$$\begin{aligned} c_{1k} &= d_{1k} , \\ c_{2k} &= -d_{2k} , \end{aligned} \qquad (9.31)$$

have expectations

$$\begin{aligned} \mathrm{E}(c_{1k}) &= \pi_1 - \pi_2 + \tau_1 - \tau_2 - \lambda_1 , \\ \mathrm{E}(c_{2k}) &= \pi_2 - \pi_1 + \tau_1 - \tau_2 + \lambda_2 . \end{aligned} \qquad (9.32)$$

Under the null hypothesis $\mathrm{H}_0 : \pi_1 = \pi_2$ and the familiar reparametrization $\lambda_1 + \lambda_2 = 0$, these expectations coincide, i.e., $\mathrm{E}(c_{1k}) = \mathrm{E}(c_{2k})$. An unbiased estimator for the difference of the period effects $\pi_d = \pi_1 - \pi_2$ is given by

$$\hat{\pi}_d = 1/2(c_{1\cdot} - c_{2\cdot}) \qquad (9.33)$$

and we get the test statistic with s_D from (9.28)

$$T_\pi = \frac{\hat{\pi}_d}{\frac{1}{2} s_D} \sqrt{\frac{n_1 n_2}{n_1 + n_2}} , \qquad (9.34)$$

which again follows a t–distribution with $(n_1 + n_2 - 2)$ degrees of freedom.

Unequal Carry–Over Effects

If the hypothesis $\lambda_1 = \lambda_2$ is rejected, the above procedure for testing $\tau_1 = \tau_2$ should not be used since it is based on biased estimators. Given $\lambda_d = \lambda_1 - \lambda_2 \neq 0$, we get

$$\mathrm{E}(\hat{\tau}_d) = \mathrm{E}\left(\frac{d_{1\cdot} - d_{2\cdot}}{2} \right) = \tau_d - \frac{\lambda_d}{2} . \qquad (9.35)$$

With

$$\hat{\lambda}_d = y_{11\cdot} + y_{12\cdot} - y_{21\cdot} - y_{22\cdot} \tag{9.36}$$

and

$$\hat{\tau}_d = 1/2(y_{11\cdot} - y_{12\cdot} - y_{21\cdot} + y_{22\cdot}) \tag{9.37}$$

an unbiased estimator $\hat{\tau}_{d|\lambda_d}$ of τ_d is given by

$$\begin{aligned}\hat{\tau}_{d|\lambda_d} &= 1/2(y_{11\cdot} - y_{12\cdot} - y_{21\cdot} + y_{22\cdot}) + 1/2(y_{11\cdot} + y_{12\cdot} - y_{21\cdot} - y_{22\cdot}) \\ &= y_{11\cdot} - y_{21\cdot}. \end{aligned} \tag{9.38}$$

The unbiased estimator of τ_d for $\lambda_d \neq 0$ is identical to the estimator of a parallel group study. The estimator is based on between–subject information of the first period and the measurements. Testing for $H_0 : \tau_d = 0$ is done following a two–sample t–test, but using the measurements of the first period only, to estimate the variance. Thus, the sample size might become too small to get significant results for the treatment effect.

Regarding the reparametrization

$$\lambda_1 + \lambda_2 = 0\,, \tag{9.39}$$

we see that the estimator $\hat{\pi}_d$ is still unbiased

$$\begin{aligned}\mathrm{E}(\hat{\pi}_d) &= \mathrm{E}\left(\frac{c_{1\cdot} - c_{2\cdot}}{2}\right) \\ &= 1/2\mathrm{E}\left(\frac{1}{n_1}\sum_{k=1}^{n_1} c_{1k} - \frac{1}{n_2}\sum_{k=1}^{n_2} c_{2k}\right) \\ &= 1/2\left(\frac{1}{n_1}\sum_{k=1}^{n_1} \mathrm{E}(c_{1k}) - \frac{1}{n_2}\sum_{k=1}^{n_2} \mathrm{E}(c_{2k})\right) \\ &= \frac{1}{2}(2\pi_1 - 2\pi_2 - (\lambda_1 + \lambda_2)) \qquad [\text{cf. } (9.32)] \\ &= \pi_d \qquad [\text{cf. } (9.39)]\,, \end{aligned}$$

and thus $\hat{\pi}_d$ is unbiased, even if $\lambda_d = \lambda_1 - \lambda_2 \neq 0$ but $\lambda_1 + \lambda_2 = 0$.

9.3.2 Analysis of Variance

Considering higher–order cross–over designs, it is useful to test the effects using F–tests obtained from an analysis of variance table. Such a table was presented by Grizzle (1965) for the special case $n_1 = n_2$. The first general table was given by Hills and Armitage (1979). The sums of squares may be derived for the 2×2 cross–over design as a simple example of a split–plot design. The subjects form the main plots while the periods are treated as the subplots at which repeated measurements are taken (cf. Section 6.8).

With this in mind, we get

$$SS_{\text{Total}} = \sum_{i=1}^{2}\sum_{j=1}^{2}\sum_{k=1}^{n_i} y_{ijk}^2 - \frac{Y_{\cdots}^2}{2(n_1+n_2)},$$

between–subjects:

$$SS_{\text{Carry-over}} = \frac{2n_1n_2}{(n_1+n_2)}(y_{1\cdot\cdot} - y_{2\cdot\cdot})^2,$$

$$SS_{\text{b-s Residual}} = \sum_{i=1}^{2}\sum_{k=1}^{n_i} \frac{Y_{i\cdot k}^2}{2} - \sum_{i=1}^{2} \frac{Y_{i\cdot\cdot}^2}{2n_i},$$

within–subjects:

$$SS_{\text{Treat}} = \frac{n_1n_2}{2(n_1+n_2)}(y_{11\cdot} - y_{12\cdot} - y_{21\cdot} + y_{22\cdot})^2,$$

$$SS_{\text{Period}} = \frac{n_1n_2}{2(n_1+n_2)}(y_{11\cdot} - y_{12\cdot} + y_{21\cdot} - y_{22\cdot})^2,$$

$$SS_{\text{w-s Residual}} = \sum_{i=1}^{2}\sum_{j=1}^{2}\sum_{k=1}^{n_i} y_{ijk}^2 - \sum_{i=1}^{2}\sum_{j=1}^{2} \frac{Y_{ij\cdot}^2}{n_i} - SS_{\text{b-s Residual}}.$$

Source	SS	df	MS	F
Between–subjects				
Carry–over	$SS_{\text{c-o}}$	1	$MS_{\text{c-o}}$	$F_{\text{c-o}}$
Residual (between–subjects)	$SS_{\text{Residual(b-s)}}$	n_1+n_2-2	$MS_{\text{Residual(b-s)}}$	
Within–subjects				
Direct treatment effect	SS_{Treat}	1	MS_{Treat}	F_{Treat}
Period effect	SS_{Period}	1	MS_{Period}	F_{Period}
Residual (within–subjects)	$SS_{\text{Residual(w-s)}}$	n_1+n_2-2	$MS_{\text{Residual(w-s)}}$	
Total	SS_{Total}	$2(n_1+n_2)-1$		

TABLE 9.2. Analysis of variance table for 2 × 2 cross–over designs (Jones and Kenward, 1989, p. 31; Hills and Armitage, 1979).

MS	$\text{E}(MS)$
$MS_{\text{c-o}}$	$[(2n_1n_2)/(n_1+n_2)](\lambda_1-\lambda_2)^2 + (2\sigma_s^2+\sigma^2)$
$MS_{\text{Residual(b-s)}}$	$(2\sigma_s^2+\sigma^2)$
MS_{Treat}	$(2n_1n_2)/(n_1+n_2)[(\tau_1-\tau_2)-(\lambda_1-\lambda_2)/2]^2+\sigma^2$
MS_{Period}	$[(2n_1n_2)/(n_1+n_2)](\pi_1-\pi_2)^2+\sigma^2$
$MS_{\text{Residual(w-s)}}$	σ^2

TABLE 9.3. E(MS).

The F–statistics are built according to Table 9.3.

Under $H_0 : \lambda_1 = \lambda_2$, the expressions $MS_{\text{c-o}}$ and $MS_{\text{Residual(b-s)}}$ have the same expectations and we use the statistic $F_{c-o} = \text{MS}_{\text{c-o}}/MS_{\text{Residual(b-s)}}$.

Assuming $\lambda_1 = \lambda_2$ and $H_0 : \tau_1 = \tau_2$, MS_{Treat} and $MS_{\text{Residual(w-s)}}$ have equal expectations σ^2. Therefore, we get $F_{\text{Treat}} = MS_{\text{Treat}}/MS_{\text{Residual(w-s)}}$.

Testing for period effects does not depend upon the assumption that $\lambda_1 = \lambda_2$ holds. Since MS_{Period} and $MS_{\text{Residual(w-s)}}$ have expectation σ^2 considering $H_0 : \pi_1 = \pi_2$, the statistic $F_{\text{Period}|H_0} = MS_{\text{Period}}/MS_{\text{Residual(w-s)}}$ follows a central F–distribution.

Example 9.1. A clinical trial is used to compare the effect of two soporifics A and B. Response is the prolongation of sleep (in minutes).

Group 1		Patient					
Period	Treatment	1	2	3	4	$Y_{1j\cdot}$	$y_{1j\cdot}$
1	A	20	40	30	20	110	27.5
2	B	30	50	40	40	160	40.0
	$Y_{1\cdot k}$	50	90	70	60	$Y_{1\cdot\cdot} =$	270
						$Y_{1\cdot\cdot}/4 =$	67.50
						$y_{1\cdot\cdot} =$	33.75
Differences	d_{1k}	-10	-10	-10	-20	$d_{1.} =$	-12.5

Group 2		Patient					
Period	Treatment	1	2	3	4	$Y_{2j\cdot}$	$y_{2j\cdot}$
1	B	30	40	20	30	120	30.0
2	A	20	50	10	10	90	22.5
	$Y_{2\cdot k}$	50	90	30	40	$Y_{2\cdot\cdot} =$	210
						$Y_{2\cdot\cdot}/4 =$	52.50
						$y_{2\cdot\cdot} =$	26.25
Differences	d_{2k}	10	-10	10	20	$d_{2.} =$	7.5

t–Tests

$H_0 : \lambda_1 = \lambda_2$ (no carry–over effect):

$$
\begin{aligned}
(9.13)\quad \hat{\lambda}_d &= \frac{Y_{1\cdot\cdot}}{4} - \frac{Y_{2\cdot\cdot}}{4} = \frac{270}{4} - \frac{210}{4} = 15, \\
(9.18)\quad 3s_1^2 &= \sum_{k=1}^{4}(Y_{1\cdot k} - \frac{Y_{1\cdot\cdot}}{n_i})^2 \\
&= (50 - 67.5)^2 + \cdots + (60 - 67.5)^2 = 875, \\
(9.18)\quad 3s_2^2 &= (50 - 52.5)^2 + \cdots + (40 - 52.5)^2 = 2075, \\
(9.17)\quad s^2 &= \frac{2950}{6} = 491.67 = 22.17^2,
\end{aligned}
$$

$$(9.19)\quad T_\lambda = \frac{15}{22.17}\sqrt{\frac{16}{8}} = 0.96\,.$$

Decision. $T_\lambda = 0.96 < 1.94 = t_{6;0.90(\text{two--sided})} \Rightarrow \mathrm{H}_0 : \lambda_1 = \lambda_2$ is not rejected. Therefore, we can go on testing the main effects.

$\mathrm{H}_0 : \tau_1 = \tau_2$ (no treatment effect).

We compute

$$\begin{aligned}
d_{1\cdot} &= \frac{-10-10-10-20}{4} = -12.5,\\
d_{2\cdot} &= \frac{10-10+10+20}{4} = 7.5,\\
(9.24)\quad \hat{\tau}_d &= 1/2(d_{1\cdot} - d_{2\cdot}) = -10,\\
3s_{1D}^2 &= \sum (d_{1k} - d_{1\cdot})^2\\
&= (-10+12.5)^2 + \cdots + (-20+12.5)^2 = 75,\\
3s_{2D}^2 &= (10-7.5)^2 + \cdots + (20-7.5)^2 = 475,\\
(9.28)\quad s_D^2 &= \frac{75+475}{6} = 9.57^2,\\
(9.29)\quad T_\tau &= \frac{-10}{9.57/2}\sqrt{\frac{4\cdot 4}{4+4}} = -2.96\,.
\end{aligned}$$

Decision. With $t_{6;0.95(\text{two-sided})} = 2.45$ and $t_{6;0.95(\text{one-sided})} = 1.94$ the hypothesis $\mathrm{H}_0 : \tau_1 = \tau_2$ is rejected one–sided, as well as two–sided, which means a significant treatment effect.

$\mathrm{H}_0 : \pi_1 = \pi_2$ (no period effect).

We calculate

$$\begin{aligned}
(9.33)\quad \hat{\pi}_d &= 1/2(c_{1\cdot} - c_{2\cdot}) = 1/2(d_{1\cdot} + d_{2\cdot})\\
&= 1/2(-12.5 + 7.5) = -2.5,\\
(9.34)\quad T_\pi &= \frac{-2.5}{9.57/2}\sqrt{2} = -0.74\,.
\end{aligned}$$

$\mathrm{H}_0 : \pi_1 = \pi_2$ cannot be rejected (one– and two–sided).

From the analysis of variance we get the same $F_{1,6} = t_6^2$ statistics.

	SS	df	MS	F	
Carry-over	225	1	225.00	$0.92 = 0.96^2$	
Residual (b-s)	1475	6	245.83		
Treatment	400	1	400.00	$8.73 = 2.96^2$	*
Period	25	1	25.00	$0.55 = 0.74^2$	
Residual (w-s)	275	6	45.83		
Total	2400	15			

$$\begin{aligned}
SS_{\text{Total}} &= 16,800 - \frac{480^2}{2 \cdot 8} = 2400, \\
SS_{\text{c-o}} &= \frac{2 \cdot 4 \cdot 4}{4+4}(33.75 - 26.25)^2 = 225, \\
SS_{\text{Residual(b-s)}} &= 1/2(50^2 + 90^2 + \cdots + 40^2) - \left(\frac{270^2}{8} - \frac{210^2}{8}\right) \\
&= \frac{32,200}{2} - \frac{117,000}{8} \\
&= 16,100 - 14,625 = 1475, \\
SS_{\text{Treat}} &= \frac{4 \cdot 4}{2(4+4)}(27.5 - 40.0 - 30.0 + 22.5)^2 \\
&= (-20)^2 = 400, \\
SS_{\text{Period}} &= (27.5 - 40.0 + 30.0 - 22.5)^2 \\
&= (-5)^2 = 25, \\
SS_{\text{Residual(w-s)}} &= 16,800 - 1/4(110^2 + 160^2 + 120^2 + 90^2) - 1475 \\
&= 16,800 - 15,050 - 1475 = 275\,.
\end{aligned}$$

9.3.3 Residual Analysis and Plotting the Data

In addition to t– and F–tests, it is often desirable to represent the data using plots. We will now describe three methods of plotting the data which will allow us to detect patients being conspicuous by their response (outliers) and interactions such as carry–over effects.

Subject profile plots are produced for each group by plotting each subject's reponse against the period label. To summarize the data, we choose a **groups–by–periods** plot in which the group–by–period means are plotted against the period labels and points which refer to the same treatment are connected. Using Example 9.1 we get the following plots.

All patients in Group 1 show increasing response when they cross–over from treatment A to treatment B. In Group 2, the profile of patient 2 (uppermost line) exhibits a decreasing response while the other three profiles show an increasing tendency.

Figure 9.4 shows that in both periods treatment B leads to higher response than treatment A (difference of means $B - A : 30 - 27.5 = 2.5$

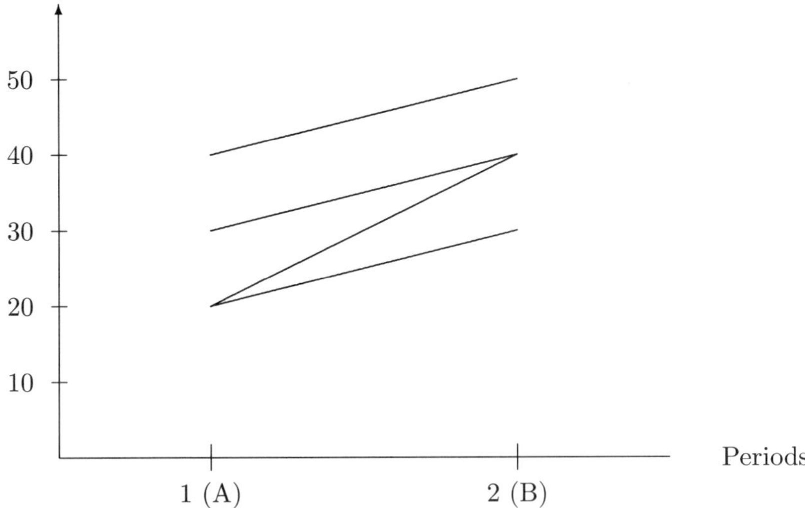

FIGURE 9.2. Individual profiles (Group 1).

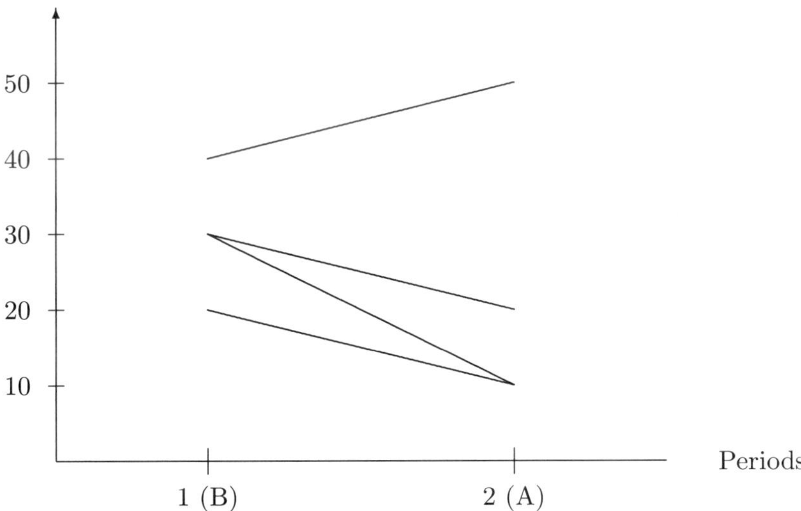

FIGURE 9.3. Individual profiles (Group 2).

for period 1; $40 - 22.5 = 17.5$ for period 2; so that $\hat{\tau}_d(B - A) = \frac{1}{2}(17.5 + 2.5) = 10 = -\hat{\tau}_d(A - B)$). It would also be possible to say that treatment A shows a slight carry–over effect that strengthens B (or B has a carry–over effect that reduces A). This difference in the treatment effects is not statistically significant according to the results we obtained from testing *treatment* × *period* interactions (= carry–over effect). Without

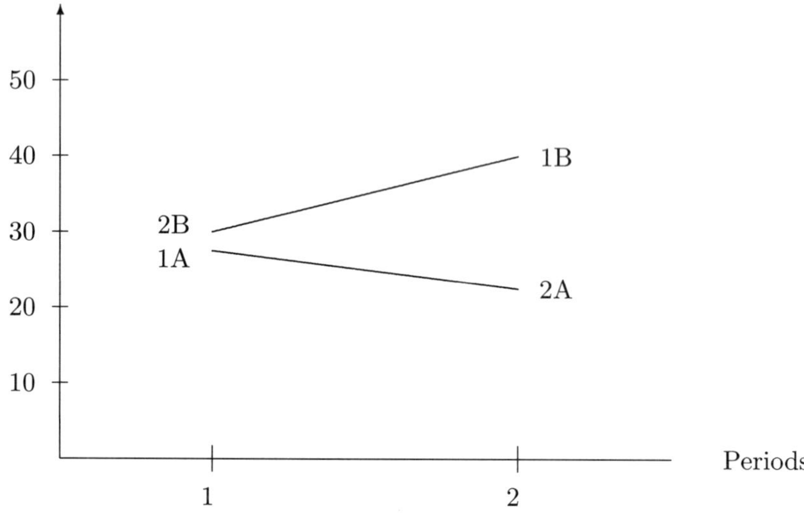

FIGURE 9.4. Group–period plots.

doubt, we can say that treatment A has lower response than treatment B in period 1 and this effect is even more pronounced in period 2. Another interesting view is given by the **differences–by–totals** plot where the subjects' differences d_{ik} are plotted against the total responses $Y_{i \cdot k}$. Plotting the pairs $(d_{ik}, Y_{i \cdot k})$ and connecting the outermost points of each group by a convex hull, we get a clear impression of carry–over and treatment effects. Since the statistic for carry–over is based on $\hat{\lambda}_d = (Y_{1..}/n_1 - Y_{2..}/n_2)$, the two hulls will be separated horizontally if $\lambda_d \neq 0$. In the same way the treatment effect based on $\hat{\tau}_d = \frac{1}{2}(d_{1.} - d_{2.})$ will manifest if the two hulls are being vertically separated.

Figure 9.5 shows vertically separated hulls indicating a treatment effect (which we already know is significant according to our tests). On the other hand, the hulls are not separated horizontally and indicate no carry–over effect.

Analysis of Residuals

The components $\hat{\epsilon}_{ijk}$ of $\hat{\epsilon} = (y - X\hat{\beta})$ are the estimated residuals which are used to check the model assumptions on the errors ϵ_{ijk}. Using appropriate plots, we can check for outliers and revise our assumptions on normal distribution and independency. The response values corresponding to unusually large standardized residuals are called outliers. A standardized residual is given by

$$r_{ijk} = \frac{\hat{\epsilon}_{ijk}}{\sqrt{\operatorname{Var}(\hat{\epsilon}_{ijk})}}, \tag{9.40}$$

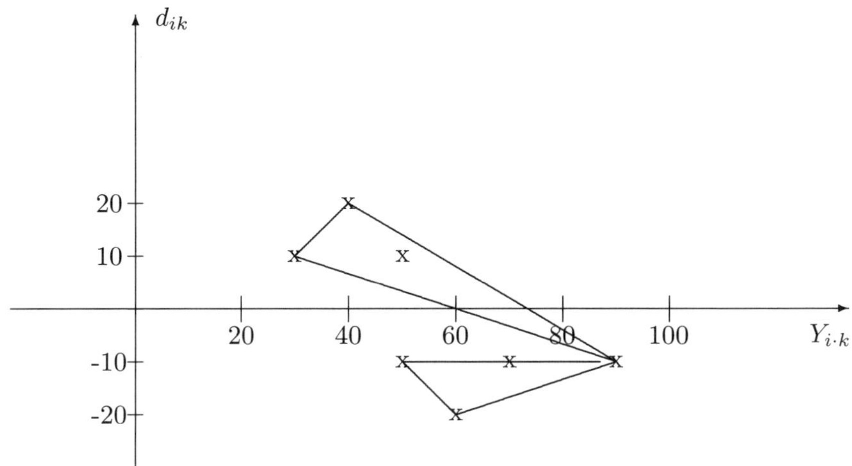

FIGURE 9.5. Difference–response–total plot to Example 9.1

with the variance factor σ^2 being estimated with $MS_{\text{Residual(w-s)}}$.

From the 2 × 2 cross–over, we get

$$\hat{y}_{ijk} = y_{i \cdot k} + y_{ij \cdot} - y_{i \cdot \cdot} \tag{9.41}$$

and

$$\text{Var}(\hat{\epsilon}_{ijk}) = \text{Var}(y_{ijk} - \hat{y}_{ijk}) = \frac{(n_i - 1)}{2n_i}\sigma^2 . \tag{9.42}$$

Then

$$r_{ijk} = \frac{\hat{\epsilon}_{ijk}}{\sqrt{MS_{\text{Residual(w-s)}}(n_i - 1)/2n_i}} . \tag{9.43}$$

This is the internally Studentized residual and follows a beta–distribution. We, however, regard r_{ijk} as $N(0,1)$–distributed and choose the two–sided quantile 2.00 (instead of $u_{0.975} = 1.96$) to test for y_{ijk} being an outlier.

Remark. If a more exact analysis is required, externally Studentized residuals should be used, since they follow the F–distribution (and can therefore be tested directly) and. additionally, are more sensitive to outliers (cf. Beckman and Trussel, 1974; Rao and Toutenburg, 1999, pp. 218–222).

Patient	Group 1 (AB)				Patient	Group 2 (BA)				
	y_{ijk}	$\hat{y}_{ijk}$	$\hat{\epsilon}_{ijk}$	r_{ijk}		y_{ijk}	$\hat{y}_{ijk}$	$\hat{\epsilon}_{ijk}$	r_{ijk}	
1	20	18.75	1.25	0.30	1	30	28.75	1.25	0.30	
2	40	38.75	1.25	0.30	2	40	48.75	−8.75	−2.10	*
3	30	28.75	1.25	0.30	3	20	18.75	1.25	0.30	
4	20	23.75	−3.75	−0.90	4	30	23.75	6.25	1.51	

Hence, patient 2 in Group 2 is an outlier.

Remark. If $\epsilon_{ijk} \sim N(0, \sigma^2)$ is not tenable, the response values are substituted by their ranks and the hypotheses are tested with the Wilcoxon–Mann–Whitney test (cf. Section 2.5) instead of using t–tests.

A detailed discussion of the various approaches for the 2×2 cross–over and, especially, their interpretations may be found in Jones and Kenward (1989, Chapter 2) and Ratkowsky et al. (1993).

Comment on the Procedure of Testing

Grizzle (1965) suggested testing carry–over effects on a quite high level of significance ($\alpha = 0.1$) first. If this leads to a significant result, then the test for treatment effects is to be based on the data of the first period only. If it is not significant, then the treatment effects are tested using the differences between the periods. This procedure has certain disadvantages. For example, Brown jr (1980) showed that this pretest is of minor efficiency in the case of real carry–over effects.

The hypothesis of no carry–over effect is very likely to be rejected even if there is a true carry–over effect. Hence, the biased test [(9.29)] (biased, because the carry–over was not recognized) is used to test for treatment differences. This test is conservative in the case of a true positive carry–over effect and therefore is insensitive to potential differences in treatments. On the other hand, this test will exceed the level of significance if there is a true negative carry–over effect (not very likely in practice, since this refers to a withdrawal effect).

If there is no true carry–over effect, the null hypothesis is very likely to be rejected erroneously ($\alpha = 0.1$) and the less efficient test using first–period data only is performed.

Brown jr (1980) concluded that this method is not very useful in testing treatment effects as it depends upon the outcome of the pretest.

Further comments are given in the Section 9.3.4.

9.3.4 Alternative Parametrizations in 2 × 2 Cross–Over

Model (9.1) was introduced as the classical approach and is labeled parametrization No. 1 using the notation of Ratkovsky, Evans and Alldredge (1993). A more general parametrization of the 2×2 cross–over design, that includes a sequence effect γ_i, is given by

$$y_{ijk} = \mu + \gamma_i + s_{ik} + \pi_j + \tau_t + \lambda_r + \epsilon_{ijk} \,, \tag{9.44}$$

with $i, j, t, r = 1, 2$ and $k = 1, \ldots, n_i$. The data are summarized in a table containing the cell means $y_{ij\cdot}$, i.e.,

		Period 1	Period 2
Sequence	1	$y_{11\cdot}$	$y_{12\cdot}$
	2	$y_{21\cdot}$	$y_{22\cdot}$

Here Sequence 1 indicates that the treatments are given in the order (AB) and Sequence 2 has the (BA) order. Using the common restrictions

$$\gamma_2 = -\gamma_1, \quad \pi_2 = -\pi_1, \quad \tau_2 = -\tau_1, \quad \lambda_2 = -\lambda_1\,, \tag{9.45}$$

and writing $\gamma_1 = \gamma$, $\pi_1 = \pi$, $\tau_1 = \tau$, $\lambda_1 = \lambda$ for brevity, we get the following equations representing the four expectations:

$$\begin{aligned}
\mu_{11} &= \mu + \gamma + \pi + \tau \\
\mu_{12} &= \mu + \gamma - \pi - \tau + \lambda, \\
\mu_{21} &= \mu - \gamma + \pi - \tau, \\
\mu_{22} &= \mu - \gamma - \pi + \tau - \lambda\,.
\end{aligned}$$

In matrix notation this is equivalent to

$$\begin{pmatrix} \mu_{11} \\ \mu_{12} \\ \mu_{21} \\ \mu_{22} \end{pmatrix} = X\beta = \begin{pmatrix} 1 & 1 & 1 & 1 & 0 \\ 1 & 1 & -1 & -1 & 1 \\ 1 & -1 & 1 & -1 & 0 \\ 1 & -1 & -1 & 1 & -1 \end{pmatrix} \begin{pmatrix} \mu \\ \gamma \\ \pi \\ \tau \\ \lambda \end{pmatrix}. \tag{9.46}$$

This (4×5)–matrix X has rank 4, so that β is only estimable if one of the parameters is removed. Various parametrizations are possible depending on which of the five parameters is removed and then confounded with the remaining ones.

Parametrization No. 1

The classical approach ignores the sequence parameter. Its expectations may therefore be represented as a submodel of (9.46) by dropping the second column of X:

$$X_1\beta_1 = \begin{pmatrix} 1 & 1 & 1 & 0 \\ 1 & -1 & -1 & 1 \\ 1 & 1 & -1 & 0 \\ 1 & -1 & 1 & -1 \end{pmatrix} \begin{pmatrix} \mu \\ \pi \\ \tau \\ \lambda \end{pmatrix}. \tag{9.47}$$

From this we get

$$X_1'X_1 = \begin{pmatrix} E & \mathbf{0} \\ \mathbf{0} & H \end{pmatrix},$$

where

$$E = 4I_2\,,$$

$$
\begin{aligned}
H &= \begin{pmatrix} 4 & -2 \\ -2 & 2 \end{pmatrix}, \quad |X_1'X_1| = 64\,, \\
(X_1'X_1)^{-1} &= \begin{pmatrix} E^{-1} & \mathbf{0} \\ \mathbf{0} & H^{-1} \end{pmatrix} \quad \text{[cf. Theorem A.4]},
\end{aligned}
$$

with $E^{-1} = \frac{1}{4}I_2$, $H^{-1} = \begin{pmatrix} 1/2 & 1/2 \\ 1/2 & 1 \end{pmatrix}$. The least squares estimate of β_1 is

$$
\hat{\beta}_1 = \begin{pmatrix} \hat{\mu} \\ \hat{\pi} \\ \hat{\tau} \\ \hat{\lambda} \end{pmatrix} = (X_1'X_1)^{-1}X_1' \begin{pmatrix} y_{11\cdot} \\ y_{12\cdot} \\ y_{21\cdot} \\ y_{22\cdot} \end{pmatrix}. \tag{9.48}
$$

We calculate

$$
\begin{aligned}
X_1' \begin{pmatrix} y_{11\cdot} \\ y_{12\cdot} \\ y_{21\cdot} \\ y_{22\cdot} \end{pmatrix} &= \begin{pmatrix} 1 & 1 & 1 & 1 \\ 1 & -1 & 1 & -1 \\ 1 & -1 & -1 & 1 \\ 0 & 1 & 0 & -1 \end{pmatrix} \begin{pmatrix} y_{11\cdot} \\ y_{12\cdot} \\ y_{21\cdot} \\ y_{22\cdot} \end{pmatrix} \\
&= \begin{pmatrix} y_{11\cdot} + y_{12\cdot} + y_{21\cdot} + y_{22\cdot} \\ y_{11\cdot} - y_{12\cdot} + y_{21\cdot} - y_{22\cdot} \\ y_{11\cdot} - y_{12\cdot} - y_{21\cdot} + y_{22\cdot} \\ y_{12\cdot} - y_{22\cdot} \end{pmatrix}. \qquad (9.49)
\end{aligned}
$$

Therefore, the least squares estimation gives

$$
\begin{aligned}
\hat{\beta}_1 &= \begin{pmatrix} \hat{\mu} \\ \hat{\pi} \\ \hat{\tau} \\ \hat{\lambda} \end{pmatrix} = (X_1'X_1)^{-1}X_1' \begin{pmatrix} y_{11\cdot} \\ y_{12\cdot} \\ y_{21\cdot} \\ y_{22\cdot} \end{pmatrix} \qquad (9.50) \\
&= \begin{pmatrix} (y_{11\cdot} + y_{12\cdot} + y_{21\cdot} + y_{22\cdot})/4 \\ (y_{11\cdot} - y_{12\cdot} + y_{21\cdot} - y_{22\cdot})/4 \\ (y_{11\cdot} - y_{21\cdot})/2 \\ (y_{11\cdot} + y_{12\cdot} - y_{21\cdot} - y_{22\cdot})/2 \end{pmatrix}, \qquad (9.51)
\end{aligned}
$$

from which we get the following results:

$$
\begin{aligned}
\hat{\mu} &= y_{\cdots}\,, && (9.52) \\
\hat{\pi} &= (y_{\cdot 1\cdot} - y_{\cdot 2\cdot})/2 = (c_{1\cdot} - c_{2\cdot})/4 = \frac{\hat{\pi}_d}{2} \quad \text{[cf. (9.33)]}\,, && (9.53) \\
\hat{\tau} &= (y_{11\cdot} - y_{21\cdot})/2 = \frac{\hat{\tau}_{d/\lambda_d}}{2} \quad \text{[cf. (8.38)]}\,, && (9.54) \\
\hat{\lambda} &= y_{1\cdot\cdot} - y_{2\cdot\cdot} = \hat{\lambda}_d/2 \quad \text{[cf. (9.13)]}\,. && (9.55)
\end{aligned}
$$

The estimators $\hat{\tau}$ and $\hat{\lambda}$ are correlated

$$
\mathrm{V}(\hat{\tau}, \hat{\lambda}) = \sigma^2 H^{-1} = \sigma^2 \begin{pmatrix} 1/2 & 1/2 \\ 1/2 & 1 \end{pmatrix},
$$

with $\rho(\hat{\tau}, \hat{\lambda}) = \frac{1}{2}/(\frac{1}{2} \cdot 1)^{1/2} = 0.707$. The estimation of $\hat{\tau}$ is always twice as accurate as the estimation of $\hat{\lambda}$, although $\hat{\tau}$ uses data of the first period only and is confounded with the difference between the two groups (sequences).

Remark. In fact, parametrization No. 1 is a three–factorial design with the main effects π, τ, and λ and with τ and λ being correlated. On the other hand, the classical approach uses the split–plot model in addition to parametrization (9.1). So it is obvious that we will get different results depending on which parametrization we use. We will demonstrate this in Example 8.2, where the four different parametrizations are applied to our data set of Example 9.1.

Parametrization No. 1(a)

If the test for no carry–over effect does not reject $H_0 : \lambda = 0$ against $H_1 : \lambda \neq 0$ using the test statistic $F_{1,df} = \hat{\lambda}_d^2 / \operatorname{Var}(\hat{\lambda}_d)$ (cf. (9.19)), our model can be reduced to the following

$$\tilde{X}_1 \tilde{\beta}_1 = \begin{pmatrix} 1 & 1 & 1 \\ 1 & -1 & -1 \\ 1 & 1 & -1 \\ 1 & -1 & 1 \end{pmatrix} \begin{pmatrix} \mu \\ \pi \\ \tau \end{pmatrix} \tag{9.56}$$

and we get the same estimators $\hat{\mu}$ [(9.52)] and $\hat{\pi}$ [(9.53)] as before, but now the estimator $\hat{\tau}$ is based on both periods' data

$$\begin{aligned} \hat{\tau} &= (y_{11\cdot} - y_{12\cdot} - y_{21\cdot} + y_{22\cdot})/4 \\ &= (d_{1\cdot} - d_{2\cdot})/4 \\ &= \hat{\tau}_d/2 \qquad [\text{cf. } (9.24)] \,. \end{aligned} \tag{9.57}$$

The results of parametrizations No. 1 and No. 1(a) are the same as the classical univariate results we obtained in Section 9.3.1 (except for a factor of 1/2 in $\hat{\pi}$, $\hat{\tau}$, and $\hat{\lambda}$). But, in addition, the dependency in estimating the treatment effect τ and the carry–over effect λ is explained.

Parametrization No. 2

In the first parametrization, the interaction *treatment* × *period* was aliased with the carry–over effect λ. We now want to parametrize this interaction directly. Dropping the sequence effect, the model of expectations is as follows:

$$\mathrm{E}(y_{ijk}) = \mu_{ij} = \mu + \pi_j + \tau_t + (\tau\pi)_{tj} \,. \tag{9.58}$$

Using effect coding, the codings of the interaction effects are just the products of the involved main effects. Therefore, we get

$$\begin{pmatrix} \mu_{11} \\ \mu_{12} \\ \mu_{21} \\ \mu_{22} \end{pmatrix} = X_2\beta_2 = \begin{pmatrix} 1 & 1 & 1 & 1 \\ 1 & -1 & -1 & 1 \\ 1 & 1 & -1 & -1 \\ 1 & -1 & 1 & -1 \end{pmatrix} \begin{pmatrix} \mu \\ \pi \\ \tau \\ (\pi\tau) \end{pmatrix} . \tag{9.59}$$

Since the column vectors are orthogonal, we easily get $(X_2'X_2) = 4I_4$ and, therefore, the parameter estimations are independent (cf. Section 6.3). The estimators are

$$\hat{\beta}_2 = \begin{pmatrix} \hat{\mu} \\ \hat{\pi} \\ \hat{\tau} \\ \widehat{(\pi\tau)} \end{pmatrix} = \begin{pmatrix} y_{\cdot\cdot\cdot} \\ \hat{\pi}_d/2 \\ (y_{11\cdot} - y_{12\cdot} - y_{21\cdot} + y_{22\cdot})/4 \\ (y_{11\cdot} + y_{12\cdot} - y_{21\cdot} - y_{22\cdot})/4 \end{pmatrix} . \tag{9.60}$$

Note that $\hat{\mu}$ and $\hat{\pi}$ are as in the first parametrization. The estimator $\hat{\tau}$ in (9.60) and the estimator $\hat{\tau}$ [(9.57)] in the reduced model (9.56) coincide. The estimator $\widehat{(\pi\tau)}$ may be written as (cf. (9.55))

$$\widehat{(\pi\tau)} = (y_{1\cdot\cdot} - y_{2\cdot\cdot})/2 = \hat{\lambda}_d/4 = \hat{\lambda}/2 , \tag{9.61}$$

and coincides—except for a factor of 1/2—with the estimation of the carry–over effect (9.55) in model (9.47). So it is obvious that there is an intrinsic aliasing between the two parameters λ and $(\pi\tau)$.

Parametrization No. 3

Supposing that a carry–over effect λ or, alternatively, an interaction effect $(\pi\tau)$ may be excluded from analysis, the model now contains only main effects. We already discussed model (9.56). Now we want to introduce the sequence effect γ as an additional main effect. With $\gamma_2 = -\gamma_1 = \gamma$, we get

$$\begin{pmatrix} \mu_{11} \\ \mu_{12} \\ \mu_{21} \\ \mu_{22} \end{pmatrix} = X_3\beta_3 = \begin{pmatrix} 1 & 1 & 1 & 1 \\ 1 & 1 & -1 & -1 \\ 1 & -1 & 1 & -1 \\ 1 & -1 & -1 & 1 \end{pmatrix} \begin{pmatrix} \mu \\ \gamma \\ \pi \\ \tau \end{pmatrix} , \tag{9.62}$$

$$\begin{aligned} (X_3'X_3) &= 4I_4 , \\ \hat{\beta}_3 &= \begin{pmatrix} \hat{\mu} \\ \hat{\gamma} \\ \hat{\pi} \\ \hat{\tau} \end{pmatrix} = 1/4X_3' \begin{pmatrix} y_{11\cdot} \\ y_{12\cdot} \\ y_{21\cdot} \\ y_{22\cdot} \end{pmatrix} \\ &= \begin{pmatrix} y_{\cdot\cdot\cdot} \\ (y_{11\cdot} + y_{12\cdot} - y_{21\cdot} - y_{22\cdot})/4 \\ (y_{11\cdot} - y_{12\cdot} + y_{21\cdot} - y_{22\cdot})/4 \\ (y_{11\cdot} - y_{12\cdot} - y_{21\cdot} + y_{22\cdot})/4 \end{pmatrix} \end{aligned} \tag{9.63}$$

$$= \begin{pmatrix} y_{\cdot\cdot\cdot} \\ (y_{1\cdot\cdot} - y_{2\cdot\cdot})/2 \\ (y_{\cdot 1\cdot} - y_{\cdot 2\cdot})/2 \\ \hat{\tau}_d/2 \end{pmatrix} . \tag{9.64}$$

The sequence effect γ is estimated using the contrast in the total response of both groups (AB) and (BA) and we see the equivalence $\hat{\gamma} = \widehat{(\pi\tau)} = \hat{\lambda}_d/4$. The period effect π is estimated using the contrast in the total response of both periods and coincides with $\hat{\pi}$ in parametrizations No. 1 (cf. (9.53)) and No. 2 (cf. (9.60)) The estimation of $\hat{\tau}$ is the same as $\hat{\tau}$ [(9.57)] in the reduced model [(9.56)] and $\hat{\tau}$ (cf. (9.60)) in parametrization No. 2. Furthermore, the estimates in $\hat{\beta}_3$ are independent, so that, e.g., $H_0 : \tau = 0$ can be tested not depending on $\gamma = \lambda_d = 0$ (in contrast to parametrization No. 1).

Parametrization No. 4

Here, the main–effects treatment and sequence and their interaction are represented in a two–factorial model (cf. Milliken and Johnson, 1984)

$$E(y_{ijk}) = \mu_{ij} = \mu + \gamma_i + \tau_t + (\gamma\tau)_{it} , \tag{9.65}$$

i.e.,

$$\begin{pmatrix} \mu_{11} \\ \mu_{12} \\ \mu_{21} \\ \mu_{22} \end{pmatrix} = X_4\beta_4 = \begin{pmatrix} 1 & 1 & 1 & 1 \\ 1 & 1 & -1 & -1 \\ 1 & -1 & -1 & 1 \\ 1 & -1 & 1 & -1 \end{pmatrix} \begin{pmatrix} \mu \\ \gamma \\ \tau \\ (\gamma\tau) \end{pmatrix} . \tag{9.66}$$

Since $X_4'X_4 = 4I_4$, the components of β_4 can be estimated independently as

$$\hat{\beta}_4 = \begin{pmatrix} \hat{\mu} \\ \hat{\gamma} \\ \hat{\tau} \\ \widehat{(\gamma\tau)} \end{pmatrix} = \begin{pmatrix} y_{\cdot\cdot\cdot} \\ (y_{1\cdot\cdot} - y_{2\cdot\cdot})/2 \\ \hat{\tau}_d/2 \\ (y_{\cdot 1\cdot} - y_{\cdot 2\cdot})/2 \end{pmatrix} . \tag{9.67}$$

Values of $\hat{\gamma}$ in parametrizations 3 and 4 are the same. Analogously, the values of $\hat{\tau}$ coincide in parametrizations 2, 3, and 4 whereas the interaction effect *sequence* × *treatment* $\widehat{(\gamma\tau)}$ refers to the period effect π in parametrizations 1, 2, and 3.

Remark. From the various parametrizations we get the following results:

(i) In parametrization No. 1, the estimators of τ and λ are correlated. In contrast to the arguments of Ratkovsky et al. (1993, pp. 89–90), the values of $E(MS)$ given in Table 9.3 are valid. $E(MS_{\text{Treat}})$ depends on $(\lambda_1 - \lambda_2) = 2\lambda$ so that testing for $H_0 : \tau = 0$ may be done either using a central t–test if $\lambda = 0$ or using a noncentral t–test if λ is known. A difficulty in the argument is certainly that τ and λ are correlated but not represented in the

	Parametrization					
	Classical	No. 1	No. 1(a)	No. 2	No. 3	No. 4
$\hat{\mu}$	$y_{...}$	$y_{...}$	$y_{...}$	$y_{...}$	$y_{...}$	$y_{...}$
$\hat{\gamma}$	—	—	—	—	$\hat{\lambda}_d/4$	$\hat{\lambda}_d/4$
$\hat{\pi}$	$\hat{\pi}_d = \frac{1}{2}(d_{1\cdot} + d_{2\cdot})$	$\hat{\pi}_d/2$	$\hat{\pi}_d/2$	$\hat{\pi}_d/2$	$\hat{\pi}_d/2$	—
$\hat{\tau}$	$\hat{\tau}_{d/\lambda_d} = y_{11\cdot} - y_{21\cdot}$	$\hat{\tau}_{d/\lambda_d}/2$	$\hat{\tau}_d/2$	$\hat{\tau}_d/2$	$\hat{\tau}_d/2$	$\hat{\tau}_d/2$
$\hat{\lambda}$	$\hat{\lambda}_d = 2(y_{1\cdot\cdot} - y_{2\cdot\cdot})$	$\hat{\lambda}_d/2$	—	—	—	—
$\widehat{(\tau\pi)}$	—	—	—	$\hat{\lambda}_d/4$	—	—
$\widehat{(\gamma\tau)}$	—	—	—	—	—	$\hat{\pi}_d/2$

TABLE 9.4. Estimators using six different parametrizations.

two–factorial hierarchy "main effect A, main effect B, and the interaction A ×B ".

(ii) In parametrization No. 2, the carry–over effect is indirectly represented as the alias effect of the interaction $(\pi\tau)$. We can use the common hierarchical test procedure, as in a two–factorial model with interaction, since the design is orthogonal. If the interaction is not significant the estimators of the main effects remain the same (in contrast to parametrization No. 1).

(iii) The analysis of data of a 2×2 cross–over design is done in two steps. In the first step, we test for carry–over using one of the parametrizations in which the carry–over effect is separable from the main effects, e.g., parametrization No. 3, and it is not surprising that the result will be the same as if we had used the sequence effect.

We consider the following experiment. We take two groups of subjects and apply the treatments in both groups in the same order (AB). If there is an interaction effect (maybe a significant carry–over effect in the classical approach of Grizzle or a significant sequence effect in parametrization No. 3 of Ratkovsky et al. (1993)), then we conclude that the two groups must consist of two different classes of subjects. There is either a difference per se between the subjects of the two groups, or treatment A shows different persistencies in the two groups. Since the latter is not very likely, it is clear that the subjects of both groups are different in their reactions. And therefore it is a sequence effect but not a carry–over effect. We try to avoid this confusion by randomizing the subjects.

Regarding the classical (AB)/(BA) design, there are two ways to interpret a significant interaction effect:

(a) either it is a true sequence effect as a result of insufficient randomization; or

(b) it is a true carry–over effect; this will be the case if there is no doubt about the randomization process.

Since the actual set of data may hardly be used to decide whether the randomization succeeded or failed, it is necessary to make a distinction before we analyze our data.

If the subjects have not been randomized, the possibility of a sequence effect should attract our attention. The F–statistics given for parametrization No. 3 are valid and do not depend upon whether the sequence effect is significant or not, because there is no natural link between a sequence effect and a treatment or a period effect.

Given the case that we did randomize our subjects, then there is no need to consider a sequence effect and, therefore, the interaction effect is to be regarded as a result of carry–over.

The carry–over effect was introduced as the persisting effect of a treatment during the subsequent period of treatment and is represented as an additive component in our model. Therefore, it is evident that the F–statistics for treatment and period effects, derived from parametrization No. 3 or from the classical approach, are no longer valid if the carry–over effect is significant.

To continue our examination, we choose one of the following alternatives:

(a) We try to test treatment effects using the data of the first period only. This might be difficult because the sample size is likely to be too small for a parallel group design. Of course we then omit the sequence effect from our analysis (because we have only this first period).

(b) A significant carry–over effect may also be regarded as a suffcient indicator that the two treatments differ in their effects. At least we can state that the two treatments have different persistencies and therefore they are not equal.

It can be assumed that Ratkovsky et al. (1993) regarded the analysis of variance tables to be read simultaneously and that the given F–statistics for carry–over, treatment, and period effects are always valid. But they are not. This is only the case if the carry–over effect was proven to be non-significant. Only with a nonsignificant carry–over effect are the expressions for treatment and period effect valid. If the label *carry–over* is replaced by the label *sequence effect*, then the ordering of tests is not important and the table is no longer misleading to readers who only just glance at the literature. The interpretation of the results must reflect this relabeling, too. Then, of course, we do not know anything about the carry–over effect which, mostly, is of more importance than a sequence effect. Using the classical approach, the analysis of variance table is valid.

(iv) From a theoretical point of view, it is interesting to extend the 2 × 2 design by three additional periods: a baseline period and two washout periods (Figure 9.6). This approach was suggested by Ratkovsky et al. (1993, Chapter 3.6), but is rarely applied because of the amount of effort.

		Period				
		1	2	3	4	5
Sequence	1	Baseline	A	Washout	B	Washout
	2	Baseline	B	Washout	A	Washout

FIGURE 9.6. Extended 2×2 cross–over design.

The linear model then contains two additional period effects and carry–over effects of first and second order. The main advantages are that all parameters are estimable, there is no dependence between treatment and carry–over effects, and we get reduced variance.

(v) Possible modifications of the 2×2 cross–over are $2 \times n$ designs like

		Period		
		1	2	3
Sequence	1	A	B	B
	2	B	A	A

		Period		
		1	2	3
Sequence	1	A	B	A
	2	B	A	B

or $n \times 2$ designs like

		Period	
		1	2
Sequence	1	A	B
	2	B	A
	3	A	A
	4	B	B

Adding baseline and washout periods may further improve these designs. A comprehensive treatment of this subject matter is given by Ratkovsky et al. (1993, Chapter 4).

Example 9.2. (Continuation of Example 9.1). The data of Example 9.1 are now analyzed with parametrizations 2, 3, and 4 using the SAS procedure GLM. In the split–plot model (classical approach) the following analysis of variance table was obtained for the data of Example 9.1 (cf. Section 9.3.2).

Source	SS	df	MS	F
Carry-over	225	1	225.00	0.92
Residual (b–s)	1475	6	245.83	
Treatment	400	1	400.00	8.73 *
Period	25	1	25.00	0.55
Residual (w–s)	275	6	45.83	
Total	2400	15		

The treatment effect was found to be significant.

Parametrization No. 1 does not take the split–plot character of the design (limited randomization) into account. Therefore, the two sums of squares

SS (b-s) and SS (w-s) are added for $SS_{\text{Residual}} = 1750$. Table 9.5 shows this result in the upper part (SS type I). The lower part (SS type II) gives the result using first–period data only, because the model contains the carry–over effect. All other parametrizations do not contain carry–over effects and the important sums of squares are found in the lower part (SS type II) of the table. We note that the following F–values coincide

Carry-over (resp., Sequence):	$F = 0.92$ (classical, No. 3, No. 4).
Treatment:	$F = 8.73$ (classical, No. 3, No. 4).
Period:	$F = 0.55$ (classical, No. 3).

The different parametrizations were calculated using the following small SAS programs.

```
proc glm;
class seq subj period treat carry;
model y = period treat carry /solution ss1 ss2;
title "Parametrization 1";
run;

proc glm;
class seq subj period treat carry;
model y = period treat treat(period) /solution ss1 ss2;
title "Parametrization 2";
run;

proc glm;
class seq subj period treat carry;
model y = seq subj(seq) period treat /solution ss1 ss2;
random subj(seq);
title "Parametrization 3";
run;

proc glm;
class seq subj period treat carry;
model y = seq subj(seq) treat seq(treat) /solution ss1 ss2;
random subj(seq);
title "Parametrization 4";
run;

data Example 8.2;
input subj seq period treat $ carry $ y @@;

cards;
1 1 1 a 0 20  1 1 2 b a 30
2 1 1 a 0 40  2 1 2 b a 50
```

```
3 1 1 a 0 30  3 1 2 b a 40
4 1 1 a 0 20  4 1 2 b a 40
1 2 1 b 0 30  1 2 2 a b 20
2 2 1 b 0 40  2 2 2 a b 50
3 2 1 b 0 20  3 2 2 a b 10
4 2 1 b 0 30  4 2 2 a b 10
run;
```

Parametrization No. 1				
Source	**df**	SS **type I**	MS	F
Periods	1	25.00	25.00	0.17
Treatments	1	400.00	400.00	2.74
Carry–over	1	225.00	225.00	1.54
Residual	12	1750.00	145.83	
	df	SS **type I**	MS	F
Treatments	1	12.50	12.50	0.09
Carry–over	1	225.00	225.00	1.54
Residual	12	1750.00	145.83	

Parametrization No. 2				
Source	*df*	SS **type I**	MS	F
Periods (P)	1	25.00	25.00	0.17
Treatments (T)	1	400.00	400.00	2.74
$P \times T$	1	225.00	225.00	1.54
Residual	12	1750.00	145.83	
	df	SS **type I**	MS	F
Treatments	1	400.00	400.00	2.74
$P \times T$	1	225.00	225.00	1.54
Residual	12	1750.00	145.83	

Parametrization No. 3				
Source	*df*	SS **type I**	MS	F
between–subjects				
Sequence	1	225.00	225.00	0.92
Residual	6	1475.00	245.83	
	df	SS **type I**	MS	F
within–subjects				
Periods	1	25.00	25.00	0.55
Treatments	1	400.00	400.00	8.73
Residual	6	275.00	45.83	

Parametrization No. 4				
Source	*df*	SS **type I**	MS	F
between–subjects				
Sequence	1	225.00	225.00	0.92
Residual	6	1475.00	245.83	
	df	SS **type I**	MS	F
within–subjects				
Treatments	1	400.00	400.00	8.73
Seq × treat.	1	25.00	25.00	0.55
Residual	6	275.00	45.83	

Table 9.5. GLM results of the four parametrizations.

9.3.5 Cross–Over Analysis Using Rank Tests

Known rank tests from other designs with two independent groups offer a nonparametric approach to analyze a cross–over trial. These tests are based on the model given in Table 8.1. However, the random effects may now follow any continuous distribution with expectation zero. The advantage of using nonparametric methods is that there is no need to assume a normal distribution. According to the difficulties mentioned above, we now assume either that there are no carry–over effects or that they are at least ignorable.

Rank Test on Treatment Differences

The null hypothesis that there are no differences between the two treatments implies that the period differences follow the same distribution

$$\mathrm{H}_0 : F_{d1}(d_{1k}) = F_{d2}(d_{2k}), \quad k = 1, \ldots, n_i \,. \tag{9.68}$$

Here F_{d1} and F_{d2} are continuous distributions with identical variances. Then the null hypothesis of no treatment effects may be tested using the Wilcoxon, Mann, and Whitney statistic (cf. Section 2.5 and Koch, 1972).

We calculate the period differences d_{1k} and d_{2k} (cf. (9.20)). These $N = (n_1 + n_2)$ differences then get ranks from 1 to N. Let

$$r_{ik}^{\phi} = [\text{rank of } d_{ik} \text{ in} \{d_{11}, \ldots, d_{1n_1}, d_{21}, \ldots, d_{2n_2}\}], \tag{9.69}$$

with $i = 1, 2,\ k = 1, \ldots, n_i$. In the case of ties we use mean ranks. For both groups (AB) and (BA), we get the sum of ranks R_1 (resp., R_2) which are used to build the test statistics U_1 (resp., U_2) [(2.38) (resp., (2.39))].

Rank Tests on Period Differences

The null hypothesis of no period differences is

$$\mathrm{H}_0 : F_{c1}(c_{1k}) = F_{c2}(c_{2k}), \quad k = 1, \ldots, n_i \,, \tag{9.70}$$

and so the distribution of the difference $c_{1k} = (y_{11k} - y_{12k})$ equals the distribution of the difference $c_{2k} = (y_{22k} - y_{21k})$. Again, F_{ci} $(i = 1, 2)$ are continuous distributions with equal variances.

The null hypothesis H_0 is then tested in the same way as H_1 in (9.68) using the Wilcoxon, Mann, and Whitney test.

9.4 2 × 2 Cross–Over and Categorical (Binary) Response

9.4.1 Introduction

In many applications, the response is categorical. This is the case in pretests when only a rough overview of possible relations is needed. Often a continuous response is not available. For example, recovering from a mental illness cannot be measured on a continuous scale, categories like "worse, constant, better" would be sufficient.

Example: Patients suffering from depression participate in two treatments A and B. Their response to each treatment is coded binary with 1 for improvement and 0 : no change. The profile of each subject is then one of the pairs $(0,0), (0,1), (1,0)$, and $(1,1)$. To summarize the data we count how often each pair occurs.

Group	(0, 0)	(0, 1)	(1, 0)	(1, 1)	Total
1 (AB)	n_{11}	n_{12}	n_{13}	n_{14}	$n_{1.}$
2 (BA)	n_{21}	n_{22}	n_{23}	n_{24}	$n_{2.}$
Total	$n_{.1}$	$n_{.2}$	$n_{.3}$	$n_{.4}$	$n_{..}$

TABLE 9.6. 2 × 2 Cross–over with binary response.

Contingency Tables and Odds Ratio

The two columns in the middle of this 2 × 4 contingency table may indicate a treatment effect. Assuming no period effect and under the null hypothesis H_0 : "no treatment effect", the two responses $n_A = (n_{13}+n_{22})$ for treatment A and $n_B = (n_{12}+n_{23})$ for treatment B have equal probabilities and follow the same binomial distribution n_A (resp., n_B) $\sim B(n_{.2}+n_{.3}; \frac{1}{2})$.

The odds ratio

$$\widehat{OR} = \frac{n_{12}n_{23}}{n_{22}n_{13}} \tag{9.71}$$

may also indicate a treatment effect.

Testing for carry–over effects is done—similar to the test statistic T_λ [(9.19)], which is based mainly on $\hat{\lambda} = Y_{1..}/n_1 - Y_{2..}/n_2$—by comparing the differences in the total response values for the profiles (0, 0) and (1, 1). Instead of differences, we choose the odds ratio

$$\widehat{OR} = \frac{n_{11}n_{24}}{n_{14}n_{21}} \tag{9.72}$$

which should equal 1 under H_0 : "no treatment × period effect". Using the 2 × 2 table

A	B
C	D

, the odds ratio is $\widehat{OR} = AD/BC$ with the following asymptotic distribution

$$\frac{(\ln(\widehat{OR}))^2}{\hat{\sigma}^2_{\ln(\widehat{OR})}} \sim \chi^2_1, \tag{9.73}$$

where

$$\hat{\sigma}^2_{\ln(\widehat{OR})} = \left(\frac{1}{A} + \frac{1}{B} + \frac{1}{C} + \frac{1}{D}\right) \tag{9.74}$$

(cf. Agresti, 1990). We can now test the significance of the two odds ratios (9.71) and (9.72).

McNemar's Test

Application of this test assumes no period effects. Only values of subjects are considered, who show a preference for one of the treatments. These subjects have either a (0, 1) or (1, 0) response profile.

There are $n_P = (n_{.2} + n_{.3})$ subjects who show a preference for one of the treatments. $n_A = (n_{13} + n_{22})$ prefer treatment A and $n_B = (n_{12} + n_{23})$ prefer treatment B.

Under the null hypothesis of no treatment effects, n_A (resp., n_B) are binomial distributed $B(n_P; \frac{1}{2})$. The hypothesis is tested using the following statistic (cf. Jones and Kenward, 1989, p. 93):

$$\chi^2_{MN} = \frac{(n_A - n_B)^2}{n_P}, \tag{9.75}$$

where χ^2_{MN} is asymptotically χ^2–distributed with one degree of freedom under the null hypothesis.

Mainland–Gart Test

Based on a logistic model, Gart (1969) proposed a test for treatment differences, which is equivalent to Fisher's exact test using the following 2×2 contingency table:

Group	(0, 1)	(1, 0)	Total
1 (AB)	n_{12}	n_{13}	$n_{12} + n_{13} = m_1$
2 (BA)	n_{22}	n_{23}	$n_{22} + n_{23} = m_2$
Total	$n_{.2}$	$n_{.3}$	$m_.$

This test is described in Jones and Kenward (1989, p. 113). Asymptotically, the hypothesis of no treatment differences may be tested using one of the common tests for 2×2 contingency tables, e.g., the χ^2–statistic

$$\chi^2 = \frac{m_.(n_{12}n_{23} - n_{13}n_{22})^2}{m_1 m_2 n_{.2} n_{.3}}. \tag{9.76}$$

This statistic follows a χ^2_1–distribution under the null hypothesis. This test and the test with $\ln(\widehat{OR})$ (cf. (9.73)) coincide.

Prescott Test

The above tests have one thing in common: subjects showing no preference for one of the treatments are discarded from the analysis. Prescott (1981) includes these subjects in his test, by means of the marginal sums $n_{1.}$ and $n_{2.}$. The following 2×3 table will be used:

Group	(0, 1)	(0, 0) or (1, 1)	(1, 0)	Total
1 (AB)	n_{12}	$n_{11}+n_{14}$	n_{13}	$n_{1.}$
2 (BA)	n_{22}	$n_{21}+n_{24}$	n_{23}	$n_{2.}$
Total	$n_{.2}$	$n_{.1}+n_{.4}$	$n_{.3}$	$n_{..}$

We first consider the difference between the first and second response. Depending on the response profile (1, 0), (0, 0), (1, 1), or (0,1), this difference takes the values +1, 0, or -1.

Assuming that treatment A is better, we expect the first group (AB) to have a higher mean difference than the second group (BA). The mean difference of the response values in Group 1 (AB) is

$$\frac{1}{n_{1.}}\sum_{k=1}^{n_{1.}}(y_{12k}-y_{11k}) = \frac{n_{12}-n_{13}}{n_{1.}} = -d_{1.} \tag{9.77}$$

and in Group 2 (BA)

$$\frac{1}{n_{2.}}\sum_{k=1}^{n_{2.}}(y_{22k}-y_{21k}) = \frac{n_{22}-n_{23}}{n_{2.}} = -d_{2.}\,. \tag{9.78}$$

Prescott's test statistic (cf. Jones and Kenward, 1989, p. 100) under the null hypothesis H_0 : *no direct treatment effect* (i.e., $E(d_{1.}-d_{2.})=0$) is

$$\chi^2(P) = [(n_{12}-n_{13})n_{..} - (n_{.2}-n_{.3})n_{1.}]^2/V \tag{9.79}$$

with

$$V = n_{1.}n_{2.}[(n_{.2}+n_{.3})n_{..} - (n_{.2}-n_{.3})^2]/n_{..}\, d. \tag{9.80}$$

Asymptotically, $\chi^2(P)$ follows the χ^2_1–distribution under H_0.

This test, however, has the disadvantage that only the hypothesis of no–treatment differences can be tested. As a uniform approach for testing all important hypotheses one could choose the approach of Grizzle, Starmer and Koch (1969).

Remark. Another, and often more efficient, method of analysis is given by loglinear models, especially models with uncorrelated two–dimensional binary response. These were examined thoroughly in recent years (cf. Chapter 7).

Example 9.3. A comparison between a placebo A and a new drug B for treating depression might have shown the following results (1 : improvement, 0 : no improvement):

Group	(0, 0)	(0, 1)	(1, 0)	(1, 1)	Total
1 (AB)	5	14	3	6	28
2 (BA)	10	7	18	10	45
Total	15	21	21	16	73

We check for H_0 : "treatment $\times$ period–effect $=$ 0" (i.e., no carry–over effect) using the odds ratio [(9.72)]

$$\widehat{OR} = \frac{5 \cdot 10}{6 \cdot 10} = 0.83 \quad \text{and} \quad \ln(\widehat{OR}) = -0.1823\,.$$

We get

$$\hat{\sigma}^2_{\ln \widehat{OR}} = 1/5 + 1/10 + 1/6 + 1/10 = 0.5667$$

and

$$\frac{(\ln(\widehat{OR}))^2}{\hat{\sigma}^2_{\ln \widehat{OR}}} = 0.06 < 3.84 = \chi^2_{1;0.95}\,,$$

so that H_0 cannot be rejected. In the same way, we get for the odds ratio [(9.71)]

$$\begin{aligned} \widehat{OR} &= \frac{14 \cdot 18}{7 \cdot 3} = 12\,, \quad \ln(\widehat{OR}) = 2.48\,, \\ \hat{\sigma}^2_{\ln \widehat{OR}} &= (1/14 + 1/18 + 1/7 + 1/3) = 0.60\,, \\ \frac{(\ln(\widehat{OR}))^2}{\hat{\sigma}^2_{\ln OR}} &= 10.24 > 3.84\,, \end{aligned}$$

and this test rejects H_0 : no–treatment effect. Since there is no carry–over effect, we can use McNemar's test

$$\begin{aligned} \chi^2_{MN} &= \frac{((3+7) - (14+18))^2}{21+21} \\ &= \frac{22^2}{42} = 11.53 > 3.84\,, \end{aligned}$$

which gives the same result. For Prescott's test we get

$$\begin{aligned} V &= 28 \cdot 45[(21+21) \cdot 73]/73 \\ &= 28 \cdot 45 \cdot 42 = 52920\,, \\ \chi^2(P) &= [(14-3) \cdot 73 - (21-21) \cdot 28]^2/V \\ &= (11 \cdot 73)^2/V = 12.28 > 3.84\,, \end{aligned}$$

and H_0 : *no–treatment effect* is also rejected.

9.4.2 Loglinear and Logit Models

In Table 9.6, we see that Group 1 (AB) and Group 2 (BA) are represented by four distinct categorical response profiles (0, 0), (0, 1), (1, 0), and (1, 1).

We assume that each row (and, therefore, each variable) is an independent observation from a multinomial distribution $M(n_{i.}; \pi_{i1}, \pi_{i2}, \pi_{i3}, \pi_{i4})$ $(i = 1, 2)$. Using the appropriate parametrizations and logit or loglinear models, we try to define a bivariate binary variable (Y_1, Y_2), which represents the four profiles and their probabilities according to the model of the 2×2 cross–over design. There are various approaches available for handling this.

Bivariate Logistic Model

Generally, Y_1 and Y_2 denote a pair of correlated binary variables. We first want to follow the approach of Jones and Kenward (1989, p. 106) who use the following bivariate logistic model according to Cox (1970) and McCullagh and Nelder (1989):

$$P(Y_1 = y_1, Y_2 = y_2) = \exp(\beta_0 + \beta_1 y_1 + \beta_2 y_2 + \beta_{12} y_1 y_2)\,, \tag{9.81}$$

with the binary response being coded with +1 and −1 in contrast to the former coding. This coding relates to the transformation $Z_i = (2Y_i - 1)$ $(i = 1, 2)$, which was used by Cox (1972a). The parameter β_0 is a scaling constant to assure us that the four probabilities sum to 1. This depends upon the other three parameters. The parameter β_{12} measures the correlation between the two variables. β_1 and β_2 depict the main effects.

The four possible observations are now put into (9.81) in order to get the joint distribution

$$\begin{aligned}
\ln P(Y_1 = 1, Y_2 = 1) &= \beta_0 + \beta_1 + \beta_2 + \beta_{12}\,,\\
\ln P(Y_1 = 1, Y_2 = -1) &= \beta_0 + \beta_1 - \beta_2 - \beta_{12}\,,\\
\ln P(Y_1 = -1, Y_2 = 1) &= \beta_0 - \beta_1 + \beta_2 - \beta_{12}\,,\\
\ln P(Y_1 = -1, Y_2 = -1) &= \beta_0 - \beta_1 - \beta_2 + \beta_{12}\,.
\end{aligned}$$

Bayes' theorem gives

$$\begin{aligned}
\frac{P(Y_1 = 1 \mid Y_2 = 1)}{P(Y_1 = -1 \mid Y_2 = 1)} &= \frac{P(Y_1 = 1, Y_2 = 1)/P(Y_2 = 1)}{P(Y_1 = -1, Y_2 = 1)/P(Y_2 = 1)}\\
&= \frac{\exp(\beta_0 + \beta_1 + \beta_2 + \beta_{12})}{\exp(\beta_0 - \beta_1 + \beta_2 - \beta_{12})}\\
&= \exp 2(\beta_1 + \beta_{12})\,.
\end{aligned}$$

We now get the logits

$$\begin{aligned}
\text{logit}[P(Y_1 = 1 \mid Y_2 = 1)] &= \ln \frac{P(Y_1 = 1 \mid Y_2 = 1)}{P(Y_1 = -1 \mid Y_2 = 1)} = 2(\beta_1 + \beta_{12})\,,\\
\text{logit}[P(Y_1 = 1 \mid Y_2 = -1)] &= \ln \frac{P(Y_1 = 1 \mid Y_2 = -1)}{P(Y_1 = -1 \mid Y_2 = -1)} = 2(\beta_1 - \beta_{12})\,,
\end{aligned}$$

and the conditional log–odds ratio

$$\text{logit}[P(Y_1 = 1 \mid Y_2 = 1)] - \text{logit}[P(Y_1 = 1 \mid Y_2 = -1)] = 4\beta_{12}\,, \tag{9.82}$$

i.e.,

$$\frac{P(Y_1 = 1 \mid Y_2 = 1)P(Y_1 = -1 \mid Y_2 = -1)}{P(Y_1 = -1 \mid Y_2 = 1)P(Y_1 = 1 \mid Y_2 = -1)} = \exp(4\beta_{12})\,. \tag{9.83}$$

This refers to the relation

$$\frac{m_{11}m_{22}}{m_{12}m_{21}} = \exp(4\lambda_{11}^{XY})$$

between the odds ratio and interaction parameter in the loglinear model (cf. Chapter 7). In the same way we get, for $i, j = 1, 2$ $(i \neq j)$,

$$\operatorname{logit}[P(Y_i = 1 \mid Y_j = y_j)] = 2(\beta_i + y_j\beta_{12})\,. \tag{9.84}$$

For a specific subject of one of the groups (AB or BA), a treatment effect exists if the response is either (1, -1) or (-1, 1). From the log–odds ratio for this combination we get

$$\operatorname{logit}[P(Y_1 = 1 \mid Y_2 = -1)] - \operatorname{logit}[P(Y_2 = 1 \mid Y_1 = -1)] = 2(\beta_1 - \beta_2)\,. \tag{9.85}$$

This is an indicator for a treatment effect within a group.

Assuming the same parameter β_{12} for both groups AB and BA, the following expression is an indicator for a period effect:

$$\begin{aligned}\operatorname{logit}[P(Y_i^{AB} = 1 \mid Y_j^{AB} = y_j)] &- \operatorname{logit}[P(Y_i^{BA} = 1 \mid Y_j^{BA} = y_j)] \\ &= 2(\beta_i^{AB} - \beta_i^{BA})\,. \end{aligned} \tag{9.86}$$

This relation is directly derived from (9.84) with an additional indexing for the two groups AB and BA. The assumption $\beta_{12}^{AB} = \beta_{12}^{BA}$ is important, i.e., identical interaction in both groups.

Logit Model of Jones and Kenward for the Classical Approach

Let y_{ijk} denote the binary response of subject k of group i in period j $(i = 1, 2,\ j = 1, 2,\ k = 1, \ldots, n_i)$. Again we choose the coding as in Table 9.6 with $y_{ijk} = 1$ denoting success and $y_{ijk} = 0$ for failure. Using logit–links we want to reparametrize the model according to Table 9.1 for the bivariate binary response (y_{i1k}, y_{i2k})

$$\operatorname{logit}(\pi_{ij}) = \ln\left(\frac{\pi_{ij}}{1 - \pi_{ij}}\right) = X\beta\,, \tag{9.87}$$

where X denotes the design matrix using effect coding for the two groups and the two periods (cf. (9.47))

$$X = \begin{pmatrix} 1 & 1 & 1 & 0 \\ 1 & -1 & -1 & 1 \\ 1 & 1 & -1 & 0 \\ 1 & -1 & 1 & -1 \end{pmatrix} \tag{9.88}$$

and $\beta = (\mu\ \pi\ \tau\ \lambda)'$ is the parameter vector using the reparametrization conditions

$$\pi = -\pi_1 = \pi_2, \quad \tau = -\tau_1 = \tau_2, \quad \lambda = -\lambda_1 = \lambda_2\,. \tag{9.89}$$

(i) For both of the two groups and the two periods of the 2×2 cross–over with binary response, the logits show the following relation to the model in Table 9.1:

$$\begin{aligned}
\text{logit } P(y_{11k}=1) &= \ln\left(\frac{P(y_{11k}=1)}{P(y_{11k}=0)}\right) = \ln\left(\frac{P(y_{11k}=1)}{1-P(y_{11k}=1)}\right)\\
&= \mu - \pi - \tau\,,\\
\text{logit } P(y_{12k}=1) &= \mu + \pi + \tau - \lambda\,,\\
\text{logit } P(y_{21k}=1) &= \mu - \pi + \tau\,,\\
\text{logit } P(y_{22k}=1) &= \mu + \pi - \tau + \lambda\,.
\end{aligned}$$

We get, for example,

$$P(y_{11k}=1) = \frac{\exp(\mu-\pi-\tau)}{1+\exp(\mu-\pi-\tau)}\,,$$

and

$$P(y_{11k}=0) = \frac{1}{1+\exp(\mu-\pi-\tau)}\,.$$

(ii) To start with, we assume that the two observations of each subject in period 1 and 2 are independent. The joint probabilities π_{ij}:

Group	(0, 0)	(0, 1)	(1, 0)	(1, 1)
1 (AB)	π_{11}	π_{12}	π_{13}	π_{14}
2 (BA)	π_{21}	π_{22}	π_{23}	π_{24}

are the product of the probabilities defined above. We introduce a normalizing constant for the case of nonresponse (0, 0) to adjust the other probabilities. The constant c_1 is chosen so that the four probabilities sum to 1 (in Group 2 this constant is c_2):

$$\left.\begin{aligned}
\pi_{11} &= P(y_{11k}=0,\ y_{12k}=0) = \exp(c_1)\,,\\
\pi_{12} &= P(y_{11k}=0,\ y_{12k}=1) = \exp(c_1+\mu+\pi+\tau-\lambda)\,,\\
\pi_{13} &= P(y_{11k}=1,\ y_{12k}=0) = \exp(c_1+\mu-\pi-\tau)\,,\\
\pi_{14} &= P(y_{11k}=1,\ y_{12k}=1) = \exp(c_1+2\mu-\lambda)\,.
\end{aligned}\right\}\ . \tag{9.90}$$

Then

$$\exp(c_1)[1+\exp(\mu+\pi+\tau-\lambda)+\exp(\mu-\pi-\tau)+\exp(2\mu-\lambda)] = 1\,,$$

will give $\exp(c_1)$.

(iii) Jones and Kenward (1989, p. 109) chose the following parametrization to represent the interaction referring to β_{12}. They introduce a new parameter σ to denote the mean interaction of both groups (i.e., $\sigma = (\beta_{12}^{AB} + \beta_{12}^{BA})/2$) and another parameter ϕ that measures the interaction difference ($\phi = (\beta_{12}^{AB} - \beta_{12}^{BA})/2$). In the logarithms of the probabilities, the model for the two groups is as follows (Table 9.7).

	Group 1	Group 2
(0, 0)	$\ln \pi_{11} = c_1 + \sigma + \phi$	$\ln \pi_{21} = c_2 + \sigma - \phi$
(0, 1)	$\ln \pi_{12} = c_1 + \mu + \pi + \tau - \lambda - \sigma - \phi$	$\ln \pi_{22} = c_2 + \mu + \pi - \tau + \lambda - \sigma + \phi$
(1, 0)	$\ln \pi_{13} = c_1 + \mu - \pi - \tau - \sigma - \phi$	$\ln \pi_{23} = c_2 + \mu - \pi + \tau - \sigma + \phi$
(1, 1)	$\ln \pi_{14} = c_1 + 2\mu - \lambda + \sigma + \phi$	$\ln \pi_{24} = c_2 + 2\mu + \lambda + \sigma - \phi$

TABLE 9.7. Logit model of Jones and Kenward.

The values of c_i and μ are somewhat difficult to interpret. The nuisance parameters σ and ϕ represent the dependency in the structure of the subjects of the two groups.

From Table 9.7 we obtain the following relations, among the parameters π, τ, and λ, and the odds ratios

$$\begin{aligned} \pi &= 1/4(\ln \pi_{12} + \ln \pi_{22} - \ln \pi_{13} - \ln \pi_{23}) \\ &= 1/4 \ln \left(\frac{\pi_{12}\pi_{22}}{\pi_{13}\pi_{23}} \right), \end{aligned} \tag{9.91}$$

$$\lambda = 1/2 \ln \left(\frac{\pi_{11}\pi_{24}}{\pi_{14}\pi_{21}} \right) \quad \text{(cf. (9.72))}, \tag{9.92}$$

$$\tau = 1/4 \ln \left(\frac{\pi_{12}\pi_{23}}{\pi_{13}\pi_{22}} \right) \quad \text{(cf. (9.71))}. \tag{9.93}$$

The null hypotheses $H_0 : \pi = 0$, $H_0 : \tau = 0$, $H_0 : \lambda = 0$ can be tested using likelihood ratio tests in the appropriate 2×2 table.

For π: $\begin{array}{|cc|} \hat{m}_{12} & \hat{m}_{13} \\ \hat{m}_{23} & \hat{m}_{22} \end{array}$

(second and third column of Table 9.6, where the second row BA is reversed to get the same order AB as the first row).

For λ: $\begin{array}{|cc|} \hat{m}_{11} & \hat{m}_{14} \\ \hat{m}_{21} & \hat{m}_{24} \end{array}$

(first and last column of Table 9.6).

For τ: $\begin{array}{|cc|} \hat{m}_{12} & \hat{m}_{13} \\ \hat{m}_{22} & \hat{m}_{23} \end{array}$

(second and third column of Table 9.6).

The estimators $\hat{m}_{ij}$ are taken from the appropriate loglinear model, corresponding to the hypothesis.

Remark. The modeling [(9.90)] of the probabilities π_{1j} of the first group (and analogously for the second group) is based on the assumption that the response of each subject is independent over the two periods. Since this assumption cannot be justified in a cross–over design, this within–subject dependency has to be introduced afterward using the parameters σ and ϕ. This guarantees the formal independency of $\ln(\hat{\pi}_{ij})$ and therefore the applicability of loglinear models. This approach, however, is critically examined by Ratkovsky et al. (1993, p. 300), who suggest the following alternative approach.

Sequence	$(1,1)$	$(1,0)$	$(0,1)$	$(0,0)$
1 (AB)	$m_{11} =$ $n_{1\cdot} P_A P_{B\mid A}$	$m_{12} =$ $n_{1\cdot} P_A(1 - P_{B\mid A})$	$m_{13} =$ $n_{1\cdot}(1 - P_A)P_{B\mid \bar{A}}$	$m_{14} =$ $n_{1\cdot}(1 - P_A)(1 - P_{B\mid \bar{A}})$
2 (BA)	$m_{21} =$ $n_{2\cdot} P_B P_{A\mid B}$	$m_{22} =$ $n_{2\cdot} P_B(1 - P_{A\mid B})$	$m_{23} =$ $n_{2\cdot}(1 - P_B)P_{A\mid \bar{B}}$	$m_{24} =$ $n_{2\cdot}(1 - P_B)(1 - P_{A\mid \bar{B}})$

TABLE 9.8. Expectations m_{ij} of the 2×4 contingency table.

Logit Model of Ratkovsky, Evans, and Alldredge (1993)

The cross–over experiment aims to analyze the relationship between the transitions $(0,1)$ and $(1,0)$ and the constant response profiles $(0,0)$ and $(1,1)$. We define the following probabilities:

(i) unconditional:

$$\begin{aligned} P_A &: \quad P(\text{success of } A), \\ P_B &: \quad P(\text{success of } B); \end{aligned}$$

(ii) conditional (conditioned on the preceding treatment):

$$\begin{aligned} P_{A\mid B} &: \quad P(\text{success of } A \mid \text{success of } B), \\ P_{A\mid \bar{B}} &: \quad P(\text{success of } B \mid \text{no success of } B); \end{aligned}$$

and, analogously, $P_{B\mid A}$ and $P_{B\mid \bar{A}}$. The contingency tables of the two groups then have the expectations m_{ij} of cell counts illustrated in Table 9.8. The proper table of observed response values is as follows (Table 9.6 transformed and using N_{ij} instead of n_{ij}):

$(1,1)$	$(1,0)$	$(0,1)$	$(0,0)$	
N_{11}	N_{12}	N_{13}	N_{14}	$n_{1\cdot}$
N_{21}	N_{22}	N_{23}	N_{24}	$n_{2\cdot}$

The loglinear model for sequence i (group, $i = 1, 2$) can then be written as follows

$$\begin{pmatrix} \ln(N_{i1}) \\ \ln(N_{i2}) \\ \ln(N_{i3}) \\ \ln(N_{i4}) \end{pmatrix} = X\beta_i + \epsilon_i \,, \tag{9.94}$$

where the vector of errors ϵ_i is such that $p \lim \epsilon_i = \mathbf{0}$. From Table 9.8, we get the design matrix for the two groups

$$X = \begin{pmatrix} 1 & 1 & 0 & 1 & 0 & 0 & 0 \\ 1 & 1 & 0 & 0 & 1 & 0 & 0 \\ 1 & 0 & 1 & 0 & 0 & 1 & 0 \\ 1 & 0 & 1 & 0 & 0 & 0 & 1 \end{pmatrix} \tag{9.95}$$

and the vectors of the parameters

$$\beta_1 = \begin{pmatrix} \ln(n_{1\cdot}) \\ \ln(P_A) \\ \ln(1-P_A) \\ \ln(P_{B|A}) \\ \ln(1-P_{B|A}) \\ \ln(P_{B|\bar{A}}) \\ \ln(1-P_{B|\bar{A}}) \end{pmatrix}, \quad \beta_2 = \begin{pmatrix} \ln(n_{2\cdot}) \\ \ln(P_B) \\ \ln(1-P_B) \\ \ln(P_{A|B}) \\ \ln(1-P_{A|B}) \\ \ln(P_{A|\bar{B}}) \\ \ln(1-P_{A|\bar{B}}) \end{pmatrix}. \tag{9.96}$$

Under the usual assumption of independent multinomial distributions $M(n_{i\cdot}, \pi_{i1}, \pi_{i2}, \pi_{i3}, \pi_{i4})$, we get the estimators of the parameters $\hat{\beta}_i$ by solving iteratively the likelihood equations using the Newton–Raphson procedure. An algorithm to solve this problem is given in Ratkovsky et al. (1993, Appendix 7.A). The authors mention that the implementation is quite difficult.

Taking advantage of the structure of Table 9.8, this difficulty can be avoided by transforming the problem (equivalently reducing it) to a standard problem that can be solved with standard software.

From Table 9.8, we get the following relations

$$\left.\begin{array}{rcl} (m_{11}+m_{12})/n_{1\cdot} & = & P_A P_{B|A} + P_A(1-P_{B|A}) = P_A\,, \\ (m_{13}+m_{14})/n_{1\cdot} & = & (1-P_A)\,, \end{array}\right\} \tag{9.97}$$

$\Rightarrow$

$$\begin{aligned} \ln(m_{11}+m_{12}) - \ln(m_{13}+m_{14}) &= \ln(P_A) - \ln(1-P_A) \\ &= \text{logit}(P_A)\,, &(9.98) \\ \ln(m_{11}) - \ln(m_{12}) &= \text{logit}(P_{B|A})\,, &(9.99) \\ \ln(m_{13}) - \ln(m_{14}) &= \text{logit}(P_{B|\bar{A}})\,, &(9.100) \end{aligned}$$

and, analogously,

$$\begin{aligned} \ln(m_{21}+m_{22}) - \ln(m_{23}+m_{24}) &= \text{logit}(P_B)\,, &(9.101) \\ \ln(m_{21}) - \ln(m_{22}) &= \text{logit}(P_{A|B})\,, &(9.102) \\ \ln(m_{23}) - \ln(m_{24}) &= \text{logit}(P_{A|\bar{B}})\,. &(9.103) \end{aligned}$$

The logits, as a measure for the various effects in the 2×2 cross–over, are developed using one of the four parametrizations given in Section 9.3.4 for the main effects and the additional effects for the within–subject correlation. To avoid overparametrization, we drop the carry–over effect λ which is represented as an alias effect anyhow, using the other interaction effects (cf. Section 9.3.4). The model of Ratkovsky et al. (1993, REA model), has the following structure.

REA Model

$$\begin{aligned}
\text{logit}(P_A) &= \mu + \gamma_1 + \pi_1 + \tau_1\,, \\
\text{logit}(P_{B|A}) &= \mu + \gamma_1 + \pi_2 + \tau_2 + \alpha_{11}\,, \\
\text{logit}(P_{B|\bar{A}}) &= \mu + \gamma_1 + \pi_2 + \tau_2 + \alpha_{10}\,, \\
\text{logit}(P_B) &= \mu + \gamma_2 + \pi_1 + \tau_2\,, \\
\text{logit}(P_{A|B}) &= \mu + \gamma_2 + \pi_2 + \tau_1 + \alpha_{21}\,, \\
\text{logit}(P_{A|\bar{B}}) &= \mu + \gamma_2 + \pi_2 + \tau_1 + \alpha_{20}\,.
\end{aligned}$$

μ, γ_i, π_i, and τ_i denote the usual parameters for the four main effects overall–mean, sequence, period, and treatment. The new parameters have the meaning:

α_{i1} is the association effect averaged over subjects of sequence i if period 1 treatment was a success; and

α_{i0} is the analog for failure .

Using the sum–to–zero conventions: for the within–subject effects, we

$\gamma = \gamma_1 = -\gamma_2$ sequence effect,

$\pi = \pi_1 = -\pi_2$ period effect,

$\tau = \tau_1 = -\tau_2$ treatment effect,

and

$\alpha_{i0} = -\alpha_{i1}$ association effect,

can represent the REA model for the two sequences as follows

$$\begin{pmatrix} \text{logit}(P_A) \\ \text{logit}(P_{B|A}) \\ \text{logit}(P_{B|\bar{A}}) \\ \text{logit}(P_B) \\ \text{logit}(P_{A|B}) \\ \text{logit}(P_{A|\bar{B}}) \end{pmatrix} = \begin{pmatrix} 1 & 1 & 1 & 1 & 0 & 0 \\ 1 & 1 & -1 & -1 & 1 & 0 \\ 1 & 1 & -1 & -1 & -1 & 0 \\ 1 & -1 & 1 & -1 & 0 & 0 \\ 1 & -1 & -1 & 1 & 0 & 1 \\ 1 & -1 & -1 & 1 & 0 & -1 \end{pmatrix} \begin{pmatrix} \mu \\ \gamma \\ \pi \\ \tau \\ \alpha_{11} \\ \alpha_{21} \end{pmatrix},$$

$$\mathbf{Logit} = X_s \beta_s\,. \tag{9.104}$$

Replacing the estimators of the logits on the left side by the relations (9.98)–(9.103), and replacing the expected counts m_{ij} by the obseverd counts N_{ij}, we get the following solutions

$$\hat{\beta}_s = X_s^{-1} \widehat{\mathbf{Logit}}\,, \tag{9.105}$$

i.e.,

$$\begin{pmatrix} \hat{\mu} \\ \hat{\gamma} \\ \hat{\pi} \\ \hat{\tau} \\ \hat{\alpha}_{11} \\ \hat{\alpha}_{21} \end{pmatrix} = \frac{1}{8} \begin{pmatrix} 2 & 1 & 1 & 2 & 1 & 1 \\ 2 & 1 & 1 & -2 & -1 & -1 \\ 2 & -1 & -1 & 2 & -1 & -1 \\ 2 & -1 & -1 & -2 & 1 & 1 \\ 0 & 4 & -4 & 0 & 0 & 0 \\ 0 & 0 & 0 & 0 & 4 & -4 \end{pmatrix} \begin{pmatrix} \widehat{\text{Logit}}(P_A) \\ \widehat{\text{Logit}}(P_{B|A}) \\ \widehat{\text{Logit}}(P_{B|\bar{A}}) \\ \widehat{\text{Logit}}(P_B) \\ \widehat{\text{Logit}}(P_{A|B}) \\ \widehat{\text{Logit}}(P_{A|\bar{B}}) \end{pmatrix} . \tag{9.106}$$

With (9.98)–(9.103) (m_{ij} replaced by N_{ij}) we get

$$\widehat{\text{Logit}}(P_A) = \ln\left(\frac{N_{11}+N_{12}}{N_{13}+N_{14}}\right), \tag{9.107}$$

$$\widehat{\text{Logit}}(P_{B|A}) = \ln\left(\frac{N_{11}}{N_{12}}\right), \tag{9.108}$$

$$\widehat{\text{Logit}}(P_{B|\bar{A}}) = \ln\left(\frac{N_{13}}{N_{14}}\right), \tag{9.109}$$

$$\widehat{\text{Logit}}(P_B) = \ln\left(\frac{N_{21}+N_{22}}{N_{23}+N_{24}}\right), \tag{9.110}$$

$$\widehat{\text{Logit}}(P_{A|B}) = \ln\left(\frac{N_{21}}{N_{22}}\right), \tag{9.111}$$

$$\widehat{\text{Logit}}(P_{A|\bar{B}}) = \ln\left(\frac{N_{23}}{N_{24}}\right). \tag{9.112}$$

In the saturated model (9.104), $\text{rank}(X_s) = 6$, so that the parameter estimates $\hat{\beta}_s$ can be derived directly from the estimated logits from (9.105). The parameter estimates in the saturated model (9.104) are

$$\begin{aligned} \hat{\alpha}_{11} &= 1/2[\widehat{\text{Logit}}(P_{B|A}) - \widehat{\text{Logit}}(P_{B|\bar{A}})] \\ &= 1/2 \ln\left(\frac{N_{11}N_{14}}{N_{12}N_{13}}\right), \end{aligned} \tag{9.113}$$

$$\hat{\alpha}_{21} = 1/2 \ln\left(\frac{N_{21}N_{24}}{N_{22}N_{23}}\right). \tag{9.114}$$

Then $\exp(2\hat{\alpha}_{11})$, for example, is the odds ratio in the 2×2 table of the AB sequence

	1	0
1	N_{11}	N_{12}
0	N_{13}	N_{14}

.

$$8\hat{\mu} = \ln\left(\frac{N_{11}+N_{12}}{N_{13}+N_{14}}\right)^2 \left(\frac{N_{11}N_{13}}{N_{12}N_{14}}\right)$$

$$
\begin{aligned}
& + \ln \left(\frac{N_{21} + N_{22}}{N_{23} + N_{24}} \right)^2 \left(\frac{N_{21} N_{23}}{N_{22} N_{24}} \right) && (9.115) \\
&= a_1 + a_2 \,, \\
8\hat{\gamma} &= a_1 - a_2 \,, && (9.116) \\
8\hat{\pi} &= \ln \left(\frac{N_{11} + N_{12}}{N_{13} + N_{14}} \right)^2 \left(\frac{N_{12} N_{14}}{N_{11} N_{13}} \right) \\
& + \ln \left(\frac{N_{21} + N_{22}}{N_{23} + N_{24}} \right)^2 \left(\frac{N_{22} N_{24}}{N_{21} N_{23}} \right) && (9.117) \\
&= a_3 + a_4 \,, \\
8\hat{\tau} &= a_3 - a_4 \,. && (9.118)
\end{aligned}
$$

The covariance matrix of $\hat{\beta}_s$ is derived considering the covariance structure of the logits in the weighted least–squares estimation (cf. Chapter 7). For the saturated model or submodels (after dropping nonsignificant parameters), the parameter estimates are given by standard software.

Ratkovsky et al. (1993, p. 310) give an example of the application of the procedure SAS PROC CATMOD. The file has to be organized according to (9.107)–(9.112) and Table 9.9 ($Y = 1$: success, $Y = 2$: failure).

Count	Y		Count in Example 9.3
$N_{11} + N_{12}$	1	$\widehat{\text{Logit}}(P_A)$	16
$N_{13} + N_{14}$	2		14
N_{11}	1	$\widehat{\text{Logit}}(P_{B\mid A})$	14
N_{12}	2		2
N_{13}	1	$\widehat{\text{Logit}}(P_{B\mid \bar{A}})$	15
N_{14}	2		9
$N_{21} + N_{22}$	1	$\widehat{\text{Logit}}(P_B)$	23
$N_{23} + N_{24}$	2		15
N_{21}	1	$\widehat{\text{Logit}}(P_{A\mid B})$	18
N_{22}	2		5
N_{23}	1	$\widehat{\text{Logit}}(P_{A\mid \bar{B}})$	4
N_{24}	2		11

TABLE 9.9. Data organization in SAS PROC CATMOD (saturated model).

Example 9.4. The efficiency of a treatment (B) compared to a placebo (A) for a mental illness is examined using a 2×2 cross–over experiment (Table 9.10). Coding is 1 : improvement and 0 : no improvement.

We first check for H_0 : "treatment × period effect = 0" using the odds ratio [(9.72)]

$$
\begin{aligned}
\widehat{OR} &= \frac{9 \cdot 18}{14 \cdot 11} = 1.05 \,, \\
\ln(\widehat{OR}) &= 0.05 \,,
\end{aligned}
$$

Group	(0, 0)	(0, 1)	(1, 0)	(1, 1)	Total
1 (AB)	9	5	2	14	30
2 (BA)	11	4	5	18	38
Total	20	9	7	32	68

TABLE 9.10. Response profiles in a 2 × 2 cross–over with binary response.

$$\begin{aligned} \hat{\sigma}^2_{\ln \widehat{OR}} &= 1/9 + 1/18 + 1/14 + 1/11 = 0.33\,, \\ \frac{(\ln(\widehat{OR}))^2}{\hat{\sigma}^2_{\ln \widehat{OR}}} &= 0.01 < 3.84\,, \end{aligned}$$

so that H_0 is not rejected. Now we can run the tests for treatment effects.

The Mainland–Gart test uses the following 2×2 table:

Group	(0, 1)	(1, 0)	Total
1 (AB)	5	2	7
2 (BA)	4	5	9
Total	9	7	16

Pearson's χ^2_1–statistic with

$$\chi^2 = \frac{16(5\cdot 5 - 2\cdot 4)^2}{9\cdot 7\cdot 7\cdot 9} = 1.17 < 3.84 = \chi^2_{1;0.95}$$

does not indicate a treatment effect (p–value: 0.2804).

The Mainland–Gart test and Fisher's exact test do test the same hypothesis but the p–values are different. Fisher's exact test (cf. Section 2.6.2) gives, for the three tables,

$$\begin{array}{|cc|}\hline 2 & 5 \\ 5 & 4 \\ \hline\end{array} \quad \begin{array}{|cc|}\hline 1 & 6 \\ 6 & 3 \\ \hline\end{array} \quad \begin{array}{|cc|}\hline 0 & 7 \\ 7 & 2 \\ \hline\end{array}$$

the following probabilities

$$\begin{aligned} P_1 &= \frac{7!\ 9!\ 7!\ 9!}{16!} \cdot \frac{1}{5!2!4!5!} = 0.2317, \\ P_2 &= \frac{2\cdot 4}{6\cdot 6} P_1 = 0.0515, \\ P_3 &= \frac{1\cdot 3}{7\cdot 7} P_2 = 0.0032\,, \end{aligned}$$

with $P = P_1 + P_2 + P_3 = 0.2364$, so that $H_0 : P((AB)) = P((BA))$ is not rejected.

Prescott's test uses the following 2×3 table:

Group	(0, 1)	(0, 0) or (1, 1)	(1, 0)	Total
(AB)	5	9 + 14	2	30
(BA)	4	11 + 18	6	38
Total	9	52	7	68

$$\begin{aligned} V &= 30 \cdot 38[(9+7) \cdot 68 - (9-7)^2]/68 \\ &= \frac{30 \cdot 38}{68}[16 \cdot 68 - 4] = 18172.94, \\ \chi^2(P) &= [(5-2) \cdot 68 - (9-7) \cdot 30]^2/V \\ &= \frac{144^2}{V} = 1.14 < 3.84\,. \end{aligned}$$

H_0 : treatment effect = 0 is not rejected.

Saturated REA Model

The analysis of the REA model using SAS gives the following table, after calling this procedure in SAS:

```
PROC CATMOD DATA = BEISPIEL 8.4;
WEIGHT COUNT;
DIRECT SEQUENCE PERIOD TREAT
ASSOC_AB ASSOC_BA;
MODEL Y = SEQUENCE PERIOD TREAT
  ASSOC_AB ASSOC_BA /
  NOGLS ML;
RUN;
```

Effect	Estimate	S.E.	Chi-Square	p–Value
INTERCEPT	0.3437	0.1959	3.08	0.0793
SEQUENCE	0.0626	0.1959	0.10	0.7429
PERIOD	-0.0623	0.1959	0.10	0.7470
TREAT	-0.2096	0.1959	1.14	0.2846
ASSOC_AB	1.2668	0.4697	7.27	0.0070 *
ASSOC_BA	1.1463	0.3862	8.81	0.0030 *

None of the main effects is significant.

Remark. The parameter estimates may be checked directly using formulas (9.113)–(9.118):

$$\begin{aligned} \hat{\mu} &= 1/8 \ln\left[\left(\frac{14+2}{9+5}\right)^2 \frac{14 \cdot 5}{9 \cdot 2}\right] + 1/8 \ln\left[\left(\frac{18+5}{11+4}\right)^2 \frac{18 \cdot 4}{11 \cdot 5}\right] \\ &= 0.2031 + 0.1406 = 0.3437, \\ \hat{\gamma} &= 0.2031 - 0.1406 = 0.0625, \\ \hat{\pi} &= 1/8 \ln\left[\left(\frac{16}{14}\right)^2 \frac{18}{70}\right] + 1/8 \ln\left[\left(\frac{23}{15}\right)^2 \frac{55}{72}\right] \\ &= -0.1364 + 0.0732 = -0.0632, \end{aligned}$$

$$\begin{aligned}
\hat{\tau} &= -0.1364 - 0.0732 = -0.2096, \\
\hat{\alpha}_{11} &= 1/2 \ln\left(\frac{9 \cdot 14}{5 \cdot 2}\right) = 1.2668, \\
\hat{\alpha}_{21} &= 1/2 \ln\left(\frac{11 \cdot 18}{4 \cdot 5}\right) = 1.1463\,.
\end{aligned}$$

Analysis via GEE1 (cf. Chapter 7)

The analysis of the data set using the GEE1 procedure of Heumann (1993) gives the following results for parametrization No. 2 (model (9.58)):

Effect	Estimates	Naive S.E.	Robust S.E.	P-Robust
INTERCEPT	0.1335	0.3569	0.3569	0.7154
TREATMENT	0.2939	0.4940	0.4940	0.5521
PERIOD	0.1849	0.4918	0.4918	0.7071
TREAT x PERIOD	-0.0658	0.7040	0.8693	0.9397

The working correlation is 0.5220. All effects are not significant.

9.5 Exercises and Questions

9.5.1 Give a description of the linear model of cross–over designs. What is its relationship to repeated measures and split–plot designs? What are the main effects and the interaction effect?

9.5.2 Review the test strategy in the 2×2 cross–over. Assuming the carry–over effect to be significant, what effect is still testable? Is this test useful?

9.5.3 What is the difference between the classical approach and the four alternative parametrizations? Describe the relationship between randomization versus carry–over effect and parallel groups versu sequence effect.

9.5.4 Consider the following 2×2 cross–over with binary response:

Group	(0, 0)	(0, 1)	(1, 0)	(1, 1)	Total
1 (AB)	n_{11}	n_{12}	n_{13}	n_{14}	$n_{1\cdot}$
2 (BA)	n_{21}	n_{22}	n_{23}	n_{24}	$n_{2\cdot}$

Which contingency tables and corresponding odds ratios are indicators for the treatment effect or treatment $\times$ period effect?

9.5.5 Review the tests of McNemar, Mainland–Gart, and Prescott (assumptions, objectives).

10

Statistical Analysis of Incomplete Data

10.1 Introduction

A basic problem in the statistical analysis of data sets is the loss of single observations, of variables, or of single values. Rubin (1976) can be regarded as the pioneer of the modern theory of *Nonresponse in Sample Surveys*. Little and Rubin (1987) and Rubin (1987) have discussed fundamental concepts for handling missing data based on decision theory and models for the mechanism of nonresponse.

Standard statistical methods have been developed to analyze rectangular data sets, i.e., to analyze a matrix

$$X = \begin{pmatrix} x_{11} & \cdots & \cdots & x_{1p} \\ \vdots & * & & \vdots \\ & & & * \\ \vdots & & * & \vdots \\ x_{n1} & \cdots & \cdots & x_{np} \end{pmatrix}.$$

The columns of the matrix X represent variables observed for each unit, and the rows of X represent units (cases, observations) of the variables. Here, data on all scales can be observed:

- interval-scaled data;
- ordinal-scaled data; and
- nominal-scaled data.

In practice, some of the observations may be missing. This fact is indicated by the symbol "*".

Examples:

- People do not always give answers to all of the items in a questionnaire. Answers may be missing at random (a question was overlooked) or not missing at random (individuals are not always willing to give detailed information concerning personal items like drinking behavior, income, sexual behavior, etc.).
- Mechanical experiments in industry (e.g., quality control by pressure) sometimes destroy the object and the response is missing. If there is a strong causal relationship between the object of the experiment and the loss of response, then it may be expected that the response is not missing at random.
- In clinical long–time studies, some individuals may not cooperate or do not participate over the whole period and drop out. In the analysis of lifetime data, these individuals are called censored. Censoring is a mechanism causing nonrandomly missing data.

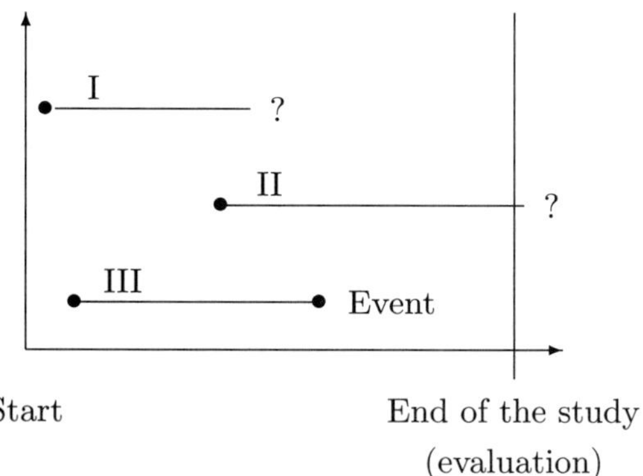

FIGURE 10.1. Censored individuals (I : drop–out and II : censored by the end point) and an individual with response (event) (III).

Statistical Methods with Missing Data

There are mainly three general approaches to handling the missing data problem in statistical analysis.

(i) Complete Case Analysis

Analyses using only complete cases confine their attention to those cases (rows of the matrix X) where all p variables are observed. Let X be rearranged according to

$$X = \begin{pmatrix} \underset{n_1,p}{X_c} \\ \underset{n_2,p}{X_*} \end{pmatrix}$$

where X_c (c : complete) is fully observed. The statistical analysis makes use of the data in X_c only. The complete case analysis tends to become inefficient if the percentage $(n_2/n) \cdot 100$ is increasing and if there are blocks in the pattern of missing data. The selection of complete cases can lead to a selectivity bias in the estimates if selection is heterogeneous with respect to the covariates. Hence, the crucial concern is whether the complete cases constitute a random subsample of X or not.

Example 10.1. Suppose that age under 60 and age over 60 are the two levels of the binary variable X (age of individuals). Assume the following situation in a lifetime data analysis:

	Start	End
< 60	100	60
> 60	100	40

The drop–out percentage is 40% and 60%, respectively. Hence, one has to test if there is a selectivity bias in estimating survivorship models and, if the tests are significant, one has to correct the estimations by adjustment methods (see, e.g., Walther and Toutenburg, 1991).

(ii) Filling In the Missing Values (Imputation for Nonresponse)

Imputation is a general and flexible alternative to the complete case analysis. The missing cells in the submatrix X_* are replaced by guesses or correlation–based predictors transforming X_* to $\hat{X}_*$. However, this method can lead to severe biases in statistical analysis, as the imputed values, in general, are different from the true but missing data. We will discuss this problem in detail in the case of regression. Sometimes, the statistician has no other choice but to fill–up the matrix X_*, especially if the percentage of complete units is too small. There are several approaches for imputation. Popular among them are the following:

- *Hot deck imputation.* Recorded units of the sample are substituted for missing data.
- *Cold deck imputation.* A missing value is replaced by a constant value, as, for example, a unit from external (or previous) samples.

- *Mean imputation.* Based on the sample of the responding units, means are substituted for the missing cells.
- *Regression (correlation) imputation.* Based on the correlative structure of the matrix X_c, missing values are replaced by predicted values from a regression of the missing item on items observed for the unit.

(iii) Model–Based Procedures

Modeling techniques are generated by factorization of the likelihood according to the observation and missing patterns. Parameters can be estimated by iterative maximum likelihood procedures starting with the complete cases. These methods are discussed in full by Little and Rubin (1987).

Multiple Imputation

The idea of multiple imputation (Rubin, 1987) is to achieve a variability of the estimate by repeated imputation and analysis of each of the so–completed data sets. The final estimate can then be calculated, for example, by taking the means.

Missing Data Mechanisms

Ignorable nonresponse. Knowledge of the mechanism for nonresponse is a central element in choosing an appropriate statistical analysis. If the mechanism is under control of the statistician, and if it generates a random subsample of the whole sample, then it may be called ignorable.

Example: Assume $Y \sim N(\mu, \sigma^2)$ to be a univariate normally distributed response variable and denote the planned whole sample by $(y_1, \ldots, y_m, y_{m+1}, \ldots, y_n)'$. Suppose that indeed only a subsample denoted by $y_{\text{obs}} = (y_1, \ldots, y_m)'$ of responses is observed and the remaining responses $y_{\text{mis}} = (y_{m+1}, \ldots, y_n)'$ are missing. If the values are missing at random (MAR), then the vector $(y_1, \ldots, y_m)'$ is a random subsample. The only disadvantage is a loss of sample size and, hence, a loss of efficiency of the unbiased estimators $\bar{y}$ and s_y^2.

Nonignorable nonresponse occurs if the probability $P(y_i \text{ observed})$ is a function of the value y_i itself, as happens, for example, in the case of censoring. In general, estimators based on nonrandom subsamples are biased.

MAR, OAR, and MCAR

Let us assume a bivariate sample of (X, Y) such that X is completely observed but that some values of Y are missing. This structure is a special case of a so–called monotone pattern of missing data.

This situation is typical for longitudinal studies or questionnaires, when one variable is known for all elements of the sample, but the other variable is unknown for some of them.

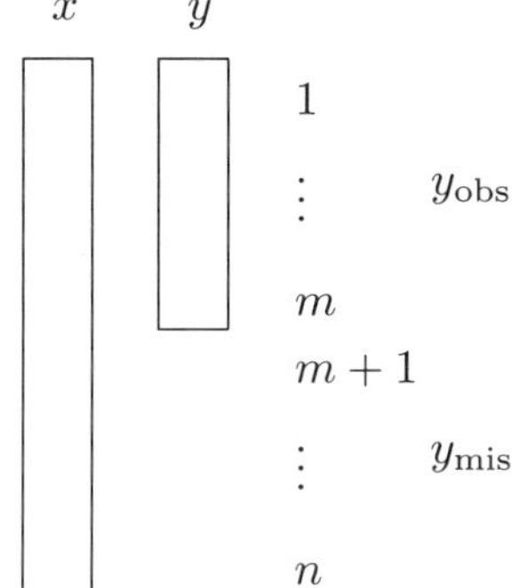

FIGURE 10.2. Monotone pattern in the bivariate case.

Examples:

X	Y
Age	Income
Placebo	Blood pressure after 28 days
Cancer	Life span

The probability of the response of Y can be dependent on X and Y in the following manner:

(i) dependent on X and Y;

(ii) dependent on X but independent of Y; and

(iii) independent of X and Y.

In case (iii) the missing data is said to be missing at random (MAR) and the observed data is said to be observed at random (OAR). Thus the missing data is said to be missing completely at random (MCAR). As a consequence, the data y_{obs} constitutes a random subsample of $y = (y_{\text{obs}}, y_{\text{mis}})'$. In case (ii) the missing data is MAR but the observed values are not necessarily a random subsample of y. However, within fixed X–levels, the y–values y_{obs} are OAR.

In case (i) the data is neither MAR nor OAR and hence, the missing data mechanism is not ignorable. In cases (ii) and (iii) the missing data mechanisms are ignorable for methods using the likelihood function. In case (iii) this is true for methods based on the sample as well.
If the conditional distribution of $Y \mid X$ has to be investigated, then MAR is sufficient to have efficient estimators. On the other hand, if the marginal distribution of Y is of interest (e.g., estimation of μ by $\bar{y}$ based on the m complete observations), then MCAR is a necessary assumption to avoid a

bias. Suppose that the joint density function of X and Y is factorized as

$$f(X,Y) = f(X)f(Y \mid X)$$

where $f(X)$ is the marginal density of X and $f(Y \mid X)$ is the conditional density of $Y \mid X$. It is obvious that analysis of $f(Y \mid X)$ has to be based on the m jointly observed data points. Estimating y_{mis} coincides with the classical prediction.

Example: Suppose that X is a categorical covariate with two categories $X = 1$ (age > 60 years) and $X = 0$ (age ≤ 60 years). Let Y be the lifetime of a denture. It may happen that the younger group of patients participates less often in the follow–ups compared to the older group. Therefore, one may expect that $P(y_{\text{obs}} \mid X = 1) > P(y_{\text{obs}} \mid X = 0)$.

10.2 Missing Data in the Response

In controlled experiments such as clinical trials, the design matrix X is fixed and the response is observed for the different factor levels of X. The analysis is done by means of analysis of variance or the common linear model and the associated test procedures (cf. Chapter 3). In this situation, it is realistic to assume that missing values occur in the response y and not in the design matrix X. This results in an unbalanced response. Even if we can assume that MCAR holds, sometimes it may be more advantageous to fill–up the vector y than to confine the analysis to the complete cases. This is the fact, for example, in factorial (cross–classified) designs with few replications.

10.2.1 Least Squares Analysis for Complete Data

Let Y be the response variable, X the (T,K)–matrix of design, and assume the linear model

$$y = X\beta + \epsilon, \quad \epsilon \sim N(\mathbf{0}, \sigma^2 \mathrm{I}). \tag{10.1}$$

The OLSE of β is given by $b = (X'X)^{-1}X'y$ and the unbiased estimator of σ^2 is given by

$$\begin{aligned} s^2 &= (y - Xb)'(y - Xb)(T-K)^{-1} \\ &= \frac{\sum_{t=1}^{T}(y_t - \hat{y}_t)^2}{T-K}. \end{aligned} \tag{10.2}$$

To test linear hypotheses of the type $R\beta = \mathbf{0}$ (R a $(J \times K)$–matrix of rank J), we use the test statistic

$$F_{J,T-K} = \frac{(Rb)'(R(X'X)^{-1}R')^{-1}(Rb)}{Js^2} \tag{10.3}$$

(cf. Sections 3.7 and 3.8).

10.2.2 Least Squares Analysis for Filled–Up Data

The following method was proposed by Yates (1933). Assume that $(T-m)$ responses in y are missing. Reorganize the data matrices according to

$$\begin{pmatrix} y_{\text{obs}} \\ y_{\text{mis}} \end{pmatrix} = \begin{pmatrix} X_c \\ X_* \end{pmatrix} \beta + \begin{pmatrix} \epsilon_c \\ \epsilon_* \end{pmatrix} . \tag{10.4}$$

The complete case estimator of β is then given by

$$b_c = (X_c' X_c)^{-1} X_c' y_{\text{obs}} \tag{10.5}$$

$(X_c : m \times K)$ and the classical predictor of the $(T-m)$–vector y_{mis} is given by

$$\hat{y}_{\text{mis}} = X_* b_c . \tag{10.6}$$

Inserting this estimator into (10.4) for y_{mis} and estimating β in the filled–up model is equivalent to minimizing the following function with respect to β (cf. (3.6))

$$\begin{aligned} S(\beta) &= \left\{ \begin{pmatrix} y_{\text{obs}} \\ \hat{y}_{\text{mis}} \end{pmatrix} - \begin{pmatrix} X_c \\ X_* \end{pmatrix} \beta \right\}' \left\{ \begin{pmatrix} y_{\text{obs}} \\ \hat{y}_{\text{mis}} \end{pmatrix} - \begin{pmatrix} X_c \\ X_* \end{pmatrix} \beta \right\} \\ &= \sum_{t=1}^{m} (y_t - x_t' \beta)^2 + \sum_{t=m+1}^{T} (\hat{y}_t - x_t' \beta)^2 \longrightarrow \min_{\beta}! \end{aligned} \tag{10.7}$$

The first sum is minimized by b_c [(10.5)]. Replacing β in the second sum by b_c equates this sum to zero (cf. (10.6)), i.e., to its absolute minimum. Therefore, the estimator b_c minimizes the error–sum–of–squares $S(\beta)$ [(10.7)] and b_c is seen to be the OLSE of β in the filled–up model.

Estimating σ^2

(i) If the data are complete, then $s^2 = \sum_{t=1}^{T} (y_t - \hat{y}_t)^2 / (T-K)$ is the correct estimator of σ^2.

(ii) If $(T-m)$ values are missing (i.e., y_{mis} in (10.4)), then

$$\hat{\sigma}^2_{\text{mis}} = \sum_{t=1}^{m} (y_t - \hat{y}_t)^2 / (m-K) \tag{10.8}$$

would be the appropriate estimator of σ^2.

(iii) On the other hand, if the missing data are filled–up according to the method of Yates, we automatically receive the estimator

$$\hat{\sigma}^2_{\text{Yates}} = \left\{ \sum_{t=1}^{m} (y_t - \hat{y}_t)^2 + \sum_{t=m+1}^{T} (\hat{y}_t - \hat{y}_t)^2 \right\} / (T-K)$$

$$= \sum_{t=1}^{m} (y_t - \hat{y}_t)^2 / (T - K) . \tag{10.9}$$

Therefore we get the relationship

$$\hat{\sigma}^2_{\text{Yates}} = \hat{\sigma}^2_{\text{mis}} \cdot \frac{m-K}{T-K} < \hat{\sigma}^2_{\text{mis}} , \tag{10.10}$$

and hence the method of Yates underestimates the variance. As a consequence of this, the confidence intervals (cf. (3.148), (3.149), and (3.164)) turn out to be too small and the test statistics (cf. (10.3)) become too large, implying that null hypotheses can be rejected more often. To ensure correct tests, the estimate of the variance and all the following statistics would have to be corrected by the factor $(T - K)/(m - K)$.

10.2.3 Analysis of Covariance—Bartlett's Method

Bartlett (1937) suggested an improvement of Yates' ANOVA, which is known as Bartlett's ANCOVA (analysis of covariance). This procedure is as follows:

(i) each missing value is replaced by an arbitrary estimate (guess): $y_{\text{mis}} \Rightarrow \hat{y}_{\text{mis}}$;

(ii) define an indicator matrix $\underset{T,(T-m)}{Z}$ as a covariate according to

$$Z = \begin{pmatrix} 0 & 0 & 0 & \cdots & 0 \\ 0 & 0 & 0 & \cdots & 0 \\ \vdots & \vdots & \vdots & & \vdots \\ 0 & 0 & 0 & \cdots & 0 \\ 1 & 0 & 0 & \cdots & 0 \\ 0 & 1 & 0 & \cdots & 0 \\ \vdots & \vdots & \vdots & & \vdots \\ 0 & 0 & 0 & \cdots & 1 \end{pmatrix} . \tag{10.11}$$

The m null vectors indicate the observed cases and the $(T - m)$–vectors e_i' indicate the missing values. This covariate Z leads to an additional parameter $\underset{(T-m),1}{\gamma}$ in the model that has to be estimated

$$\begin{aligned} \begin{pmatrix} y_{\text{obs}} \\ \hat{y}_{\text{mis}} \end{pmatrix} &= X\beta + Z\gamma + \epsilon \\ &= (X, Z) \begin{pmatrix} \beta \\ \gamma \end{pmatrix} + \epsilon . \end{aligned} \tag{10.12}$$

The OLSE of the parameter vector $\begin{pmatrix} \beta \\ \gamma \end{pmatrix}$ is found by minimizing the error-sum-of-squares

$$S(\beta, \gamma) = \sum_{t=1}^{m} (y_t - x_t'\beta - \mathbf{0}'\gamma)^2 + \sum_{t=m+1}^{T} (\hat{y}_t - x_t'\beta - e_t'\gamma)^2. \quad (10.13)$$

The first term is minimal for $\hat{\beta} = b_c$ [(10.5)], whereas the second term becomes minimal (equating to zero) for $\hat{\gamma} = \hat{y}_{\text{mis}} - X_* b_c$. Hence, the sum total is minimal for $(b_c, \hat{\gamma})'$, and so

$$\begin{pmatrix} b_c \\ \hat{y}_{\text{mis}} - X_* b_c \end{pmatrix} \quad (10.14)$$

is the OLSE of $\begin{pmatrix} \beta \\ \gamma \end{pmatrix}$ in the model (10.12). Choosing the guess $\hat{y}_{\text{mis}} = X_* b_c$ (as in Yates' method), we get $\hat{\gamma} = 0$. Both methods lead to the complete case OLSE b_c as an estimate of β. Introducing the additional parameter γ (which is not of any statistical interest) has one advantage: the degree of freedom in estimating σ^2 in model (10.12) is now T minus the number of estimated parameters, i.e., $T - K - (T - m) = m - K$, and is hence correct. Therefore Bartlett's ANCOVA leads to $\hat{\sigma}^2 = \hat{\sigma}^2_{\text{mis}}$ (cf. (10.8)), an unbiased estimator of σ^2.

10.3 Missing Values in the X–Matrix

In econometric models, other than in experimental design in biology or pharmacy, the matrix X is not fixed but contains observations of exogeneous variables. Hence X may be a matrix of random variables, and missing observations can occur. In general, we may assume the following structure of the data

$$\begin{pmatrix} y_{\text{obs}} \\ y_{\text{mis}} \\ y^*_{\text{obs}} \end{pmatrix} = \begin{pmatrix} X_{\text{obs}} \\ X^*{}_{\text{obs}} \\ X_{\text{mis}} \end{pmatrix} \beta + \epsilon\,. \quad (10.15)$$

Estimation of y_{mis} corresponds to the prediction problem. The classical prediction is equivalent to the method of Yates. Based on these arguments, we may confine ourselves to the substructure

$$\begin{pmatrix} y_{\text{obs}} \\ y^*_{\text{obs}} \end{pmatrix} = \begin{pmatrix} X_{\text{obs}} \\ X_{\text{mis}} \end{pmatrix} \beta + \epsilon \quad (10.16)$$

of (10.15) and change the notation as follows:

$$\begin{pmatrix} y_c \\ y_* \end{pmatrix} = \begin{pmatrix} X_c \\ X_* \end{pmatrix} \beta + \begin{pmatrix} \epsilon_c \\ \epsilon_* \end{pmatrix}, \quad \begin{pmatrix} \epsilon_c \\ \epsilon_* \end{pmatrix} \sim (\mathbf{0}, \sigma^2 \mathrm{I}). \quad (10.17)$$

The submodel

$$y_c = X_c\beta + \epsilon_c \tag{10.18}$$

stands for the completely observed data (c : complete), and we have $y_c : m \times 1$, $X_c : m \times K$, and $\text{rank}(X_c) = K$. Assume that X is nonstochastic. If not, we would use conditional expectations.

The other submodel

$$y_* = X_*\beta + \epsilon_* \tag{10.19}$$

is of dimension $(T - m) = J$. The vector y_* is observed completely. In the matrix X_* some observations are missing. The notation X_* will underline that X_* is partially incomplete, in contrast to the matrix X_{mis}, which is completely missing. Combining both of the submodels in model (10.17) corresponds to the so–called mixed model. Therefore, it seems to be natural to use the method of mixed estimation.

The optimal estimator of β in model (10.17) is given by the mixed estimator (cf. Rao and Toutenburg, 1999, Chapter 5)

$$\begin{aligned}\hat{\beta}(X_*) &= (X_c'X_c + X_*'X_*)^{-1}(X_c'y_c + X_*'y_*) \\ &= b_c + S_c^{-1}X_*'(I_J + X_*S_c^{-1}X_*')^{-1}(y_* - X_*b_c),\end{aligned} \tag{10.20}$$

where

$$b_c = (X_c'X_c)^{-1}X_c'y_c \tag{10.21}$$

is the OLSE in the complete case submodel (10.18) and

$$S_c = X_c'X_c. \tag{10.22}$$

The covariance matrix of $\hat{\beta}(X_*)$ is

$$\text{V}(\hat{\beta}(X_*)) = \sigma^2(S_c + S_*)^{-1} \tag{10.23}$$

with

$$S_* = X_*'X_*. \tag{10.24}$$

The mixed estimator (10.20) is not operational though, due to the fact that X_* is partially unknown.

10.3.1 Missing Values and Loss of Efficiency

Before we discuss the different methods for estimating missing values, let us study the consequences of confining the analysis to the complete case model [(10.18)]. Our measure to compare $\hat{\beta}_c$ and $\hat{\beta}(X_*)$ is the scalar risk

$$R(\hat{\beta}, \beta, S_c) = \text{tr}\{S_c\,\text{V}(\hat{\beta})\}, \tag{10.25}$$

which coincides with the MSE–III risk. From Theorem A.3(iii) we have the identity

$$(S_c + X_*'X_*)^{-1} = S_c^{-1} - S_c^{-1}X_*'(\mathrm{I}_J + X_*S^{-1}X_*')^{-1}X_*S_c^{-1}. \tag{10.26}$$

Applying this we get the risk of $\hat\beta(X_*)$ as

$$\begin{aligned}\sigma^{-2}R(\hat\beta(X_*),\beta,S_c) &= \mathrm{tr}\{S_c(S_c+S_*)^{-1}\}\\ &= K - \mathrm{tr}\{(I_J + B'B)^{-1}B'B\},\end{aligned} \tag{10.27}$$

where $B = S_c^{-1/2}X_*'$.

The $(J \times J)$–matrix $B'B$ is nonnegative definite with rank $(B'B) = J^*$. If $\mathrm{rank}(X_*) = J < K$ holds, then $J^* = J$ and hence $B'B > 0$.

Let $\lambda_1 \geq \ldots \geq \lambda_J \geq 0$ denote the eigenvalues of B, let $\Lambda = \mathrm{diag}(\lambda_1,\ldots,\lambda_J)$, and let P denote the matrix of orthogonal eigenvectors. Then we have (Theorem A.11) $B'B = P\Lambda P'$ and

$$\begin{aligned}\mathrm{tr}\{(I_J+B'B)^{-1}B'B\} &= \mathrm{tr}\{P(I_J+\Lambda)^{-1}P'P\Lambda P'\}\\ &= \mathrm{tr}\{(I_J+\Lambda)^{-1}\Lambda\}\\ &= \sum_{i=1}^{J}\frac{\lambda_i}{1+\lambda_i}.\end{aligned} \tag{10.28}$$

The MSE–III risk of b_c is

$$\sigma^{-2}R(b_c,\beta,S_c) = \mathrm{tr}\{S_cS_c^{-1}\} = K. \tag{10.29}$$

Using the MSE–III criterion, we may conclude that

$$R(b_c,\beta,S_c) - R(\hat\beta(X_*),\beta,S_c) = \sum\frac{\lambda_i}{1+\lambda_i} \geq 0, \tag{10.30}$$

and, hence, that $\hat\beta(X_*)$ is superior to b_c. We want to continue the comparison according to a different criterion, which compares the size of the risks instead of their differences.

Definition 10.1. The relative efficiency of an estimator $\hat\beta_1$, compared to another estimator $\hat\beta_2$, is defined as the following ratio

$$\mathrm{eff}(\hat\beta_1,\hat\beta_2,A) = \frac{R(\hat\beta_2,\beta,A)}{R(\hat\beta_1,\beta,A)}. \tag{10.31}$$

$\hat\beta_1$ is said to be less efficient than $\hat\beta_2$ if

$$\mathrm{eff}(\hat\beta_1,\hat\beta_2,A) \leq 1.$$

Using (10.27)–(10.29) we find

$$\mathrm{eff}(b_c,\hat\beta(X_*),S_c) = 1 - \frac{1}{K}\sum\frac{\lambda_i}{1+\lambda_i} \leq 1. \tag{10.32}$$

The relative efficiency of the complete case estimator b_c, compared to the mixed estimator in the full model (10.17), is smaller than or equal to one

$$\max\left[0, 1 - \frac{J}{K}\frac{\lambda_1}{1+\lambda_1}\right] \leq \text{eff}(b_c, \hat{\beta}(X_*), S_c) \leq 1 - \frac{J}{K}\frac{\lambda_J}{1+\lambda_J} \leq 1. \quad (10.33)$$

Examples:

(i) Let $X_* = X_c$, so that in the full model the design matrix X_c is used twice. Then $B'B = X_c S_c^{-1} X_c'$ is idempotent of rank $J = K$. Therefore, we have $\lambda_i = 1$ (Theorem A.36(i)) and hence

$$\text{eff}(b_c, \hat{\beta}(X_c), S_c) = 1/2. \quad (10.34)$$

(ii) $J = 1$ (one row of X is incomplete). Then $X_* = x_*'$ becomes a $(1 \times K)$–vector and $B'B = x_*' S_c^{-1} x_*$ becomes a scalar. Let $\mu_1 \geq \ldots \geq \mu_K > 0$ be the eigenvalues of S_c and let $\Gamma = (\gamma_1, \ldots, \gamma_K)$ be the matrix of the corresponding orthogonal eigenvectors.

Therefore, we may write $\hat{\beta}(x_*)$ as

$$\hat{\beta}(x_*) = (S_c + x_* x_*')^{-1}(X_c' y_c + x_* y_*) \quad (10.35)$$

and observe that

$$\mu_1^{-1} x_*' x_* \leq x_*' S_c^{-1} x_* = \sum \mu_j^{-1}(x_*'\gamma_j)^2 \leq \mu_K^{-1} x_*' x_* \,. \quad (10.36)$$

According to (10.32), the relative efficiency becomes

$$\text{eff}(b_c, \hat{\beta}(x_*), S_c) = 1 - \frac{1}{K}\frac{x_*' S_c^{-1} x_*}{1 + x_*' S_c^{-1} x_*} = 1 - \frac{1}{K}\frac{\sum \mu_j^{-1}(x_*'\gamma_j)^2}{1 + \sum \mu_j^{-1}(x_*'\gamma_j)^2} \leq 1 \quad (10.37)$$

and, hence,

$$1 - \frac{\mu_1 \mu_K^{-1} x_*' x_*}{K(\mu_1 + x_*' x_*)} \leq \text{eff}(b_c, \hat{\beta}(x_*), S_c) \leq 1 - \frac{x_*' x_*}{K(\mu_1 \mu_K^{-1})(\mu_K + x_*' x_*)} \,. \quad (10.38)$$

The relative efficiency of b_c in comparison to $\hat{\beta}(x_*)$ is dependent on the vector x_* (or rather its quadratic norm $x_*' x_*$), as well as on the eigenvalues of the matrix S_c, especially on the so–called condition number μ_1/μ_K and the span $(\mu_1 - \mu_K)$ between the largest and smallest eigenvalues.

Let $x_* = g\gamma_i$ $(i = 1, \ldots, K)$, where g is a scalar and define $M = \text{diag}(\mu_1, \ldots, \mu_K)$. For these x_*–vectors, which are parallel to the eigenvectors of S_c, the quadratic risk of the estimators $\hat{\beta}(g\gamma_i)$ becomes

$$\begin{aligned} \sigma^{-2} R(\hat{\beta}(g\gamma_i), \beta, S_c) &= \text{tr}\{\Gamma M \Gamma' (\Gamma M \Gamma' + g^2 \gamma_i \gamma_i')^{-1}\} \\ &= K - 1 + \frac{\mu_i}{\mu_i + g^2} \,. \end{aligned} \quad (10.39)$$

Hence, the relative efficiency of b_c reaches its maximum if x_* is parallel to γ_1 (eigenvector corresponding to the maximum eigenvalue μ_1). Therefore, the loss in efficiency by removing one row x_* is minimal for $x_* = g\gamma_1$ and maximum for $x_* = g\gamma_K$. This fact corresponds to the result of Silvey (1969), namely, that the goodness–of–fit of the OLSE can be improved, if additional observations are taken in the direction which was most imprecise. This is just the direction of the eigenvector corresponding to the minimal eigenvalue μ_K of S_c.

10.3.2 Standard Methods for Incomplete X–Matrices

(i) Complete Case Analysis

The idea of the first method is to confine the analysis to the completely observed submodel [(10.18)]. The corresponding estimator of β is $b_c = S_c^{-1}X_c'y_c$ [(10.21)], which is unbiased and has the covariance matrix $\mathrm{V}(b_c) = \sigma^2 S_c^{-1}$. Using the estimator b_c is only feasible for a small percentage of missing or incomplete rows in X_*, i.e., for $[(T-m)/T]\cdot 100\%$ at the most, and assumes that MAR holds. The assumption of MAR may not be tenable if, for instance, too many rows in X_* are parallel to the eigenvector γ_K corresponding to the eigenvalue μ_K of S_c.

(ii) Zero–Order Regression (ZOR)

This method by Weisberg (1980), also called the method of sample means, replaces a missing value x_{ij} of the jth regressor X_j by the sample mean of the observed values of X_j. Denote the index sets of the missing values of X_j by

$$\Phi_j = \{i : x_{ij} \text{ missing}\}, \quad j = 1, \ldots, K, \tag{10.40}$$

and let M_j be the number of elements in Φ_j. Then for j fixed, any missing value x_{ij} in X_* is replaced by

$$\hat{x}_{ij} = \bar{x}_j = \frac{1}{T-M_j}\sum_{i\notin\Phi_j} x_{ij}. \tag{10.41}$$

This method may be recommended, as long as the sample mean is a good estimator for the mean of the jth column. If, somehow, the data in the jth column are trended or follows a growth curve, then $\bar{x}_j$ is not a good estimator and, hence, replacing missing values by $\bar{x}_j$ may cause a bias. If all the missing values x_{ij} are replaced by the corresponding column means $\bar{x}_j$ $(j = 1, \ldots, K)$, then the matrix X_* results in a—now completely known—matrix $X_{(1)}$. Hence, an operationalized version of the mixed model [(10.17)] is

$$\begin{pmatrix} y_c \\ y_* \end{pmatrix} = \begin{pmatrix} X_c \\ X_{(1)} \end{pmatrix}\beta + \begin{pmatrix} \epsilon \\ \epsilon_{(1)} \end{pmatrix}. \tag{10.42}$$

For the vector of errors $\epsilon_{(1)}$, we have

$$\epsilon_{(1)} = (X_* - X_{(1)})\beta + \epsilon_* \tag{10.43}$$

with

$$\epsilon_{(1)} \sim \{(X_* - X_{(1)})\beta, \sigma^2 \mathrm{I}_J\} \tag{10.44}$$

and $J = (T - m)$.

In general, replacing missing values can result in a biased mixed model, since $(X_* - X_{(1)}) \neq \mathbf{0}$ holds. If X is a matrix of stochastic regressor variables, then, at the most, one may expect that $\mathrm{E}(X_* - X_{(1)}) = \mathbf{0}$ holds.

(iii) First–Order Regression (FOR)

This term comprises a set of methods, which make use of the structure of the matrix X by setting up additional regressions. Based on the index sets Φ_j in (10.40), the dependence of each column x_j ($j = 1, \dots, K$, j fixed) on the other columns is modeled according to the following relationship

$$x_{ij} = \theta_{0j} + \sum_{\substack{\mu=1 \\ \mu \neq j}}^{K} x_{i\mu}\theta_{\mu j} + u_{ij}, \quad i \notin \Phi = \bigcup_{j=1}^{K} \Phi_j. \tag{10.45}$$

The missing values x_{ij} in X_* are estimated and replaced by

$$\hat{x}_{ij} = \hat{\theta}_{0j} + \sum_{\substack{\mu=1 \\ \mu \neq j}}^{K} x_{i\mu}\hat{\theta}_{\mu j} \quad (i \in \Phi_j). \tag{10.46}$$

(iv) Correlation Methods for Stochastic X

In the case of stochastic regressors $X_1, \dots, X_K$ (or $X_2, \dots, X_K$, if $X_1 = \mathbf{1}$), the vector β is estimated by solving the normal equations

$$\mathbf{Cov}(x_i, x_j)\hat{\beta} = \mathbf{Cov}(x_i, y) \quad (i, j = 1, \dots, K), \tag{10.47}$$

where $\mathbf{Cov}(x_i, x_j)$ is the $(K \times K)$–sample covariance matrix. The (i, j)th element of $\mathbf{Cov}(x_i, x_j)$ is calculated from the pairwise observed elements of the variables X_i and X_j. Similarly, $\mathbf{Cov}(x_i, y)$ makes use of pairwise observed elements of x_i and y. Since this method frequently leads to unsatisfactory results, we will not deal with this method any further. Based on simulation studies, Haitovsky (1968) concludes that in most situations the complete case estimator b_c is superior to the correlation method.

Maximum–Likelihood Estimates of Missing Values

Suppose that the errors are normally distributed, i.e., $\epsilon \sim N(\mathbf{0}, \sigma^2 \mathrm{I}_T)$. Moreover, assume a so–called monotone pattern of missing values, which enables a factorization of the likelihood (cf. Little and Rubin, 1987). We confine ourselves to the most simple case and assume that the matrix X_* is completely unobserved. This requires a model which contains no constant. Then X_*, in the mixed model (10.17), may be treated as an unknown parameter. The loglikelihood corresponding to the estimators of the unknown parameters β, σ^2, and the "parameter" X_* may be written as

$$\begin{aligned} \ln L(\beta, \sigma^2, X_*) &= -\frac{n}{2}\ln(2\pi) - \frac{n}{2}\ln(\sigma^2) \\ &- \frac{1}{2\sigma^2}(y_c - X_c\beta, y_* - X_*\beta)' \begin{pmatrix} y_c - X_c\beta \\ y_* - X_*\beta \end{pmatrix}. \end{aligned} \tag{10.48}$$

Differentiating with respect to β, σ^2, and X_* leads to the following normal equations

$$\frac{1}{2}\frac{\partial \ln L}{\partial \beta} = \frac{1}{2\sigma^2}\{X_c'(y_c - X_c\beta) + X_*'(y_* - X_*\beta)\} = \mathbf{0}, \tag{10.49}$$

$$\begin{aligned} \frac{\partial \ln L}{\partial \sigma^2} &= \frac{1}{2\sigma^2}\{-n + \frac{1}{\sigma^2}(y_c - X_c\beta)'(y_c - X_c\beta) \\ &+ \frac{1}{\sigma^2}(y_* - X_*\beta)'(y_* - X_*\beta)\} = 0 \end{aligned} \tag{10.50}$$

and

$$\frac{\partial \ln L}{\partial X_*} = \frac{1}{2\sigma^2}(y_* - X_*\beta)\beta' = \mathbf{0}. \tag{10.51}$$

This results in the ML estimators for β and σ^2:

$$\hat{\beta} = b_c = S_c^{-1} X_c' y_c\,, \tag{10.52}$$

$$\hat{\sigma}^2 = \frac{1}{m}(y_c - X_c b_c)'(y_c - X_c b_c), \tag{10.53}$$

which are only based on the complete submodel (10.18). Hence, the ML estimator $\hat{X}_*$ is solution (cf. (10.36) with $\hat{\beta} = b_c$) of

$$y_* = \hat{X}_* b_c. \tag{10.54}$$

Only if $K = 1$, the solution is unique

$$\hat{x}_* = \frac{y_*}{b_c}\,, \tag{10.55}$$

where $b_c = (x_c' x_c)^{-1} x_c' y_c$ (cf. Kmenta, 1971). For $K > 1$, a $(J \times (K-1))$–fold set of solutions $\hat{X}_*$ exists. If any solution $\hat{X}_*$ of (10.39) is substituted for X_* in the mixed model, i.e.,

$$\begin{pmatrix} y_c \\ y_* \end{pmatrix} = \begin{pmatrix} X_c \\ \hat{X}_* \end{pmatrix}\beta + \begin{pmatrix} \epsilon_c \\ \epsilon_* \end{pmatrix}, \tag{10.56}$$

then the following identity holds

$$\begin{aligned} \hat{\beta}(\hat{X}_*) &= (S_c + \hat{X}_*'\hat{X}_*)^{-1}(X_c'y_c + \hat{X}_*'y_*) \\ &= (S_c + \hat{X}_*'\hat{X}_*)^{-1}(S_c\beta + X_c'\epsilon_c + \hat{X}_*'\hat{X}_*\beta + \hat{X}_*'\hat{X}_* S_c^{-1} X_c'\epsilon_c) \end{aligned}$$

$$\begin{aligned} &= \beta + (S_c + \hat{X}_*'\hat{X}_*)^{-1}(S_c + \hat{X}_*'\hat{X}_*)S_c^{-1}X_c'\epsilon_c \\ &= \beta + S_c^{-1}X_c'\epsilon_c \\ &= b_c\,. \end{aligned} \tag{10.57}$$

Remark. The OLSE $\hat{\beta}(\hat{X}_*)$ in the model filled up with the ML estimator $\hat{X}_*$ equals the OLSE b_c in the submodel with the incomplete observations. This is true for other monotone patterns as well.

On the other hand, if the pattern is not monotone, then the ML equations have to be solved by iterative procedures as, for example, the EM algorithm by Dempster, Laird and Rubin (1977) (cf. algorithms by Oberhofer and Kmenta, 1974).

Further discussions of the problem of estimating missing values can be found in Little and Rubin (1987), Weisberg (1980) and Toutenburg (1992a, Chapter 8). Toutenburg, Heumann, Fieger and Park (1995) propose a unique solution of the normal equation (10.49) according to

$$\min_{\hat{X}_*,\lambda}\{|S_c + \hat{X}_*'\hat{X}_*|^{-1} - 2\lambda'(y_* - \hat{X}_*b_c)\}. \tag{10.58}$$

The solution is

$$\hat{X}_* = \frac{y_*y_c'X_c}{y_c'x_X S_c^{-1}X_c'y_c}\,. \tag{10.59}$$

10.4 Adjusting for Missing Data in 2×2 Cross–Over Designs

In Chapter 9, procedures for testing a 2×2 cross–over design were introduced for continuous response. In practice, small sample sizes are an important factor for the employment of the cross–over design. Hence, for studies of this kind, it is especially important to use all available information and to include the data of incomplete observations in the analysis as well.

10.4.1 Notation

We assume that data are only missing for the second period of treatment. Moreover, we assume that the response (y_{i1k}, y_{i2k}) of group i is ordered, so that the first m_i pairs represent the complete data sets. The last $(n_i - m_i)$ pairs are then the incomplete pairs of response. The first m_i values of the response of period j, which belong to complete observation pairs of group i, are now stacked in the vector

$$y_{ij}' = (y_{ij1}, \ldots, y_{ijm_i})\,. \tag{10.60}$$

Those observations of the first period which are assigned to incomplete response pairs are denoted by

$$y_{i1}^{*\prime} = (y_{i1(m_i+1)}, \ldots, y_{i1n_i}) \tag{10.61}$$

for group i. The $(m \times 2)$–data matrix Y of the complete data and the $((n-m) \times 1)$–vector y_1^* of the incomplete data can now be written as

$$Y = \begin{pmatrix} y_{11} & y_{12} \\ y_{21} & y_{22} \end{pmatrix}, \quad y_1^* = \begin{pmatrix} y_{11}^* \\ y_{21}^* \end{pmatrix}, \tag{10.62}$$

with $m = m_1 + m_2$ and $n = n_1 + n_2$. Additionally, we assume that

$$\begin{aligned} (y_{i1k}, y_{i2k}) &\overset{\text{i.i.d.}}{\sim} N((\mu_{i1}, \mu_{i2}), \Sigma) \quad \text{for } k = 1, \ldots, m_i\,, \\ y_{i1k} &\overset{\text{i.i.d.}}{\sim} N(\mu_{i1}, \sigma_{11}^2) \qquad \text{for } k = m_i + 1, \ldots, n_i\,. \end{aligned} \tag{10.63}$$

Here Σ denotes the covariance matrix

$$\Sigma = \begin{pmatrix} \sigma_{11} & \sigma_{12} \\ \sigma_{21} & \sigma_{22} \end{pmatrix} \tag{10.64}$$

with

$$\sigma_{jj'} = \mathrm{Cov}(y_{ijk}, y_{ij'k}) \tag{10.65}$$

and, hence, $\sigma_{11} = \mathrm{Var}(y_{i1k})$ and $\sigma_{22} = \mathrm{Var}(y_{i2k})$. The correlation coefficient ρ can now be written as

$$\rho = \frac{\sigma_{12}}{\sqrt{\sigma_{11}\sigma_{22}}}\,. \tag{10.66}$$

Additionally, we assume that the rows of the matrix Y are independent of the rows of the vector y_1^*. The entire sample can now be described by the two vectors $u' = (y_{11}', y_{21}', y_1^{*\prime})$ and $v' = (y_{12}', y_{22}')$. Hence, the $(n \times 1)$–vector u represents the observations of the first period and the $(m \times 1)$–vector v those of the second period. Since we interpret the observed response pairs as independent realizations of a random sample of a bivariate normal distribution, we can express the density function of (u, v) as the product of the marginal density of u and the conditional density of v given u. The density function of u is

$$f_u = \left(\frac{1}{\sqrt{2\pi\sigma_{11}}}\right)^n \exp\left(-\frac{1}{2\sigma_{11}} \sum_{i=1}^{2} \sum_{k=1}^{n_i} (y_{i1k} - \mu_{i1})^2\right) \tag{10.67}$$

and the conditional density of v given u is

$$\begin{aligned} f_{v|u} &= \left(1/\sqrt{2\pi\sigma_{22}(1-\rho^2)}\right)^m \\ &\quad \cdot \exp\left(-1/2\sigma_{22}(1-\rho^2) \sum_{i=1}^{2} \sum_{k=1}^{m_i} (y_{i2k} - \mu_{i2} - (\rho\sqrt{\sigma_{22}}/\sigma_{11})(y_{i1k} - \mu_{i1}))^2\right). \end{aligned} \tag{10.68}$$

The joint density function $f_{u,v}$ of (u, v) is now

$$f_{u,v} = f_u f_{v|u} . \tag{10.69}$$

10.4.2 Maximum Likelihood Estimator (Rao, 1956)

We now estimate the unknown parameters $\mu_{11}, \mu_{21}, \mu_{12}$, and μ_{22}, as well as the unknown components $\sigma_{jj'}$ of the covariance matrix Σ. The loglikelihood is $\ln L = \ln f_u + \ln f_{v|u}$ with

$$\ln f_u = -\frac{n}{2}\ln(2\pi\sigma_{11}) - \frac{1}{2\sigma_{11}}\sum_{i=1}^{2}\sum_{k=1}^{n_i}(y_{i1k} - \mu_{i1})^2 \tag{10.70}$$

and

$$\begin{aligned} \ln f_{v|u} = & - m/2\ln(2\pi\sigma_{22}(1-\rho^2)) \\ & - 1/(2\sigma_{22}(1-\rho^2))\sum_{i=1}^{2}\sum_{k=1}^{m_i}\left(y_{i2k} - \mu_{i2} - \rho\sqrt{\sigma_{22}/\sigma_{11}}(y_{i1k} - \mu_{i1})\right)^2 . \end{aligned} \tag{10.71}$$

Let us introduce the following notation

$$\sigma^* = \sigma_{22}(1-\rho^2) , \tag{10.72}$$

$$\beta = \rho\sqrt{\frac{\sigma_{22}}{\sigma_{11}}} , \tag{10.73}$$

$$\mu_{i2}^* = \mu_{i2} - \beta\mu_{i1} . \tag{10.74}$$

Equation (10.71) can now be transformed, and we get

$$\ln f_{v|u} = -(m/2)\ln(2\pi\sigma^*) - (1/2\sigma^*)\sum_{i=1}^{2}\sum_{k=1}^{m_i}(y_{i2k} - \mu_{i2}^* - \beta y_{i1k})^2 . \tag{10.75}$$

This leads to a factorization of the loglikelihood into the two terms (10.70) and (10.75), where no two of the unknown parameters $\mu_{11}, \mu_{21}, \mu_{12}^*, \mu_{22}^*, \sigma_{11}$, σ^*, and β show up in one summand at the same time. Hence maximization of the loglikelihood can be done independently for the unknown parameters and we find the maximum–likelihood estimates

$$\left.\begin{aligned} \hat{\mu}_{i1} &= y_{i1\cdot}^{(n_i)} , \\ \hat{\mu}_{i2} &= y_{i2\cdot}^{(m_i)} + \hat{\beta}\left(\hat{\mu}_{i1} - y_{i1\cdot}^{(m_i)}\right) , \\ \hat{\beta} &= \frac{s_{12}}{s_{11}} , \\ \hat{\sigma}_{11} &= 1/n\sum_{i=1}^{2}\sum_{k=1}^{n_i}(y_{i1k} - \hat{\mu}_{i1})^2 , \\ \hat{\sigma}_{22} &= s_{22} + \hat{\beta}^2(\hat{\sigma}_{11} - s_{11}) , \\ \hat{\sigma}_{12} &= \hat{\beta}\hat{\sigma}_{11} . \end{aligned}\right\} \tag{10.76}$$

If we write

$$
\begin{aligned}
y_{ij\cdot}^{(c)} &= \frac{1}{a}\sum_{k=1}^{a} y_{ijk}, \\
s_{jj'} &= \frac{1}{m_1+m_2}\sum_{i=1}^{2}\sum_{k=1}^{m_i}\left(y_{ijk}-y_{ij\cdot}^{(m_i)}\right)\left(y_{ij'k}-y_{ij'\cdot}^{(m_i)}\right), \quad (10.77)
\end{aligned}
$$

then $\hat{\beta}$ and $\hat{y}_{ij\cdot}^{(c)}$ are independent for $a = n_i, m_i$. Consequently, the covariance matrix $\Gamma_i = ((\gamma_{i,uv}))$ of $(\hat{\mu}_{i1}, \hat{\mu}_{i2})$ is

$$
\Gamma_i = \begin{pmatrix} \sigma_{11}/n_i & \sigma_{12}/n_i \\ \sigma_{12}/n_i & [\sigma_{22} + \left(1-\frac{m_i}{n_i}\right)\sigma_{11}\left(\mathrm{Var}(\hat{\beta})-\beta^2\right)]/m_i \end{pmatrix} \quad (10.78)
$$

with

$$
\mathrm{Var}(\hat{\beta}) = \mathrm{E}\left(\mathrm{Var}(\hat{\beta}|y_1)\right) = \frac{\sigma_{22}(1-\hat{\rho}^2)}{\sigma_{11}(m-4)}, \quad (10.79)
$$

$$
\hat{\rho} = \hat{\beta}\sqrt{\frac{\hat{\sigma}_{11}}{\hat{\sigma}_{22}}}. \quad (10.80)
$$

10.4.3 Test Procedures

We now develop test procedures for large and small sample sizes and formulate the hypotheses $\mathrm{H}_0^{(1)}$: no interaction, $\mathrm{H}_0^{(2)}$: no treatment effect, and $\mathrm{H}_0^{(3)}$: no effect of the period:

$$
\begin{aligned}
\mathrm{H}_0^{(1)}: \theta_1 &= \mu_{11}+\mu_{12}-\mu_{21}-\mu_{22} = 0, \quad (10.81) \\
\mathrm{H}_0^{(2)}: \theta_2 &= \mu_{11}-\mu_{12}-\mu_{21}+\mu_{22} = 0, \quad (10.82) \\
\mathrm{H}_0^{(3)}: \theta_2 &= \mu_{11}-\mu_{12}+\mu_{21}-\mu_{22} = 0. \quad (10.83)
\end{aligned}
$$

Large Samples

The estimates (10.76) lead to the maximum–likelihood estimate $\hat{\theta}_1$ of θ_1. For large sample sizes m_1 and m_2, the distribution of Z_1, defined by

$$
Z_1 = \frac{\hat{\theta}_1}{\sqrt{\sum_{i=1}^{2}(\tilde{\gamma}_{i,11}+2\tilde{\gamma}_{i,12}+\tilde{\gamma}_{i,22})}}, \quad (10.84)
$$

can be approximated by the $N(0,1)$–distribution if $\mathrm{H}_0^{(1)}$ holds. Here $\tilde{\gamma}_{i,uv}$ denote the estimates of the elements of the covariance matrix Γ_i. These are found by replacing $\hat{\sigma}_{11}$ [(10.76)] and $s_{jj'}$ [(10.77)] by their unbiased estimates

$$
\tilde{\sigma}_{11} = \frac{n}{n-2}\hat{\sigma}_{11}, \quad (10.85)
$$

$$\tilde{s}_{jj'} = \frac{m}{m-2} s_{jj'} . \tag{10.86}$$

The maximum–likelihood estimate $\hat{\theta}_2$ for θ_2 is derived from the estimates in (10.76). The test statistic Z_2, given by

$$Z_2 = \frac{\hat{\theta}_2}{\sqrt{\sum_{i=1}^{2} (\tilde{\gamma}_{i,11} - 2\tilde{\gamma}_{i,12} + \tilde{\gamma}_{i,22})}} , \tag{10.87}$$

is approximatively $N(0,1)$–distributed for large samples m_1 and m_2 under $H_0^{(2)}$. Analogously, we find the distribution of the test statistic Z_3:-

$$Z_3 = \frac{\hat{\theta}_3}{\sqrt{\sum_{i=1}^{2} (\tilde{\gamma}_{i,11} - 2\tilde{\gamma}_{i,12} + \tilde{\gamma}_{i,22})}} \tag{10.88}$$

and construct the maximum–likelihood estimate $\hat{\theta}_3$ for θ_3.

Small Samples

For small sample sizes m_1 and m_2, Rao (1956) suggests approximating the distribution of Z_1 by a t–distribution with $v_1 = \frac{1}{2}(n + m - 5)$ degrees of freedom. The choice of v_1 degrees of freedom is explained as follows: The estimates of the variances σ_{11} and σ^* ($\hat{\sigma}^* = s_{22} - \hat{\beta} s_{12}$) are based on $(n-2)$ and $(n-3)$ degrees of freedom, and their mean is $v_1 = \frac{1}{2}(n+m-5)$. If there are no missing values in the second period ($n = m$), then a t–distribution with $(n-2)$ degrees of freedom should be chosen. This test then corresponds to the previously introduced test based on T_λ [(9.19)].

Rao chooses a t–distribution with $v_2 = (m-2)$ degrees of freedom for the approximation of the distribution of Z_2 and Z_3. Morrison (1973) constructs a test for a comparison of the means of a bivariate normal distribution for missing values in one variable at the most. Morrison derives the test statistic from the maximum–likelihood estimate and specifies its distribution as a t–distribution, where the degrees of freedom are only dependent on the number of completely observed response pairs. These tests are equivalent to the tests in Section 9.3.1 if no data are missing.

Example 10.2. In Example 10.1, patient 2 in Group 2 was identified as an outlier. We now want to check to what extent the estimates of the effects vary when the observation of this patient in the second period is excluded from the analysis. We reorganize the data so that patient 2 in Group 2 comes last.

Group 1		Group 2	
A	B	B	A
20	30	30	20
40	50	20	10
30	40	30	10
20	40	40	—

Summarizing in matrix notation (cf. (10.62)), we have

$$Y = \left(\begin{array}{c|c} 20 & 30 \\ 40 & 50 \\ 30 & 40 \\ 20 & 40 \\ \hline 30 & 20 \\ 20 & 10 \\ 30 & 10 \end{array}\right), \quad y_1^* = (40) \,. \tag{10.89}$$

The unbiased estimates are calculated with $n_1 = 4, n_2 = 4, m_1 = 4$, and $m_2 = 3$ by inserting (10.85) and (10.86) in (10.76). We calculate

$$\begin{aligned}
y_{11\cdot}^{(n_1)} &= 1/4\,(20 + 40 + 30 + 20) = 27.50, \\
y_{11\cdot}^{(m_1)} &= 1/4\,(20 + 40 + 30 + 20) = 27.50, \\
y_{12\cdot}^{(m_1)} &= 1/4\,(30 + 50 + 40 + 40) = 40.00, \\
y_{21\cdot}^{(n_2)} &= 1/4\,(30 + 20 + 30 + 40) = 30.00, \\
y_{21\cdot}^{(m_2)} &= 1/3\,(30 + 20 + 30) = 26.67, \\
y_{22\cdot}^{(m_1)} &= 1/3\,(20 + 10 + 10) = 13.33,
\end{aligned}$$

and

$$\begin{aligned}
\tilde{s}_{11} &= \frac{1}{7-2}\left[(20 - 27.50)^2 + \cdots + (20 - 27.50)^2\right. \\
&\quad \left. + (30 - 26.67)^2 + \cdots + (30 - 26.67)^2\right] = 68.33, \\
\tilde{s}_{22} &= \frac{1}{7-2}\left[(30 - 40.00)^2 + \cdots + (40 - 40.00)^2\right. \\
&\quad \left. + (20 - 13.33)^2 + (10 - 13.33)^2 + (10 - 13.33)^2\right] = 53.33, \\
\tilde{s}_{12} &= \frac{1}{7-2}\left[(20 - 27.50)(30 - 40) + \cdots + (20 - 27.50)(40 - 40)\right. \\
&\quad \left. + (30 - 26.67)(20 - 13.33) + \cdots + (30 - 26.67)(10 - 13.33)\right] \\
&= 46.67, \\
\tilde{s}_{21} &= \tilde{s}_{12}.
\end{aligned}$$

With

$$\hat{\beta} = \frac{\tilde{s}_{12}}{\tilde{s}_{11}} = \frac{53.33}{68.33} = 0.68$$

we find

$$\begin{aligned}
\hat{\mu}_{11} &= y_{11\cdot}^{(n_1)} = 27.50, \\
\hat{\mu}_{21} &= y_{21\cdot}^{(n_2)} = 30.00, \\
\hat{\mu}_{12} &= 40.00 + 0.68 \cdot (27.50 - 27.50) = 40.00, \\
\hat{\mu}_{22} &= 13.33 + 0.68 \cdot (30.00 - 26.67) = 15.61,
\end{aligned}$$

and with

$$\begin{aligned}
\tilde{\sigma}_{11} &= \frac{1}{8-2}\left[(20-27.50)^2 + \cdots + (20-27.50)^2\right. \\
&\quad \left. + (30-30)^2 + \cdots + (30-30)^2\right] = 79.17, \\
\tilde{\sigma}_{22} &= 53.33 + 0.68^2 \cdot (79.17 - 68.33) = 58.39, \\
\tilde{\sigma}_{12} &= 0.68 \cdot 79.17 = 54.07, \\
\tilde{\sigma}_{21} &= \tilde{\sigma}_{12}\,,
\end{aligned}$$

we get

$$\begin{aligned}
\hat{\rho} &= 0.68 \cdot \sqrt{\frac{79.17}{58.39}} = 0.80 \qquad \text{[cf. (10.80)]}, \\
\widehat{\operatorname{Var}(\hat{\beta})} &= \frac{58.39 \cdot (1 - 0.80^2)}{79.17 \cdot (7-4)} = 0.09 \qquad \text{[cf. (10.79)]}.
\end{aligned}$$

We now determine the two covariance matrices [(10.78)]

$$\begin{aligned}
\Gamma_1 &= \begin{pmatrix} 79.17/4 & 54.07/4 \\ 54.07/4 & [58.39 + \left(1 - \frac{4}{4}\right) \cdot 79.17 \cdot \left(0.09 - 0.68^2\right)]/4 \end{pmatrix} \\
&= \begin{pmatrix} 19.79 & 13.52 \\ 13.52 & 14.60 \end{pmatrix}, \\
\Gamma_2 &= \begin{pmatrix} 19.79 & 13.52 \\ 13.52 & 16.98 \end{pmatrix}.
\end{aligned}$$

Finally, our test statistics are

$$\begin{aligned}
&\text{interaction:} \quad Z_1 = 21.89/11.19 = 1.96 \qquad \text{[5 degrees of freedom]}, \\
&\text{treatment:} \quad Z_2 = -26.89/4.13 = -6.50 \qquad \text{[5 degrees of freedom]}, \\
&\text{period:} \quad Z_3 = 1.89/4.13 = 0.46 \qquad \text{[5 degrees of freedom]}.
\end{aligned}$$

The following table shows a comparison with the results of the analysis of the complete data set:

	Complete			Incomplete		
	t	df	p–Value	t	df	p–Value
Carry-over	0.96	6	0.376	1.96	5	0.108
Treatment	-2.96	6	0.026	-6.50	5	0.001
Period	0.74	6	0.488	0.46	5	0.667

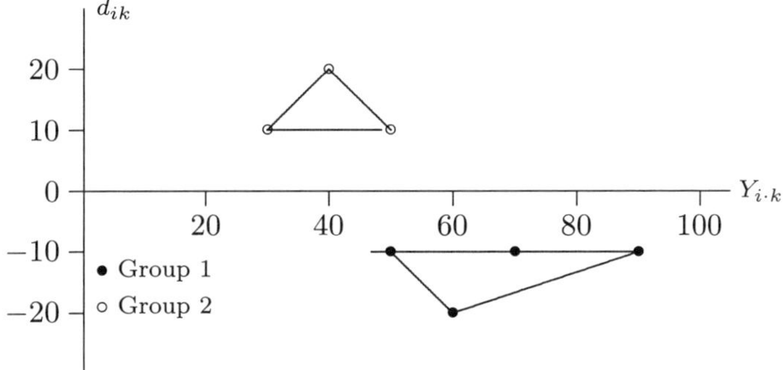

FIGURE 10.3. Difference–response–total plot of the incomplete data set.

An interesting result is that by excluding the second observation of patient 2, the treatment effect achieves an even higher level of significance of $p = 0.001$ (compared to $p = 0.026$ before). However, the carry–over effect of $p = 0.108$ is now very close to the limit of significance of $p = 0.100$ proposed by Grizzle. This is easily seen in the difference–response–total plot (Figure 10.3), which shows a clear separation of the covering, in the horizontal as well as the vertical direction (cf. Figure 8.5).

10.5 Missing Categorical Data

The procedures which have been introduced so far are all based on the linear regression model [(10.1)] with one continuous endogeneous variable Y. In many applications however, this assumption does not hold. Often Y is defined as a binary response variable and hence has a binomial distribution. Because of this, statistical analysis of incompletely observed categorical data demands different procedures than those previously described. For a clear and understandable representation of the different procedures, a three–dimensional contingency table is chosen where only one of the three categorical variables is assumed to be observed incompletely.

10.5.1 Introduction

Let Y be a binary outcome variable and let X_1 and X_2 be two covariates with J and K categories. The contingency table is thus of the dimension

$2 \times J \times K$. We assume that only X_2 is observed incompletely. The response of the covariate X_2 is indicated by an additional variable

$$R_2 = \begin{cases} 1 & \text{if } X_2 \text{ is not missing}, \\ 0 & \text{if } X_2 \text{ is missing} . \end{cases} \tag{10.90}$$

This leads to a new random variable

$$Z_2 = \begin{cases} X_2 & \text{if } R_2 = 1, \\ K+1 & \text{if } R_2 = 0 . \end{cases} \tag{10.91}$$

Assume that Y is related to X_1 and X_2 by the logistic model, a generalized linear model with logit link. This model assesses the effects of the covariates X_1 and X_2 on the outcome variable Y.

Let $\mu_{i|jk} = P(Y = i \mid X_1 = j, X_2 = k)$ be the conditional distribution of the binary variable Y, given the values of the covariates X_1 and X_2. The logistic model without interaction is

$$\ln\left(\frac{\mu_{1|jk}}{1 - \mu_{1|jk}}\right) = \beta_0 + \beta_{1j} + \beta_{2k} \tag{10.92}$$

or

$$\mu_{1|jk} = \frac{\exp(\beta_0 + \beta_{1j} + \beta_{2k})}{1 + \exp(\beta_0 + \beta_{1j} + \beta_{2k})} . \tag{10.93}$$

The parameters β_{1j} and β_{2k} describe the effect of the jth category of X_1 and the kth category of X_2 on the outcome variable Y. The parameter vector $\beta' = (\beta_0, \beta_{11}, \ldots, \beta_{1J}, \beta_{21}, \ldots, \beta_{2K})$ is estimated by the maximum–likelihood approach.

10.5.2 *Maximum Likelihood Estimation in the Complete Data Case*

Let $\pi^*_{ijk} = P(Y = i, X_1 = j, X_2 = k)$ be the joint distribution of the three variables for the complete data case and define

$$\begin{aligned} \gamma_{k|j} &= P(X_2 = k \mid X_1 = j), \\ \tau_j &= P(X_1 = j) . \end{aligned} \tag{10.94}$$

This parametrization allows a factorization of the joint distribution of Y, X_1, and X_2:

$$\begin{aligned} \pi^*_{ijk} &= \mu_{i|jk}\, \gamma_{k|j}\, \tau_j \\ &= (\mu_{1|jk})^i \, (1 - \mu_{1|jk})^{1-i}\, \gamma_{k|j}\, \tau_j . \end{aligned} \tag{10.95}$$

The contribution of a single observation with the values $Y = i, X_1 = j$, and $X_2 = k$ to the loglikelihood is

$$\ln\left((\mu_{1|jk})^i \, (1 - \mu_{1|jk})^{1-i} \right) + \ln \gamma_{k|j} + \ln \tau_j . \tag{10.96}$$

Hence, the loglikelihood is additive in the parameters and can be maximized independently for β, γ and τ. The maximum–likelihood estimate of β results from maximizing the loglikelihood of the entire sample

$$l_n^*(\beta) = \sum_{i=0}^{1} \sum_{j=1}^{J} \sum_{k=1}^{K} n_{ijk}^* \, l^*(\beta; i, j, k) \tag{10.97}$$

with

$$l^*(\beta; , i, j, k) = \ln \left(\left(\mu_{1|jk}\right)^i \left(1 - \mu_{1|jk}\right)^{1-i} \right),$$

where n_{ijk}^* is the number of elements with $Y = i, X_1 = j$, and $X_2 = k$. However, these equations are nonlinear in β and, hence, the maximization task involves an iterative method. A standard procedure for nonlinear optimization is the Newton–Raphson method or one of its variants, like the Fisher–scoring method.

10.5.3 Ad–Hoc Methods

Complete Case Analysis

Similar to the previously described situation with continuous variables, the complete case analysis is a standard approach for incomplete categorical data as well: the incompletely observed cases are eliminated from the data set. This reduced sample can now be analyzed by the maximum–likelihood approach for completely observed contingency tables (cf. Section 10.5.2).

Filling the Contingency Table

Unlike imputation methods that fill up the gaps in the data set (cf. Section 10.1), the filling method by Vach and Blettner (1991) fills up the cells of the contingency table. This is done by distributing the elements with a missing value of X_2, i.e., with the value $Z_2 = K+1$, to the other cells, dependent on the (known) values of Y and X_1.

Let n_{ijk} be the number of elements with the values $Y = i, X_1 = j$, and $Z_2 = k$, i.e., the cell counts of the $[2 \times J \times (K+1)]$–contingency table. The filled–up contingency table is then

$$n_{ijk}^{\text{FILL}} = n_{ijk} + n_{ijK+1} \frac{n_{ijk}}{\sum_{k=1}^{K} n_{ijk}} . \tag{10.98}$$

To this new $(2 \times J \times K)$ table, the maximum–likelihood procedure for completely observed contingency tables is applied, according to Section 10.5.2.

10.5.4 Model–Based Methods

Maximum–Likelihood Estimation in the Incomplete Data Case

Let $\pi_{ijk} = P(Y=i, X_1=j, Z_2=k)$ be the joint distribution of the variables Y, X_1, and Z_2, and define

$$q_{ijk} = P(R_2=1 \mid Y=i, X_1=j, X_2=k)\,. \tag{10.99}$$

The parametrization [(10.94) and (10.99)] enables a decomposition of the joint distribution (cf. Vach and Schumacher, 1993, p. 355). However, we have to distinguish between the case that the value of X_2 is known

$$\begin{aligned}\pi_{ijk} &= P(Y=i, X_1=j, Z_2=k)\\ &= P(Y=i, X_1=j, X_2=k, R_2=1)\\ &= P(R_2=1 \mid Y=i, X_1=j, X_2=k)\, P(Y=i \mid X_1=j, X_2=k)\\ &\quad \times P(X_2=k \mid X_1=j)\, P(X_1=j)\\ &= q_{ijk}\left(\mu_{1|jk}\right)^i \left(1-\mu_{1|jk}\right)^{1-i} \gamma_{k|j}\, \tau_j\,. \end{aligned} \tag{10.100}$$

and the case that the value of X_2 is missing, i.e., $k = K+1$:

$$\begin{aligned}\pi_{ijK+1} &= P(Y=i, X_1=j, Z_2=K+1)\\ &= P(Y=i, X_1=j, R_2=0)\\ &= P(R_2=0 \mid Y=i, X_1=j)\, P(Y=i \mid X_1=j)\, P(X_1=j)\\ &= \Big(\sum_{k=1}^{K} P(R_2=0 \mid Y=i, X_1=j, X_2=k)\, P(Y=i \mid X_1=j, X_2=k)\\ &\qquad \times P(X_2=k \mid X_1=j)\Big)\, P(X_1=j)\\ &= \Big(\sum_{k=1}^{K} (1-q_{ijk})\left(\mu_{1|jk}\right)^i \left(1-\mu_{1|jk}\right)^{1-i} \gamma_{k|j}\Big)\, \tau_j\,. \end{aligned} \tag{10.101}$$

Note that this distribution, unlike the complete data case, is dependent on the parameter q. Furthermore, the loglikelihood is not additive in the parameters β, γ, τ, and q and, hence, cannot be maximized separately for the parameters.

If the missing values are missing at random (MAR), then the missing probability is independent of the true value k of X_2, i.e.,

$$P(R_2=1 \mid Y=i, X_1=j, X_2=k) \equiv P(R_2=1 \mid Y=i, X_1=j) \tag{10.102}$$

and thus $q_{ijk} \equiv q_{ij}$. For the joint distribution of Y, X_1, and Z_2 (cf. (10.100) and (10.101)) this leads to

$$\pi_{ijk} = q_{ij} \left(\mu_{1|jk}\right)^i \left(1 - \mu_{1|jk}\right)^{1-i} \gamma_{k|j}\, \tau_j \tag{10.103}$$

for $k = 1, \ldots, K$ and to

$$\pi_{ijK+1} = (1 - q_{ij}) \left(\sum_{k=1}^{K} \left(\mu_{1|jk}\right)^i \left(1 - \mu_{1|jk}\right)^{1-i} \gamma_{k|j}\right) \tau_j \tag{10.104}$$

for $k = K+1$.

The contribution of a single element to the loglikelihood under the MAR assumption is now

$$\ln\, q_{ij} + \ln \left(\left(\mu_{1|jk}\right)^i \left(1 - \mu_{1|jk}\right)^{1-i} \right) + \ln\, \gamma_{k|j} + \ln\, \tau_j \tag{10.105}$$

for $k = 1, \ldots, K$ and

$$\ln\,(1 - q_{ij}) + \ln \left(\sum_{k=1}^{K} \left(\mu_{1|jk}\right)^i \left(1 - \mu_{1|jk}\right)^{1-i} \gamma_{k|j} \right) + \ln\, \tau_j \tag{10.106}$$

for $k = K+1$.

The loglikelihood disintegrates into three summands; hence, maximizing the loglikelihood for β can now be done independently of q. If the value of X_2 is missing, it is impossible to split the second summand depending on β and γ any further. Hence, the maximum–likelihood estimation of β requires joint maximization of the following loglikelihood for (β, γ), where γ is regarded as a nuisance parameter,

$$l_n^{\mathrm{ML}}(\beta, \gamma) = \sum_{i=0}^{1} \sum_{j=1}^{J} \sum_{k=1}^{K+1} n_{ijk}\, l^{\mathrm{ML}}(\beta, \gamma\,; i, j, k) \tag{10.107}$$

with

$$l^{\mathrm{ML}}\left(\beta, \gamma; i, j, k\right) = \begin{cases} \ln \left(\left(\mu_{1|jk}\right)^i \left(1 - \mu_{1|jk}\right)^{1-i} \right) + \ln\, \gamma_{k|j} & \text{for } k = 1, \ldots, K, \\ \ln \left(\sum_{k=1}^{K} \left(\mu_{1|jk}\right)^i \left(1 - \mu_{1|jk}\right)^{1-i} \gamma_{k|j} \right) & \text{for } k = K+1\,, \end{cases}$$

where n_{ijk} is the number of elements with $Y = i$, $X_1 = j$, and $Z_2 = k$.

Analogously to the complete data case, the computation of the estimates of β and γ requires an iterative procedure such as the Fisher–scoring method. Let $\theta = (\beta, \gamma)$. The iteration step of the Fisher–scoring method is

$$\theta^{(t+1)} = \theta^{(t)} + \left(I_{\theta\theta}^{\mathrm{ML}}(\theta^{(t)}, \hat{\tau}^n, \hat{q}^n)\right)^{-1} S_n^{\mathrm{ML}}(\theta^{(t)})\,, \tag{10.108}$$

with the score function

$$S_n^{\mathrm{ML}}(\theta) = \frac{1}{n} \frac{\partial}{\partial \theta}\, l_n^{\mathrm{ML}}(\theta) \tag{10.109}$$

and the information matrix

$$I_\theta^{\mathrm{ML}}(\theta, \tau, q) = -\operatorname{E}_{\theta,\tau,q}\left(\frac{\partial^2}{\partial\theta\,\partial\theta'}\, l^{\mathrm{ML}}\left(\beta; Y, X_1, Z_2\right)\right). \tag{10.110}$$

Pseudo–Maximum–Likelihood Estimation (PML)

In order to simplify the computation of the maximum–likelihood estimate of the regression parameter β, the nuisance parameter γ may be estimated from the observed values of X_1 and Z_2 and inserted into the loglikelihood, instead of joint iterative estimation along with β. A possible estimate (cf. Pepe and Fleming, 1991) is

$$\hat{\gamma}_{k|j} = \frac{n_{+jk}}{\sum_{k=1}^{K} n_{+jk}}. \tag{10.111}$$

This estimate is only consistent for γ under very strict assumptions for the missing mechanism. Vach and Schumacher (1993), p. 356, suggest applying this estimate to the filled up contingency table of the filling method (cf. Section 10.5.3)

$$\widetilde{\gamma}_{k|j} = \frac{n_{+jk}^{\mathrm{FILL}}}{\sum_{k=1}^{K} n_{+jk}^{\mathrm{FILL}}} = \frac{n_{0jk}\frac{\sum_{k=1}^{K+1} n_{0jk}}{\sum_{k=1}^{K} n_{0jk}} + n_{1jk}\frac{\sum_{k=1}^{K+1} n_{1jk}}{\sum_{k=1}^{K} n_{1jk}}}{\sum_{k=1}^{K+1} n_{+jk}}. \tag{10.112}$$

This estimate is consistent for γ if the MAR assumption holds. PML estimation of β is now achieved by iterative maximization of the following loglikelihood:

$$l_n^{\mathrm{PML}}(\beta) = \sum_{i=0}^{1}\sum_{j=1}^{J}\sum_{k=1}^{K+1} n_{ijk}\, l^{\mathrm{PML}}(\beta, \widetilde{\gamma}; i, j, k) \tag{10.113}$$

with

$$l^{\mathrm{PML}}\left(\beta, \widetilde{\gamma}; i, j, k\right) = \begin{cases} \ln\left(\left(\mu_{1|jk}\right)^{i}\left(1-\mu_{1|jk}\right)^{1-i}\right) & \text{for } k = 1, \ldots, K, \\ \ln\left(\left(\sum_{k=1}^{K}\mu_{1|jk}\widetilde{\gamma}_{k|j}\right)^{i}\left(1-\sum_{k=1}^{K}\mu_{1|jk}\widetilde{\gamma}_{k|j}\right)^{1-i}\right), & k = K+1. \end{cases}$$

10.6 Exercises and Questions

10.6.1 What is a selectivity bias and what is meant by drop–out in long–term studies?

10.6.2 Name the essential methods for imputation and describe them.

10.6.3 Explain the missing data mechanisms MAR, OAR, and MCAR by means of a bivariate sample.

10.6.4 Describe the OLS methods of Yates and Bartlett. What is the difference?

10.6.5 Assume that in a regression model values in the matrix X are missing and are to be replaced. Which methods may be used? Explain the effect on the unbiasedness of the final estimator $\hat{\beta}$.

Appendix A
Matrix Algebra

There are numerous books on matrix albegra which contain results useful for the discussion of linear models. See, for instance, books by Graybill (1961), Mardia et al. (1979), Searle (1982), Rao (1973), Rao and Mitra (1971), Rao and Rao (1998) to mention a few. We collect in this Appendix some of the important results for ready reference. Proofs are not generally given. References to original sources are given wherever necessary.

A.1 Introduction

Definition A.1. An $(m \times n)$–matrix A is a rectangular array of elements in m rows and n columns.

In the context of the material treated in this book and in this Appendix the elements of a matrix are taken as real numbers.

We refer to an $(m \times n)$–matrix of type (or order) $m \times n$ and indicate this by writing $A : m \times n$ or $\underset{m,n}{A}$.

Let a_{ij} be the element in the ith row and the jth column of A. Then A may be represented as

$$A = \begin{pmatrix} a_{11} & a_{12} & \cdots & a_{1n} \\ a_{21} & a_{22} & \cdots & a_{2n} \\ \vdots & \vdots & & \cdots \\ a_{m1} & a_{m2} & \cdots & a_{mn} \end{pmatrix} = (a_{ij}).$$

A matrix with $n = m$ rows and columns is called a square matrix. A square matrix, having zeros as elements below (above) the diagonal, is called an upper (lower) triangular matrix.

Let A and B be two matrices with the same dimensions, i.e., with the same number of rows m and columns n. Then the sum of the matrices $A \pm B$ is defined element by element, i.e.,

$$A \pm B = \begin{pmatrix} a_{11} \pm b_{11} & a_{12} \pm b_{12} & \dots & a_{1n} \pm b_{1n} \\ a_{21} \pm b_{21} & a_{22} \pm b_{22} & \dots & a_{2n} \pm b_{2n} \\ \vdots & \vdots & & \vdots \\ a_{m1} \pm b_{m1} & a_{m2} \pm b_{m2} & \dots & a_{mn} \pm b_{mn} \end{pmatrix}.$$

Also an element–by–element operation is the multiplication of a matrix with a scalar. Therefore $\nu A = \nu \cdot a_{ij} \; \forall i = 1, \dots, m, j = 1, \dots, n$.

Definition A.2. The transpose $A' : n \times m$ of a matrix $A : m \times n$ is given by interchanging the rows and columns of A. Thus

$$A' = (a_{ji}).$$

Then we have the following rules:

$$(A')' = A, \quad (A+B)' = A' + B', \quad (AB)' = B'A.'$$

Definition A.3. A square matrix is called symmetric, if $A' = A$.

Example A.1. Let x be a random vector with an expectation vector $\mathrm{E}(x) = \mu$. Then the covariance matrix of x is defined by

$$\mathrm{cov}(x) = \mathrm{E}(x - \mu)(x - \mu)' .$$

Any covariance matrix is symmetric.

Definition A.4. An $(m \times 1)$–matrix a is said to be an m–vector and is written as a column

$$a = \begin{pmatrix} a_1 \\ \vdots \\ a_m \end{pmatrix}.$$

Definition A.5. A $(1 \times n)$–matrix a' is said to be a row vector

$$a' = (a_1, \dots, a_n).$$

Hence, a matrix $A : m \times n$ may be written, alternatively, as

$$A = (a_{(1)}, \dots, a_{(n)}) = \begin{pmatrix} a_1' \\ \vdots \\ a_m' \end{pmatrix}$$

with

$$a_{(j)} = \begin{pmatrix} a_{1j} \\ \vdots \\ a_{mj} \end{pmatrix}, \quad a_i = \begin{pmatrix} a_{i1} \\ \vdots \\ a_{in} \end{pmatrix}.$$

Definition A.6. The $(n \times 1)$–row vector $(1, \ldots, 1)'$ is denoted by $\mathbf{1}'_n$ or $\mathbf{1}'$.

Definition A.7. The matrix $A : m \times m$ with $a_{ij} = 1$ (for all i, j) is given the symbol J_m, i.e.,

$$J_m = \begin{pmatrix} 1 & \ldots & 1 \\ \vdots & & \vdots \\ 1 & \vdots & 1 \end{pmatrix} = \mathbf{1}_m \mathbf{1}'_m .$$

Definition A.8. The n–vector

$$e_i = (0, \ldots, 0, 1, 0, \ldots, 0)',$$

whose ith component is one and whose remaining components are zero, is called the ith unit vector.

Definition A.9. A $(n \times n)$–matrix, with elements 1 on the main diagonal and zeros off the diagonal, is called the identity matrix I_n.

Definition A.10. A square matrix $A : n \times n$, with zeros in the off diagonal, is called a diagonal matrix. We write

$$A = \operatorname{diag}(a_{11}, \ldots, a_{nn}) = \operatorname{diag}(a_{ii}) = \begin{pmatrix} a_{11} & & 0 \\ & \ddots & \\ 0 & & a_{nn} \end{pmatrix}.$$

Definition A.11. A matrix A is said to be partitioned if its elements are arranged in submatrices.

Examples are

$$\underset{m,n}{A} = \left(\underset{m,r}{A_1}, \underset{m,s}{A_2} \right) \quad \text{with} \quad r + s = n$$

or

$$\underset{m,n}{A} = \begin{pmatrix} \underset{r,n-s}{A_{11}} & \underset{r,s}{A_{12}} \\ \underset{m-r,n-s}{A_{21}} & \underset{m-r,s}{A_{22}} \end{pmatrix}.$$

For partitioned matrices we get the transpose as

$$A' = \begin{pmatrix} A'_1 \\ A'_2 \end{pmatrix}, \quad A' = \begin{pmatrix} A'_{11} & A'_{21} \\ A'_{12} & A'_{22} \end{pmatrix},$$

respectively.

A.2 Trace of a Matrix

Definition A.12. Let $a_{11}, \ldots, a_{nn}$ be the elements on the main diagonal of a square matrix $A : n \times n$. Then the trace of A is defined as the sum

$$\mathrm{tr}(A) = \sum_{i=1}^{n} a_{ii}.$$

Theorem A.1. Let A and B be square $(n \times n)$–matrices and let c be a scalar factor. Then we have the following rules:

(i) $\mathrm{tr}(A \pm B) = \mathrm{tr}(A) \pm \mathrm{tr}(B)$.

(ii) $\mathrm{tr}(A') = \mathrm{tr}(A)$.

(iii) $\mathrm{tr}(cA) = c\,\mathrm{tr}(A)$.

(iv) $\mathrm{tr}(AB) = \mathrm{tr}(BA)$.

(v) $\mathrm{tr}(AA') = \mathrm{tr}(A'A) = \sum_{i,j} a_{ij}^2$.

(vi) If $a = (a_1, \ldots, a_n)'$ is an n–vector, then its squared norm may be written as

$$\| a \|^2 = a'a = \sum_{i=1}^{n} a_i^2 = \mathrm{tr}(aa').$$

Note: The rules (iv) and (v) also hold for the cases $A : n \times m$ and $B : m \times n$.

A.3 Determinant of a Matrix

Definition A.13. Let $n > 1$ be a positive integer. The determinant of a square matrix $A : n \times n$ is defined by

$$|A| = \sum_{i=1}^{n} (-1)^{i+j} a_{ij} |M_{ij}| \quad \text{(for any } j,\ j \text{ fixed)},$$

with $|M_{ij}|$ being the minor of the element a_{ij}. $|M_{ij}|$ is the determinant of the remaining $[(n-1) \times (n-1)]$–matrix when the ith row and the jth column of A are deleted. $A_{ij} = (-1)^{i+j} |M_{ij}|$ is called the cofactor of a_{ij}.

Example A.2.
$n = 2$:

$$|A| = a_{11}a_{22} - a_{12}a_{21} .$$

$n = 3$: First column ($j = 1$) fixed:

$$A_{11} = (-1)^2 \begin{vmatrix} a_{22} & a_{23} \\ a_{32} & a_{33} \end{vmatrix},$$

$$
\begin{aligned}
A_{21} &= (-1)^3 \begin{vmatrix} a_{12} & a_{13} \\ a_{32} & a_{33} \end{vmatrix}, \\
A_{31} &= (-1)^4 \begin{vmatrix} a_{12} & a_{13} \\ a_{22} & a_{23} \end{vmatrix}, \\
\Rightarrow |A| &= a_{11}A_{11} + a_{21}A_{21} + a_{31}A_{31} .
\end{aligned}
$$

Note: As an alternative, we may fix a row and develop the determinant of A according to

$$|A| = \sum_{j=1}^{n} (-1)^{i+j} a_{ij} |M_{ij}| \quad \text{(for any } i,\ i \text{ fixed)}.$$

Definition A.14. A square matrix A is said to be regular or nonsingular if $|A| \neq 0$. Otherwise A is said to be singular.

Theorem A.2. Let A and B be $(n \times n)$–square matrices and let c be a scalar. Then we have:

(i) $|A'| = |A|$.

(ii) $|cA| = c^n |A|$.

(iii) $|AB| = |A||B|$.

(iv) $|A^2| = |A|^2$.

(v) If A is diagonal or triangular, then

$$|A| = \prod_{i=1}^{n} a_{ii}.$$

(vi) For $D = \begin{pmatrix} \underset{n,n}{A} & \underset{n,m}{C} \\ \underset{m,n}{O} & \underset{m,m}{B} \end{pmatrix}$ we have

$$\begin{vmatrix} A & C \\ O & B \end{vmatrix} = |A||B|,$$

and, analogously,

$$\begin{vmatrix} A' & O' \\ C' & B' \end{vmatrix} = |A||B|.$$

(vii) If A is partitioned with $A_{11} : p \times p$ and $A_{22} : q \times q$ square and nonsingular, then

$$
\begin{aligned}
\begin{vmatrix} A_{11} & A_{12} \\ A_{21} & A_{22} \end{vmatrix} &= |A_{11}||A_{22} - A_{21}A_{11}^{-1}A_{12}| \\
&= |A_{22}||A_{11} - A_{12}A_{22}^{-1}A_{21}|.
\end{aligned}
$$

Proof. Define the following matrices

$$Z_1 = \begin{pmatrix} I & -A_{12}A_{22}^{-1} \\ 0 & I \end{pmatrix} \quad and \quad Z_2 = \begin{pmatrix} I & 0 \\ -A_{22}^{-1}A_{21} & I \end{pmatrix},$$

where $|Z_1| = |Z_2| = 1$ by (vi). Then we have

$$Z_1 A Z_2 = \begin{pmatrix} A_{11} - A_{12}A_{22}^{-1}A_{21} & 0 \\ 0 & A_{22} \end{pmatrix}$$

and [using (iii) and (iv)]

$$|Z_1 A Z_2| = |A| = |A_{22}||A_{11} - A_{12}A_{22}^{-1}A_{21}|.$$

(viii) $\begin{vmatrix} A & x \\ x' & c \end{vmatrix} = |A|(c - x'A^{-1}x)$ where x is an $(n,1)$–vector.

Proof. Use (vii) with A instead of A_{11} and c instead of A_{22}.

(ix) Let $B : p \times n$ and $C : n \times p$ be any matrices and let $A : p \times p$ be a nonsingular matrix. Then

$$\begin{aligned} |A + BC| &= |A||I_p + A^{-1}BC| \\ &= |A||I_n + CA^{-1}B|. \end{aligned}$$

Proof. The first relationship follows from (iii) and

$$(A + BC) = A(I_p + A^{-1}BC),$$

immediately.

The second relationship is a consequence of (vii) applied to the matrix

$$\begin{aligned} \begin{vmatrix} I_p & -A^{-1}B \\ C & I_n \end{vmatrix} &= |I_p||I_n + CA^{-1}B| \\ &= |I_n||I_p + A^{-1}BC|. \end{aligned}$$

(x) $|A + aa'| = |A|(1 + a'A^{-1}a)$, if A is nonsingular.

(xi) $|I_p + BC| = |I_n + CB|$, if $B : (p, n)$ and $C : (n, p)$.

A.4 Inverse of a Matrix

Definition A.15. The inverse of a square matrix $A : n \times n$ is written as A^{-1}. The inverse exists if and only if A is nonsingular. The inverse A^{-1} is unique and characterized by

$$AA^{-1} = A^{-1}A = I.$$

Theorem A.3. If all the inverses exist we have:

(i) $(cA)^{-1} = c^{-1}A^{-1}$.

(ii) $(AB)^{-1} = B^{-1}A^{-1}$.

(iii) If $A : p \times p$, $B : p \times n$, $C : n \times n$, and $D : n \times p$, then

$$(A + BCD)^{-1} = A^{-1} - A^{-1}B(C^{-1} + DA^{-1}B)^{-1}DA^{-1}.$$

(iv) If $1 + b'A^{-1}a \neq 0$, then we get, from (iii),

$$(A + ab')^{-1} = A^{-1} - \frac{A^{-1}ab'A^{-1}}{1 + b'A^{-1}a}.$$

(v) $|A^{-1}| = |A|^{-1}$.

Theorem A.4 (Inverse of a Partitioned Matrix).
For partitioned regular A:

$$A = \begin{pmatrix} E & F \\ G & H \end{pmatrix},$$

where $E : (n_1 \times n_1)$, $F : (n_1 \times n_2)$, $G : (n_2 \times n_1)$, and $H : (n_2 \times n_2)$ $(n_1 + n_2 = n)$ are such that E and $D = H - GE^{-1}F$ are regular, the partitioned inverse is given by

$$A^{-1} = \begin{pmatrix} E^{-1}(I + FD^{-1}GE^{-1}) & -E^{-1}FD^{-1} \\ -D^{-1}GE^{-1} & D^{-1} \end{pmatrix} = \begin{pmatrix} A^{11} & A^{12} \\ A^{21} & A^{22} \end{pmatrix}.$$

Proof. Check that the product of A and A^{-1} reduces to the identity matrix, i.e.,

$$AA^{-1} = A^{-1}A = I.$$

A.5 Orthogonal Matrices

Definition A.16. A square matrix $A : n \times n$ is said to be orthogonal if $AA' = I = A'A$. For orthogonal matrices we have:

(i) $A' = A^{-1}$.

(ii) $|A| = \pm 1$.

(iii) Let $\delta_{ij} = 1$ for $i = j$ and 0 for $i \neq j$, denote the Kronecker symbol. Then the row vectors a_i and the column vectors $a_{(i)}$ of A satisfy the conditions

$$a_i'a_j = \delta_{ij}\,, \quad a_{(i)}'a_{(j)}' = \delta_{ij}\,.$$

(iv) AB is orthogonal, if A and B are orthogonal.

Theorem A.5. For $A : n \times n$ and $B : n \times n$ symmetric, there exists an orthogonal matrix H such that $H'AH$ and $H'BH$ become diagonal if and only if A and B commute, i.e.,

$$AB = BA.$$

A.6 Rank of a Matrix

Definition A.17. The rank of $A : m \times n$ is the maximum number of linearly independent rows (or columns) of A. We write $\text{rank}(A) = p$.

Theorem A.6 (Rules for Ranks).

(i) $0 \leq \text{rank}(A) \leq \min(m, n)$.

(ii) $\text{rank}(A) = \text{rank}(A')$.

(iii) $\text{rank}(A + B) \leq \text{rank}(A) + \text{rank}(B)$.

(iv) $\text{rank}(AB) \leq \min\{\text{rank}(A), \text{rank}(B)\}$.

(v) $\text{rank}(AA') = \text{rank}(A'A) = \text{rank}(A) = \text{rank}(A')$.

(vi) For $B : m \times m$ and $C : n \times n$ regular, we have $\text{rank}(BAC) = \text{rank}(A)$.

(vii) For $A : n \times n$, $\text{rank}(A) = n$ if and only if A is regular.

(viii) If $A = \text{diag}(a_i)$, then $\text{rank}(A)$ equals the number of the $a_i \neq 0$.

A.7 Range and Null Space

Definition A.18.

(i) The range $\mathcal{R}(A)$ of a matrix $A : m \times n$ is the vector space spanned by the column vectors of A, that is,

$$\mathcal{R}(A) = \left\{ z : z = Ax = \sum_{i=1}^{n} a_{(i)} x_i, \ x \in \mathbb{R}^n \right\} \subset \mathbb{R}^m ,$$

where $a_{(1)}, \ldots, a_{(n)}$ are the column vectors of A.

(ii) The null space $\mathcal{N}(A)$ is the vector space defined by

$$\mathcal{N}(A) = \{x \in \Re^n \text{ and } Ax = 0\} \subset \Re^n .$$

Theorem A.7.

(i) $\text{rank}(A) = \dim \mathcal{R}(A)$, where $\dim V$ denotes the number of basis vectors of a vector space V.

(ii) $\dim \mathcal{R}(A) + \dim \mathcal{N}(A) = n$.

(iii) $\mathcal{N}(A) = \{\mathcal{R}(A')\}^{\perp}$.
($V^{\perp}$ the orthogonal complement of a vector space V defined by $V^{\perp} = \{x : x'y = 0 \text{ for all } y \in V\}$).

(iv) $\mathcal{R}(AA') = \mathcal{R}(A)$.

(v) $\mathcal{R}(AB) \subseteq \mathcal{R}(A)$ for any A and B.

(vi) For $A \geq 0$ and any B, $\mathcal{R}(BAB') = \mathcal{R}(BA)$.

A.8 Eigenvalues and Eigenvectors

Definition A.19. If $A : p \times p$ is a square matrix, then

$$q(\lambda) = |A - \lambda I|$$

is a pth–order polynomial in λ. The p roots $\lambda_1, \ldots, \lambda_p$ of the characteristic equation $q(\lambda) = |A - \lambda I| = 0$ are called eigenvalues or characteristic roots of A.

The eigenvalues possibly may be complex numbers. Since $|A - \lambda_i I| = 0$, $A - \lambda_i I$ is a singular matrix. Hence, there exists a nonzero vector $\gamma_i \neq 0$ satisfying $(A - \lambda_i I)\gamma_i = 0$, i.e.,

$$A\gamma_i = \lambda_i \gamma_i.$$

γ_i is called a (right) eigenvector of A for the eigenvalue λ_i. If λ_i is complex, then γ_i may have complex components. An eigenvector γ with real components is called standardized if $\gamma'\gamma = 1$.

Theorem A.8.

(i) If x and y are nonzero eigenvectors of A for λ_i and α and β are any real numbers, then $\alpha x + \beta y$ is also an eigenvector for λ_i, i.e.,

$$A(\alpha x + \beta y) = \lambda_i(\alpha x + \beta y).$$

Thus the eigenvectors for any λ_i span a vector space which is called an eigenspace of A for λ_i.

(ii) The polynomial $q(\lambda) = |A - \lambda I|$ has the normal form in terms of the roots

$$q(\lambda) = \prod_{i=1}^{p} (\lambda_i - \lambda).$$

Hence, $q(0) = \prod_{i=1}^{p} \lambda_i$ and

$$|A| = \prod_{i=1}^{p} \lambda_i.$$

(iii) Matching the coefficients of λ^{n-1} in $q(\lambda) = \prod_{i=1}^{p}(\lambda_i - \lambda)$ and $|A - \lambda I|$ gives

$$\mathrm{tr}(A) = \sum_{i=1}^{p} \lambda_i.$$

(iv) Let $C : p \times p$ be a regular matrix. Then A and CAC^{-1} have the same eigenvalues λ_i. If γ_i is an eigenvector for λ_i, then $C\gamma_i$ is an eigenvector of CAC^{-1} for λ_i.

Proof. As C is nonsingular, it has an inverse C^{-1} with $CC^{-1} = I$. We have $|C^{-1}| = |C|^{-1}$ and

$$\begin{aligned} |A - \lambda I| &= |C||A - \lambda C^{-1}C||C^{-1}| \\ &= |CAC^{-1} - \lambda I|. \end{aligned}$$

Thus, A and CAC^{-1} have the same eigenvalues. Let $A\gamma_i = \lambda_i \gamma_i$ and multiply from the left by C:

$$CAC^{-1}C\gamma_i = (CAC^{-1})(C\gamma_i) = \lambda_i(C\gamma_i).$$

(v) The matrix $A + \alpha I$ with α a real number has the eigenvalues $\tilde{\lambda}_i = \lambda_i + \alpha$ and the eigenvectors of A and $A + \alpha I$ coincide.

(vi) Let λ_1 denote any eigenvalue of $A : p \times p$ with eigenspace H of dimension r. If k denotes the multiplicity of λ_1 in $q(\lambda)$, then

$$1 \leq r \leq k.$$

Remark.

(a) For symmetric matrices A we have $r = k$.

(b) If A is not symmetric, then it is possible that $r < k$.

Example A.3. $A = \begin{pmatrix} 0 & 1 \\ 0 & 0 \end{pmatrix}$, $A \neq A'$

$$|A - \lambda I| = \begin{vmatrix} -\lambda & 1 \\ 0 & -\lambda \end{vmatrix} = \lambda^2 = 0.$$

The multiplicity of the eigenvalue $\lambda_{1,2} = 0$ is $k = 2$.

The eigenvectors for $\lambda = 0$ are $\gamma = \alpha \begin{pmatrix} 1 \\ 0 \end{pmatrix}$ and generate an eigenspace of dimension 1.

(c) If for any particular eigenvalue λ, $\dim(H) = r = 1$, then the standardized eigenvector for λ is unique (up to the sign).

Theorem A.9. Let $A : n \times p$ and $B : p \times n$, with $n \geq p$, be any two matrices. Then, from Theorem A.2(vii)

$$\begin{vmatrix} -\lambda I_n & -A \\ B & I_p \end{vmatrix} = (-\lambda)^{n-p}|BA - \lambda I_p| = |AB - \lambda I_n|.$$

Hence the n eigenvalues of AB are equal to the p eigenvalues of BA plus the eigenvalue 0 with multiplicity $n - p$. Suppose that $x \neq 0$ is an eigenvector of AB for any particular $\lambda \neq 0$. Then $y = Bx$ is an eigenvector of BA for this λ and we have $y \neq 0$, too.

Corollary. A matrix $A = aa'$ with $a \neq 0$ has the eigenvalues 0 and $\lambda = a'a$ and the eigenvector a.

Corollary. The nonzero eigenvalues of AA' are equal to the nonzero eigenvalues of $A'A$.

Theorem A.10. If A is symmetric, then all the eigenvalues are real.

A.9 Decomposition of Matrices

Theorem A.11 (Spectral Decomposition Theorem).
Any symmetric matrix $A : (p \times p)$ can be written as

$$A = \Gamma \Lambda \Gamma' = \sum \lambda_i \gamma_{(i)} \gamma'_{(i)} ,$$

where $\Lambda = \text{diag}(\lambda_1, \ldots, \lambda_p)$ is the diagonal matrix of the eigenvalues of A and $\Gamma = (\gamma_{(1)}, \ldots, \gamma_{(p)})$ is the matrix of the standardized eigenvectors $\gamma_{(i)}$. Γ is orthogonal

$$\Gamma \Gamma' = \Gamma' \Gamma = I .$$

Theorem A.12. Suppose A is symmetric and $A = \Gamma \Lambda \Gamma'$. Then:

(i) A and Λ have the same eigenvalues with the same multiplicity.

(ii) From $A = \Gamma \Lambda \Gamma'$ we get $\Lambda = \Gamma' A \Gamma$.

(iii) If $A : p \times p$ is a symmetric matrix, then for any integer n, $A^n = \Gamma \Lambda^n \Gamma'$ and $\Lambda^n = \text{diag}(\lambda_i^n)$. If the eigenvalues of A are positive, then we can define the rational powers

$$A^{r/s} = \Gamma \Lambda^{r/s} \Gamma' \quad \text{with } \Lambda^{r/s} = \text{diag}(\lambda_i^{r/s})$$

for integers $s > 0$ and r.

Important special cases are when $\lambda_i > 0$

$$A^{-1} = \Gamma \Lambda^{-1} \Gamma' \quad \text{with } \Lambda^{-1} = \text{diag}(\lambda_i^{-1}),$$

the symmetric square root decomposition of A is when $\lambda_i \geq 0$

$$A^{1/2} = \Gamma \Lambda^{1/2} \Gamma' \quad \text{with } \Lambda^{1/2} = \text{diag}(\lambda_i^{1/2})$$

and, if $\lambda_i > 0$,

$$A^{-1/2} = \Gamma \Lambda^{-1/2} \Gamma' \quad \text{with } \Lambda^{-1/2} = \text{diag}(\lambda_i^{-1/2}).$$

(iv) For any square matrix A the rank of A equals the number of nonzero eigenvalues.

Proof. According to Theorem A.6(vi) we have $\text{rank}(A) = \text{rank}(\Gamma \Lambda \Gamma') = \text{rank}(\Lambda)$. But $\text{rank}(\Lambda)$ equals the number of nonzero λ_i's.

(v) A symmetric matrix A is uniquely determined by its distinct eigenvalues and the corresponding eigenspaces. If the distinct eigenvalues λ_i are ordered as $\lambda_1 \geq \cdots \geq \lambda_p$, then the matrix Γ is unique (up to sign).

(vi) $A^{1/2}$ and A have the same eigenvectors. Hence, $A^{1/2}$ is unique.

(vii) Let $\lambda_1 \geq \lambda_2 \geq \cdots \geq \lambda_k > 0$ be the nonzero eigenvalues and let $\lambda_{k+1} = \cdots = \lambda_p = 0$. Then we have

$$A = (\Gamma_1 \Gamma_2) \begin{pmatrix} \Lambda_1 & 0 \\ 0 & 0 \end{pmatrix} \begin{pmatrix} \Gamma_1' \\ \Gamma_2' \end{pmatrix} = \Gamma_1 \Lambda_1 \Gamma_1'$$

with $\Lambda_1 = \text{diag}(\lambda_1, \ldots, \lambda_k)$ and $\Gamma_1 = (\gamma_{(1)}, \ldots, \gamma_{(k)})$, whereas $\Gamma_1' \Gamma_1 = I_k$ holds so that Γ_1 is column–orthogonal.

(viii) A symmetric matrix A is of rank 1 if and only if $A = aa'$ where $a \neq 0$.

Proof. If $\text{rank}(A) = \text{rank}(\Lambda) = 1$, then $\Lambda = \begin{pmatrix} \lambda & 0 \\ 0 & 0 \end{pmatrix}$, $A = \lambda \gamma \gamma' = aa'$ with $a = \sqrt{\lambda} \gamma$. If $A = aa'$, then by Theorem A.6(iv) we have $\text{rank}(A) = \text{rank}(a) = 1$.

Theorem A.13 (Singular Value Decomposition of a Rectangular Matrix). Let A be a rectangular $(n \times p)$–matrix of rank r. Then we have

$$\underset{n,p}{A} = \underset{n,r}{U} \; \underset{r,r}{L} \; \underset{r,p}{V'}$$

with $U'U = I_r$, $V'V = I_r$ and $L = \text{diag}(l_1, \ldots, l_r)$, $l_i > 0$.

For a proof, see Rao (1973), p. 42.

Theorem A.14. If $A : p \times q$ has $\text{rank}(A) = r$, then A contains at least one nonsingular (r, r)–submatrix X, such that A has the so–called normal presentation

$$\underset{p,q}{A} = \begin{pmatrix} \underset{r,r}{X} & \underset{r,q-r}{Y} \\ \underset{p-r,r}{Z} & \underset{p-r,q-r}{W} \end{pmatrix}.$$

All square submatrices of type $(r + s, r + s)$ with $(s \geq 1)$ are singular.

Proof. As $\text{rank}(A) = \text{rank}(X)$ holds, the first r rows of (X, Y) are linearly independent. Then the $(p - r)$–rows (Z, W) are linear combinations of (X, Y) i.e., there exists a matrix F such that

$$(Z, W) = F(X, Y).$$

Analogously, there exists a matrix H satisfying

$$\begin{pmatrix} Y \\ W \end{pmatrix} = \begin{pmatrix} X \\ Z \end{pmatrix} H.$$

Hence, we get $W = FY = FXH$ and

$$A = \begin{pmatrix} X & Y \\ Z & W \end{pmatrix} = \begin{pmatrix} X & XH \\ FX & FXH \end{pmatrix}$$
$$= \begin{pmatrix} I \\ F \end{pmatrix} X(I, H)$$
$$= \begin{pmatrix} X \\ FX \end{pmatrix} (I, H) = \begin{pmatrix} I \\ F \end{pmatrix} (X, XH).$$

As X is nonsingular, the inverse X^{-1} exists. Then we obtain $F = ZX^{-1}$, $H = X^{-1}Y$, $W = ZX^{-1}Y$, and

$$A = \begin{pmatrix} X & Y \\ Z & W \end{pmatrix} = \begin{pmatrix} I \\ ZX^{-1} \end{pmatrix} X(I, X^{-1}Y)$$
$$= \begin{pmatrix} X \\ Z \end{pmatrix} (I, X^{-1}Y)$$
$$= \begin{pmatrix} I \\ ZX^{-1} \end{pmatrix} (X\ Y).$$

Theorem A.15 (Full Rank Factorization).

(i) If $A : p \times q$ has $\text{rank}(A) = r$, then A may be written as

$$\underset{p,q}{A} = \underset{p,r}{K} \ \underset{r,q}{L}$$

with K of full column rank r and L of full row rank r.

Proof. Theorem A.14.

(ii) If $A : p \times q$ has $\text{rank}(A) = p$, then A may be written as

$$A = M(I, H) \quad \text{where } M : p \times p \text{ is regular.}$$

Proof. Theorem A.15(i).

A.10 Definite Matrices and Quadratic Forms

Definition A.20. Suppose $A : n \times n$ is symmetric and $x : n \times 1$ is any vector. Then the quadratic form in x is defined as the function

$$Q(x) = x'Ax = \sum_{i,j} a_{ij} x_i x_j .$$

Clearly $Q(0) = 0$.

Definition A.21. The matrix A is called positive definite (p.d.) if $Q(x) > 0$ for all $x \neq 0$. We write $A > 0$.

Note: If $A > 0$, then $(-A)$ is called negative definite.

Definition A.22. The quadratic form $x'Ax$ (and the matrix A, also) is called positive semidefinite (p.s.d.), if $Q(x) \geq 0$ for all x and $Q(x) = 0$ for at least one $x \neq 0$.

Definition A.23. The quadratic form $x'Ax$ and A) is called nonnegative definite (n.n.d.), if it is either p.d. or p.s.d., i.e., if $x'Ax \geq 0$ for all x. If A is n.n.d., we write $A \geq 0$.

Theorem A.16. Let the $(n \times n)$–matrix $A > 0$. Then:

(i) A has all eigenvalues $\lambda_i > 0$.

(ii) $x'Ax > 0$ for any $x \neq 0$.

(iii) A is nonsingular and $|A| > 0$.

(iv) $A^{-1} > 0$.

(v) $\text{tr}(A) > 0$.

(vi) Let $P : n \times m$ be of $\text{rank}(P) = m \leq n$. Then $P'AP > 0$ and, in particular, $P'P > 0$, choosing $A = I$.

(vii) Let $P : n \times m$ be of $\text{rank}(P) < m \leq n$. Then $P'AP \geq 0$ and $P'P \geq 0$.

Theorem A.17. Let $A : n \times n$ and $B : n \times n$ be such that $A > 0$ and $B : n \times n \geq 0$. Then:

(i) $C = A + B > 0$.

(ii) $A^{-1} - (A + B)^{-1} \geq 0$.

(iii) $|A| \leq |A + B|$.

Theorem A.18. Let $A \geq 0$. Then:

(i) $\lambda_i \geq 0$.

(ii) $\text{tr}(A) \geq 0$.

(iii) $A = A^{1/2}A^{1/2}$ with $A^{1/2} = \Gamma\Lambda^{1/2}\Lambda'$.

(iv) For any matrix $C : n \times m$ we have $C'AC \geq 0$.

(v) For any matrix C we have $C'C \geq 0$ and $CC' \geq 0$.

Theorem A.19. For any matrix $A \geq 0$ we have $0 \leq \lambda_i \leq 1$ if and only if $(I - A) \geq 0$.

Proof. Write the symmetric matrix A in its spectral form as $A = \Gamma\Lambda\Gamma'$. Then we have

$$(I - A) = \Gamma(I - \Lambda)\Gamma' \geq 0$$

if and only if

$$\Gamma'\Gamma(I - \Lambda)\Gamma'\Gamma = I - \Lambda \geq 0.$$

(a) If $I - \Lambda \geq 0$, then for the eigenvalues of $I - A$ we have $1 - \lambda_i \geq 0$, i.e., $0 \leq \lambda_i \leq 1$.

(b) If $0 \leq \lambda_i \leq 1$, then for any $x \neq 0$:

$$x'(I - \Lambda)x = \sum x_i^2(1 - \lambda_i) \geq 0,$$

i.e., $I - \Lambda \geq 0$.

Theorem A.20 (Theobald, 1974).
Let $D : n \times n$ be symmetric. Then $D \geq 0$ if and only if $\text{tr}\{CD\} \geq 0$ for all $C \geq 0$.

Proof. D is symmetric, so that

$$D = \Gamma\Lambda\Gamma' = \sum \lambda_i \gamma_i \gamma_i'$$

and, hence,

$$\begin{aligned} tr\{CD\} &= tr\left\{\sum \lambda_i C \gamma_i \gamma_i'\right\} \\ &= \sum \lambda_i \gamma_i' C \gamma_i. \end{aligned}$$

(a) Let $D \geq 0$ and, hence, $\lambda_i \geq 0$ for all i. Then $tr(CD) \geq 0$ if $C \geq 0$.

(b) Let $\text{tr}\{CD\} \geq 0$ for all $C \geq 0$. Choose $C = \gamma_i \gamma_i'$ ($i = 1, \ldots, n$, i fixed) so that

$$\begin{aligned} 0 \leq \text{tr}\{CD\} &= \text{tr}\left\{\gamma_i \gamma_i' \left(\sum_j \lambda_j \gamma_j \gamma_j'\right)\right\} \\ &= \lambda_i \ (i = 1, \ldots, n) \end{aligned}$$

and $D = \Gamma\Lambda\Gamma' \geq 0$.

Theorem A.21. Let $A : n \times n$ be symmetric with eigenvalues $\lambda_1 \geq \ldots \geq \lambda_n$. Then

$$\sup_x \frac{x'Ax}{x'x} = \lambda_1, \quad \inf_x \frac{x'Ax}{x'x} = \lambda_n.$$

Proof. See Rao (1973), p. 62.

Theorem A.22. Let $A : n \times r = (A_1, A_2)$, with A_1 of order $n \times r_1$ and A_2 of order $n \times r_2$ and $\text{rank}(A) = r = r_1 + r_2$.

Define the orthogonal projectors $M_1 = A_1(A_1'A_1)^{-1}A_1'$ and $M = A(A'A)^{-1}A'$. Then

$$M = M_1 + (I - M_1)A_2(A_2'(I - M_1)A_2)^{-1}A_2'(I - M_1).$$

Proof. M_1 and M are symmetric idempotent matrices fulfilling $M_1 A_1 = 0$ and $MA = 0$. Using Theorem A.4 for partial inversion of $A'A$, i.e.,

$$(A'A)^{-1} = \begin{pmatrix} A_1'A_1 & A_1'A_2 \\ A_2'A_1 & A_2'A_2 \end{pmatrix}^{-1},$$

and using the special form of the matrix D defined in Theorem A.4, i.e.,

$$D = A_2'(I - M_1)A_2,$$

straightforward calculation concludes the proof.

Theorem A.23. Let $A : n \times m$, with $\text{rank}(A) = m \le n$ and $B : m \times m$, be any symmetric matrix. Then

$$ABA' \ge 0 \quad \text{if and only if} \quad B \ge 0.$$

Proof. (i) $B \ge 0 \Rightarrow ABA' \ge 0$ for all A.

(ii) Let $\text{rank}(A) = m \le n$ and assume $ABA' \ge 0$, so that $x'ABA'x \ge 0$ for all $x \in E^n$.
We have to prove that $y'By \ge 0$ for all $y \in E^m$. As $\text{rank}(A) = m$, the inverse $(A'A)^{-1}$ exists. Setting $z = A(A'A)^{-1}y$, we have $A'z = y$ and $y'By = z'ABA'z \ge 0$ so that $B \ge 0$.

Definition A.24. Let $A : n \times n$ and $B : n \times n$ be any matrices. Then the roots $\lambda_i = \lambda_i^B(A)$ of the equation

$$|A - \lambda B| = 0$$

are called the eigenvalues of A in the metric of B. For $B = I$ we obtain the usual eigenvalues defined in Definition A.19 (cf. Dhrymes (1978)).

Theorem A.24. Let $B > 0$ and $A \ge 0$. Then $\lambda_i^B(A) \ge 0$.
Proof. $B > 0$ is equivalent to $B = B^{1/2}B^{1/2}$ with $B^{1/2}$ nonsingular and unique (Theorem A.12(iii)). Then we may write

$$0 = |A - \lambda B| = |B^{1/2}|^2 |B^{-1/2}AB^{-1/2} - \lambda I|$$

and $\lambda_i^B(A) = \lambda_i^I(B^{-1/2}AB^{-1/2}) \ge 0$, as $B^{-1/2}AB^{-1/2} \ge 0$.

Theorem A.25 (Simultaneous Diagonalization).
Let $B > 0$ and $A \ge 0$ and denote by $\Lambda = \text{diag}(\lambda_i^B(A))$ the diagonal matrix of the eigenvalues of A in the metric of B. Then there exists a nonsingular matrix W such that

$$B = W'W \quad \text{and} \quad A = W'\Lambda W.$$

Proof. From the proof of Theorem A.24 we know that the roots $\lambda_i^B(A)$ are the usual eigenvalues of the matrix $B^{-1/2}AB^{-1/2}$. Let X be the matrix of the corresponding eigenvectors:

$$B^{-1/2}AB^{-1/2}X = X\Lambda,$$

i.e.,

$$A = B^{1/2} X \Lambda X' B^{1/2} = W' \Lambda W$$

with $W' = B^{1/2} X$ regular and

$$B = W'W = B^{1/2} X X' B^{1/2} = B^{1/2} B^{1/2}.$$

Theorem A.26. Let $A > 0$ (or $A \geq 0$) and $B > 0$. Then

$$B - A > 0 \quad \text{if and only if} \quad \lambda_i^B(A) < 1.$$

Proof. Using Theorem A.25 we may write

$$B - A = W'(I - \Lambda)W,$$

i.e.,

$$\begin{aligned} x'(B - A)x &= x'W'(I - \Lambda)Wx \\ &= y'(I - \Lambda)y \\ &= \sum (1 - \lambda_i^B(A)) y_i^2 \end{aligned}$$

with $y = Wx$, W regular and, hence, $y \neq 0$ for $x \neq 0$. Then $x'(B - A)x > 0$ holds if and only if

$$\lambda_i^B(A) < 1.$$

Theorem A.27. Let $A > 0$ (or $A \geq 0$) and $B > 0$. Then

$$A - B \geq 0$$

if and only if

$$\lambda_i^B(A) \leq 1.$$

Proof. Similar to Theorem A.26.

Theorem A.28. Let $A > 0$ and $B > 0$. Then

$$B - A > 0 \quad \text{if and only if} \quad A^{-1} - B^{-1} > 0.$$

Proof. From Theorem A.25 we have

$$B = W'W, \quad A = W' \Lambda W.$$

Since W is regular we have

$$B^{-1} = W^{-1} {W'}^{-1}, \quad A^{-1} = W^{-1} \Lambda^{-1} {W'}^{-1},$$

i.e.,

$$A^{-1} - B^{-1} = W^{-1}(\Lambda^{-1} - I){W'}^{-1} > 0,$$

as $\lambda_i^B(A) < 1$ and, hence, $\Lambda^{-1} - I > 0$.

Theorem A.29. Let $B - A > 0$. Then $|B| > |A|$ and $\mathrm{tr}(B) > \mathrm{tr}(A)$.

If $B - A \geq 0$, then $|B| \geq |A|$ and $\mathrm{tr}(B) \geq \mathrm{tr}(A)$.
Proof. From Theorem A.25 and Theorem A.2(iii), (v) we get

$$\begin{aligned} |B| &= |W'W| = |W|^2, \\ |A| &= |W'\Lambda W| = |W|^2|\Lambda| = |W|^2 \prod \lambda_i^B(A), \end{aligned}$$

i.e.,

$$|A| = |B| \prod \lambda_i^B(A).$$

For $B - A > 0$ we have $\lambda_i^B(A) < 1$, i.e., $|A| < |B|$.
For $B - A \geq 0$ we have $\lambda_i^B(A) \leq 1$, i.e., $|A| \leq |B|$.

$B - A > 0$ implies $\mathrm{tr}(B - A) > 0$, and $\mathrm{tr}(B) > \mathrm{tr}(A)$. Analogously, $B - A \geq 0$ implies $\mathrm{tr}(B) \geq \mathrm{tr}(A)$.

Theorem A.30 (Cauchy–Schwarz Inequality).
Let x and y be real vectors of the same dimension. Then

$$(x'y)^2 \leq (x'x)(y'y),$$

with equality if and only if x and y are linearly dependent.

Theorem A.31. Let x and y be n–vectors and $A > 0$. Then we have the following results:

(i) $(x'Ay)^2 \leq (x'Ax)(y'Ay)$.

(ii) $(x'y)^2 \leq (x'Ax)(y'A^{-1}y)$.

Proof. (i) $A \geq 0$ is equivalent to $A = BB$ with $B = A^{1/2}$ (Theorem A.18(iii)). Let $Bx = \tilde{x}$ and $By = \tilde{y}$. Then (i) is a consequence of Theorem A.30.

(ii) $A > 0$ is equivalent to $A = A^{1/2}A^{1/2}$ and $A^{-1} = A^{-1/2}A^{-1/2}$. Let $A^{1/2}x = \tilde{x}$ and $A^{-1/2}y = \tilde{y}$, then (ii) is a consequence of Theorem A.30.

Theorem A.32. Let $A > 0$ and let T be any square matrix. Then:

(i) $\sup_{x \neq 0}(x'y)^2/x'Ax = y'A^{-1}y$.

(ii) $\sup_{x \neq 0}(y'Tx)^2/x'Ax = y'TA^{-1}T'y$.

Proof. Use Theorem A.31(ii).

Theorem A.33. Let $I : n \times n$ be the identity matrix and a an n–vector. Then

$$I - aa' \geq 0 \quad \text{if and only if} \quad a'a \leq 1.$$

Proof. The matrix aa' is of rank 1 and $aa' \geq 0$. The spectral decomposition is $aa' = C\Lambda C'$ with $\Lambda = \text{diag}(\lambda, 0, \ldots, 0)$ and $\lambda = a'a$. Hence, $I - aa' = C(I - \Lambda)C' \geq 0$ if and only if $\lambda = a'a \leq 1$ (see Theorem A.19).

Theorem A.34. Assume $MM' - NN' \geq 0$. Then there exists a matrix H such that $N = MH$.

Proof. (Milliken and Akdeniz, 1977). Let $M(n,r)$ of $\text{rank}(M) = s$ and let x be any vector $\in \mathcal{R}(I - MM^-)$, implying $x'M = 0$ and $x'MM'x = 0$. As NN' and $MM' - NN'$ (by assumption) are n.n.d., we may conclude that $x'NN'x \geq 0$ and

$$x'(MM' - NN')x = -x'NN'x \geq 0,$$

so that $x'NN'x = 0$ and $x'N = 0$. Hence, $N \subset \mathcal{R}(M)$ or, equivalently, $N = MH$ for some matrix $H(r,k)$.

Theorem A.35. Let A be an $(n \times n)$–matrix and assume $(-A) > 0$. Let a be an n–vector. In the case of $n \geq 2$, the matrix $A + aa'$ is never n.n.d.

Proof. (Guilkey and Price, 1981). The matrix aa' is of rank ≤ 1. In the case of $n \geq 2$ there exists a nonzero vector w such that $w'aa'w = 0$ implying $w'(A + aa')w = w'Aw < 0$.

A.11 Idempotent Matrices

Definition A.25. A square matrix A is called idempotent if it satisfies

$$A^2 = AA = A.$$

An idempotent matrix A is called an orthogonal projector if $A = A'$. Otherwise, A is called an oblique projector.

Theorem A.36. Let $A : n \times n$ be idempotent with $\text{rank}(A) = r \leq n$. Then we have:

(i) The eigenvalues of A are 1 or 0.

(ii) $\text{tr}(A) = \text{rank}(A) = r$.

(iii) If A is of full rank n, then $A = I_n$.

(iv) If A and B are idempotent and if $AB = BA$, then AB is also idempotent.

(v) If A is idempotent and P is orthogonal, then PAP' is also idempotent.

(vi) If A is idempotent, then $I - A$ is idempotent and

$$A(I - A) = (I - A)A = 0.$$

Proof. (i) The characteristic equation

$$Ax = \lambda x$$

multiplied by A gives

$$AAx = Ax = \lambda Ax = \lambda^2 x.$$

Multiplication of both the equations by x' then yields

$$x'Ax = \lambda x'x = \lambda^2 x'x,$$

i.e.,

$$\lambda(\lambda - 1) = 0.$$

(ii) From the spectral decomposition

$$A = \Gamma\Lambda\Gamma'$$

we obtain

$$\operatorname{rank}(A) = \operatorname{rank}(\Lambda) = \operatorname{tr}(\Lambda) = r,$$

where r is the number of characteristic roots with value 1.

(iii) Let $\operatorname{rank}(A) = \operatorname{rank}(\Lambda) = n$, then $\Lambda = I_n$ and

$$A = \Gamma\Lambda\Gamma' = I_n.$$

(iv)–(vi) follow from the definition of an idempotent matrix.

A.12 Generalized Inverse

Definition A.26. Let A be an $(m \times n)$–matrix. Then a matrix $A^- : n \times m$ is said to be a generalized inverse (g–inverse) of A if

$$AA^-A = A$$

holds.

Theorem A.37. A generalized inverse always exists although it is not unique in general.

Proof. Assume $\operatorname{rank}(A) = r$. According to Theorem A.13 we may write

$$\underset{m,n}{A} = \underset{m,r}{U} \; \underset{r,r}{L} \; \underset{r,n}{V'}$$

with $U'U = I_r$ and $V'V = I_r$ and

$$L = \operatorname{diag}(l_1, \ldots, l_r), \quad l_i > 0.$$

Then

$$A^- = V \begin{pmatrix} L^{-1} & X \\ Y & Z \end{pmatrix} U',$$

where X, Y, and Z are arbitrary matrices (of suitable dimensions), is a g–inverse.

Using Theorem A.14, i.e.,

$$A = \begin{pmatrix} X & Y \\ Z & W \end{pmatrix}$$

with X nonsingular, we have

$$A^- = \begin{pmatrix} X^{-1} & 0 \\ 0 & 0 \end{pmatrix}$$

as a special g–inverse.

For details on g–inverses, the reader is referred to Rao and Mitra (1971).

Definition A.27 (Moore–Penrose Inverse). A matrix A^+ satisfying the following conditions is called a Moore–Penrose inverse of A:

$$\begin{array}{llll} (i) & AA^+A = A; & (ii) & A^+AA^+ = A^+; \\ (iii) & (A^+A)' = A^+A; & (iv) & (AA^+)' = AA^+. \end{array}$$

A^+ is unique.

Theorem A.38. For any matrix $A : m \times n$ and any g–inverse $A^- : m \times n$ we have:

(i) A^-A and AA^- are idempotent.

(ii) $\text{rank}(A) = \text{rank}(AA^-) = \text{rank}(A^-A)$.

(iii) $\text{rank}(A) \leq \text{rank}(A^-)$.

Proof. (i) Using the definition of the g–inverse:

$$(A^-A)(A^-A) = A^-(AA^-A) = A^-A.$$

(ii) According to Theorem A.6(iv) we get

$$\text{rank}(A) = \text{rank}(AA^-A) \leq \text{rank}(A^-A) \leq \text{rank}(A),$$

i.e., $\text{rank}(A^-A) = \text{rank}(A)$. Analogously, we see that $\text{rank}(A) = \text{rank}(AA^-)$.

(iii) $\text{rank}(A) = \text{rank}(AA^-A) \leq \text{rank}(AA^-) \leq \text{rank}(A^-)$.

Theorem A.39. Let A be an $(m \times n)$–matrix. Then:

(i) A regular $\Rightarrow A^+ = A^{-1}$.

(ii) $(A^+)^+ = A$.

(iii) $(A^+)' = (A')^+$.

(iv) $\text{rank}(A) = \text{rank}(A^+) = \text{rank}(A^+A) = \text{rank}(AA^+)$.

(v) A an orthogonal projector $\Rightarrow A^+ = A$.

(vi) $\text{rank}(A) : m \times n = m. \Rightarrow A^+ = A'(AA')^{-1}$ and $AA^+ = I_m$.

(vii) $\text{rank}(A) : m \times n = n. \Rightarrow A^+ = (A'A)^{-1}A'$ and $A^+A = I_n$.

(viii) If $P : m \times m$ and $Q : n \times n$ are orthogonal $\Rightarrow$ $(PAQ)^+ = Q^{-1}A^+P^{-1}$.

(ix) $(A'A)^+ = A^+(A')^+$ and $(AA')^+ = (A')^+A^+$.

(x) $A^+ = (A'A)^+A' = A'(AA')^+$.

Theorem A.40 (Baksalary et al., 1983). Let $M : n \times n \geq 0$ and $N : m \times n$ be any matrices. Then

$$M - N'(NM^+N')^+N \geq 0$$

if and only if

$$\mathcal{R}(N'NM) \subset \mathcal{R}(M).$$

Theorem A.41. Let A be any square $(n \times n)$–matrix and let a be an n–vector with $a \notin \mathcal{R}(A)$. Then a g–inverse of $A + aa'$ is given by

$$\begin{aligned} (A + aa')^- &= A^- - \frac{A^-aa'U'U}{a'U'Ua} \\ &\quad - \frac{VV'aa'A^-}{a'VV'a} + \phi \frac{VV'aa'U'U}{(a'U'Ua)(a'VV'a)}, \end{aligned}$$

with A^- any g–inverse of A and

$$\phi = 1 + a'A^-a, \quad U = I - AA^-, \quad V = I - A^-A.$$

Proof. Straightforward by checking $AA^-A = A$.

Theorem A.42. Let A be a square $(n \times n)$–matrix. Then we have the following results:

(i) Assume a and b to be vectors with $a, b \in \mathcal{R}(A)$ and let A be symmetric. Then the bilinear form $a'A^-b$ is invariant to the choice of A^-.

(ii) $A(A'A)^-A'$ is invariant to the choice of $(A'A)^-$.

Proof. (i) $a, b \in \mathcal{R}(A) \;\Rightarrow\; a = Ac$ and $b = Ad$.

Using the symmetry of A gives

$$\begin{aligned} a'A^-b &= c'A'A^-Ad \\ &= c'Ad. \end{aligned}$$

(ii) Using the row–wise representation of A as $A = \begin{pmatrix} a_1' \\ \vdots \\ a_n' \end{pmatrix}$ gives

$$A(A'A)^-A' = (a_i'(A'A)^-a_j).$$

As $A'A$ is symmetric, we may conclude then: (i) that all bilinear forms $a_i'(A'A)a_j$ are invariant to the choice of $(A'A)^-$ and, hence, (ii) is proved.

Theorem A.43. Let $A : n \times n$ be symmetric, $a \in \mathcal{R}(A)$, $b \in \mathcal{R}(A)$, and assume $1 + b'A^+a \neq 0$. Then

$$(A + ab')^+ = A^+ - \frac{A^+ab'A^+}{1 + b'A^+a}.$$

Proof. Straightforward, using Theorems A.41 and A.42.

Theorem A.44. Let $A : n \times n$ be symmetric, a an n–vector, and $\alpha > 0$ any scalar. Then the following statements are equivalent:

(i) $\alpha A - aa' \geq 0$.

(ii) $A \geq 0$, $a \in \mathcal{R}(A)$, and $a'A^-a \leq \alpha$, with A^- being any g–inverse of A.

Proof. (i) $\Rightarrow$ (ii) $\alpha A - aa' \geq 0 \Rightarrow \alpha A = (\alpha A - aa') + aa' \geq 0 \Rightarrow A \geq 0$. Using Theorem A.12 for $\alpha A - aa' \geq 0$ we have $\alpha A - aa' = BB$ and, hence,

$$\begin{aligned} & \alpha A = BB + aa' = (B, a)(B, a)' \\ \Rightarrow \quad & \mathcal{R}(\alpha A) = \mathcal{R}(A) = \mathcal{R}(B, a) \\ \Rightarrow \quad & a \in \mathcal{R}(A) \\ \Rightarrow \quad & a = Ac \quad \text{with} \quad c \in E^n. \\ \Rightarrow \quad & a'A^-a = c'Ac. \end{aligned}$$

As $\alpha A - aa' \geq 0 \;\Rightarrow\; x'(\alpha A - aa')x \geq 0$ for any vector x. Choosing $x = c$ we have

$$\alpha c'Ac - c'aa'c = \alpha c'Ac - (c'Ac)^2 \geq 0$$

$$\Rightarrow c'Ac \leq \alpha.$$

(ii) $\Rightarrow$ (i) Let $x \in E^n$ be any vector. Then, using Theorem A.30

$$\begin{aligned} x'(\alpha A - aa')x &= \alpha x'Ax - (x'a)^2 \\ &= \alpha x'Ax - (x'Ac)^2 \\ &\geq \alpha x'Ax - (x'Ax)(c'Ac) \end{aligned}$$

$$\Rightarrow \quad x'(\alpha A - aa')x \geq (x'Ax)(\alpha - c'Ac).$$

In (ii) we have assumed $A \geq 0$ and $c'Ac = a'A^-a \leq \alpha$. Hence, $\alpha A - aa' \geq 0$.

Remark: This theorem is due to Baksalary et al. (1983).

Theorem A.45. For any matrix A we have

$$A'A = 0 \quad \text{if and only if } A = 0.$$

Proof. (i) A=0 $\Rightarrow \quad A'A = 0$.
(ii) Let $A'A = 0$ and let $A = (a_{(1)}, \ldots, a_{(n)})$ be the column–wise presentation. Then

$$A'A = (a'_{(i)}a_{(j)}) = 0,$$

so that all the elements on the diagonal are zero: $a'_{(i)}a_{(i)} = 0 \;\Rightarrow\; a_{(i)} = 0$ and $A = 0$.

Theorem A.46. Let $X \neq 0$ be an $(m \times n)$–matrix and let A be an $(n \times n)$ matrix. Then

$$X'XAX'X = X'X \quad \Rightarrow \quad XAX'X = X \quad \text{and} \quad X'XAX' = X'.$$

Proof. As $X \neq 0$ and $X'X \neq 0$, we have

$$\begin{aligned} X'XAX'X - X'X &= (X'XA - I)X'X = 0 \quad \Rightarrow \\ & (X'XA - I) = 0 \quad \Rightarrow \\ 0 &= (X'XA - I)(X'XAX'X - X'X) \\ &= (X'XAX' - X')(XAX'X - X) = Y'Y\,, \end{aligned}$$

so that (by Theorem A.45) $Y = 0$ and, hence, $XAX'X = X$.

Corollary. Let $X \neq 0$ be an (m, n)–matrix and let A and b be (n, n)–matrices. Then

$$AX'X = BX'X \longleftrightarrow AX' = BX'.$$

Theorem A.47 (Albert's Theorem). Let $A = \begin{pmatrix} A_{11} & A_{12} \\ A_{21} & A_{22} \end{pmatrix}$ be symmetric. Then:

(a) $A \geq 0$ if and only if:

(i) $A_{22} \geq 0$;
(ii) $A_{21} = A_{22}A_{22}^{-}A_{21}$;
(iii) $A_{11} \geq A_{12}A_{22}^{-}A_{21}$.

((ii) and (iii) are invariant of the choice of A_{22}^{-}).

(b) $A > 0$ if and only if:

(i) $A_{22} > 0$;
(ii) $A_{11} > A_{12}A_{22}^{-1}A_{21}$.

Proof. (Bekker and Neudecker, 1989)

(a) Assume $A \geq 0$.

(i) $A \geq 0 \;\Rightarrow\; x'Ax \geq 0$ for any x. Choosing $x' = (0', x_2')$,
$\Rightarrow x'Ax = x_2'A_{22}x_2 \geq 0$ for any $x_2 \Rightarrow A_{22} \geq 0$.

(ii) Let $B' = (0, I - A_{22}A_{22}^{-}) \;\Rightarrow$

$$\begin{aligned} B'A &= ((I - A_{22}A_{22}^{-})A_{21}, A_{22} - A_{22}A_{22}^{-}A_{22}) \\ &= ((I - A_{22}A_{22}^{-})A_{21}, 0) \end{aligned}$$

and

$$B'AB = B'A^{1/2}A^{1/2}B = 0 \Rightarrow B'A^{1/2} = 0 \text{ (Theorem A.45)}$$

$$\Rightarrow \quad B'A^{1/2}A^{1/2} = B'A = 0$$
$$\Rightarrow \quad (I - A_{22}A_{22}^{-})A_{21} = 0.$$

This proves (ii).

(iii) Let $C' = (I, -(A_{22}^{-}A_{21})')$. As $A \geq 0 \Rightarrow$

$$\begin{aligned} 0 \leq C'AC &= A_{11} - A_{12}(A_{22}^{-})'A_{21} - A_{12}A_{22}^{-}A_{21} \\ &\quad + A_{12}(A_{22}^{-})'A_{22}A_{22}^{-}A_{21} \\ &= A_{11} - A_{12}A_{22}^{-}A_{21} \end{aligned}$$

(as A_{22} is symmetric, we have $(A_{22}^{-})' = A_{22}$).

Assume now (i), (ii), and (iii). Then

$$D = \begin{pmatrix} A_{11} - A_{12}A_{22}^{-}A_{21} & 0 \\ 0 & A_{22} \end{pmatrix} \geq 0,$$

as the submatrices are n.n.d. by (i) and (ii). Hence,

$$A = \begin{pmatrix} I & A_{12}(A_{22}^{-}) \\ 0 & I \end{pmatrix} D \begin{pmatrix} I & 0 \\ A_{22}^{-}A_{21} & I \end{pmatrix} \geq 0.$$

(b) Proof as in (a) if A_{22}^{-} is replaced by A_{22}^{-1}.

Theorem A.48. If $A : n \times n$ and $B : n \times n$ are symmetric, then:

(a) $0 \leq B \leq A$ if and only if:

(i) $A \geq 0$;
(ii) $B = AA^{-}B$;
(iii) $B \geq BA^{-}B$.

(b) $0 < B < A$ if and only if $0 < A^{-1} < B^{-1}$.

Proof. Apply Theorem A.47 to the matrix $\begin{pmatrix} B & B \\ B & A \end{pmatrix}$.

Theorem A.49. Let A be symmetric and let $c \in \mathcal{R}(A)$. Then the following statements are equivalent:

(i) $\text{rank}(A + cc') = \text{rank}(A)$.

(ii) $\mathcal{R}(A + cc') = \mathcal{R}(A)$.

(iii) $1 + c'A^{-}c \neq 0$.

Corollary. Assume (i) or (ii) or (iii) to hold, then

$$(A + cc')^{-} = A^{-} - \frac{A^{-}cc'A^{-}}{1 + c'A^{-}c}$$

for any choice of A^{-}.

Corollary. Assume (i) or (ii) or (iii) to hold, then

$$\begin{aligned} c'(A+cc')^{-}c &= c'A^{-}c - \frac{(c'A^{-}c)^2}{1+c'A^{-}c} \\ &= 1 - \frac{1}{1+c'A^{-}c}. \end{aligned}$$

Moreover, as $c \in \mathcal{R}(A+cc')$, this is seen to be invariant for the special choice of the g–inverse.

Proof. $c \in \mathcal{R}(A) \Leftrightarrow AA^{-}c = c \Rightarrow$

$$\mathcal{R}(A+cc') = \mathcal{R}(AA^{-}(A+cc')) \subset \mathcal{R}(A).$$

Hence, (i) and (ii) become equivalent. Consider the following product of matrices

$$\begin{pmatrix} 1 & 0 \\ c & A+cc' \end{pmatrix} \begin{pmatrix} 1 & -c \\ 0 & I \end{pmatrix} \begin{pmatrix} 1 & 0 \\ -A^{-}c & I \end{pmatrix} = \begin{pmatrix} 1+c'A^{-}c & -c \\ 0 & A \end{pmatrix}.$$

The left–hand side has the rank

$$1 + \text{rank}(A+cc') = 1 + \text{rank}(A)$$

(see (i) or (ii)). The right–hand side has the rank $1 + \text{rank}(A)$ if and only if $1 + c'A^{-}c \neq 0$.

Theorem A.50. Assume $A : n \times n$ to be a symmetric and nonsingular matrix and assume $c \notin \mathcal{R}(A)$. Then we have:

(i) $c \in \mathcal{R}(A+cc')$.

(ii) $\mathcal{R}(A) \subset \mathcal{R}(A+cc')$.

(iii) $c'(A+cc')^{-}c = 1$.

(iv) $A(A+cc')^{-}A = A$.

(v) $A(A+cc')^{-}c = 0$.

Proof. As A is assumed to be nonsingular, the equation $Al = 0$ has a nontrivial solution $l \neq 0$ which may be standardized as $l/(c'l)$, such that $c'l = 1$. Then we have $c = (A+cc')l \in \mathcal{R}(A+cc')$ and, hence, (i) is proved. Relation (ii) holds as $c \notin \mathcal{R}(A)$. Relation (i) is seen to be equivalent to

$$(A+cc')(A+cc')^{-}c = c.$$

Therefore (iii) follows:

$$\begin{aligned} c'(A+cc')^{-}c &= l'(A+cc')(A+cc')^{-}c \\ &= l'c = 1. \end{aligned}$$

From

$$\begin{aligned} c &= (A+cc')(A+cc')^{-}c \\ &= A(A+cc')^{-}c + cc'(A+cc')^{-}c \\ &= A(A+cc')^{-}c + c \end{aligned}$$

we have (v). (iv) is a consequence of the general definition of a g–inverse and of (iii) and (iv):

$$\begin{aligned} A + cc' &= (A+cc')(A+cc')^-(A+cc') \\ &= A(A+cc')^-A \\ &\quad +cc'(A+cc')^-cc' \quad [= cc' \text{ using (iii)}] \\ &\quad +A(A+cc')^-cc' \quad [= 0 \text{ using (v)}] \\ &\quad +cc'(A+cc')^-A \quad [= 0 \text{ using (v)}]. \end{aligned}$$

Theorem A.51. We have $A \geq 0$ if and only if:

(i) $A + cc' \geq 0$.

(ii) $(A+cc')(A+cc')^-c = c$.

(iii) $c'(A+cc')^-c \leq 1$.

Assume $A \geq 0$, then:

(a) $c = 0 \Leftrightarrow c'(A+cc')^-c = 0$.

(b) $c \in \mathcal{R}(A) \Leftrightarrow c'(A+cc')^-c < 1$.

(c) $c \notin \mathcal{R}(A) \Leftrightarrow c'(A+cc')^-c = 1$.

Proof. $A \geq 0$ is equivalent to

$$0 \leq cc' \leq A + cc'.$$

Straightforward application of Theorem A.48 gives (i)–(iii).

(a) $A \geq 0 \;\Rightarrow\; A + cc' \geq 0$. Assume $c'(A+cc')^-c = 0$ and replace c by (ii) $\Rightarrow$

$$\begin{aligned} c'(A+cc')^-(A+cc')(A+cc')^-c &= 0 \Rightarrow \\ (A+cc')(A+cc')^-c &= 0 \end{aligned}$$

as $(A+cc') \geq 0$. Assuming $c = 0 \;\Rightarrow\; c'(A+cc')c = 0$.

(b) Assume $A \geq 0$ and $c \in \mathcal{R}(A)$, and use Theorem A.49 $\Rightarrow$

$$c'(A+cc')^-c = 1 - \frac{1}{1 + c'A^-c} < 1.$$

The opposite direction of (b) is a consequence of (c).

(c) Assume $A \geq 0$ and $c \notin \mathcal{R}(A)$, and use Theorem A.50(iii) $\Rightarrow$

$$c'(A+cc')^-c = 1.$$

The opposite direction of (c) is a consequence of (b).

Note: The proofs of Theorems A.47–A.51 are given in Bekker and Neudecker (1989).

Theorem A.52. The linear equation $Ax = a$ has a solution if and only if

$$a \in \mathcal{R}(A) \quad \text{or} \quad AA^- a = a$$

for any g–inverse A.

If this condition holds, then all solutions are given by

$$x = A^- a + (I - A^- A)w,$$

where w is an arbitrary m–vector. Further $q'x$ has a unique value for all solutions of $Ax = a$ if and only if $q'A^- A = q'$, or $q \in \mathcal{R}(A')$.

For a proof see Rao (1973), p. 25.

A.13 Projections

Consider the range space $\mathcal{R}(A)$ of the matrix $A : m \times n$ with rank r. Then there exists $\mathcal{R}(A)^\perp$ which is the orthogonal complement of $\mathcal{R}(A)$ with dimension $m - r$. Any vector $x \in \Re^m$ has the unique decomposition

$$x = x_1 + x_2, \quad X_1 \in \mathcal{R}(A), \quad \text{and} \quad x_2 \in \mathcal{R}(A)^\perp,$$

of which the component x is called the orthogonal projection of x on $\mathcal{R}(A)$. The component x_1 can be computed as Px where

$$P = A(A'A)^- A'$$

which is called the projection operator on $\mathcal{R}(A)$. Note that P is unique for any choice of the g–inverse $(A'A)^-$.

Theorem A.53. For any $P : n \times n$, the following statements are equivalent:

(i) P is an orthogonal projection operator.

(ii) P is symmetric and idempotent.

For proofs and other details the reader is referred to Rao (1973) and Rao and Mitra (1971).

Theorem A.54. Let X be a matrix of order $T \times K$ with rank $r < K$ and let $U : (K - r) \times K$ be such that $\mathcal{R}(X') \cap \mathcal{R}(U') = \{0\}$.

Then:

(i) $X(X'X + U'U)^{-1}U' = 0$.

(ii) $X'X(X'X + U'U)^{-1}X'X = X'X$, i.e., $(X'X + U'U)^{-1}$ is a g–inverse of $X'X$.

(iii) $U'U(X'X + U'U)^{-1}U'U = U'U$, i.e., $(X'X + U'U)^{-1}$ is also a g–inverse of $U'U$.

(iv) $U(X'X + U'U)^{-1}U'u = u$ if $u \in \mathcal{R}(U)$.

Proof. Since $X'X + U'U$ is of full rank, there exists a matrix A such that

$$\begin{aligned} & (X'X + U'U)A = U' \\ \Rightarrow\quad & X'XA = U' - U'UA \quad \Rightarrow \quad XA = 0 \text{ and } U' = U'UA \end{aligned}$$

since $\mathcal{R}(X')$ and $\mathcal{R}(U')$ are disjoint.

(i):

$$X(X'X + U'U)^{-1}U' = X(X'X + U'U)^{-1}(X'X + U'U)A = XA = 0$$

(ii):

$$\begin{aligned} & X'X(X'X + U'U)^{-1}(X'X + U'U - U'U) \\ = \; & X'X - X'X(X'X + U'U)^{-1}U'U = X'X\,. \end{aligned}$$

The result (iii) follows on the same lines as result (ii).

(iv):

$$U(X'X + U'U)^{-1}U'u = U(X'X + U'U)^{-1}U'Ua = Ua = u$$

since $u \in \mathcal{R}(U)$.

A.14 Functions of Normally Distributed Variables

Let $x' = (x_1, \ldots, x_p)$ be a p–dimensional random vector. Then x is p–dimensional normally distributed with expectation vector μ and covariance matrix $\Sigma > 0$, i.e., $x \sim N_p(\mu, \Sigma)$, if the joint density is

$$f(x; \mu, \Sigma) = \{(2\pi)^p |\Sigma|\}^{-1/2} \exp\{-1/2(x-\mu)'\Sigma^{-1}(x-\mu)\}.$$

Theorem A.55. Assume $x \sim N_p(\mu, \Sigma)$, and $A : p \times p$ and $b : p \times 1$ nonstochastic. Then

$$y = Ax + b \sim N_q(A\mu + b, A\Sigma A') \quad \text{with } q = \text{rank}(A).$$

Theorem A.56. If $x \sim N_p(0, I)$, then

$$x'x \sim \chi^2_p$$

(central χ^2–distribution with p degrees of freedom).

Theorem A.57. If $x \sim N_p(\mu, I)$, then

$$x'x \sim \chi^2_p(\lambda)$$

has a noncentral χ^2–distribution with a noncentrality parameter

$$\lambda = \mu'\mu = \sum_{i=1}^{p} \mu_i^2.$$

Theorem A.58. If $x \sim N_p(\mu, \Sigma)$, then:

(i) $x'\Sigma^{-1}x \sim \chi^2_p(\mu'\Sigma^{-1}\mu)$.

(ii) $(x-\mu)'\Sigma^{-1}(x-\mu) \sim \chi^2_p$.

Proof. $\Sigma > 0 \Rightarrow \Sigma = \Sigma^{1/2}\Sigma^{1/2}$ with $\Sigma^{1/2}$ regular and symmetric. Hence,

$$\Sigma^{-1/2}x = y \sim N_p(\Sigma^{-1/2}\mu, I) \Rightarrow x'\Sigma^{-1}x = y'y \sim \chi^2_p(\mu'\Sigma^{-1}\mu)$$

and

$$(x-\mu)'\Sigma^{-1}(x-\mu) = (y - \Sigma^{-1/2}\mu)'(y - \Sigma^{-1/2}\mu) \sim \chi^2_p.$$

Theorem A.59. If $Q_1 \sim \chi^2_m(\lambda)$ and $Q_2 \sim \chi^2_n$, and Q_1 and Q_2 are independent, then:

(i) The ratio

$$F = \frac{Q_1/m}{Q_2/n}$$

has a noncentral $F_{m,n}(\lambda)$–distribution.

(ii) If $\lambda = 0$, then $F \sim F_{m,n}$, the central F–distribution.

(iii) If $m = 1$, then $\sqrt{F}$ has a noncentral $t_n(\sqrt{\lambda})$–distribution or a central t_n–distribution if $\lambda = 0$.

Theorem A.60. If $x \sim N_p(\mu, I)$ and $A : p \times p$ is a symmetric idempotent matrix with $\operatorname{rank}(A) = r$, then

$$x'Ax \sim \chi^2_r(\mu'A\mu).$$

Proof. We have $A = P\Lambda P'$ (Theorem A.11) and without loss of generality (Theorem A.36(i)) we may write $\Lambda = \begin{pmatrix} I_r & 0 \\ 0 & 0 \end{pmatrix}$, i.e., $P'AP = \Lambda$ with P orthogonal. Let $P = \begin{pmatrix} \underset{p,r}{P_1} & \underset{p,(p-r)}{P_2} \end{pmatrix}$ and

$$P'x = y = \begin{pmatrix} y_1 \\ y_2 \end{pmatrix} = \begin{pmatrix} P_1'x \\ P_2'x \end{pmatrix}.$$

Therefore

$$\begin{aligned} y &\sim N_p(P'\mu, I_p) && \text{(Theorem A.55),} \\ y_1 &\sim N_r(P_1'\mu, I_r), && \text{and} \\ y_1'y_1 &\sim \chi^2_r(\mu'P_1P_1'\mu) && \text{(Theorem A.57).} \end{aligned}$$

As P is orthogonal, we have

$$\begin{aligned} A &= (PP')A(PP') = P(P'AP)P \\ &= (P_1 \; P_2)\begin{pmatrix} I_r & 0 \\ 0 & 0 \end{pmatrix}\begin{pmatrix} P_1' \\ P_2' \end{pmatrix} = P_1P_1' \end{aligned}$$

and, therefore,

$$x'Ax = x'P_1P_1'x = y_1'y_1 \sim \chi_r^2(\mu'A\mu).$$

Theorem A.61. Assume $x \sim N_p(\mu, I)$, $A : p \times p$ an idempotent of rank r, and $B : p \times n$ any matrix.

Then the linear form Bx is independent of the quadratic form $x'Ax$ if and only if $AB = 0$.

Proof. Let P be the matrix as in Theorem A.60. Then $BPP'AP = BAP = 0$, as $BA = 0$ was assumed. Let $BP = D = (D_1, D_2) = (BP_1, BP_2)$, then

$$BPP'AP = (D_1, D_2)\begin{pmatrix} I_r & 0 \\ 0 & 0 \end{pmatrix} = (D_1, 0) = (0, 0),$$

so that $D_1 = 0$. This gives

$$Bx = BPP'x = Dy = (0, D_2)\begin{pmatrix} y_1 \\ y_2 \end{pmatrix} = D_2y_2$$

where $y_2 = P_2'x$. Since P is orthogonal and, hence, regular we may conclude that all the components of $y = P'x$ are independent $\Rightarrow$ $Bx = D_2y_2$ and $x'Ax = y_1'y_1$ are independent.

Theorem A.62. Let $x \sim N_p(0, I)$ and assume A and B to be idempotent $p \times p$ matrices with $\text{rank}(A) = r$ and $\text{rank}(B) = s$. Then the quadratic forms $x'Ax$ and $x'Bx$ are independent if and only if $BA = 0$.

Proof. If we use P from Theorem A.60 and set $C = P'BP$ (C symmetric) we get, with the assumption $BA = 0$,

$$\begin{aligned} CP'AP &= P'BPP'AP \\ &= P'BAP = 0. \end{aligned}$$

Using

$$\begin{aligned} C &= \begin{pmatrix} P_1 \\ P_2 \end{pmatrix} B(P_1'\ P_2') \\ &= \begin{pmatrix} C_1 & C_2 \\ C_2' & C_3 \end{pmatrix} = \begin{pmatrix} P_1BP_1' & P_1BP_2' \\ P_2BP_1' & P_2BP_2' \end{pmatrix} \end{aligned}$$

this relation may be written as

$$CP'AP = \begin{pmatrix} C_1 & C_2 \\ C_2' & C_3 \end{pmatrix}\begin{pmatrix} I_r & 0 \\ 0 & 0 \end{pmatrix} = \begin{pmatrix} C_1 & 0 \\ C_2' & 0 \end{pmatrix} = 0\,.$$

Therefore, $C_1 = 0$ and $C_2 = 0$,

$$\begin{aligned} x'Bx &= x'(PP')B(PP')x \\ &= x'P(P'BP)P'x \\ &= x'PCP'x \\ &= (y_1', y_2')\begin{pmatrix} 0 & 0 \\ 0 & C_3 \end{pmatrix}\begin{pmatrix} y_1 \\ y_2 \end{pmatrix} = y_2'C_3y_2\,. \end{aligned}$$

As shown in Theorem A.60, we have $x'Ax = y_1'y_1$ and, therefore, the quadratic forms $x'Ax$ and $x'Bx$ are independent.

A.15 Differentiation of Scalar Functions of Matrices

Definition A.28. If $f(X)$ is a real function of an $m \times n$ matrix $X = (x_{ij})$, then the partial differential of f with respect to X is defined as the $(m \times n)$–matrix of partial differentials $\partial f / \partial x_{ij}$:

$$\frac{\partial f(X)}{\partial X} = \begin{pmatrix} \partial f/\partial x_{11} & \dots & \partial f/\partial x_{1n} \\ \vdots & & \vdots \\ \partial f/\partial x_{m1} & \dots & \partial f/\partial x_{mn} \end{pmatrix}.$$

Theorem A.63. Let x be an n–vector and A a symmetric $(n \times n)$–matrix. Then

$$\frac{\partial}{\partial x} x'Ax = 2Ax.$$

Proof.

$$\begin{aligned} x'Ax &= \sum_{r,s=1}^{n} a_{rs} x_r x_s \,, \\ \frac{\partial f}{\partial x_i} x'Ax &= \sum_{\substack{s=1 \\ (s \neq i)}}^{n} a_{is} x_s + \sum_{\substack{r=1 \\ (r \neq i)}}^{n} a_{ri} x_r + 2a_{ii} x_i \\ &= 2 \sum_{s=1}^{n} a_{is} x_s \qquad (\text{as } a_{ij} = a_{ji}) \\ &= 2a_i'x \qquad (a_i' \text{: } i\text{th row vector of } A). \end{aligned}$$

According to Definition A.28 we get

$$\frac{\partial x'Ax}{\partial x} = \begin{pmatrix} \partial/\partial x_1 \\ \vdots \\ \partial/\partial x_n \end{pmatrix} (x'Ax) = 2 \begin{pmatrix} a_1' \\ \vdots \\ a_n' \end{pmatrix} x = 2Ax.$$

Theorem A.64. If x is an n–vector, y an m–vector, and C an $(n \times m)$–matrix, then

$$\frac{\partial}{\partial C} x'Cy = xy'.$$

Proof.

$$\begin{aligned} x'Cy &= \sum_{r=1}^{m}\sum_{s=1}^{n} x_s c_{sr} y_r, \\ \partial/(\partial c_{k\lambda}) x'Cy &= x_k y_\lambda \quad \text{(the } (k,\lambda)\text{th element of } xy'), \\ \partial/\partial C x'Cy &= (x_k y_\lambda) = xy'. \end{aligned}$$

Theorem A.65. Let x be a K–vector, A a symmetric $(T \times T)$–matrix, and C a $(T \times K)$–matrix. Then

$$\frac{\partial}{\partial C} x'C'ACx = 2ACxx'.$$

Proof. We have

$$\begin{aligned} x'C' &= \left(\sum_{i=1}^{K} x_i c_{1i}, \ldots, \sum_{i=1}^{K} x_i c_{Ti} \right), \\ \frac{\partial}{\partial c_{k\lambda}} &= (0, \ldots, 0, x_\lambda, 0, \ldots, 0) \quad (x_\lambda \text{ is an element of the } k\text{th column}). \end{aligned}$$

Using the product rule yields

$$\frac{\partial}{\partial c_{k\lambda}} x'C'ACx = \left(\frac{\partial}{\partial c_{k\lambda}} x'C' \right) ACx + x'C'A \left(\frac{\partial}{\partial c_{k\lambda}} Cx \right).$$

Since

$$x'C'A = \left(\sum_{t=1}^{T}\sum_{i=1}^{K} x_i c_{ti} a_{t1}, \ldots, \sum_{t=1}^{T}\sum_{i=1}^{K} x_i c_{ti} a_{Tt} \right)$$

we get

$$\begin{aligned} x'C'A \left(\frac{\partial}{\partial c_{k\lambda}} Cx \right) &= \sum_{t,i} x_i x_\lambda c_{ti} a_{kt} \\ &= \sum_{t,i} x_i x_\lambda c_{ti} a_{tk} \quad \text{(as } A \text{ is symmetric)} \\ &= \left(\frac{\partial}{\partial c_{k\lambda}} x'C' \right) ACx. \end{aligned}$$

But $\sum_{t,i} x_i x_\lambda c_{ti} a_{tk}$ is just the (k,λ)th element of the matrix $ACxx'$.

Theorem A.66. Assume $A = A(x)$ to be an $(n \times n)$–matrix, where its elements $a_{ij}(x)$ are real functions of a scalar x. Let B be an $(n \times n)$–matrix, such that its elements are independent of x. Then

$$\frac{\partial}{\partial x} \operatorname{tr}(AB) = \operatorname{tr}\left(\frac{\partial A}{\partial x} B \right).$$

Proof.

$$\begin{aligned}\operatorname{tr}(AB) &= \sum_{i=1}^{n}\sum_{j=1}^{n} a_{ij}b_{ji},\\ \frac{\partial}{\partial x}\operatorname{tr}(AB) &= \sum_{i}\sum_{j}\frac{\partial a_{ij}}{\partial x}b_{ji}\\ &= \operatorname{tr}\left(\frac{\partial A}{\partial x}B\right),\end{aligned}$$

where $\partial A/\partial x = \partial a_{ij}/\partial x$.

Theorem A.67. For the differential of the trace we have the following rules:

	y	$\partial y/\partial X$
(i)	$\operatorname{tr}(AX)$	A'
(ii)	$\operatorname{tr}(X'AX)$	$(A + A')X$
(iii)	$\operatorname{tr}(XAX)$	$X'A + A'X'$
(iv)	$\operatorname{tr}(XAX')$	$X(A + A')$
(v)	$\operatorname{tr}(X'AX')$	$AX' + X'A$
(vi)	$\operatorname{tr}(X'AXB)$	$AXB + A'XB'$

Differentiation of Inverse Matrices

Theorem A.68. Let $= T(x)$ be a regular matrix, such that its elements depend on a scalar x. Then

$$\frac{\partial T^{-1}}{\partial x} = -T^{-1}\frac{\partial T}{\partial x}T^{-1}.$$

Proof. We have $T^{-1}T = I$, $\partial I/\partial x = 0$,

$$\frac{\partial(T^{-1}T)}{\partial x} = \frac{\partial T^{-1}}{\partial x}T + T^{-1}\frac{\partial T}{\partial x} = 0.$$

Theorem A.69. For nonsingular X we have

$$\begin{aligned}\frac{\partial \operatorname{tr}(AX^{-1})}{\partial X} &= -(X^{-1}AX^{-1})',\\ \frac{\partial \operatorname{tr}(X^{-1}AX^{-1}B)}{\partial X} &= -(X^{-1}AX^{-1}BX^{-1} + X^{-1}BX^{-1}AX^{-1})'.\end{aligned}$$

Proof. Use Theorems A.67, A.68 and the product rule.

Differentiation of a Determinant

Theorem A.70. For a nonsingular matrix Z we have:

(i) $\partial/\partial Z\,|Z| = |Z|(Z')^{-1}$.

(ii) $\partial/\partial Z\,\log|Z| = (Z')^{-1}$.

A.16 Miscellaneous Results, Stochastic Convergence

Theorem A.71 (Kronecker Product). Let $A : m \times n = (a_{ij})$ and $B : p \times q = (b_{rs})$ be any matrices. Then the Kronecker product of A and B is defined as

$$\underset{mp,nq}{C} = \underset{m,n}{A} \otimes \underset{p,q}{B} = \begin{pmatrix} a_{11}B & a_{12}B & \cdots & a_{1n}B \\ \vdots & \vdots & & \cdots \\ a_{m1}B & a_{m2}B & \cdots & a_{mn}B \end{pmatrix}$$

and the following rules hold:

(i) $c(A \otimes B) = (cA) \otimes B = A \otimes (cB)$ (c a scalar).

(ii) $A \otimes (B \otimes C) = (A \otimes B) \otimes C$.

(iii) $A \otimes (B + C) = (A \otimes B) + (A \otimes C)$.

(iv) $(A \otimes B)' = A' \otimes B'$.

Theorem A.72 (Tschebyschev's Inequality). For any n–dimensional random vector X and a given scalar $\epsilon > 0$ we have

$$P\{|X| \geq \epsilon\} \leq \frac{E\,|X|^2}{\epsilon^2}\,.$$

Proof. Let $F(x)$ be the joint distribution function of $X = (x_1, \ldots, x_n)$. Then

$$\begin{aligned} E|x|^2 &= \int |x|^2 \, \mathrm{d}F(x) \\ &= \int_{\{x:|x|\geq\epsilon\}} |x|^2 \, \mathrm{d}F(x) + \int_{\{x:|x|<\epsilon\}} |x|^2 \, \mathrm{d}F(x) \\ &\geq \epsilon^2 \int_{\{x:|x|\geq\epsilon\}} \mathrm{d}F(x) = \epsilon^2 P\{|x| \geq \epsilon\}\,. \end{aligned}$$

Definition A.29. Let $\{x(t)\}$, $t = 1, 2, \ldots$, be a multivariate stochastic process.

(i) Weak convergence
If

$$\lim_{t\to\infty} P\{|x(t) - \tilde{x}| \geq \delta\} = 0,$$

where $\delta > 0$ is any given scalar and $\tilde{x}$ is a finite vector, then $\tilde{x}$ is called the probability limit of $\{x(t)\}$ and we write

$$p \lim x = \tilde{x}.$$

(ii) Strong convergence
Assume that $\{x(t)\}$ is defined on a probability space (Ω, Σ, P). Then $\{x(t)\}$ is said to be strongly convergent to $\tilde{x}$, i.e.,

$$\{x(t)\} \rightarrow \tilde{x} \quad \text{almost sure (a.s.)}$$

if there exists a set $T \in \Sigma$, $P(T) = 0$, and $x_\omega(t) \rightarrow \tilde{x}_\omega$, as $T \rightarrow \infty$, for each $\omega \in \Omega - T$ (M.M. Rao, 1984, p. 45).

Theorem A.73 (Slutsky's Theorem). (i) If $p\lim x = \tilde{x}$, then $\lim_{t\to\infty} E\{x(t)\} = \bar{E}(x) = \tilde{x}$.

(ii) If c is a vector of constants, then $p\lim c = c$.

(iii) (Slutsky's Theorem) If $p\lim x = \tilde{x}$ and $y = f(x)$ is any continuous vector function of x, then $p\lim y = f(\tilde{x})$.

(iv) If A and B are random matrices, then, when the following limits exist,

$$p\lim(AB) = p\lim A)(p\lim B)$$

and

$$p\lim(A^{-1}) = (p\lim A)^{-1}.$$

(v) If $p\lim\left[\sqrt{T}(x(t) - Ex(t))\right]'\left[\sqrt{T}(x(t) - Ex(t))\right] = V$, then the asymptotic covariance matrix is

$$\bar{V}(x, x) = \bar{E}\left[x - \bar{E}(x)\right]'\left[x - \bar{E}(x)\right] = T^{-1}V.$$

Definition A.30. If $\{x(t)\}, t = 1, 2, \ldots$, is a multivariate stochastic process statisfying

$$\lim_{t\to\infty} E|x(t) - \tilde{x}|^2 = 0,$$

then $\{x(t)\}$ is called convergent in the quadratic mean, and we write

$$\text{l.i.m. } x = \tilde{x}\, d.$$

Theorem A.74. If l.i.m. $x = \tilde{x}$, then $p\lim x = \tilde{x}$.

Proof. Using Theorem A.72 we get

$$0 \leq \lim_{t\to\infty} P(|x(t) - \tilde{x}| \geq \epsilon) \leq \lim_{t\to\infty} \frac{E|x(t) - \tilde{x}|^2}{\epsilon^2} = 0.$$

Theorem A.75. If l.i.m. $(x(t) - Ex(t)) = 0$ and $\lim_{t\to\infty} Ex(t) = c$, then $p\lim x(t) = c$.

Proof.

$$\begin{aligned}
\lim_{t\to\infty} P(|x(t) - c| \geq \epsilon) &\leq \epsilon^{-2} \lim_{t\to\infty} E|x(t) - c|^2 \\
&= \epsilon^{-2} \lim_{t\to\infty} E|x(t) - Ex(t) + Ex(t) - c|^2
\end{aligned}$$

$$\begin{aligned} &= \epsilon^{-2} \lim_{t\to\infty} E|x(t) - Ex(t)|^2 + \epsilon^{-2} \lim_{t\to\infty} |Ex(t) - c|^2 \\ &\quad + 2\epsilon^{-2} \lim_{t\to\infty} \{(Ex(t) - c)'(x(t) - Ex(t))\} \\ &= 0 . \end{aligned}$$

Theorem A.76. l.i.m. $x = c$ if and only if

$$\text{l.i.m.}(x(t) - Ex(t)) = 0 \text{ and } \lim_{t\to\infty} Ex(t) = c .$$

Proof. As in Theorem A.75, we may write

$$\begin{aligned} \lim_{t\to\infty} E|x(t) - c|^2 &= \lim_{t\to\infty} E|x(t) - Ex(t)|^2 \\ &\quad + \lim_{t\to\infty} |Ex(t) - c|^2 \\ &\quad + 2 \lim_{t\to\infty} E(Ex(t) - c)'(x(t) - Ex(t)) \\ &= 0. \end{aligned}$$

Theorem A.77. Let $x(t)$ be an estimator of a parameter vector θ. Then we have the result

$$\lim_{t\to\infty} Ex(t) = \theta \quad \text{if} \quad \text{l.i.m.}(x(t) - \theta) = 0 .$$

That is, $x(t)$ is an asymptotically unbiased estimator for θ if $x(t)$ converges to θ in the quadratic mean.

Proof. Use Theorem A.76.

Appendix B
Theoretical Proofs

In this Appendix the reader will find proofs of theoretical results which we decided to put in the appendix. It is structured in accordance with the chapters of the book.

B.1 The Linear Regression Model

Proof 1 (Theorem (3.1)). Let $Ax = a$ have a solution. Then at least one vector x_0 exists, with $Ax_0 = a$. As $AA^-A = A$ for every g–inverse, we obtain

$$a = Ax_0 = AA^-Ax_0 = AA^-(Ax_0) = AA^-a\,,$$

which is just (3.12).

Now let (3.12) be true, i.e., $AA^-a = a$. Then A^-a is a solution of (3.11). Assume now that (3.11) is solvable. To prove (3.13), we have to show:

(i) that $A^-a + (I - A^-A)w$ is always a solution of (3.11) (w arbitrary); and

(ii) that every solution x of $Ax = a$ may be represented by (3.13).

Part (i) follows by insertion of the general solution, also making use of $A(I - A^-A) = \mathbf{0}$:

$$A[A^-a + (I - A^-A)w] = AA^-a = a\,.$$

To prove (ii) we choose $w = x_0$, where x_0 is a solution of the linear equation, i.e., $Ax_0 = a$. Then we have

$$\begin{aligned} A^- a + (I - A^- A)x_0 &= A^- a + x_0 - A^- Ax_0 \\ &= A^- a + x_0 - A^- a \\ &= x_0 \,, \end{aligned}$$

thus concluding the proof.

Proof 2 (Theorem (3.2)). We have to start by the following corollary:

Corollary. The set of equations

$$AXB = C \tag{B.1}$$

where $A : m \times n$, $B : p \times q$, $C : m \times q$, and $X : n \times p$ have a solution X if and only if

$$AA^- CB^- B = C \,, \tag{B.2}$$

where A^- and B^- are arbitrary g–inverses of A and B.

If X is of full rank, i.e., $\text{rank}(X) = p = K$, then we have $(X'X)^- = (X'X)^{-1}$ and the normal equations are uniquely solvable by

$$b = (X'X)^{-1} X'y \,. \tag{B.3}$$

If, more generally, $\text{rank}(X) = p < K$, then the solutions of the normal equations span the same hyperplane as Xb, i.e., for two solutions b and b^* we have

$$Xb = Xb^* \,. \tag{B.4}$$

This result is easy to prove: If b and b^*, are solutions to the normal equations, we have

$$X'Xb = X'y \quad \text{and} \quad X'Xb^* = X'y \,.$$

Accordingly, we have, for the difference of the above equations,

$$X'X(b - b^*) = 0 \,,$$

which entails

$$X(b - b^*) = 0 \quad \text{or} \quad Xb = Xb^* \,.$$

Moreover, by (B.4), the two sums of squared errors are given by

$$S(b) = (y - Xb)'(y - Xb) = (y - Xb^*)'(y - Xb^*) = S(b^*) \,.$$

Thus Theorem B.3 has been proven.

Proof 3 (Theorem (3.3)). As $\mathcal{R}(X)$ is of dimension p, an orthonormal basis $v_1, \ldots, v_p$ exists. Furthermore, we may represent the $(T \times 1)$–vector y as

$$y = \sum_{i=1}^{p} a_i v_i + \left(y - \sum_{i=1}^{p} a_i v_i \right) = c + d, \tag{B.5}$$

where $a_i = y'v_i$.

As

$$v_j'd = v_j'y - \sum_i a_i v_j' v_i = a_j - \sum_i a_i \delta_{ij} = 0 \tag{B.6}$$

(δ_{ij} denotes the Kronecker symbol), we have $c \perp d$, i.e., we have $c \in \mathcal{R}(X)$ and $d \in \mathcal{R}(X)^{\perp}$, such that y has been decomposed into two orthogonal components. This decomposition is unique as can easily be shown.

We have to show now that $c = Xb = \Theta_0$.

It follows from $c - \Theta \in \mathcal{R}(X)$ that

$$(y - c)'(c - \Theta) = d'(c - \Theta) = 0\,. \tag{B.7}$$

Considering $y - \Theta = (y - c) + (c - \Theta)$, we get

$$\begin{aligned} \tilde{S}(\Theta) = (y - \Theta)'(y - \Theta) &= (y - c)'(y - c) + (c - \Theta)'(c - \Theta) \\ &\quad + 2(y - c)'(c - \Theta) \\ &= (y - c)'(y - c) + (c - \Theta)'(c - \Theta)\,. \end{aligned} \tag{B.8}$$

$\tilde{S}(\Theta)$ reaches its minimum on $\mathcal{R}(X)$ for the choice $\Theta = c$. As $\tilde{S}(\Theta) = S(\beta)$ we find b to be the optimum $c = \Theta_0 = Xb$.

Proof 4 (Theorem (3.4)). Following Theorem 3.3, we have

$$\begin{aligned} \Theta_0 = c &= \sum_i a_i v_i = \sum_i v_i (y' v_i) \\ &= \sum_i v_i (v_i' y) \\ &= (v_1, \ldots, v_p)(v_1, \ldots, v_p)' y \\ &= BB'y \quad [B = (v_1, \ldots, v_p)] \\ &= Py\,, \end{aligned} \tag{B.9}$$

where P is obviously symmetric and idempotent.

We have to make use of the following lemma, which will be stated without proof.

Lemma. *A symmetric and idempotent $(T \times T)$–matrix P of rank $p \leq T$ represents the orthogonal projection matrix of $\mathcal{R}^T$ on a p–dimensional vector space $V = \mathcal{R}(P)$.*

(i) Determination of P if rank$(X) = K$.
The rows of B constitute an orthonormal basis of $\mathcal{R}(X) = \{\Theta : \Theta = X\beta\}$. But $X = BC$, with a regular matrix C, as the columns of X

also form a basis of $\mathcal{R}(X)$.

Thus

$$\begin{aligned} P = BB' &= XC^{-1}C'^{-1}X' = X(C'C)^{-1}X' \\ &= X(C'B'BC)^{-1}X' \quad [\text{as } B'B = I] \\ &= X(X'X)^{-1}X', \end{aligned} \tag{B.10}$$

and we finally get

$$\Theta_0 = Py = X(X'X)^{-1}X'y = Xb. \tag{B.11}$$

(ii) Determination of P if rank$(X) = p < K$.
The normal equations have a unique solution, if X is of full column rank K. A method of deriving unique solutions, if $\text{rank}(X) = p < K$, is based on imposing additional linear restrictions, which enable the identification of β.

We introduce only the general strategy by using Theorem 3.4; further details will be given in Section 3.5.

Let R be a $[((K-p) \times K)]$–matrix with $\text{rank}(R) = K - p$ and define the matrix $D = \begin{pmatrix} X \\ R \end{pmatrix}$.

Let r be a known $((K-p) \times 1)$–vector. If $\text{rank}(D) = K$, then X and R are *complementary matrices*. The matrix R represents $(K-p)$ additional linear restrictions on β (reparametrization), as it will be assumed that

$$R\beta = r. \tag{B.12}$$

Minimization of $S(\beta)$, subject to these exact linear restrictions $R\beta = r$, requires the minimization of the function

$$Q(\beta, \lambda) = S(\beta) + 2\lambda'(R\beta - r), \tag{B.13}$$

where λ stands for a $[((K-p) \times 1)]$–vector of Lagrangian multipliers. The corresponding normal equations are given by (cf. Theorem A.63–A.67)

$$\left.\begin{aligned} \frac{1}{2}\frac{\partial Q(\beta,\lambda)}{\partial \beta} &= X'X\beta - X'y + R'\lambda = \mathbf{0}, \\ \frac{1}{2}\frac{\partial Q(\beta,\lambda)}{\partial \lambda} &= R\beta - r = \mathbf{0}. \end{aligned}\right\} \tag{B.14}$$

If $r = \mathbf{0}$, we can prove the following theorem (cf. Seber (1966), p. 16):

Theorem B.1. Under the exact linear restrictions $R\beta = r$ with $\text{rank}(R) = K - p$ and $\text{rank}(D) = K$ we can state:

(i) The orthogonal projection matrix of $\mathcal{R}^T$ on $\mathcal{R}(X)$ is of the form

$$P = X(X'X + R'R)^{-1}X'. \tag{B.15}$$

(ii) The conditional ordinary least–squares estimator of β is given by

$$b(R,r) = (X'X + R'R)^{-1}(X'y + R'r)\,. \tag{B.16}$$

Proof. We start with the proof of part (i).

From the assumptions we conclude that for every $\Theta \in \mathcal{R}(X)$ a β exists, such that $\Theta = X\beta$ *and* $R\beta = r$ are valid. β is unique, as $\operatorname{rank}(D) = K$. In other words, for every $\Theta \in \mathcal{R}(X)$, the $[((T+K-p)\times 1)]$–vector is

$$\begin{pmatrix}\Theta\\ R\end{pmatrix} \in \mathcal{R}(D), \text{ therefore } \begin{pmatrix}\Theta\\ r\end{pmatrix} = D\beta \text{ (and } \beta \text{ is unique)}\,.$$

If we make use of Theorem 3.4, then we get the projection matrix of $\mathcal{R}^{T+K-p}$ on $\mathcal{R}(D)$ as

$$P^* = D(D'D)^{-1}D'\,. \tag{B.17}$$

As the projection P^* maps every element of $\mathcal{R}(D)$ onto itself we have, for every $\Theta \in \mathcal{R}(X)$,

$$\begin{aligned}\begin{pmatrix}\Theta\\ r\end{pmatrix} &= D(D'D)^{-1}D'\begin{pmatrix}\Theta\\ r\end{pmatrix}\\ &= \begin{pmatrix} X(D'D)^{-1}X' & X(D'D)^{-1}R' \\ R(D'D)^{-1}X' & R(D'D)^{-1}R'\end{pmatrix}\begin{pmatrix}\Theta\\ r\end{pmatrix},\end{aligned} \tag{B.18}$$

i.e.,

$$\Theta = X(D'D)^{-1}X'\Theta + X(D'D)^{-1}R'r\,, \tag{B.19}$$

$$r = R(D'D)^{-1}X'\Theta + R(D'D)^{-1}R'r\,. \tag{B.20}$$

Equations (B.19) and (B.20) hold for every $\Theta \in \mathcal{R}(X)$ and for all $r = R\beta \in \mathcal{R}(R)$. If we choose in (B.12) $r = 0$, then (B.19) and (B.20) specialize to

$$\Theta = X(D'D)^{-1}X'\Theta\,, \tag{B.21}$$

$$0 = R(D'D)^{-1}X'\Theta\,. \tag{B.22}$$

From (B.22) it follows that

$$\mathcal{R}(X(D'D)^{-1}R') \perp \mathcal{R}(X) \tag{B.23}$$

and as $\mathcal{R}(X(D'D)^{-1}R') = \{\Theta : \Theta = X\tilde{\beta} \text{ with } \tilde{\beta} = (D'D)^{-1}R'\beta\}$ it holds that

$$\mathcal{R}(X(D'D)^{-1}R') \subset \mathcal{R}(X)\,, \tag{B.24}$$

such that, finally,

$$X(D'D)^{-1}R' = 0 \tag{B.25}$$

(see also Tan, 1971).

The matrices $X(D'D)^{-1}X'$ and $R(D'D)^{-1}R'$ are idempotent (symmetry is evident):

$$\begin{aligned}
& X(D'D)^{-1}X'X(D'D)^{-1}X' \\
= \; & X(D'D)^{-1}(X'X + R'R - R'R)(D'D)^{-1}X' \\
= \; & X(D'D)^{-1}(X'X + R'R)(D'D)^{-1}X' - X(D'D)^{-1}R'R(D'D)^{-1}X' \\
= \; & X(D'D)^{-1}X' ,
\end{aligned}$$

as $D'D = X'X + R'R$ and (B.25) are valid.

The idempotency of $R(D'D)^{-1}R'$ can be shown in a similar way. $D'D$ and $(D'D)^{-1}$ are both positive definite (see Theorems A.16 and A.17). $R(D'D)^{-1}R'$ is positive definite (Theorem A.16(vi)) and thus regular since $\text{rank}(R) = K - p$. But there exists only one idempotent and regular matrix, namely, the identity matrix (Theorem A.36(iii))

$$R(D'D)^{-1}R' = I , \tag{B.26}$$

such that (B.20) is equivalent to $r = r$. As $P = X(D'D)^{-1}X'$ is idempotent, it represents the orthogonal projection matrix of $\mathcal{R}^T$ on a vector space $V \subset \mathcal{R}^T$ (see the lemma following Theorem 3.4).

With (B.21) we have $\mathcal{R}(X) \subset V$. But the reverse proposition is also true (see Theorem A.7(iv), (v)):

$$V = \mathcal{R}(X(D'D)^{-1}X') \subset \mathcal{R}(X) , \tag{B.27}$$

such that $V = \mathcal{R}(X)$, which proves (i).

(ii): We will solve the normal equations (B.14). With $R\beta = r$ it also holds that $R'R\beta = R'r$. Inserting the latter identity into the first equation of (B.14) yields

$$(X'X + R'R)\beta = X'y + R'r - R'\lambda .$$

Multiplication with $(D'D)^{-1}$ from the left yields

$$\beta = (D'D)^{-1}(X'y + R'r) - (D'D)^{-1}R'\lambda .$$

If we use the second equation of (B.14), (B.25), and (B.26), and then multiply by R from the left we get

$$R\beta = R(D'D)^{-1}(X'y + R'r) - R(D'D)^{-1}R'\lambda = r - \lambda , \tag{B.28}$$

from which $\hat{\lambda} = 0$ follows.

The solution of the normal equations is therefore given by

$$\hat{\beta} = b(R, r) = (X'X + R'R)^{-1}(X'y + R'r) \tag{B.29}$$

which proves (ii).

Proof 5 (Theorem (3.11)). $\mathrm{r}(\hat{\beta}, \beta)$ has to be minimized with respect to C under the restriction

$$CX = \begin{pmatrix} c_1' \\ \vdots \\ c_K' \end{pmatrix} X = \begin{pmatrix} e_1' \\ \vdots \\ e_K' \end{pmatrix} = I_K \,,$$

i.e.,

$$\min_C [\mathrm{tr}\{XCC'X'\} \mid CX - I = 0] \,.$$

This problem may be reformulated in terms of Lagrangian multipliers as

$$\min_{C_i, \lambda_i} \left[\mathrm{tr}\{XCC'X'\} - 2 \sum_{i=1}^{K} \lambda_i'(c_i'X - e_i')' \right] . \tag{B.30}$$

The $(K \times 1)$–vectors λ_i of Lagrangian multipliers may be contained in the matrix

$$\Lambda = \begin{pmatrix} \lambda_1' \\ \vdots \\ \lambda_K' \end{pmatrix} . \tag{B.31}$$

Differentiation of (B.30) with respect to C and Λ yields (Theorems A.63–A.67) the normal equations

$$X'XC - \Lambda X' = \mathbf{0} \,, \tag{B.32}$$

$$CX - I = \mathbf{0} \,. \tag{B.33}$$

The matrix $X'X$ is regular since $\mathrm{rank}(X) = K$. Premultiplication of (B.32) with $(X'X)^{-1}$ leads to

$$C = (X'X)^{-1}\Lambda X' \,,$$

from which we have (using (B.33))

$$CX = (X'AX)^{-1}\Lambda(X'X) = I_K \,,$$

namely,

$$\hat{\Lambda} = I_K \,.$$

Therefore, the optimum matrix is

$$\hat{C} = (X'X)^{-1}X' \,.$$

The actual linear unbiased estimator is given by

$$\hat{\beta}_{\mathrm{opt}} = \hat{C}y = (X'X)^{-1}X'y \,, \tag{B.34}$$

and coincides with the descriptive or empirical OLS estimator b. The estimator b is unbiased since

$$\hat{C}X = (X'X)^{-1}X'X = I_K \,, \tag{B.35}$$

(see (3.47)) and has the $(K \times K)$–covariance matrix

$$\begin{aligned} \mathrm{V}(b) = V_b &= \mathrm{E}(b-\beta)(b-\beta)' \\ &= \mathrm{E}\{(X'X)^{-1}X'\epsilon\epsilon'X(X'X)^{-1}\} \\ &= \sigma^2(X'X)^{-1}. \end{aligned} \tag{B.36}$$

Proof 6 (Theorem (3.12)). The equivalence is a direct consequence from the definition of definiteness. We will prove *(a)*.

Let $\tilde{\beta} = \tilde{C}y$ be an arbitrary unbiased estimator. Define, without loss of generality,

$$\tilde{C} = \hat{C} + D = (X'X)^{-1}X' + D.$$

Unbiasedness of $\tilde{\beta}$ requires that (3.47) is fulfilled:

$$\tilde{C}X = \hat{C}X + DX = I.$$

In view of (B.35) it is necessary that

$$DX = \mathbf{0}.$$

For the covariance matrix of $\tilde{\beta}$ we get

$$\begin{aligned} V_{\tilde{\beta}} &= \mathrm{E}(\tilde{C}y-\beta)(\tilde{C}y-\beta)' \\ &= \mathrm{E}(\tilde{C}\epsilon)(\epsilon'\tilde{C}') \\ &= \sigma^2[(X'X)^{-1}X'+D][X(X'X)^{-1}+D'] \\ &= \sigma^2[(X'X)^{-1}+DD'] \\ &= V_b + \sigma^2 DD' \geq V_b. \end{aligned}$$

Corollary. Let $V_{\tilde{\beta}} - V_b \geq 0$. Denote by $\mathrm{Var}(b_k)$ and $\mathrm{Var}(\tilde{\beta}_k)$ the main diagonal elements of V_b and $V_{\tilde{\beta}}$. Then the following inequality holds for the components of the two vectors $\tilde{\beta}$ and b:

$$\mathrm{Var}(\tilde{\beta}_i) - \mathrm{Var}(b_i) \geq 0 \quad (i = 1, \ldots, K). \tag{B.37}$$

Proof. From $V_{\tilde{\beta}} - V_b \geq 0$ we have $a'(V_{\tilde{\beta}} - V_b)a \geq 0$ for arbitrary vectors a, such that for the vectors, $e_i' = (0\ldots010\ldots0)$ with 1 at the ith position. Let A be an arbitrary symmetric matrix such that $e_i'Ae_i = a_{ii}$. Then the ith diagonal element of $V_{\tilde{\beta}} - V_b$ is just (B.37).

Proof 7 (Theorem (3.14)). Let $\tilde{d} = c'y$ be an arbitrary linear unbiased estimator of d, where c is a $(T \times 1)$–vector. Without loss of generality we set

$$c' = a'(X'X)^{-1}X' + \tilde{c}'.$$

The unbiasedness of $\tilde{d}$ requires that

$$c'X = a',$$

i.e.,

$$a'(X'X)^{-1}X'X + \tilde{c}'X = a'$$

and, therefore,

$$\tilde{c}'X = 0\,. \tag{B.38}$$

Using (3.94) we get

$$\begin{aligned} \tilde{d} - d &= a'\beta + a'(X'X)^{-1}X'\epsilon + \tilde{c}'\epsilon - a'\beta \\ &= a'(X'X)^{-1}X'\epsilon + \tilde{c}'\epsilon = c'\epsilon\,. \end{aligned}$$

The variance of $\tilde{d}$ is given by

$$\begin{aligned} \operatorname{Var}(\tilde{d}) &= \mathrm{E}(\tilde{d} - d)^2 = c'\,\mathrm{E}(\epsilon\epsilon')c = \sigma^2 c'c \\ &= \sigma^2[a'(X'X)^{-1}X' + \tilde{c}'][X(X'X)^{-1}a + \tilde{c}] \\ &= a'V_{b_0}a + \sigma^2\tilde{c}'\tilde{c}\,. \end{aligned}$$

As $\tilde{c}'\tilde{c} \geq 0$, the variance of $\tilde{d}$ will be minimized if $\tilde{c} = 0$. The estimator $c'y = a'(X'X)^{-1}X'y = a'b_0$ is therefore the best estimator among all linear unbiased estimators in the sense of a minimum variance.

Proof 8. We may use the corollary following Theorem 3.1.

The condition of unbiasedness is a condition on the matrix C, namely,

$$CX = I\,.$$

The latter equation is solvable with respect to C if and only if (B.1) holds, i.e., $X^-X = I_K$. With the help of Theorem A.38(ii), we know that $\operatorname{rank}(X^-X) = \operatorname{rank}(X)$ and $\operatorname{rank}(X) = p < K$. On the other hand, $\operatorname{rank}(I_K) = K$. Thus $(X^-X) = I_K$ cannot be valid so that $CX = I$ is not solvable.

Proof 9 (Theorem (3.15)). The proof consists of three parts.

(a) $b(R)$ is unbiased.

With $R\beta = 0$ we also have $R'R\beta = 0$ (Theorems A.45 and A.46), such that

$$\begin{aligned} \mathrm{E}(b(R)) &= (X'X + R'R)^{-1}X'X\beta \\ &= (X'X + R'R)^{-1}(X'X + R'R)\beta = \beta\,. \end{aligned}$$

$b(R)$ fulfills the restriction

$$Rb(R) = R(X'X + R'R)^{-1}X'y = 0 \quad (\text{compare } (B.25))\,.$$

(b) We immediately get

$$b(R) - \beta = (D'D)^{-1}X'\epsilon$$

and, therefore,

$$\begin{aligned} V_{b(R)} &= \mathrm{E}\{(D'D)^{-1}X'\epsilon\epsilon'X(D'D)^{-1}\} \\ &= \sigma^2(D'D)^{-1}X'X(D'D)^{-1}. \end{aligned}$$

(c) We now have to prove that $b(R)$ is the best linear conditionally unbiased estimator of β under the restriction $R\beta = 0$, i.e., the best linear unbiased estimator in model (3.75). (A somewhat different way of proof is given by Tan (1971) who deals with multivariate models using generalized inverses.)
Model (3.75) is then of the form

$$\begin{pmatrix} y \\ 0 \end{pmatrix} \begin{pmatrix} X \\ R \end{pmatrix} \beta + \begin{pmatrix} \epsilon \\ 0 \end{pmatrix}, \tag{B.39}$$

or in new symbols ($\tilde{T} = T + K - p$) of the form

$$\underset{\tilde{T}\times 1}{\tilde{y}} = \underset{\tilde{T}\times K}{D} \underset{K\times 1}{\beta} + \underset{\tilde{T}\times 1}{\tilde{\epsilon}} . \tag{B.40}$$

We have $\mathrm{E}(\tilde{\epsilon}) = 0$, $\mathrm{E}(\epsilon\epsilon') = V = \begin{pmatrix} \sigma^2 I & 0 \\ 0 & 0 \end{pmatrix}$, and $\mathrm{rank}(D) = K$, such that the model is singular. The estimator $b(R)$ is still linear in $\tilde{y}$:

$$\begin{aligned} b(R) &= (D'D)^{-1}X'y = (D'D)^{-1}(X'y + R'0) \\ &= (D'D)^{-1}D'\tilde{y} = C\tilde{y} \quad (C \text{ is a } K \times \tilde{T}\text{–matrix}). \end{aligned} \tag{B.41}$$

Since $b(R)$ is conditionally unbiased, we have

$$CD = I. \tag{B.42}$$

Let $\tilde{\beta} = \tilde{C}\tilde{y} + d$ be an arbitrary unbiased estimator of β in model (B.39).

Without loss of generality, we write

$$\tilde{C} = C + F \quad \text{with} \quad F = (F_1, F_2), \tag{B.43}$$

where $C = (D'D)^{-1}D'$ is the matrix from (B.41), F_1 is a $(K \times T)$–matrix, and F_2 is a $[(K \times (K-p))]$–matrix. Unbiasedness of $\tilde{\beta}$ in model (B.39) requires that

$$\mathrm{E}(\tilde{\beta}) = \tilde{C}D\beta + d = \beta \quad \text{for all } \beta,$$

from which we have $d = 0$ by choosing $\beta = 0$. A necessary condition for unbiasedness is thus given by

$$\begin{aligned} \tilde{C}D\beta &= CD\beta + FD\beta \\ &= CD\beta + F_1X\beta + F_2R\beta \\ &= \beta + F_1X\beta = \beta \quad [R\beta = 0 \text{ and } (B.42)] \end{aligned}$$

and, thus,

$$F_1X = 0. \tag{B.44}$$

It follows that

$$\begin{aligned}\tilde{\beta} - \beta &= (C+F)D\beta + (C+F)\tilde{\epsilon} - \beta \\ &= (C+F)\tilde{\epsilon} = \tilde{C}\tilde{\epsilon}\end{aligned}$$

and we can express the covariance matrix of $\tilde{\beta}$ in the following form:

$$\begin{aligned}\mathrm{V}_{\tilde{\beta}} = \mathrm{E}(\tilde{\beta}-\beta)(\tilde{\beta}-\beta)' &= \tilde{C}V\tilde{C}' \\ &= (C+F)V(C'+F') \\ &= CVC' + FVF' + FVC' + CVF' .\end{aligned}$$

Furthermore, we have (with $\mathrm{E}(\tilde{\epsilon}\tilde{\epsilon}') = V$, compare (B.40))

$$\begin{aligned}CVC' &= V_{b(R)} , \\ FVF' &= (F_1, F_2)\begin{pmatrix} \sigma^2 I & 0 \\ 0 & 0 \end{pmatrix}\begin{pmatrix} F_1' \\ F_2' \end{pmatrix} = \sigma^2 F_1 F_1' ,\end{aligned}$$

where $\sigma^2 F_1 F_1'$ is nonnegative definite [Theorem A.18 (v)].

For mixed products it holds that

$$\begin{aligned}FVC' &= (F_1, F_2)\begin{pmatrix} \sigma^2 I & 0 \\ 0 & 0 \end{pmatrix}\begin{pmatrix} X \\ R \end{pmatrix}(D'D)^{-1} \\ &= F_1 X (D'D)^{-1} = 0 \quad [\text{by (B.44)}] \end{aligned} \tag{B.45}$$

Finally, we get

$$V_{\tilde{\beta}} - V_{b(R)} = \sigma^2 F_1 F_1' \geq 0 \tag{B.46}$$

and the asserted optimality of $b(R)$ has been proven. Therefore, $b(R)$ is a Gauss–Markov estimator of β in model (B.39).

Proof 10 (Testing Linear Hypotheses, Case $s > 0$). Let

$$X\begin{pmatrix} G \\ R \end{pmatrix}^{-1} = \underset{T\times K}{\tilde{X}} = \left(\underset{T\times s}{\tilde{X}_1} , \underset{T\times(K-s)}{\tilde{X}_2} \right)$$

and

$$\underset{s\times 1}{\tilde{\beta}_1} = G\beta, \qquad \underset{(K-s)\times 1}{\tilde{\beta}_2} = R\beta .$$

Then the model could be rewritten as

$$y = X\beta + \epsilon = \tilde{X}_1\tilde{\beta}_1 + \tilde{X}_2\tilde{\beta}_2 + \epsilon .$$

Proof 11 (Testing Linear Hypotheses, Distribution of F). In what follows, we will determine F and its distribution for the two special cases of the general linear hypothesis.

Distribution of F

Case 1: $s = 0$
The ML estimators under H_0 (3.96) are given by

$$\hat{\beta} = \beta^* \text{ and } \hat{\sigma}^2_\omega = \frac{1}{T}(y - X\beta^*)'(y - X\beta^*). \tag{B.47}$$

The ML estimators over Ω are available from Theorem 3.18:

$$\hat{\beta} = b \quad \text{and} \quad \hat{\sigma}^2_\Omega = \frac{1}{T}(y - Xb)'(y - Xb). \tag{B.48}$$

Subsequent modifications then yield

$$\left.\begin{aligned}
& b - \beta^* = (X'X)^{-1}X'(y - X\beta^*), \\
& (b - \beta^*)'X'X = (y - X\beta^*)'X, \\
& y - Xb = (y - X\beta^*) - X(b - \beta^*), \\
& (y - Xb)'(y - Xb) = (y - X\beta^*)'(y - X\beta^*) \\
& \qquad + (b - \beta^*)'X'X(b - \beta^*) \\
& \qquad - 2(y - X\beta^*)'X(b - \beta^*) \\
& \quad = (y - X\beta^*)'(y - X\beta^*) \\
& \qquad - (b - \beta^*)'X'X(b - \beta^*).
\end{aligned}\right\} \tag{B.49}$$

It follows that

$$T(\hat{\sigma}^2_\omega - \hat{\sigma}^2_\Omega) = (b - \beta^*)'X'X(b - \beta^*), \tag{B.50}$$

and we now have the test statistic

$$F = \frac{(b - \beta^*)'X'X(b - \beta^*)}{(y - Xb)'(y - Xb)} \cdot \frac{T - K}{K}. \tag{B.51}$$

Numerator:
The following statements hold:

$$b - \beta^* = (X'X)^{-1}X'[\epsilon + X(\beta - \beta^*)] \qquad \text{[by (B.49)]},$$

$$\tilde{\epsilon} = \epsilon + X(\beta - \beta^*) \sim N(X(\beta - \beta^*), \sigma^2 I) \qquad \text{[Theorem A.82]},$$

$$X(X'X)^{-1}X' \quad \text{idempotent and of rank } K$$

$$(b - \beta^*)'X'X(b - \beta^*) = \tilde{\epsilon}'X(X'X)^{-1}X'\tilde{\epsilon}$$

$$\sim \sigma^2\chi^2_K(\sigma^{-2}(\beta - \beta^*)'X'X(\beta - \beta^*)) \qquad \text{[Theorem A.57]}$$

and $\sim \sigma^2\chi^2_K$ under H_0.

Denominator:

$$\left.\begin{array}{lr} (y - Xb)'(y - Xb) = (T - K)s^2 = \epsilon' M\epsilon & \text{[by (3.62)]}, \\ M = I - X(X'X)^{-1}X' \quad \text{idempotent of rank } T - K & \text{[A.36(vi)]}, \\ \epsilon' M\epsilon \sim \sigma^2\chi^2_{T-K} & \text{[Theorem A.60]}. \end{array}\right\} \tag{B.52}$$

We have

$$MX(X'X)^{-1}X' = \mathbf{0} \quad \text{[Theorem A.36(vi)]}, \tag{B.53}$$

such that the numerator and denominator are independently distributed (Theorem A.62).

Thus (Theorem A.59) the ratio F exhibits the following properties:

- F is distributed as $F_{K,T-K}(\sigma^{-2}(\beta - \beta^*)'X'X(\beta - \beta^*))$ under H_1; and
- F is distributed as central $F_{K,T-K}$ under $\mathrm{H}_0 : \beta = \beta^*$.

If we denote by $F_{m,n,1-q}$ the $(1 - q)$–quantile of $F_{m,n}$ (i.e., $P(F \leq F_{m,n,1-q}) = 1 - q$), then we may derive a uniformly most powerful test, given a fixed level of significance α (cf. Lehmann, 1986, p. 372):

$$\left.\begin{array}{l} \text{region of acceptance of } \mathrm{H}_0 : 0 \leq F \leq F_{K,T-K,1-\alpha}, \\ \text{critical area of } \mathrm{H}_0 : F > F_{K,T-K,1-\alpha}. \end{array}\right\} \tag{B.54}$$

A selection of critical values is provided in Appendix C.

Case 2: $s > 0$
Next we consider a decomposition of the model in order to determine the

ML estimators under H_0 (3.97) and compare them with the corresponding ML estimators over Ω. Let

$$\beta' = \left(\underset{1\times s}{\beta_1'}\,, \quad \underset{1\times(K-s)}{\beta_2'} \right) \tag{B.55}$$

and, respectively,

$$y = X\beta + \epsilon = X_1\beta_1 + X_2\beta_2 + \epsilon\,. \tag{B.56}$$

We set

$$\tilde{y} = y - X_2 r. \tag{B.57}$$

Since $\operatorname{rank}(X) = K$, we have

$$\operatorname{rank}\underset{T\times s}{(X_1)} = s, \quad \operatorname{rank}\underset{T\times(K-s)}{(X_2)} = K - s, \tag{B.58}$$

such that the inverse matrices $(X_1'X_1)^{-1}$ and $(X_2'X_2)^{-1}$ exist.

The ML estimators under H_0 are then given by

$$\hat{\beta}_2 = r, \quad \hat{\beta}_1 = (X_1'X_1)^{-1}X_1'\tilde{y}, \tag{B.59}$$

and

$$\hat{\sigma}_\omega^2 = \frac{1}{T}(\tilde{y} - X_1\hat{\beta}_1)'(\tilde{y} - X_1\hat{\beta}_1). \tag{B.60}$$

Separation of b

It can easily be seen that

$$\begin{aligned} b &= (X'X)^{-1}X'y \\ &= \begin{pmatrix} X_1'X_1 & X_1'X_2 \\ X_2'X_1 & X_2'X_2 \end{pmatrix}^{-1} \begin{pmatrix} X_1'y \\ X_2'y \end{pmatrix}. \end{aligned} \tag{B.61}$$

Making use of the formulas for the inverse of a partitioned matrix yields (Theorem A.4)

$$\begin{pmatrix} (X_1'X_1)^{-1}[I + X_1'X_2D^{-1}X_2'X_1(X_1'X_1)^{-1}] & -(X_1'X_1)^{-1}X_1'X_2D^{-1} \\ -D^{-1}X_2'X_1(X_1'X_1)^{-1} & D^{-1} \end{pmatrix}, \tag{B.62}$$

where

$$D = X_2'M_1X_2 \tag{B.63}$$

and

$$M_1 = I - X_1(X_1'X_1)^{-1}X_1' = I - P_{X_1}. \tag{B.64}$$

M_1 is (analogously to M) idempotent and of rank $T - s$, furthermore, we have $M_1X_1 = \mathbf{0}$. The $[(K-s)\times(K-s)]$–matrix

$$D = X_2'X_2 - X_2'X_1(X_1'X_1)^{-1}X_1'X_2 \tag{B.65}$$

is symmetric and regular, as the normal equations are uniquely solvable.

The components b_1 and b_2 of b are then given by

$$b = \begin{pmatrix} b_1 \\ b_2 \end{pmatrix} = \begin{pmatrix} (X_1'X_1)^{-1}X_1'y - (X_1'X_1)^{-1}X_1'X_2D^{-1}X_2'M_1y \\ D^{-1}X_2'M_1y \end{pmatrix}. \tag{B.66}$$

Various relations immediately become apparent from (B.66)

$$\left.\begin{aligned} b_2 &= D^{-1}X_2'M_1y, \\ b_1 &= (X_1'X_1)^{-1}X_1'(y - X_2b_2), \\ b_2 - r &= D^{-1}X_2'M_1(y - X_2r) \\ &= D^{-1}X_2'M_1\tilde{y} \\ &= D^{-1}X_2'M_1(\epsilon + X_2(\beta_2 - r)), \end{aligned}\right\} \tag{B.67}$$

$$\left.\begin{aligned} b_1 - \hat{\beta}_1 &= (X_1'X_1)^{-1}X_1'(y - X_2b_2 - \tilde{y}) \\ &= -(X_1'X_1)^{-1}X_1'X_2(b_2 - r) \\ &= -(X_1'X_1)^{-1}X_1'X_2D^{-1}X_2'M_1\tilde{y}. \end{aligned}\right\} \tag{B.68}$$

Decomposition of $\hat{\sigma}_\Omega^2$

We write (using symbols u and v)

$$\begin{aligned} (y - Xb) &= (y - X_2r - X_1\hat{\beta}_1) - \Big(X_1(b_1 - \hat{\beta}_1) + X_2(b_2 - r)\Big) \\ &= u - v. \end{aligned} \tag{B.69}$$

Thus, we may decompose the ML estimator $T\hat{\sigma}_\Omega^2 = (y - Xb)'(y - Xb)$ as

$$(y - Xb)'(y - Xb) = u'u + v'v - 2u'v. \tag{B.70}$$

We have

$$u = y - X_2r - X_1\hat{\beta}_1 = \tilde{y} - X_1(X_1'X_1)^{-1}X_1'\tilde{y} = M_1\tilde{y}, \tag{B.71}$$

$$u'u = \tilde{y}'M_1\tilde{y}, \tag{B.72}$$

$$\begin{aligned} v &= X_1(b_1 - \hat{\beta}_1) + X_2(b_2 - r) \\ &= -X_1(X_1'X_1)^{-1}X_1'X_2D^{-1}X_2'M_1\tilde{y} \quad \text{[by (B.67)]} \\ &\quad + X_2D^{-1}X_2'M_1\tilde{y} \quad \text{[by (B.68)]} \\ &= M_1X_2D^{-1}X_2'M_1\tilde{y}\,, \end{aligned} \tag{B.73}$$

$$\begin{aligned} v'v &= \tilde{y}'M_1X_2D^{-1}X_2'M_1\tilde{y} \\ &= (b_2 - r)'D(b_2 - r)\,, \end{aligned} \tag{B.74}$$

$$u'v = v'v\,. \tag{B.75}$$

Summarizing, we may state

$$\begin{aligned}(y - Xb)'(y - Xb) &= u'u - v'v \\ &= (\tilde{y} - X_1\hat{\beta}_1)'(\tilde{y} - X_1\hat{\beta}_1) - (b_2 - r)'D(b_2 - r)\end{aligned} \tag{B.76}$$

or

$$T(\hat{\sigma}_\omega^2 - \hat{\sigma}_\Omega^2) = (b_2 - r)'D(b_2 - r)\,. \tag{B.77}$$

Hence, for Case 2: $s > 0$, we get

$$F = \frac{(b_2 - r)'D(b_2 - r)}{(y - Xb)'(y - Xb)} \frac{T - K}{K - s}\,. \tag{B.78}$$

Distribution of F

Numerator:
We use the following relations:

$A = M_1 X_2 D^{-1} X_2' M_1$ is idempotent,

$$\begin{aligned}\text{rank}(A) = \text{tr}(A) &= \text{tr}\{(M_1 X_2 D^{-1})(X_2' M_1)\} \\ &= \text{tr}\{(X_2' M_1)(M_1 X_2 D^{-1})\} \qquad \text{[Theorem A.1(iv)]} \\ &= \text{tr}(I_{K-s}) = K - s,\end{aligned}$$

$$b_2 - r = D^{-1} X_2' M_1 \tilde{\epsilon} \quad \text{[by (B.67)]},$$

$$\tilde{\epsilon} = \epsilon + X_2(\beta_2 - r) \sim N(X_2(\beta_2 - r), \sigma^2 I), \quad \text{[Theorem A.55]},$$

$$(b_2 - r)'D(b_2 - r) = \tilde{\epsilon}' A \tilde{\epsilon} \sim \sigma^2 \chi^2_{K-s}(\sigma^{-2}(\beta_2 - r)'D(\beta_2 - r)) \tag{B.79}$$

[Theorem A.57] and

$$\sim \sigma^2 \chi^2_{K-s} \quad \text{under } \text{H}_0. \tag{B.80}$$

Denominator:
The denominator is equal in both cases, i.e., with $P_X = X(X'X)^{-1}X'$, we have

$$(y - Xb)'(y - Xb) = \epsilon'(I - P_X)\epsilon \sim \sigma^2 \chi^2_{T-K}. \tag{B.81}$$

Since

$$(I - P_X)X = (I - P_X)(X_1, X_2) = ((I - P_X)X_1, (I - P_X)X_2) = (0, 0) \tag{B.82}$$

we find

$$(I - P_X)M_1 = (I - P_X) \tag{B.83}$$

and

$$(I - P_X)A = (I - P_X)M_1 X_2 D^{-1} X_2' M_1 = \mathbf{0}, \tag{B.84}$$

such that the numerator and denominator of F (B.78) are independently distributed ([Theorem A.62]). Hence ([see also Theorem A.59]), the test statistic F is distributed under H_1 as $F_{K-s,T-K}(\sigma^{-2}(\beta_2 - r)'D(\beta_2 - r))$ and as central $F_{K-s,T-K}$ under H_0.

Proof 12 (Theorem (3.20)). Let

$$R_X^2 - R_{X_1}^2 = \frac{RSS_{X_1} - RSS_X}{SYY},$$

such that the assertion (3.161) is equivalent to

$$RSS_{X_1} - RSS_X \geq 0.$$

Since

$$\begin{aligned} RSS_X &= (y - Xb)'(y - Xb) \\ &= y'y + b'X'Xb - 2b'X'y \\ &= y'y - b'X'y \end{aligned} \tag{B.85}$$

and, analogously,

$$RSS_{X_1} = y'y - \hat{\beta}_1' X_1' y$$

where

$$b = (X'X)^{-1}X'y$$

and

$$\hat{\beta}_1 = (X_1'X_1)^{-1}X_1'y$$

are OLS estimators in the full model and in the submodel, we have

$$RSS_{X_1} - RSS_X = b'X'y - \hat{\beta}_1' X_1' y. \tag{B.86}$$

Now we have, with (B.61)–(B.67),

$$\begin{aligned} b'X'y &= (b_1', b_2') \begin{pmatrix} X_1'y \\ X_2'y \end{pmatrix} \\ &= (y' - b_2'X_2')X_1(X_1'X_1)^{-1}X_1'y + b_2'X_2'y \\ &= \hat{\beta}_1'X_1'y + b_2'X_2'M_1y \quad \text{(cf. (B.76))}. \end{aligned}$$

Thus, (B.86) becomes

$$\begin{aligned} RSS_{X_1} - RSS_X &= b_2'X_2'M_1y \\ &= y'M_1X_2D^{-1}X_2'M_1y \geq 0, \end{aligned} \tag{B.87}$$

such that (3.161) is proven.

Proof 13 (Transformation for General Linear Regression). The matrices W and W^{-1} may be decomposed [see also Theorem A.12(iii)] as

$$W = MM \quad \text{and} \quad W^{-1} = NN, \tag{B.88}$$

where $M = W^{1/2}$ and $N = W^{-1/2}$ are nonsingular. We transform the model (3.166) by premultiplication with N:

$$Ny = NX\beta + N\epsilon \tag{B.89}$$

and set

$$Ny = \tilde{y}\,, \quad NX = \tilde{X}\,, \quad N\epsilon = \tilde{\epsilon}\,. \tag{B.90}$$

Then it holds

$$\mathrm{E}(\tilde{\epsilon}) = \mathrm{E}(N\epsilon) = 0, \quad \mathrm{E}(\tilde{\epsilon}\tilde{\epsilon}') = \mathrm{E}(N\epsilon\epsilon' N) = \sigma^2 I\,, \tag{B.91}$$

such that the transformed model $\tilde{y} = \tilde{X}\beta + \tilde{\epsilon}$ obeys all assumptions of the classical regression model. The OLS estimator of β in this model is of the form

$$\begin{aligned} b &= (\tilde{X}'\tilde{X})^{-1}\tilde{X}'\tilde{y} \\ &= (X'NN'X)^{-1}X'NN'y \\ &= (X'W^{-1}X)^{-1}X'W^{-1}y\,. \end{aligned} \tag{B.92}$$

Proof 14 (Smallest Variance for Aitken Estimator). Let $\tilde{\beta} = \tilde{C}y$ be an arbitrary linear unbiased estimator of β. We set

$$\tilde{C} = \hat{C} + D \tag{B.93}$$

with

$$\hat{C} = S^{-1}X'W^{-1}\,. \tag{B.94}$$

The unbiasedness of $\tilde{\beta}$ leads to the condition $DX = 0$, such that $\hat{C}WD = 0$. Therefore, we get, for the covariance matrix,

$$\begin{aligned} V_{\tilde{\beta}} &= \mathrm{E}(\tilde{C}\epsilon\epsilon'\tilde{C}') \\ &= \sigma^2(\hat{C} + D)W(\hat{C}' + D') \\ &= \sigma^2\hat{C}W\hat{C}' + \sigma^2 DWD' \\ &= V_b + \sigma^2 DWD'\,, \end{aligned} \tag{B.95}$$

such that $V_{\tilde{\beta}} - V_b = \sigma^2 D'WD$ is nonnegative definite (Theorem A.18(v)).

Proof 15 (Estimation of σ^2). Here we have

$$\begin{aligned} \hat{\epsilon} = y - X\hat{\beta} &= (I - X(X'AX)^{-1}X'A)\epsilon\,, \\ (T-K)\hat{\sigma}^2 &= \hat{\epsilon}'\hat{\epsilon} \\ &= \mathrm{tr}\{(I - X(X'AX)^{-1}X'A)\epsilon\epsilon'(I - AX(X'AX)^{-1}X')\}\,, \\ \mathrm{E}(\hat{\sigma^2})(T-K) &= \sigma^2\,\mathrm{tr}(W - X(X'AX)^{-1}X'A) \\ &\quad + \mathrm{tr}\{\sigma^2 X(X'AX)^{-1}X'A(I - 2W) + XV_{\hat{\beta}}X'\}\,. \end{aligned} \tag{B.96}$$

If we choose the standardization $\mathrm{tr}(W) = T$, then the first term in (B.96) becomes $(T-K)$ (Theorem A.1). In the case $\hat{\beta} = (X'X)^{-1}X'y$ (i.e., $A = I$), we get

$$\begin{aligned} \mathrm{E}(\hat{\sigma}^2) &= \sigma^2 + \frac{\sigma^2}{T-K}\,\mathrm{tr}[X(X'X)^{-1}X'(I-W)] \\ &= \sigma^2 + \frac{\sigma^2}{T-K}(K - \mathrm{tr}[(X'X)^{-1}X'WX])\,. \end{aligned} \tag{B.97}$$

Proof 16 (Decomposition of P). Assume that X is partitioned as $X = (X_1, X_2)$ with $X_1 : T \times p$ and $\mathrm{rank}(X_1) = p$, $X_2 : T \times (K-p)$ and $\mathrm{rank}(X_2) = K - p$. Let $P_1 = X_1(X_1'X_1)^{-1}X_1'$ be the (idempotent) prediction matrix for X_1, and let $W = (I - P_1)X_2$ be the projection of the columns of X_2 onto the orthogonal complement of X_1. Then the matrix $P_2 = W(W'W)^{-1}W'$ is the prediction matrix for W, and P can be expressed as (using Theorem A.45)

$$P = P_1 + P_2 \tag{B.98}$$

or

$$X(X'X)^{-1}X' = X_1(X_1'X_1)^{-1}X_1' + (I-P_1)X_2[X_2'(I-P_1)X_2]^{-1}X_2'(I-P_1)\,. \tag{B.99}$$

Equation (B.98) shows that the prediction matrix P can be decomposed into the sum of two (or more) prediction matrices. Applying the decomposition (B.99) to the linear model, including a dummy variable, i.e., $y = 1\alpha + X\beta + \epsilon$, we obtain

$$P = \frac{11'}{T} + \tilde{X}(\tilde{X}'\tilde{X})^{-1}\tilde{X}' = P_1 + P_2 \tag{B.100}$$

and

$$p_{ii} = \frac{1}{T} + \tilde{x}_i'(\tilde{X}'\tilde{X})^{-1}\tilde{x}_i\,, \tag{B.101}$$

where $\tilde{X} = (x_{ij} - \bar{x}_i)$ is the matrix of the mean–corrected x–values. This is seen as follows. Application of (B.99) to $(1, X)$ gives

$$P_1 = 1(1'1)^{-1}1' = \frac{11'}{T} \tag{B.102}$$

and

$$\begin{aligned} W = (I - P_1)X &= X - 1\left(\frac{1}{T}1'X\right) \\ &= X - (1\bar{x}_1, 1\bar{x}_2, \ldots, 1\bar{x}_K) \\ &= (x_1 - \bar{x}_1, \ldots, x_K - \bar{x}_K)\,. \end{aligned} \tag{B.103}$$

Since $\tilde{X}'1 = 0$ and hence $P_2 1 = 0$, we get, from (B.100),

$$P1 = 1\frac{T}{T} + 0 = 1\,. \tag{B.104}$$

Proof 17 (Property (ii)). Since P is nonnegative definite, we have $x'Px \geq 0$ for all x and, especially, for $x_{ij} = (0, \ldots, 0, x_i, 0, x_j, 0, \ldots, 0)'$, where x_i and x_j occur at the ith and jth positions ($i \neq j$). This gives

$$x'_{ij} P x_{ij} = (x_i, x_j) \begin{pmatrix} p_{ii} & p_{ij} \\ p_{ji} & p_{jj} \end{pmatrix} \begin{pmatrix} x_i \\ x_j \end{pmatrix} \geq 0 .$$

Therefore, $P_{ij} = \begin{pmatrix} p_{ii} & p_{ij} \\ p_{ji} & p_{jj} \end{pmatrix}$ is nonnegative definite, and hence its determinant is nonnegative

$$|P_{ij}| = p_{ii} p_{jj} - p_{ij}^2 \geq 0 .$$

Proof 18 (Property (iv)). Analogous to (ii), using $I - P$ instead of P leads to (3.198).

We have

$$p_{ii} + \frac{\hat{\epsilon}_i^2}{\hat{\epsilon}'\hat{\epsilon}} \leq 1 . \tag{B.105}$$

Proof. Let $Z = (X, y)$, $P_X = X(X'X)^{-1}X'$, and $P_Z = Z(Z'Z)^{-1}Z'$. Then (B.99) and (3.181) imply

$$\begin{aligned} P_Z &= P_X + \frac{(I - P_X) y y' (I - P_X)}{y'(I - P_X) y} \\ &= P_X + \frac{\hat{\epsilon}\hat{\epsilon}'}{\hat{\epsilon}'\hat{\epsilon}} . \end{aligned} \tag{B.106}$$

Hence we find that the ith diagonal element of P_Z is equal to $p_{ii} + \hat{\epsilon}_i^2 / \hat{\epsilon}'\hat{\epsilon}$. If we now use (3.192), then (B.105) follows.

Proof 19 (p_{ij} in Multiple Regression). The proof is straightforward by using the spectral decomposition of $X'X = \Gamma \Lambda \Gamma'$ and the definition of p_{ij} and p_{ii} (cf. (3.182)), i.e.,

$$\begin{aligned} p_{ij} &= x_i'(X'X)^{-1} x_j = x_i' \Gamma \Lambda^{-1} \Gamma' x_j \\ &= \sum_{r=1}^{K} \lambda_r^{-1} x_i' \gamma_r x_j' \gamma_r \\ &= \|x_i\| \, \|x_j\| \sum \lambda_r^{-1} \cos\theta_{ir} \cos\theta_{jr} , \end{aligned}$$

where $\|x_i\| = (x_i' x_i)^{1/2}$ is the norm of the vector x_i.

Proof 20 (Likelihood–Ratio Test Statistic). Applying relationship (B.99) we obtain

$$(X, e_i)[(X, e_i)'(X, e_i)]^{-1}(X, e_i)' = P + \frac{(I - P) e_i e_i' (I - P)}{e_i'(I - P) e_i} . \tag{B.107}$$

The left–hand side may be interpreted as the prediction matrix $P_{(i)}$ when the ith observation is omitted. Therefore, we may conclude that

$$\begin{aligned} SSE(H_1) &= (T-K-1)s^2_{(i)} = y'_{(i)}(I-P_{(i)})y_{(i)} \\ &= y'\left(I - P - \frac{(I-P)e_ie_i'(I-P)}{e_i'(I-P)e_i}\right)y \\ &= SSE(H_0) - \frac{\hat{\epsilon}_i^2}{1-p_{ii}} \end{aligned} \tag{B.108}$$

holds, where we have made use of the following relationships: $(I-P)y = \hat{\epsilon}$ and $e_i'\hat{\epsilon} = \hat{\epsilon}_i$ and, moreover, $e_i'Ie_i = 1$ and $e_i'Pe_i = p_{ii}$.

Proof 21 (Andrews–Pregibon Statistic). Define $Z = (X, y)$ and consider the partitioned matrix

$$Z'Z = \begin{pmatrix} X'X & X'y \\ y'X & y'y \end{pmatrix}. \tag{B.109}$$

Since $\text{rank}(X'X) = K$, we get (cf. Theorem A.2(vii))

$$\begin{aligned} |Z'Z| &= |X'X||y'y - y'X(X'X)^{-1}X'y| \\ &= |X'X|(y'(I-P)y) \\ &= |X'X|(T-K)s^2 . \end{aligned} \tag{B.110}$$

Analogously, defining $Z_{(i)} = (X_{(i)}, y_{(i)})$, we get

$$|Z'_{(i)}Z_{(i)}| = |X'_{(i)}X_{(i)}|(T-K-1)s^2_{(i)}. \tag{B.111}$$

Therefore the ratio (3.224) becomes

$$\frac{|Z'_{(i)}Z_{(i)}|}{|Z'Z|}. \tag{B.112}$$

Proof 22 (Another Notation of the Andrews–Pregibon statistic). Using

$$Z'_{(i)}Z_{(i)} = Z'Z - z_iz_i'$$

with $z_i = (x_i', y_i)$ and Theorem A.2(x), we obtain

$$\begin{aligned} |Z'_{(i)}Z_{(i)}| &= |Z'Z - z_iz_i'| \\ &= |Z'Z|(1 - z_i'(Z'Z)^{-1}z_i) \\ &= |Z'Z|(1-p_{zii}) . \end{aligned}$$

Proof 23 (Lemma 3.25). Using Theorem A.3(iv),

$$\begin{aligned} (X'X)^{-1} &= (X'_{(i)}X_{(i)} + x_ix_i')^{-1} \\ &= (X'_{(i)}X_{(i)})^{-1} - \frac{(X'_{(i)}X_{(i)})^{-1}x_ix_i'(X'_{(i)}X_{(i)})^{-1}}{1+t_{ii}}, \end{aligned}$$

where

$$t_{ii} = x_i'(X'_{(i)}X_{(i)})^{-1}x_i .$$

We have

$$P = X(X'X)^{-1}X'$$
$$= \begin{pmatrix} X_{(i)} \\ x_i' \end{pmatrix} \left((X_{(i)}'X_{(i)})^{-1} - \frac{(X_{(i)}'X_{(i)})^{-1}x_i x_i'(X_{(i)}'X_{(i)})^{-1}}{1+t_{ii}} \right) (X_{(i)}' x_i)$$

and

$$Py = X(X'X)^{-1}X'y$$
$$= \begin{pmatrix} X_{(i)}\hat{\beta}_{(i)} - 1/(1+t_{ii})(X_{(i)}'(X_{(i)}'X_{(i)})^{-1}x_i x_i'\hat{\beta}_{(i)} - X_{(i)}'(X_{(i)}'X_{(i)})^{-1}x_i y_i) \\ 1/(1+t_{ii})(x_i'\hat{\beta}_{(i)} + t_{ii}y_i) \end{pmatrix}.$$

Since

$$(I-P)e_i = \frac{1}{1+t_{ii}} \begin{pmatrix} -X_{(i)}(X_{(i)}'X_{(i)})^{-1}x_i \\ 1 \end{pmatrix}$$

and

$$||(I-P)e_i||^2 = \frac{1}{1+t_{ii}},$$

we get

$$\tilde{e}_i\tilde{e}_i{}'y = \frac{1}{1+t_{ii}} \begin{pmatrix} X_{(i)}'(X_{(i)}'X_{(i)})^{-1}x_i x_i'\hat{\beta}_{(i)} - X_{(i)}'(X_{(i)}'X_{(i)})^{-1}x_i y_i \\ -x_i'\hat{\beta}_{(i)} + y_i \end{pmatrix}.$$

Therefore,

$$X(X'X)^{-1}X'y + \tilde{e}_i\tilde{e}_i{}'y = \begin{pmatrix} X_{(i)}\hat{\beta}_{(i)} \\ y_i \end{pmatrix}.$$

Proof 24 (Lemma 3.26). Using the fact that

$$\begin{pmatrix} X'X & X'e_i \\ e_i'X & e_i'e_i \end{pmatrix} - 1$$
$$= \begin{pmatrix} (X'X)^{-1} + (X'X)^{-1}X'e_iHe_i'He_i'X(X'X)^{-1} & -(X'X)^{-1}X'e_iH \\ -He_i'X(X'X)^{-1} & H \end{pmatrix}$$

where

$$H = (e_i'e_i - e_i'X(X'X)^{-1}Xe_i)^{-1}$$
$$= (e_i'(I-P)e_i)^{-1}$$
$$= \frac{1}{||Qe_i||^2},$$

we can show that $P(X, e_i)$, the projection matrix onto the column space of (X, e_i), becomes

$$P(X, e_i) = (X \quad e_i) \begin{pmatrix} X'X & X'e_i \\ e_i'X & e_i'e_i \end{pmatrix}^{-1} \begin{pmatrix} X' \\ e_i' \end{pmatrix}$$

$$\begin{aligned} &= P + \frac{(I-P)e_i e_i'(I-P)}{||Qe_i||^2} \\ &= P + \tilde{e}_i \tilde{e}_i'. \end{aligned}$$

Therefore

$$\begin{aligned} \hat{y}(\lambda) &= X(X'X)^{-1}X'y + \lambda e_i e_i' y \\ &= \hat{y}(0) + \lambda(P(X, e_i) - P)y \\ &= \hat{y}(0) + \lambda(\hat{y}(1) - \hat{y}(0)) \\ &= \lambda \hat{y}(1) + (1-\lambda)\hat{y}(0) \end{aligned}$$

and property (ii) can be proved by the fact that

$$\begin{aligned} \hat{\epsilon}(\lambda) &= y - \hat{y}(\lambda) \\ &= y - \hat{y}(0) - \lambda(\hat{y}(1) - \hat{y}(0)) \\ &= \hat{\epsilon} - \lambda(\hat{y}(1) - \hat{y}(0)). \end{aligned}$$

B.2 Single–Factor Experiments with Fixed and Random Effects

Proof 25 (OLS Estimate for $s = 2$). The multiplication of (4.11), by rows, with (4.12) yields

$$\begin{aligned} \hat{\mu} &= \frac{n_1 n_2 (1+n) Y_{\cdot\cdot} - n_1 n_2 Y_{1\cdot} - n_1 n_2 Y_{2\cdot}}{n_1 n_2 n^2} \\ &= \frac{n Y_{\cdot\cdot}}{n^2} = \frac{Y_{\cdot\cdot}}{n} = y_{\cdot\cdot}\,, \\ \hat{\alpha}_1 &= \frac{-n_1 n_2 Y_{\cdot\cdot} + n_2(n(1+n_2) - n_2) Y_{1\cdot} - n_1 n_2 (n-1) Y_{2\cdot}}{n_1 n_2 n^2} \\ &= -\frac{Y_{\cdot\cdot}}{n^2} + \frac{n + n n_2 - n_2}{n_1 n^2} Y_{1\cdot} - \frac{n-1}{n^2}(Y_{\cdot\cdot} - Y_{1\cdot}) \\ &= Y_{1\cdot}\left(\frac{n + n n_2 - n_2 + n n_1 - n_1}{n_1 n^2}\right) - Y_{\cdot\cdot}\left(\frac{1 - 1 + n}{n^2}\right) \\ &= \frac{Y_{1\cdot}}{n_1} - \frac{Y_{\cdot\cdot}}{n} = y_{1\cdot} - y_{\cdot\cdot} \end{aligned}$$

and, analogously,

$$\hat{\alpha}_2 = y_{2\cdot} - y_{\cdot\cdot}\,.$$

Proof 26 (Proof of the F–Distribution of $F_{1,n-s}$). We first start proving with the denomiator.

(i) Denominator
First, we derive a representation of MS_{Error} as a quadratic form in the total error vector ϵ (cf. (4.4)).

With (4.2) and (4.42) we have

$$\begin{aligned} y_{ij} - y_{i\cdot} &= \epsilon_{ij} - \epsilon_{i\cdot}\,, (\text{all } i, j), \\ \epsilon_i - \mathbf{1}_{n_i}\epsilon_{i\cdot} &= \epsilon_i - \frac{1}{n_i}\mathbf{1}_{n_i}\mathbf{1}'_{n_i}\epsilon_i \\ &= \left(I_{n_i} - \frac{1}{n_i}\mathbf{1}_{n_i}\mathbf{1}'_{n_i}\right)\epsilon_i \\ &= Q_i\epsilon_i\,, \end{aligned} \tag{B.113}$$

$$\begin{aligned} \begin{pmatrix} \epsilon_1 \\ \vdots \\ \epsilon_s \end{pmatrix} - \begin{pmatrix} \mathbf{1}_{n_1}\epsilon_{1\cdot} \\ \vdots \\ \mathbf{1}_{n_s}\epsilon_{s\cdot} \end{pmatrix} &= \begin{pmatrix} Q_1 & & \mathbf{0} \\ & \ddots & \\ \mathbf{0} & & Q_s \end{pmatrix} \epsilon \\ &= \text{diag}(Q_1, \ldots, Q_s)\epsilon \\ &= Q\epsilon\,. \end{aligned} \tag{B.114}$$

The matrices $Q_i = I_{n_i} - 1/n_i\mathbf{1}_{n_i}\mathbf{1}'_{n_i}$ are symmetric

$$Q_i = Q'_i\,,$$

hence, we have

$$Q = Q'\,.$$

Furthermore, Q_i is idempotent

$$\begin{aligned} Q_i^2 &= I_{n_i} + \frac{1}{n_i^2}\mathbf{1}_{n_i}\mathbf{1}'_{n_i}\mathbf{1}_{n_i}\mathbf{1}'_{n_i} - \frac{2}{n_i}\mathbf{1}_{n_i}\mathbf{1}'_{n_i} \\ &= Q_i\,, \end{aligned}$$

with $\text{rank}(Q_i) = \text{tr}(Q_i) = n_i - 1$. Hence, Q is idempotent as well, with $\text{rank}(Q) = \sum \text{rank}(Q_i) = n - s$.

This yields the following representation:

$$MS_{\text{Error}} = \frac{1}{n-s}\,\epsilon' Q\epsilon\,. \tag{B.115}$$

(ii) Numerator
We have

$$y = \begin{pmatrix} y_{1\cdot} \\ \vdots \\ y_{s\cdot} \end{pmatrix} = \begin{pmatrix} \mu + \alpha_1 + \epsilon_{1\cdot} \\ \vdots \\ \mu + \alpha_s + \epsilon_{s\cdot} \end{pmatrix}. \tag{B.116}$$

Under

$$\text{H}_0 : c'\mu = c' \begin{pmatrix} \mu + \alpha_1 \\ \vdots \\ \mu + \alpha_s \end{pmatrix} = 0 \tag{B.117}$$

we have

$$c'y = c' \begin{pmatrix} \epsilon_{1\cdot} \\ \vdots \\ \epsilon_{s\cdot} \end{pmatrix} = c'\epsilon \tag{B.118}$$

with

$$\begin{aligned} \epsilon &= \begin{pmatrix} 1/n_1 \mathbf{1}'_{n_1} & & \mathbf{0}' \\ & \ddots & \\ \mathbf{0}' & & 1/n_s \mathbf{1}'_{n_s} \end{pmatrix} \epsilon \\ &= \operatorname{diag}(D'_1, \ldots, D'_s)\epsilon \\ &= D'\epsilon . \end{aligned} \tag{B.119}$$

Hence, the numerator of F [(4.58)] can also be presented as a quadratic form in ϵ according to

$$\frac{(c'y)^2}{\sum c_i^2/n_i} = \frac{1}{\sum c_i^2/n_i} \epsilon' Dcc'D'\epsilon . \tag{B.120}$$

The matrix of this quadratic form is symmetric and idempotent:

$$\left(\frac{1}{\sum c_i^2/n_i} Dcc'D' \right)^2 = \frac{1}{\sum c_i^2/n_i} Dcc'D' . \tag{B.121}$$

We check this for $s = 2$. We have

$$\begin{aligned} Dcc'D' &= \begin{pmatrix} 1/n_1 \mathbf{1}_{n_1} & \mathbf{0} \\ \mathbf{0} & 1/n_2 \mathbf{1}_{n_2} \end{pmatrix} \begin{pmatrix} c_1 \\ c_2 \end{pmatrix} (c_1 \; c_2) \begin{pmatrix} 1/n_1 \mathbf{1}'_{n_1} & \mathbf{0}' \\ \mathbf{0}' & 1/n_2 \mathbf{1}'_{n_2} \end{pmatrix} \\ &= \begin{pmatrix} c_1^2/n_1^2 \mathbf{1}_{n_1} \mathbf{1}'_{n_1} & (c_1 c_2)/(n_1 n_2) \mathbf{1}_{n_1} \mathbf{1}'_{n_2} \\ (c_1 c_2)/(n_1 n_2) \mathbf{1}_{n_2} \mathbf{1}'_{n_1} & c_2^2/n_2^2 \mathbf{1}_{n_2} \mathbf{1}'_{n_2} \end{pmatrix} \end{aligned}$$

and, hence,

$$(Dcc'D')^2 = \left(\frac{c_1^2}{n_1} + \frac{c_2^2}{n_2} \right) (Dcc'D') .$$

From this the idempotence follows (cf. (B.121)). Furthermore, we have (cf. A.36(ii))

$$\operatorname{rank} \left(\frac{Dcc'D'}{\sum c_i^2/n_i} \right) = \operatorname{tr} \left(\frac{Dcc'D'}{\sum c_i^2/n_i} \right) = 1 ,$$

since $\operatorname{tr}(\mathbf{1}_{n_i} \mathbf{1}'_{n_i}) = n_i$.

(iii) Independence of numerator and denominator

The numerator and denominator of F from (4.58) are quadratic forms in ϵ with idempotent matrices, hence they have a χ_1^2–distribution, or χ_{n-s}^2–distribution, respectively. According to Theorem A.61, their ratio has an

$F_{1,n-s}$–distribution if

$$\frac{1}{\sum c_i^2/n_i} QDcc'D' = \mathbf{0}. \tag{B.122}$$

As can easily be seen, we have

$$QD = \begin{pmatrix} Q_1D_1 & & \mathbf{0} \\ & \ddots & \\ \mathbf{0} & & Q_sD_s \end{pmatrix} \tag{B.123}$$

and

$$\begin{aligned} Q_iD_i &= \left(I_{n_i} - \frac{1}{n_i}\mathbf{1}_{n_i}\mathbf{1}'_{n_i}\right)\frac{1}{n_i}\mathbf{1_{n_i}} \\ &= \frac{1}{n_i}\mathbf{1}_{n_i} - \frac{1}{n_i}\mathbf{1}_{n_i} = \mathbf{0}\,. \end{aligned}$$

Hence

$$QD = \mathbf{0}$$

and (B.122) holds.

Appendix C
Distributions and Tables

x	0.0	0.02	0.04	0.06	0.08
0.0	0.3989	0.3989	0.3986	0.3982	0.3977
0.2	0.3910	0.3894	0.3876	0.3857	0.3836
0.4	0.3814	0.3653	0.3621	0.3589	0.3555
0.6	0.3332	0.3292	0.3251	0.3209	0.3166
0.8	0.2897	0.2850	0.2803	0.2756	0.2709
1.0	0.2419	0.2371	0.2323	0.2275	0.2226
1.2	0.1942	0.1895	0.1849	0.1804	0.1758
1.4	0.1497	0.1456	0.1415	0.1374	0.1334
1.6	0.1109	0.1074	0.1039	0.1006	0.0973
1.8	0.0789	0.0761	0.0734	0.0707	0.0681
2.0	0.0539	0.0519	0.0498	0.0478	0.0459
2.2	0.0355	0.0339	0.0325	0.0310	0.0296
2.4	0.0224	0.0213	0.0203	0.0194	0.0184
2.6	0.0136	0.0167	0.0122	0.0116	0.0110
2.8	0.0059	0.0075	0.0071	0.0067	0.0063
3.0	0.0044	0.0024	0.0012	0.0006	0.0003

TABLE C.1. Density function $\phi(x)$ of the $N(0,1)$–distribution.

u	0.00	0.01	0.02	0.03	0.04
0.0	0.500000	0.503989	0.507978	0.511966	0.515953
0.1	0.539828	0.543795	0.547758	0.551717	0.555670
0.2	0.579260	0.583166	0.587064	0.590954	0.594835
0.3	0.617911	0.621720	0.625516	0.629300	0.633072
0.4	0.655422	0.659097	0.662757	0.666402	0.670031
0.5	0.691462	0.694974	0.698468	0.701944	0.705401
0.6	0.725747	0.729069	0.732371	0.735653	0.738914
0.7	0.758036	0.761148	0.764238	0.767305	0.770350
0.8	0.788145	0.791030	0.793892	0.796731	0.799546
0.9	0.815940	0.818589	0.821214	0.823814	0.826391
1.0	0.841345	0.843752	0.846136	0.848495	0.850830
1.1	0.864334	0.866500	0.868643	0.870762	0.872857
1.2	0.884930	0.886861	0.888768	0.890651	0.892512
1.3	0.903200	0.904902	0.906582	0.908241	0.909877
1.4	0.919243	0.920730	0.922196	0.923641	0.925066
1.5	0.933193	0.934478	0.935745	0.936992	0.938220
1.6	0.945201	0.946301	0.947384	0.948449	0.949497
1.7	0.955435	0.956367	0.957284	0.958185	0.959070
1.8	0.964070	0.964852	0.965620	0.966375	0.967116
1.9	0.971283	0.971933	0.972571	0.973197	0.973810
2.0	0.977250	0.977784	0.978308	0.978822	0.979325
2.1	0.982136	0.982571	0.982997	0.983414	0.983823
2.2	0.986097	0.986447	0.986791	0.987126	0.987455
2.3	0.989276	0.989556	0.989830	0.990097	0.990358
2.4	0.991802	0.992024	0.992240	0.992451	0.992656
2.5	0.993790	0.993963	0.994132	0.994297	0.994457
2.6	0.995339	0.995473	0.995604	0.995731	0.995855
2.7	0.996533	0.996636	0.996736	0.996833	0.996928
2.8	0.997445	0.997523	0.997599	0.997673	0.997744
2.9	0.998134	0.998193	0.998250	0.998305	0.998359
3.0	0.998650	0.998694	0.998736	0.998777	0.998817

TABLE C.2. Distribution function $\Phi(u)$ of the $N(0,1)$–distribution.

u	0.05	0.06	0.07	0.08	0.09
0.0	0.519939	0.523922	0.527903	0.531881	0.535856
0.1	0.559618	0.563559	0.567495	0.571424	0.575345
0.2	0.598706	0.602568	0.606420	0.610261	0.614092
0.3	0.636831	0.640576	0.644309	0.648027	0.651732
0.4	0.673645	0.677242	0.680822	0.684386	0.687933
0.5	0.708840	0.712260	0.715661	0.719043	0.722405
0.6	0.742154	0.745373	0.748571	0.751748	0.754903
0.7	0.773373	0.776373	0.779350	0.782305	0.785236
0.8	0.802337	0.805105	0.807850	0.810570	0.813267
0.9	0.828944	0.831472	0.833977	0.836457	0.838913
1.0	0.853141	0.855428	0.857690	0.859929	0.862143
1.1	0.874928	0.876976	0.879000	0.881000	0.882977
1.2	0.894350	0.896165	0.897958	0.899727	0.901475
1.3	0.911492	0.913085	0.914657	0.916207	0.917736
1.4	0.926471	0.927855	0.929219	0.930563	0.931888
1.5	0.939429	0.940620	0.941792	0.942947	0.944083
1.6	0.950529	0.951543	0.952540	0.953521	0.954486
1.7	0.959941	0.960796	0.961636	0.962462	0.963273
1.8	0.967843	0.968557	0.969258	0.969946	0.970621
1.9	0.974412	0.975002	0.975581	0.976148	0.976705
2.0	0.979818	0.980301	0.980774	0.981237	0.981691
2.1	0.984222	0.984614	0.984997	0.985371	0.985738
2.2	0.987776	0.988089	0.988396	0.988696	0.988989
2.3	0.990613	0.990863	0.991106	0.991344	0.991576
2.4	0.992857	0.993053	0.993244	0.993431	0.993613
2.5	0.994614	0.994766	0.994915	0.995060	0.995201
2.6	0.995975	0.996093	0.996207	0.996319	0.996427
2.7	0.997020	0.997110	0.997197	0.997282	0.997365
2.8	0.997814	0.997882	0.997948	0.998012	0.998074
2.9	0.998411	0.998462	0.998511	0.998559	0.998605
3.0	0.998856	0.998893	0.998930	0.998965	0.998999

TABLE C.3. Distribution function $\Phi(u)$ of the $N(0,1)$–distribution.

	Level of significance α					
df	0.99	0.975	0.95	0.05	0.025	0.01
1	0.0001	0.001	0.004	3.84	5.02	6.62
2	0.020	0.051	0.103	5.99	7.38	9.21
3	0.115	0.216	0.352	7.81	9.35	11.3
4	0.297	0.484	0.711	9.49	11.1	13.3
5	0.554	0.831	1.15	11.1	12.8	15.1
6	0.872	1.24	1.64	12.6	14.4	16.8
7	1.24	1.69	2.17	14.1	16.0	18.5
8	1.65	2.18	2.73	15.5	17.5	20.1
9	2.09	2.70	3.33	16.9	19.0	21.7
10	2.56	3.25	3.94	18.3	20.5	23.2
11	3.05	3.82	4.57	19.7	21.9	24.7
12	3.57	4.40	5.23	21.0	23.3	26.2
13	4.11	5.01	5.89	22.4	24.7	27.7
14	4.66	5.63	6.57	23.7	26.1	29.1
15	5.23	6.26	7.26	25.0	27.5	30.6
16	5.81	6.91	7.96	26.3	28.8	32.0
17	6.41	7.56	8.67	27.6	30.2	33.4
18	7.01	8.23	9.39	28.9	31.5	34.8
19	7.63	8.91	10.1	30.1	32.9	36.2
20	8.26	9.59	10.9	31.4	34.2	37.6
25	11.5	13.1	14.6	37.7	40.6	44.3
30	15.0	16.8	18.5	43.8	47.0	50.9
40	22.2	24.4	26.5	55.8	59.3	63.7
50	29.7	32.4	34.8	67.5	71.4	76.2
60	37.5	40.5	43.2	79.1	83.3	88.4
70	45.4	48.8	51.7	90.5	95.0	100.4
80	53.5	57.2	60.4	101.9	106.6	112.3
90	61.8	65.6	69.1	113.1	118.1	124.1
100	70.1	74.2	77.9	124.3	129.6	135.8

TABLE C.4. Quantiles of the χ^2–distribution.

	Levels of significance α (one–sided)			
	0.05	0.025	0.01	0.005
	Levels of significance α (two–sided)			
df	0.10	0.05	0.02	0.01
1	6.31	12.71	31.82	63.66
2	2.92	4.30	6.97	9.92
3	2.35	3.18	4.54	5.84
4	2.13	2.78	3.75	4.60
5	2.01	2.57	3.37	4.03
6	1.94	2.45	3.14	3.71
7	1.89	2.36	3.00	3.50
8	1.86	2.31	2.90	3.36
9	1.83	2.26	2.82	3.25
10	1.81	2.23	2.76	3.17
11	1.80	2.20	2.72	3.11
12	1.78	2.18	2.68	3.05
13	1.77	2.18	2.65	3.01
14	1.76	2.14	2.62	2.98
15	1.75	2.13	2.60	2.95
16	1.75	2.12	2.58	2.92
17	1.74	2.11	2.57	2.90
18	1.73	2.10	2.55	2.88
19	1.73	2.09	2.54	2.86
20	1.73	2.09	2.53	2.85
30	1.70	2.04	2.46	2.75
40	1.68	2.02	2.42	2.70
60	1.67	2.00	2.39	2.66
∞	1.64	1.96	2.33	2.58

TABLE C.5. Quantiles of the t–distribution.

					df_1				
df_2	1	2	3	4	5	6	7	8	9
1	161	200	216	225	230	234	237	239	241
2	18.51	19.00	19.16	19.25	19.30	19.33	19.36	19.37	19.38
3	10.13	9.55	9.28	9.12	9.01	8.94	8.88	8.84	8.81
4	7.71	6.94	6.59	6.39	6.26	6.16	6.09	6.04	6.00
5	6.61	5.79	5.41	5.19	5.05	4.95	4.88	4.82	4.78
6	5.99	5.14	4.76	4.53	4.39	4.28	4.21	4.15	4.10
7	5.59	4.74	4.35	4.12	3.97	3.87	3.79	3.73	3.68
8	5.32	4.46	4.07	3.84	3.69	3.58	3.50	3.44	3.39
9	5.12	4.26	3.86	3.63	3.48	3.37	3.29	3.23	3.18
10	4.96	4.10	3.71	3.48	3.33	3.22	3.14	3.07	3.02
11	4.84	3.98	3.59	3.36	3.20	3.09	3.01	2.95	2.90
12	4.75	3.88	3.49	3.26	3.11	3.00	2.92	2.85	2.80
13	4.67	3.80	3.41	3.18	3.02	2.92	2.84	2.77	2.72
14	4.60	3.74	3.34	3.11	2.96	2.85	2.77	2.70	2.65
15	4.54	3.68	3.29	3.06	2.90	2.79	2.70	2.64	2.59
20	4.35	3.49	3.10	2.87	2.71	2.60	2.52	2.45	2.40
30	4.17	3.32	2.92	2.69	2.53	2.42	2.34	2.27	2.21

TABLE C.6. Quantiles of the F_{df_1,df_2}–distribution with df_1 and df_2 degrees of freedom ($\alpha = 0.05$).

				df_1				
df_2	10	11	12	14	16	20	24	30
1	242	243	244	245	246	248	249	250
2	19.39	19.40	19.41	19.42	19.43	19.44	19.45	19.46
3	8.78	8.76	8.74	8.71	8.69	8.66	8.64	8.62
4	5.96	5.93	5.91	5.87	5.84	5.80	5.77	5.74
5	4.74	4.70	4.68	4.64	4.60	4.56	4.53	4.50
6	4.06	4.03	4.00	3.96	3.92	3.87	3.84	3.81
7	3.63	3.60	3.57	3.52	3.49	3.44	3.41	3.38
8	3.34	3.31	3.28	3.23	3.20	3.15	3.12	3.08
9	3.13	3.10	3.07	3.02	2.98	2.93	2.90	2.86
10	2.97	2.94	2.91	2.86	2.82	2.77	2.74	2.70
11	2.86	2.82	2.79	2.74	2.70	2.65	2.61	2.57
12	2.76	2.72	2.69	2.64	2.60	2.54	2.50	2.46
13	2.67	2.63	2.60	2.55	2.51	2.46	2.42	2.38
14	2.60	2.56	2.53	2.48	2.44	2.39	2.35	2.31
15	2.55	2.51	2.48	2.43	2.39	2.33	2.29	2.25
20	2.35	2.31	2.28	2.23	2.18	2.12	2.08	2.04
30	2.16	2.12	2.00	2.04	1.99	1.93	1.89	1.84

TABLE C.7. Quantiles of the F_{df_1,df_2}–distribution with df_1 and df_2 degrees of freedom ($\alpha = 0.05$).

References

Agresti, A. (1990). *Categorical Data Analysis*, Wiley.

Aitchison, J., and Silvey, S. D. (1958). Maximum likelihood estimation of parameters subject to restraints, *Annals of Mathematical Statistics* **29**: 813–828.

Albert, A. (1972). *Regression and the Moore-Penrose Pseudoinverse*, Academic Press.

Algina, J. (1995). An improved general approximation test for the man effect in a spli-plot design, *British Journal of Mathematical and Statistical Psychology* **48**: 149–160.

Algina, J. (1997). Generalization of improved general approximation tests to split-plot designs with multiple between-subjects factors and/or multiple within-subjects factors, *British Journal of Mathematical and Statistical Psychology* **50**: 243–252.

Amemiya, T. (1985). *Advanced Econometrics*, Basil Blackwell.

Andrews, D. F., and Pregibon, D. (1978). Finding outliers that matter, *Journal of the Royal Statistical Society, Series B* **40**: 85–93.

Baksalary, J. K., Kala, R., and Klaczynski, K. (1983). The matrix inequality $M \geq B^*MB$, *Linear Algebra and Its Applications* **54**: 77–86.

Bartlett, M. S. (1937). Some examples of statistical methods of research in agriculture and applied botany, *Journal of the Royal Statistical Society, Series B* **4**: 137–170.

Beckman, R. J., and Trussel, H. J. (1974). The distribution of an arbitrary Studentized residual and the effects of updating in multiple regression, *Journal of the American Statistical Association* **69**: 199–201.

Bekker, P. A., and Neudecker, H. (1989). Albert's theorem applied to problems of efficiency and MSE superiority, *Statistica Neerlandica* **43**: 157–167.

Belsley, D. A., Kuh, E., and Welsch, R. E. (1980). *Regression Diagnostics*, Wiley.

Birch, M. W. (1963). Maximum likelihood in three-way contingency tables, *Journal of the Royal Statistical Society, Series B* **25**: 220–233.

Bishop, Y. M. M., Fienberg, S. E., and Holland, P. W. (1975). *Discrete Multivariate Analysis: Theory and Practice*, MIT Press.

Boik, R. J. (1981). A priori tests in repeated measures designs: Effects of non-sphericity, *Psychometrica* **46**(3): 241–255.

Bosch, K. (1992). *Statistik-Taschenbuch*, Oldenbourg.

Box, G. E. P. (1949). A general distribution theory for a class of likelihood criteria, *Biometrics* **36**: 317–346.

Brook, R. J., and Arnold, G. C. (1985). *Applied Regression Analysis and Experimental Design*, Dekker.

Brown jr, B. W. (1980). The crossover experiment for clinical trials, *Biometrics* **36**: 69–79.

Brownie, C., and Boos, D. D. (1994). Type i error robustness of anova and anova on ranks when the number of treatments is large, *Biometrics* **50**: 542–549.

Brzeskwiniewicz, H., and Wagner, W. (1991). Covariance analysis for split-plot and split-block designs, *The American Statistician* **46**: 155–162.

Büning, H., and Trenkler, G. (1978). *Nichtparametrische statistische Methoden*, de Gruyter.

Burdick, R. (1994). Using confidence intervals to test variance components, *Journal of Quality Technology* **28**: 30–30.

Campbell, S. L., and Meyer, C. D. (1979). *Generalized Inverses of Linear Transformations*, Pitman.

Chatterjee, S., and Hadi, A. S. (1988). *Sensitivity Analysis in Linear Regression*, Wiley.

Christensen, R. (1990). *Log-linear Models*, Springer.

Cochran, W. G., and Cox, G. M. (1950). *Experimental Designs*, Wiley.

Cochran, W. G., and Cox, G. M. (1957). *Experimental Designs*, Wiley.

Cook, R. D. (1977). Detection of influential observations in linear regression, *Technometrics* **19**: 15–18.

Cook, R. D., and Weisberg, S. (1982). *Residuals and Influence in Regression*, Chapman and Hall.

Cook, R. D., and Weisberg, S. (1989). Regression diagnostics with dynamic graphics, *Technometrics* **31**: 277–291.

Cox, D. R. (1970). *The Analysis of Binary Data*, Chapman and Hall.

Cox, D. R. (1972a). The analysis of multivariate binary data, *Applied Statistics* **21**: 113–120.

Cox, D. R. (1972b). Regression models and life-tables (with discussion), *Journal of the Royal Statistical Society, Series B* **34**: 187–202.

Cox, D. R., and Snell, E. J. (1968). A general definition of residuals, *Journal of the Royal Statistical Society, Series B* **30**: 248–275.

Crowder, M. J., and Hand, D. J. (1990). *Analysis of Repeated Measures*, Chapman and Hall.

Cureton, E. E. (1967). The normal approximation to the signed-rank sampling distribution when zero differences are present., *Journal of the American Statistical Association* **62**: 1068–1069.

Dean, A., and Voss, D. (1998). *Design and Analysis of Experiments*, Springer.

Deming, W. E., and Stephan, F. F. (1940). On a least squares adjustment of sampled frequency table when the expected marginal totals are known, *Annals of Mathematical Statistics* **11**: 427–444.

Dempster, A. P., Laird, N. M., and Rubin, D. B. (1977). Maximum likelihood from incomplete data via the EM algorithm, *Journal of the Royal Statistical Society, Series B* **43**: 1–22.

Dhrymes, P. J. (1978). *Indroductory Econometrics*, Springer.

Diggle, P. J., Liang, K.-Y., and Zeger, S. L. (1994). *Analysis of Longitudinal Data*, Chapman and Hall.

Doksum, K. A., and Gasko, M. (1990). On a correspondence between models in binary regression analysis and in survival analysis, *International Statistical Review* **58**: 243–252.

Draper, N. R., and Pukelsheim, F. (1996). An overview of design of experiments, *Statistical Papers* **37**: 1–32.

Draper, N. R., and Smith, H. (1966). *Applied Regression Analysis*, Wiley.

Duncan, D. B. (1975). t-tests and intervals for comparisons suggested by the data, *Biometrics* **31**: 339–359.

Dunn, O. J. (1964). Multiple comparisons using rank sums, *Technometrics* **6**: 241–252.

Dunn, O. J., and Clark, V. A. (1987). *Applied statistics: Analysis of variance and regression*, Wiley.

Dunnett, C. W. (1955). A multiple comparison procedure for comparing treatments with a control, *Journal of the American Statistical Association* **50**: 1096–1121.

Dunnett, C. W. (1964). New tables for multiple comparisons with a control, *Biometrics* **20**: 482–491.

Fahrmeir, L., and Hamerle, A. (eds.) (1984). *Multivariate statistische Verfahren*, de Gruyter.

Fahrmeir, L., and Kaufmann, H. (1985). Consistency and asymptotic normality of the maximum likelihood estimator in generalized linear models, *Annals of Statistics* **13**: 342–368.

Fahrmeir, L., and Tutz, G. (2001). *Multivariate statistical modelling based on generalized linear models*, Springer.

Fitzmaurice, G. M., Laird, N. M., and Rotnitzky, A. G. (1993). Regression models for discrete longitudinal responses, *Statistical Science* **8**(3): 284–309.

Fleiss, J. L. (1989). A critique of recent research in the two-treatment crossover design, *Controlled Clinical Trials* **10**: 237–243.

Friedman, M. (1937). The use of ranks to avoid the assumption of normality implicit in the analysis of variance, *Journal of the American Statistical Association* **32**: 675–701.

Gail, M. H., and Simon, R. (1985). Testing for qualitative interactions between treatment effects and patient subsets, *Biometrics* **41**: 361–372.

Gart, J. J. (1969). An exact test for comparing matched proportions in crossover designs, *Biometrika* **56**(1): 75–80.

Gibbons, J. D. (1976). *Nonparametric Methods for Quantitative Analysis*, American Series In Mathematical And Management Sciences.

Girden, E. R. (1992). *ANOVA – repeated measures*, Sage Publications.

Glonek, G. V. F. (1996). A class of regression models for multivariate categorical responses, *Biometrika* **83**(1): 15–28.

Goldberger, A. S. (1964). *Econometric Theory*, Wiley.

Graybill, F. A. (1961). *An introduction to linear statistical models, Volume I*, McGraw-Hill.

Greenhouse, S. W., and Geisser, S. (1959). On methods in the analysis of profile data, *Psychometrika* **24**(2): 95–112.

Grieve, A. P. (1982). The two-period changeover design in clinical trials (letter to the editor), *Biometrics* **38**: 517–517.

Grieve, A. P. (1990). Crossover versus parallel designs, *Statistical methodology in the pharmaceutical sciences.*

Grizzle, J. E. (1965). The two-period change-over design and its use in clinical trials, *Biometrics* **21**: 467–480.

Grizzle, J. E., Starmer, F. C., and Koch, G. G. (1969). Analysis of categorical data by linear models, *Biometrics* **25**: 489–504.

Guilkey, D. K., and Price, J. M. (1981). On comparing restricted least squares estimators, *Journal of Econometrics* **15**: 397–404.

Haaland, P. D. (1989). *Experimental design in biotechnology*, Dekker.

Haitovsky, Y. (1968). Missing data in regression analysis, *Journal of the Royal Statistical Society, Series B* **34**: 67–82.

Hamerle, A., and Tutz, G. (1989). *Diskrete Modelle zur Analyse von Verweildauern und Lebenszeiten*, Campus.

Harwell, M., and Serlin, R. (1994). An empirical study of five multivariate tests for the single factor repeated measures model, *Computational Statistics and Data Analysis* **26**: 605–618.

Hays, W. L. (1988). *Statistics*, Holt, Rinehart and Winston.

Heagerty, P. J., and Zeger, S. L. (1996). Marginal regression models for clustered ordinal measurements, *Journal of the American Statistical Association* **91**(435): 1024–1036.

Hemelrijk, J. (1952). Note on wilcoxon's two-sample test when ties are present., *Annals of Mathematical Statistics* **23**: 133–135.

Heumann, C. (1993). GEE1-procedure for categorical correlated response, *Technical report*, Ludwigstr. 33, 80535 München, Germany.

Heumann, C. (1998). *Likelihoodbasierte marginale Regressionsmodelle für korrelierte kategoriale Daten*, Peter Lang Europäischer Verlag der Wissenschaften.

Heumann, C., and Jacobsen, M. (1993). *LOGGY 1.0 – Ein Programm zur Analyse von loglinearen Modellen*, C. Heumann, Ludwig-Richter-Str. 3, 85221 Dachau.

Heumann, C., Jacobsen, M., and Toutenburg, H. (1993). Rechnergestützte grafische Analyse von ordinalen Kontingenztafeln – eine Alternative zum Pareto-Prinzip, *Technical report*.

Hills, M., and Armitage, P. (1979). The two-period cross-over clinical trial, *British Journal of Clinical Pharmacology* **8**: 7–20.

Hochberg, Y., and Tamhane, A. C. (1987). *Multiple Comparison Procedures*, Wiley.

Hocking, R. R. (1973). A discussion of the two-way mixed models, *The American Statistician* **27**(4): 148–152.

Hollander, M., and Wolfe, D. A. (1973). *Nonparametric statistical methods*, Wiley.

Huynh, H., and Feldt, L. S. (1970). Conditions under which mean square ratios in repeated measurements designs have exact F-distribution, *Journal of the American Statistical Association* **65**: 1582–1589.

Huynh, H., and Mandeville, G. K. (1979). Validity conditions in repeated measures designs, *Psychological Bulletin* **86**(5): 964–973.

Ishihawa, K. (1976). *Guide to quality control*, Unipub.

Johnston, J. (1972). *Econometric methods*, McGraw-Hill.

Johnston, J. (1984). *Econometric Methods*, McGraw-Hill.

Jones, B., and Kenward, M. G. (1989). *Design and Analysis of Crossover Trials*, Chapman and Hall.

Judge, G. G., Griffiths, W. E., Hill, R. C., and Lee, T.-C. (1980). *The Theory and Practice of Econometrics*, Wiley.

Judge, G. G., Griffiths, W. E., Hill, R. C., Lütkepohl, H., and Lee, T.-C. (1985). *The theory and practice of econometrics*, Wiley.

Karim, M., and Zeger, S. L. (1988). GEE: A SAS macro for longitudinal analysis, *Technical report*, Baltimore, MD.

Kastner, C., Fieger, A., and Heumann, C. (1997). MAREG and WinMAREG—a tool for marginal regression models, *Computational Statistics and Data Analysis* **24**(2): 235–241.

Kmenta, J. (1971). *Elements of Econometrics*, Macmillan.

Koch, G. G. (1969). Some aspects of the statistical analysis of split-plot experiments in completely randomized layouts, *Journal of the American Statistical Association* **64**: 485–505.

Koch, G. G. (1972). The use of nonparametric methods in the analysis of the two period change-over design, *Biometrics* **28**: 577–584.

Koch, G. G., Landis, R. J., Freeman, J. L., Freeman, D. H., and Lehnen, R. G. (1977). A general methodology for the analysis of experiments with repeated measurements of categorical data, *Biometrics* **33**: 133–158.

Kres, H. V. (1983). *Statistical tables for multivariate analysis*, Springer.

Kruskal, W. H., and Wallis, W. A. (1952). Use of ranks in one-criterion variance analysis, *Journal of the American Statistical Association* **47**: 583–621.

Lang, J. B., and Agresti, A. (1994). Simultaneously modeling joint and marginal distributions of multivariate categorical responses, *Journal of the American Statistical Association* **89**(426): 625–632.

Larsen, W. A., and McCleary, S. J. (1972). The use of partial residual plots in regression analysis, *Technometrics* **14**: 781–790.

Lawless, J. F. (1982). *Statistical Models and Methods for Lifetime Data*, Wiley.

Lehmacher, W. (1987). *Verlaufskurven und Crossover*, Springer.

Lehmacher, W. (1991). Analysis of the crossover design in the presence of residual effects, *Statistics in Medicine* **10**: 891–899.

Lehmacher, W., and Wall, K. D. (1978). A new nonparametric approach to the comparison of k independent samples of response curves, *Biometrical Journal* **20**(3): 261–273.

Lehmann, E. L. (1986). *Testing Statistical Hypotheses*, Wiley.

Liang, K.-Y., and Zeger, S. L. (1986). Longitudinal data analysis using generalized linear models, *Biometrika* **73**: 13–22.

Liang, K.-Y., and Zeger, S. L. (1989). A class of logistic regression models for multivariate binary time series, *Journal of the American Statistical Association* **84**(406): 447–451.

Liang, K.-Y., and Zeger, S. L. (1993). Regression analysis for correlated data, *Annual Review of Public Health* **14**: 43–68.

Liang, K.-Y., Zeger, S. L., and Qaqish, B. (1992). Multivariate regression analysis for categorical data, *Journal of the Royal Statistical Society, Series B* **54**: 3–40.

Lienert, G. A. (1986). *Verteilungsfreie Methoden in der Biostatistik*, Hain.

Lipsitz, S. R., Laird, N. M., and Harrington, D. P. (1991). Generalized estimating equations for correlated binary data: Using the odds ratio as a measure of association, *Biometrika* **78**: 153–160.

Little, R. J. A., and Rubin, D. B. (1987). *Statistical Analysis with Missing Data*, Wiley.

Mardia, K. V., Kent, J. T., and Bibby, J. M. (1979). *Multivariate Analysis*, Academic Press.

Mauchly, J. W. (1940). Significance test for sphericity of a normal n-variate distribution, *Annals of Mathematical Statistics* **11**: 204–209.

McCullagh, P., and Nelder, J. A. (1989). *Generalized Linear Models*, Chapman and Hall.

McCulloch, C. E., and Searle, S. R. (2000). *Gereralized, Linear and Mixed Models*, Wiley.

McElroy, F. W. (1967). A necessary and sufficient condition that ordinary least-squares estimators be best linear unbiased, *Journal of the American Statistical Association* **62**: 1302–1304.

McFadden, D. (1974). Conditional logit analysis of qualitative choice, *Frontiers in econometrics.*

Michaelis, J. (1971). Schwellenwerte des Friedman-Tests, *Biometrische Zeitschrift* **13**: 118–122.

Miller Jr., R. G. (1981). *Simultaneous statistical inference*, Springer.

Milliken, G. A., and Akdeniz, F. (1977). A theorem on the difference of the generalized inverse of two nonnegative matrices, *Communications in Statistics, Part A—Theory and Methods* **6**: 73–79.

Milliken, G. A., and Johnson, D. E. (1984). *Analysis of messy data. Volume 1: Designed experiments*, Van Nostrand Reinhold.

Mitzel, H. C., and Games, P. A. (1981). Circularity and multiple comparisons in repeated measure designs, *British Journal of Mathematical and Statistical Psychology* **34**: 253–259.

Molenberghs, G., and Lesaffre, E. (1994). Marginal modeling of correlated ordinal data using a multivariate Plackett distribution, *Journal of the American Statistical Association* **89**(426): 633–644.

Montgomery, D. C. (1976). *Design and analysis of experiments*, Wiley.

Morrison, D. F. (1973). A test for equality of means of correlated variates with missing data on one response, *Biometrika* **60**: 101–105.

Morrison, D. F. (1983). *Applied Linear Statistical Methods*, Prentice Hall.

Nelder, J. A., and Wedderburn, R. W. M. (1972). Generalized linear models, *Journal of the Royal Statistical Society, Series A* **135**: 370–384.

Neter, J., Wassermann, W., and Kutner, M. H. (1990). *Applied Linear Statistical Models*, Irwin.

Oberhofer, W., and Kmenta, J. (1974). A general procedure for obtaining maximum likelihood estimates in generalized regression models, *Econometrica* **42**: 579–590.

Park, S. H., Kim, Y. H., and Toutenburg, H. (1992). Regression diagnostics for removing an observation with animating graphics, *Statistical Papers* **33**: 227–240.

Pepe, M. S., and Fleming, T. R. (1991). A nonparametric method for dealing with mismeasured covariate data, *Journal of the American Statistical Association* **86**: 108–113.

Petersen, R. G. (1985). *Design and analysis of experiments*, Dekker.

Pollock, D. S. G. (1979). *The Algebra of Econometrics*, Wiley.

Pratt, J. W. (1959). Remarks on zeros and ties in the wilcoxon signed rank procedures., *Journal of the American Statistical Association* **54**: 655–667.

Prentice, R. L. (1988). Correlated binary regression with covariates specific to each binary observation, *Biometrics* **44**: 1033–1048.

Prentice, R. L., and Zhao, L. P. (1991). Estimating equations for parameters in means and covariances of multivariate discrete and continuous responses, *Biometrics* **47**: 825–839.

Prescott, R. J. (1981). The comparison of success rates in cross-over trials in the presence of an order effect, *Applied Statistics* **30**(1): 9–15.

Puri, M. L., and Sen, P. K. (1971). *Nonparametric methods in multivariate analysis*, Wiley.

Rao, C. R. (1956). Analysis of dispersion with incomplete observations on one of the characters, *Journal of the Royal Statistical Society, Series B* (18): 259–264.

Rao, C. R. (1973). *Linear Statistical Inference and Its Applications*, Wiley.

Rao, C. R. (1988). Methodology based on the L_1-norm in statistical inference, *Sankhya, Series A* **50**: 289–313.

Rao, C. R., and Mitra, S. K. (1971). *Generalized Inverse of Matrices and Its Applications*, Wiley.

Rao, C. R., and Rao, M. B. (1998). *Matrix Algebra and Its Applications to Statistics and Econometrics*, World Scientific.

Rao, C. R., and Toutenburg, H. (1999). *Linear Models: Least Squares and Alternatives*, Springer.

Ratkovsky, D. A., Evans, M. A., and Alldredge, J. R. (1993). *Cross-over experiments. Design, analysis and application.*, Dekker.

Rosner, B. (1984). Multivariate methods in ophtalmology with application to paired-data situations, *Biometrics* **40**: 1025–1035.

Rouanet, H., and Lepine, D. (1970). Comparison between treatments in a repeated-measurement design: ANOVA and multivariate methods, *British Journal of Mathematical and Statistical Psychology* **23**(2): 147–163.

Roy, S. N. (1953). On a heuristic method of test construction and its use in multivariate analysis, *Annals of Mathematical Statistics* **24**: 220–238.

Roy, S. N. (1957). *Some aspects of multivariate analysis*, Wiley.

Rubin, D. B. (1976). Inference and missing data, *Biometrika* **63**: 581–592.

Rubin, D. B. (1987). *Multiple Imputation for Nonresponse in Sample Surveys*, Wiley.

Sachs, L. (1974). *Angewandte Statistik: Planung und Auswertung, Methoden und Modelle*, Springer.

Scheffé, H. (1953). A method for judging all contrasts in the analysis of variance, *Biometrika* **40**: 87–104.

Scheffé, H. (1956). A 'mixed model' for the analysis of variance, *Annals of Mathematical Statistics* **27**: 23–26.

Scheffé, H. (1959). *The Analysis of Variance*, Wiley.

Schneeweiß, H. (1990). *Ökonometrie*, Physica.

Searle, S. R. (1982). *Matrix Algebra Useful for Statistics*, Wiley.

Searle, S. R., Casella, G., and McCulloch, C. E. (1992). *Variance components*, Wiley.

Seber, G. A. F. (1966). *The linear hypothesis: a general theory*, Griffin.

Silvey, S. D. (1969). Multicollinearity and imprecise estimation, *Journal of the Royal Statistical Society, Series B* **35**: 67–75.

Snedecor, G. W., and Cochran, W. G. (1967). *Statistical Methods*, Ames: Iowa State University Press.

Tan, W. Y. (1971). Note on an extension of the GM-theorem to multivariate linear regression models, *SIAM Journal on Applied Mathematics* **1**: 24–28.

Theobald, C. M. (1974). Generalizations of mean square error applied to ridge regression, *Journal of the Royal Statistical Society, Series B* **36**: 103–106.

Timm, N. H. (1975). *Multivariate analysis with applications in education and psychology*, Brooks/Cole Publishing Company.

Toutenburg, H. (1992a). *Lineare Modelle*, Physica.

Toutenburg, H. (1992b). *Moderne nichtparametrische Verfahren der Risikoanalyse*, Physica.

Toutenburg, H., Heumann, C., Fieger, A., and Park, S. H. (1995). Missing values in regression: Mixed and weighted mixed estimation, *Statistical Sciences, Procedings of the 2nd Gauss Symposium, Munich 1983.*

Toutenburg, H., Toutenburg, S., and Walther, W. (1991). *Datenanalyse und Statistik für Zahnmediziner*, Hanser.

Toutenburg, H., and Walther, W. (1992). Statistische Behandlung unvollständiger Datensätze, *Deutsche Zahnärztliche Zeitschrift* **47**: 104–106.

Toutenburg, S. (1977). *Eine Methode zur Berechnung des Betreungsgrades in der prothetischen und konservierenden Zahnmedizin auf der Basis von Arbeitsablaufstudien, Arbeitszeitmessungen und einer Morbiditätsstudie*, PhD thesis.

Trenkler, G. (1981). *Biased Estimators in the Linear Regression Model*, Hain.

Tukey, J. W. (1953). The problem of multiple comparisons, *Technical report.*

Vach, W., and Blettner, M. (1991). Biased estimation of the odds ratio in case-control studies due to the use of ad-hoc methods of correcting for missing values in confounding variables, *American Journal of Epidemiology* **134**: 895–907.

Vach, W., and Schumacher, M. (1993). Logistic regression with incompletely observed categorial covariates: A comparison of three approaches, *Biometrika* **80**: 353–362.

Waller, R. A., and Duncan, D. B. (1972). A bayes rule for the symmetric multiple comparison problem, *Journal of the American Statistical Association* **67**: 253–255.

Walther, W. (1992). Ein Modell zur Erfassung und statistischen Bewertung klinischer Therapieverfahren—entwickelt durch Evaluation des Pfeilerverlustes bei Konuskronenersatz, *Habilitationsschrift.*

Walther, W., and Toutenburg, H. (1991). Datenverlust bei klinischen Studien, *Deutsche Zahnärztliche Zeitschrift* **46**: 219–222.

Wedderburn, R. W. M. (1974). Quasi-likelihood functions, generalized linear models, and the Gauss-Newton method, *Biometrika* **61**: 439–447.

Wedderburn, R. W. M. (1976). On the existence and uniqueness of the maximum likelihood estimates for certain generalized linear models, *Biometrika* **63**: 27–32.

Weerahandi, S. (1995). Anova under unequal error variances, *Biometrics* **51**: 589–599.

Weisberg, S. (1980). *Applied Linear Regression*, Wiley.

Wilks, S. S. (1932). Moments and distributions of estimates of population parameters from fragmentary samples, *Annals of Mathematical Statistics* **3**: 163–195.

Wilks, S. S. (1938). The large-sample distribution of the likelihood ratio for testing composite hypotheses, *Annals of Mathematical Statistics* **9**: 60–62.

Woolson, R. F. (1987). *Statistical methods for the analysis of biomedical data*, Wiley.

Wu, C. F. J., and Hamada, M. (2000). *Experiments: Planning, Analysis and Parameter Design Optimization*, Wiley.

Yates, F. (1933). The analysis of replicated experiments when the field results are incomplete, *Empire Journal of Experimental Agriculture* **1**: 129–142.

Zhao, L. P., and Prentice, R. L. (1990). Correlated binary regression using a generalized quadratic model, *Biometrika* **77**: 642–648.

Zhao, L. P., Prentice, R. L., and Self, S. G. (1992). Multivariate mean parameter estimation by using a partly exponential model, *Journal of the Royal Statistical Society, Series B* **54**(3): 805–811.

Zimmermann, H., and Rahlfs, W. (1978). Testing hypotheses in the two period change-over with binary data, *Biometrical Journal* **20**(2): 133–141.

Index

Springer Texts in Statistics *(continued from page ii)*

McPherson: Applying and Interpreting Statistics: A Comprehensive Guide, Second Edition
Mueller: Basic Principles of Structural Equation Modeling: An Introduction to LISREL and EQS
Nguyen and Rogers: Fundamentals of Mathematical Statistics: Volume I: Probability for Statistics
Nguyen and Rogers: Fundamentals of Mathematical Statistics: Volume II: Statistical Inference
Noether: Introduction to Statistics: The Nonparametric Way
Nolan and Speed: Stat Labs: Mathematical Statistics Through Applications
Peters: Counting for Something: Statistical Principles and Personalities
Pfeiffer: Probability for Applications
Pitman: Probability
Rawlings, Pantula and Dickey: Applied Regression Analysis
Robert: The Bayesian Choice: From Decision-Theoretic Foundations to Computational Implementation, Second Edition
Robert and Casella: Monte Carlo Statistical Methods
Rose and Smith: Mathematical Statistics with *Mathematica*
Santner and Duffy: The Statistical Analysis of Discrete Data
Saville and Wood: Statistical Methods: The Geometric Approach
Sen and Srivastava: Regression Analysis: Theory, Methods, and Applications
Shao: Mathematical Statistics
Shorack: Probability for Statisticians
Shumway and Stoffer: Time Series Analysis and Its Applications
Terrell: Mathematical Statistics: A Unified Introduction
Timm: Applied Multivariate Analysis
Toutenburg: Statistical Analysis of Designed Experiments, Second Edition
Whittle: Probability via Expectation, Fourth Edition
Zacks: Introduction to Reliability Analysis: Probability Models and Statistical Methods